D0745448

Palaeomagnetism and Tectonics of the Mediterranean Region

Geological Society Special Publications

Series Editor A.J. FLEET

GEOLOGICAL SOCIETY SPECIAL PUBLICATION NO. 105

Palaeomagnetism and Tectonics of the Mediterranean Region

EDITED BY

A. MORRIS and D. H. TARLING
Department of Geological Sciences, University of Plymouth, UK

1996
Published by
The Geological Society
London

THE GEOLOGICAL SOCIETY

The Society was founded in 1807 as the Geological Society of London and is the oldest geological society in the world. It received its Royal Charter in 1825 for the purpose of 'investigating the mineral structure of the Earth'. The Society is Britain's national society for geology with a membership of 7500 (1993). It has countrywide coverage and approximately 1000 members reside overseas. The Society is responsible for all aspects of the geological sciences including professional matters. The Society has its own publishing house, which produces the Society's international journals, books and maps, and which acts as the European distributor for publication of the American Association of Petroleum Geologists, SEPM and the Geological Society of America.

Fellowship is open to those holding a recognized honours degree in geology or cognate subject and who have at least two years' relevant postgraduate experience, or who have not less than six years' relevant experience in geology or a cognate subject. A Fellow who has not less than five years' relevant postgraduate experience in the practice of geology may apply for validation and, subject to approval, may be able to use the designatory letters C Geol (Chartered Geologist).

Further information about the Society is available from the Membership Manager, The Geological Society, Burlington House, Piccadilly, London W1V 0JU, UK. The Society is a Registered Charity No. 210161.

Published by The Geological Society from:
The Geological Society Publishing House
Unit 7
Brassmill Enterprise Centre
Brassmill Lane
Bath BA1 3JN
UK
(*Orders*: Tel 01225 445046
            Fax 01225 442836)

First published 1996

British Library Cataloguing in Publication Data

A catalogue record for this book is available from the British Library

ISBN 1–897799–55–1

Typeset by Type Study, Scarborough, UK

Printed by Alden Press, Osney Mead, Oxford, UK

Distributors

USA
  AAPG Bookstore
  PO Box 979
  Tulsa
  OK 74101–0979
  USA
  (*Orders*: Tel (918) 584–2555
             Fax (918) 560–2652)

Australia
  Australian Mineral Foundation
  63 Conyngham Street
  Glenside
  South Australia 5065
  Australia
  (*Orders*: Tel (08) 379–0444
             Fax (08) 379–4634)

India
  Affiliated East-West Press pvt Ltd
  G-1/16 Ansari Road
  New Delhi 110 002
  India
  (*Orders*: Tel (11) 327–9113
             Fax (11) 326–0538)

Japan
  Kanda Book Trading Co.
  Tanikawa Building
  3–2 Kanda Surugadai
  Chiyoda-Ku
  Tokyo 101
  Japan
  (*Orders*: Tel (03) 3255–3497
             Fax (03) 3255–3495)

Contents

Preface

The Mediterranean region represents a complex mosaic of continental, microcontinental and ophiolitic terranes, whose overall evolution has been controlled by relative movements between the African and Eurasian plates. Deciphering the sequence of tectonic events in this region can be likened to attempting to reconstruct all the pictures in a stack of jigsaw puzzles when 90% of the pieces are missing (and the remaining 10% are no longer their original shape!). Palaeomagnetic studies have played an important part in unravelling this 3D puzzle. The palaeomagnetic technique provides quantitative constraints on our reconstruction, since it can tell us which way each of the remaining pieces should be oriented (using magnetic declinations), their relative position with respect to the top of each picture (using magnetic inclinations), and in some cases which piece belongs to which picture (using magnetic dating). Perhaps more importantly, palaeomagnetism can also tell us something about the processes which led to the present confusion in our puzzle box.

This volume illustrates the increasingly diverse range of tectonic, magnetostratigraphic, volcanological and archaeological problems being addressed through palaeomagnetic research within the Mediterranean realm. The 33 papers herein span the full width of the Mediterranean basin and present results from Permian to Quaternary rocks. Together they provide a snap-shot of the current state of palaeomagnetic research in the region. As such, they form an intermediate step in solving our 3D puzzle and are not intended to represent the final set of pictures.

ANTONY MORRIS
DON H. TARLING

Palaeomagnetism and tectonics of the Mediterranean region: an introduction

A. MORRIS & D. H. TARLING

Department of Geological Sciences, University of Plymouth, Drake Circus, Plymouth, PL4 8AA, UK

The Mediterranean region has been one of the most intensely studied segments of the Alpine–Himalayan chain. Geological and geophysical studies have shown that the region represents a mosaic of microcontinental and ophiolitic terranes, resulting from a sequence of strike-slip and closure movements between the African and Eurasian margins of the Tethyan Ocean. Numerous early palaeomagnetic investigations suggested that many of these terranes underwent important tectonic rotations with respect to the major continents, e.g. the Iberian peninsula (Van der Voo 1969), Sardinia and Corsica (Westphal 1977), the Ionian Islands (Laj *et al.* 1982), and the Troodos ophiolite (Moores & Vine 1971). These rotations occur in a variety of geological settings, ranging from those active during oceanic crustal genesis to those associated with the late stages of continental deformation. More recent palaeomagnetic studies have provided an increasingly detailed picture of rotational deformation from the Atlantic margin to eastern Turkey. Tectonic rotations are now recognized on all scales from that of microplates down to individual thrust sheets. The papers contained in this volume, cover the full range of tectonic, magnetostratigraphic and archaeomagnetic problems currently being addressed in the region. Together they form a comprehensive review of an exciting and challenging field of research. To increase the usefulness of the volume to non-palaeomagnetists, a full glossary of palaeomagnetic and rock magnetic terms is provided.

Since the tectonic evolution of the Mediterranean Tethyan belt has been controlled by the relative motion history of the African and Eurasian plates, this introduction begins with a brief examination of the most commonly used apparent polar wander paths (APWPs) for these plates and the kinematic framework derived from the Atlantic Ocean spreading record. Interpretation of palaeomagnetic results obtained in the Mediterranean must be consistent with the plate tectonic framework defined by these data.

The African and Eurasian apparent polar wander paths

A full discussion of the myriad APWPs proposed for the African and Eurasian plates is beyond the scope of this brief review. Instead we restrict discussion to a comparison of the two most widely adopted sets of APWPs, those of Westphal *et al.* (1986) and Besse & Courtillot (1991). A fundamental problem encountered in constructing an African APWP is the relative lack of reliable Mesozoic poles from Africa itself. This problem is usually overcome by using palaeomagnetic data from other continents rotated into African coordinates.

The approach adopted by Westphal *et al.* (1986) was to construct pure Eurasian and African curves using what they considered to be the most reliable data for stable Eurasia and stable Africa (i.e. north of the Alpine belt and south of the South Atlas fault respectively). They then used the kinematics for the Atlantic bordering plates given by Savostin *et al.* (1986) to transfer other APWPs to Eurasia, calculated a mean synthetic APWP, and finally transferred this path to Africa. The resulting polar wander reference curves are shown in Figs 1a,b and 2a,b. A similar procedure was followed by Besse & Courtillot (1991), who used a more limited dataset of higher quality data which excluded some of the Soviet poles used by Westphal *et al.* (1986). Pure Eurasian, North American and Indian APWPs were transferred to the African reference frame and combined with African data to define a master APWP for Africa. This was then transferred back to the other reference frames to obtain master APWPs for the other continents. The Eurasian and African APWPs of Besse & Courtillot (1991) are shown in Fig. 1c,d and Fig. 2c,d respectively. It should be noted that uncertainties in poles from a single continent will be transferred to the final synthetic APWPs for all other continents by the process adopted by both Westphal *et al.* (1986) and Besse & Courtillot (1991). Potential problems with some of the

From Morris, A. & Tarling, D. H. (eds), 1996, *Palaeomagnetism and Tectonics of the Mediterranean Region*, Geological Society Special Publication No. 105, pp. 1–18.

1

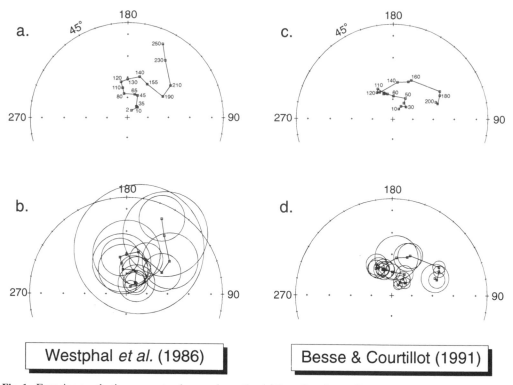

Fig. 1. Eurasian synthetic apparent polar wander paths. (**a**) Late Permian to Quaternary path of Westphal *et al.* (1986); (**b**) As (a) but showing α_{95} cones of confidence associated with each mean pole position; (**c**) Early Jurassic to Late Miocene path of Besse & Courtillot (1991); (**d**) As (c) but showing α_{95} cones of confidence associated with each mean pole position. Numbers refer to age of poles in Ma. Equal angle projections with minimum latitudes of 45°.

Jurassic North American poles used by Besse & Courtillot (1991) are discussed by Channell (this volume). In addition, the final curves are critically dependent on the plate kinematic model used, although Besse & Courtillot (1991) believe that errors incurred during APWP transfer are unlikely to exceed 1 or 2°.

A more intuitive way of examining these data is to plot the declination and palaeolatitude expected from the Eurasian and African paths at any particular locality as a function of time. An example is shown in Fig. 3 for an arbitrary location near Rome (42°N, 13°E). In this case the resulting reference curves show the declination and palaeolatitude which would be expected at Rome if the Italian peninsula had remained rigidly attached either to Eurasia or to Africa. Deviations between any observed data from the expected values are a measure of the relative palaeolatitudinal displacements or rotations of Italy with respect to stable Eurasia or Africa. This type of plot can be used to graphically show the affinity of sampled units to

the major bounding plates by comparing palaeolatitudes calculated from site mean inclinations to the expected values.

For the Tertiary period the differences between values derived from the Westphal *et al.* (1986) and Besse & Courtillot (1991) APWPs are within the bounds of errors expected in any palaeomagnetic study. Choice of reference curve for Tertiary directions, therefore, has little effect on inferred displacements and rotations. Significant differences in declinations expected from the African APWPs are observed in the Early Cretaceous segment. These could result in a 10° difference in inferred rotation angles depending upon choice of reference path. Of greater importance for palaeogeographic reconstructions are differences in expected African palaeolatitudes for the Mid-Jurassic, which differ by nearly 15° in the case of a locality near Rome. The African palaeolatitude expected at this locality from the path of Westphal *et al.* (1986) is nearly coincident with the Eurasian palaeolatitude expected from the path of Besse

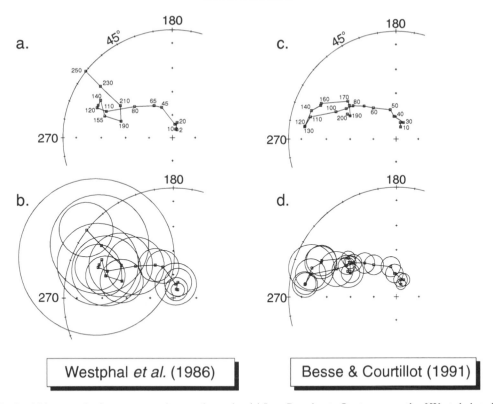

Fig. 2. African synthetic apparent polar wander paths. (**a**) Late Permian to Quaternary path of Westphal *et al.* (1986) (**b**) As (a) but showing α_{95} cones of confidence associated with each mean pole position; (**c**) Early Jurassic to Late Miocene path of Besse & Courtillot (1991); (**d**) As (c) but showing α_{95} cones of confidence associated with each mean pole position. Numbers refer to age of poles in Ma. Equal angle projections with minimum latitudes of 45°.

& Courtillot (1991). Mid-Jurassic results from Italy could potentially be interpreted as showing either African or Eurasian affinities of Italy, depending upon choice of reference path.

The relative motion history of the African and Eurasian plates

The first comprehensive kinematic model of the relative motion of the Atlantic-bordering continents based on the identification and fit of magnetic anomalies in the North and Central Atlantic oceans was made by Pitman & Talwani (1972). Their reconstruction suggested that Africa moved past Eurasia in a sinistral sense during the period from the Late Triassic to the Late Cretaceous. Dextral relative motion then occurred between the Late Cretaceous and the earliest Oligocene. However, more recent correlations by Livermore & Smith (1983), Savostin *et al.* (1986) and Dewey *et al.* (1989), based on more accurate sea-floor spreading data and

robust methods of analysis, do not support a change in the sense of relative motion at this time. Instead, these studies indicate a smooth sinistral strike-slip motion of Africa relative to Eurasia, which changes to NE-directed compression in the Late Cretaceous in response to opening of the North Atlantic Ocean (Fig. 4). The reconstructed positions of Africa and Eurasia for the Oxfordian, Santonian–Campanian boundary, and Tortonian according to Savostin *et al.* (1986) are shown in Fig. 5, along with superimposed palaeolatitudes calculated from the APWPs of Westphal *et al.* (1986). These clearly show the strike-slip and closure phases of the relative motion path and the progressive destruction of the Tethyan Ocean. Major events in the geodynamic evolution of the Mediterranean Tethys can be related elegantly to this kinematic framework. For example, the Mid-Jurassic phase of ophiolite emplacement in the Greek region coincides with the rapid eastward shift of Africa relative to

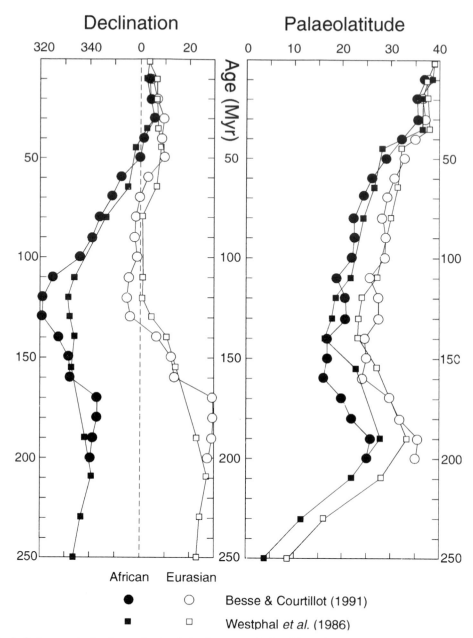

Fig. 3. Reference declination and palaeolatitude curves calculated from the apparent polar wander paths of Figs 1 & 2 for an arbitrary locality near Rome (42°N; 13°E). Solid symbols show African data; open symbols show Eurasian data; circles show values derived from the APWPs of Besse & Courtillot (1991); squares show values derived from the APWPs of Westphal *et al.* (1986). Note the close agreement between the two sets of paths over the last 100 Ma.

Eurasia at about 170 Ma, produced by initiation of spreading in the central Atlantic (Robertson & Dixon 1984). Compression and ophiolite emplacement in the Late Cretaceous in Turkey and Early to Mid-Tertiary in the Central Mediterranean are linked to the Late Cretaceous and Tertiary convergent phase (Robertson & Grasso 1995).

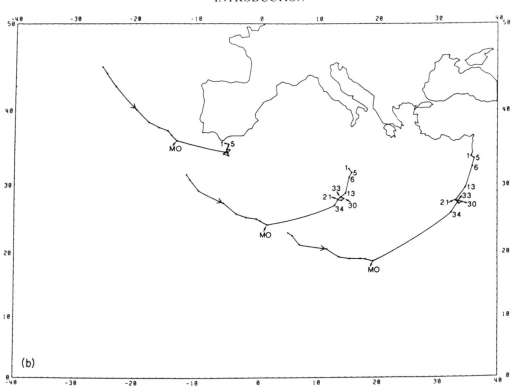

(b)

Fig. 4. Flow lines depicting the motion of three points on the present day African coastline with respect to Europe, based on magnetic anomaly fitting (from Dewey *et al*. 1989). The first points on the paths correspond to 175 Ma, whereas the points labelled MO correspond to 118 Ma. This relative motion path is essentially the same as that proposed by Savostin (1986), which was used in the construction of Fig. 5.

Overview of papers in the volume

This volume is organized into five sections. The first three address tectonic and magnetostratigraphic applications of palaeomagnetism and are presented in geographic order from west to east. The subdivisions used are somewhat arbitrary, and results from any one region can not be considered in isolation. For example, the Apulian platform of Italy, discussed in the Central Mediterranean section, is part of the foreland to the Hellenide orogenic system, discussed in the Eastern Mediterranean section. The final sections deal with volcanological applications and archaeomagnetism. The locations of field areas discussed by the various authors are shown in Fig. 6. We summarize below the main conclusions of the papers in each section.

Western Mediterranean

The tectonics of the western Mediterranean have been controlled by interaction between the Iberian, Eurasian and the African plates. The Iberian Meseta (Fig. 7) was consolidated in the Hercynian orogeny during the Late Carboniferous (Van der Voo 1993). Palaeomagnetic analyses of Permo-Carboniferous formations show that it underwent a 35° anticlockwise rotation relative to Europe (Van der Voo 1969). This rotation terminated before the Late Cretaceous and was driven by opening of the Bay of Biscay. To the south of the Iberian Meseta lies the highly arcuate Alpine chain of the Betic-Rif arc (Fig. 7). The first three papers concern the tectonic evolution of the Betic segment of this chain.

The Betic Cordillera is divided into Internal and External Zones. The Internal Zone comprises a series of metamorphosed Palaeozoic and Triassic rocks which were deformed in the Palaeogene as a result of African–Eurasian convergence. The External Zone consists of unmetamorphosed Mesozoic continental margin sequences, which were folded and thrust towards the NW onto the Iberian Meseta during the Early to Mid-Miocene as a result of

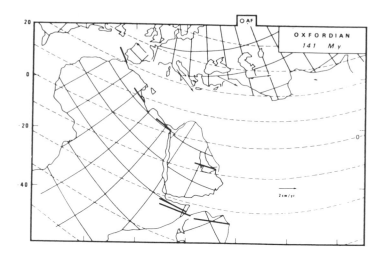

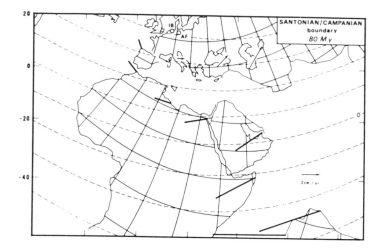

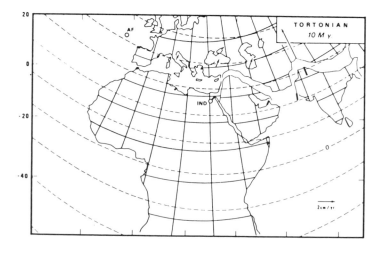

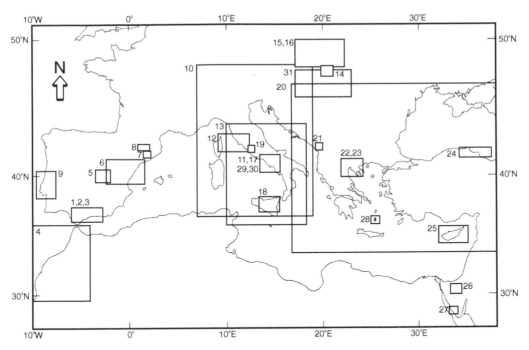

Fig. 6. Areas in the Mediterranean region discussed by different authors in this volume: 1, Kirker &
McClelland; 2, Villalaín *et al.*; 3, Feinberg, Saddiqi & Michard (Feinberg *et al. a*); 4, Khattach *et al.*; 5, Rey *et
al.*; 6, Juárez *et al.*; 7, Garcés *et al.*; 8, Keller & Gehring; 9, Pereira *et al.*; 10, Channell; 11, Iorio, Nardi *et al.*
(Iorio *et al. a*); 12, Mattei *et al.*; 13, Fedi *et al.*; 14, Márton & Márton; 15, Krs *et al.*; 16, Houša *et al.*; 17, Iorio,
Tarling *et al.* (Iorio *et al. b*); 18, McClelland *et al.*; 19, Florindo & Sagnotti; 20, Robertson *et al.*; 21, Mauritsch
et al.; 22, Kondopoulou *et al.*; 23, Feinberg, Edel *et al.* (Feinberg *et al. b*); 24, Piper *et al.*; 25, Morris; 26,
Kafafy *et al.*; 27, Abdeldayem & Tarling; 28, Bardot *et al.*; 29, de' Gennaro *et al.*; 30, Incoronato; 31, Márton.
Papers by Scalera *et al.* and Evans are not restricted to specific study areas.

convergence with the Internal Zones. Previous
palaeomagnetic studies in the External Betics
(e.g. Platzman & Lowrie 1992; Allerton *et al.*
1993) have found a consistent pattern of clockwise vertical axis rotations related to dextral
shear during deformation. Within the western
part of the External Zone, **Kirker & McClelland**
have carried out a high resolution palaeomagnetic study employing some innovative techniques to provide constraints on timing of
magnetization and on rotation mechanisms.
Their field area forms part of the Sierra del
Endrinal thrust system, where previous sampling by Platzman & Lowrie (1992) identified an
anomalously high clockwise rotation. **Kirker &
McClelland** stress the importance of integrated

structural and palaeomagnetic analyses, and
eschew the application of simple tilt corrections
about present bedding strikes in such tectonically complex settings. Instead, they use a
comparison of bedding/remanence angles with
appropriate reference directions to distinguish
pre-, syn- and post-folding remanence components. Intermediate unblocking temperature
components are found to represent a Miocene
syn-deformation remagnetization, whereas high
unblocking temperature components are primary Late Jurassic magnetizations. The net
tectonic rotation at each site was then calculated, being defined as the axis and angle of
rotation which simultaneously restores the
present bedding pole to the vertical and the

Fig. 5. Reconstructed positions of Eurasia, Africa, Arabia and India (from Savostin *et al.* 1986). (**a**)
Oxfordian, Late Jurassic (141 Ma); (**b**) Santonian–Campanian boundary, Late Cretaceous (80 Ma); (**c**)
Tortonian, Late Miocene (10 Ma). Palaeolatitudes calculated from the APWPs of Westphal *et al.* (1986) are
superimposed in 10° intervals. Present latitudes and longitudes on the continents are shown at 10° intervals.
Note the sinistral movement of Africa with respect to Eurasia between (a) and (b), and the almost complete
closure of the Tethys Ocean between (b) and (c).

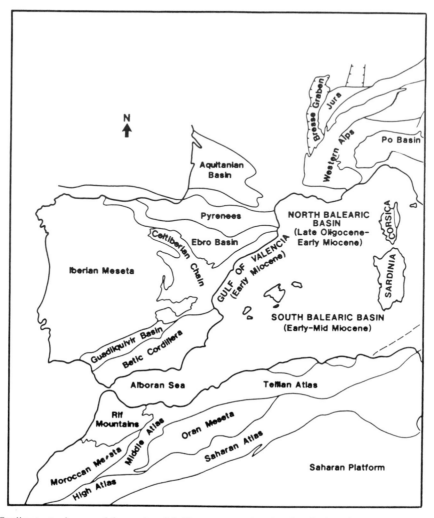

Fig. 7. Outline tectonic map of the western Mediterranean region showing the location of major units (modified from Dewey *et al.* 1989). Not all the units named are discussed by authors in the 'Western Mediterranean' section of the volume.

mean remanence to the appropriate reference direction. This construction avoids the errors which may be introduced by arbitrarily dividing the deformation at a site into a tilt and a vertical axis rotation. **Kirker & McClelland** then go on to deconstruct these net tectonic rotations by referring to the sequence of structural events proposed for the region. The results document a two phase rotation history involving large clockwise rotation on gently-dipping thrust faults and subsequent folding. No purely vertical axis rotations or simple tilting are required to explain the anomalous declinations. The implications of this study for future palaeomagnetic

analyses in complex structural terrains are significant.

The Neogene remagnetization of the Upper Jurassic limestones of the western Betics is also discussed by **Villalaín *et al*.** They show that the secondary component has similar unblocking temperature and coercivity spectra to the pre-folding (presumed Jurassic) component, and could easily be misinterpreted as the primary magnetization. **Villalaín *et al*.** suggest that some of the data previously obtained by Platzman & Lowrie (1992) may be more easily explained if the characteristic magnetization is re-interpreted as the Neogene remagnetization.

They point out that such remagnetizations can provide tectonically useful information, provided that careful field tests are performed. This point is taken up by **Feinberg** *et al.* **(a)**, who show how tectonic constraints can be obtained from remagnetized ultrabasic massifs, even when accurate tilt corrections are *impossible* to define. These authors have sampled the Ronda peridotites within the metamorphic Internal Zone of the Betics. The characteristic remanence postdates folding of the internal foliation within the peridotites, whereas the presence of both normal and reverse polarity components is evidence of an extended period of magnetization acquisition. Remagnetization probably took place during progressive exhumation and cooling following a Late Oligocene–Early Miocene metamorphic event. By modelling the effects of geologically reasonable amounts of tilting on the appropriate Early Miocene reference direction, **Feinberg** *et al.* **(a)** demonstrate that observed easterly declinations result from post-metamorphic clockwise rotation of the peridotites, rather than from significant tilting. Consideration of additional data from the Beni-Bousera peridotite massif to the south of the Alboran Sea leads **Feinberg** *et al.* **(a)** to conclude that opposing rotations of the Betic and Rifian peridotites occurred during the Early Miocene as a result of extensional collapse of Alboran thickened crust.

To the south of the Rifian segment of the Betic-Rif chain, the Hercynian basement of the Moroccan Meseta domain is separated from stable Africa by the High Atlas and Anti-Atlas fold belts (Fig. 7). **Khattach** *et al.* review available palaeomagnetic data from Palaeozoic rocks of the Meseta and Anti-Atlas and again find evidence for a widespread remagnetization event. In this case, resetting of radiometric ages to around 290–300 Ma strongly supports a late Hercynian (Late Carboniferous to Permian) age for the remagnetization. Essentially identical directions are observed throughout the coastal Meseta, central Meseta and Anti-Atlas regions, indicating no significant relative tectonic rotations or palaeolatitudinal changes between the major terranes south of the Rif since the Permian.

The *Iberian* Meseta is bounded to the NE by the major intraplate mountain belt of the Iberian Range (or Celtiberian Chain; Fig. 7), which formed in a Palaeogene compressive regime by tectonic inversion of an early Mesozoic rift basin. The range is subdivided into two branches separated by Tertiary grabens: the Aragonese branch to the NE and the Castilian branch to the SW. **Rey** *et al.* have analysed five sections of

continental and marginal marine sediments within the Castilian branch to determine a Middle Triassic magnetostratigraphy. While these sequences have not been pervasively remagnetized, detailed analyses show that the NRM has a complex structure including a primary Triassic component, a PTRM produced during uplift, and a Tertiary–Recent CRM overprint carried by goethite and authigenic hematite. This young overprint is not entirely removed by thermal demagnetization experiments, and introduces a slight bias in the resulting Triassic magnetostratigraphy. **Rey** *et al.* cancel out this bias by averaging data from successive adjacent magnetozones. The tectonics of the Aragonese branch are examined by **Juárez** *et al.*, who attempt to determine whether palaeomagnetic data from the Iberian Range should be incorporated into the Iberian APWP. Oxfordian pelagic limestones are found to contain high unblocking temperature primary components of both normal and reverse polarity. The primary directions are consistent with the expected Jurassic direction for Iberia, and reflect the 35° regional anticlockwise rotation. A consistent, *pre-folding*, normal polarity secondary magnetization of probable Early Cretaceous age is also present. The angular difference between this and the Jurassic component indicates that 15° of anticlockwise rotation of Iberia was complete by the Early Cretaceous. The southernmost section sampled, however, appears to have undergone a 45–50° clockwise rotation with respect to the other sections after remagnetization. Although declination errors introduced by simple tilt corrections can not be completely excluded, **Juárez** *et al.* suggest that this clockwise rotation relates to dextral shear along a major NW-SE trending basement fault. They conclude that inclusion of Iberian Range data in the Iberian APWP is not always justified.

Between the Iberian Range and the Pyrenees lie the Ebro basin, Catalan Coastal Ranges, and the emerged margins of the Valencia Trough of the western Mediterranean (Fig. 7). The latter feature is a Late Oligocene to Recent sedimentary basin characterized by highly attenuated continental crust (Banda & Santanach 1992). Late Oligocene to Early Burdigalian rifting of the Catalan margin during opening of the Valencia Trough resulted in formation of the Vallés-Penedés half-graben (Bartrina *et al.* 1992), in which a variety of alluvial and shallow marine sediments were deposited. The fine-grained facies have been studied by **Garcés** *et al.* who examine the influence of sedimentary facies on the degree of inclination shallowing. They demonstrate that finely-laminated distal alluvial facies are the least reliable recorders of geomag-netic field inclination, whereas reworked and massive facies yield more reliable data. The degree of inclination bias can be quantified using anisotropy of magnetic susceptibility measure-ments, and a linear correlation is found between $\ln(P)$ and $\ln(\tan(I_{NRM}))$, where P is the degree of anisotropy and I_{NRM} is the mean remanence inclination. This relationship provides a method for correcting observed inclinations, by linear regression to find the value of I_{NRM} correspond-ing to $P = 1$ (i.e. the case of no sediment anisotropy). The origin of the shallower-than-expected inclinations frequently encountered in Tertiary sedimentary, volcanic and plutonic units in the Mediterranean realm is the subject of on-going debate. Alternative explanations in-clude northward palaeolatitudinal movements (e.g. Beck & Schermer 1994) and departures from a geocentric axial dipole field (Westphal *et al.* 1993). It is clear from the analysis of **Garcés *et al.***, however, that thorough testing for facies control on magnetic inclination should be per-formed before resorting to tectonic or geomag-netic explanations for shallow inclinations in sedimentary units.

The major Alpine orogenic belt of the Pyrenees was primarily formed by Cretaceous to Early Neogene collision between the Iberian plate and Eurasia. It consists of two foreland belts and a central Axial Zone, from which thrust sheets diverge and are displaced to both north and south, thereby accommodating con-siderable shortening. Palaeomagnetic studies in the Pyrenees show the importance of vertical axis rotations during the collisional phase (e.g. McClelland & McCaig 1989; Dinarés *et al.* 1992). In a study of the SE Pyrenees, **Keller & Gehring** demonstrate that the relatively neg-lected Neogene *post-collisional* phase also in-volved significant rotations. By applying a geometrical correction to compensate for the effect of post-magnetization plate motion, Keller & Gehring isolate the effect of intra-plate deformation on the observed magnetization directions. The data are then used to reconstruct the Palaeogene thrust geometry in the SE Pyrenees. Removal of the effects of the post-collisional deformation leads to a reduced estimate of the amount of shortening due to plate collision. The authors conclude that the small component of post-collisional deformation has strongly influenced the present-day ge-ometry of the orogenic belt.

In the final paper in this section, **Pereira *et al.*** describe results obtained from igneous rocks within the Lusitanian Basin of Portugal, which forms the western margin of the Iberian Meseta.

Two phases of magmatic activity affected the basin. Palaeomagnetic directions from the ig-neous rocks of the older Upper Jurassic to Early Cretaceous episode reflect the Early to Late Cretaceous 35° anticlockwise rotation of the Iberian plate. A more complex distribution of directions is observed in the younger Late Cretaceous to Eocene intrusions, and is at-tributed by Pereira *et al.* to localized rotations about both vertical and horizontal axes (prob-ably during the Miocene).

Central Mediterranean and Carpathians

The papers in this section are divided into those which address regional tectonics and those which describe magnetostratigraphic appli-cations of palaeomagnetism. The tectonic de-bate in the Central Mediterranean has been dominated by discussion of the motion history of the Adria microplate, and the extent to which it may be considered a fixed promontary of the African plate. Early palaeomagnetic studies (e.g. VandenBerg *et al.* 1978) suggested that Adria (peninsular Italy) experienced an Eocene anticlockwise rotation with respect to Africa. This interpretation is fraught with uncertainty, however, since the reference APWP for Africa is ill-defined and localised anticlockwise rotations of thrust sheets are frequently observed. **Chan-nell** provides an up-to-date review of palaeo-magnetic data relevant to this problem. Attention is drawn to uncertainties in some of the North American reference poles which are commonly rotated to African coordinates and used to define the African APWP. **Channell** selects a restricted set of six Jurassic–Cretaceous poles from North America which are then rotated to provide a revised African APWP for this time interval. Permian to Cretaceous poles from the Southern Alps and Cretaceous poles from Istria, Iblei, Gargano and Apulia (Fig. 8) are consistent with the African APWP. This implies no significant rotation of Adria with respect to Africa. In contrast, pole paths from the Umbria–Marche region of Italy and the Transdanubian Central Range of Hungary (Fig. 8) are African in shape but are displaced from the African path. These discrepancies are attributed by **Channell** to local anticlockwise rotations of thrust sheets. Further constraints on the palaeogeography are provided by data from the Northern Calcareous Alps, which are shown to be detached from the Southern Alps during the Late Triassic and Jurassic.

The difficulties in obtaining reliable palaeo-magnetic data from the weakly magnetized sedimentary facies so prevalent throughout the

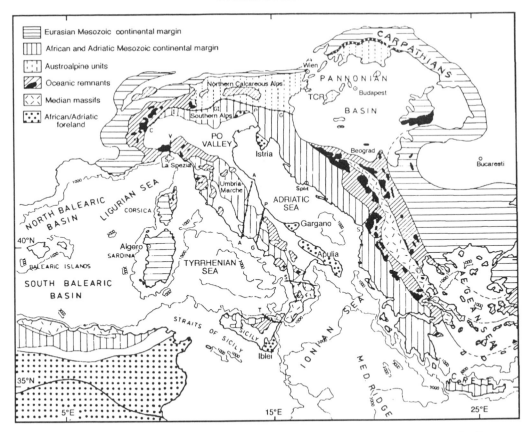

Fig. 8. Outline tectonic map of the central Mediterranean region showing the location of major units (from Channell this volume). Not all the units named are discussed by authors in the 'Central Mediterranean & Carpathians' section of the volume. TCR, Transdanubian Central Range of the Pannonian Basin.

Mediterranean realm is exemplified by the paper by **Iorio** *et al.* (*a*). These authors have attempted to document relative rotations between the Monte Matese block of the Southern Apennines and Apulia. Many of the Miocene localities visited were found to be affected by post-depositional disturbances, while magnetic rem-anences at the majority of sites sampled were found to be dominated by recent field direction overprints. Tectonically meaningful data were obtained at eight sites (representing only 5% of the total sample collection) within Lower to Middle Miocene carbonates at Pescorosito. Previous data from Lower Cretaceous car-bonates just to the east indicate a 40° difference in declination with respect to coeval data from Gargano. This angle is essentially the same as the difference between the new Pescorosito mean direction and the modelled Miocene reference direction, suggesting that much of the anticlockwise rotation documented at the Lower Cretaceous sites took place since at most Middle

Miocene times. **Iorio** *et al.* (*a*) conclude that other tectonic rotations within the Southern Apennines chain are likely to be equally young.

The Neogene deformation of Adria is also examined by **Mattei** *et al.* who show that the Tyrrhenian margin of Italy experienced no significant rotation since the Late Miocene, and may be considered as a non-rotating hinterland for the widespread rotational deformation of the Apennines. They suggest that rotations ob-served in the Apennines can not be linked to post-Tortonian opening of the Tyrrhenian Sea. Their preferred model involves initial thrust-related rotations within the external zones, eastward migration of the deformation front, and finally extensional collapse and rifting of the western sector. These new data from the Tyrrhenian margin lead to further subdivision of the Italian peninsula into rotated and non-rotated domains. An alternative to palaeomag-netic documentation of this pattern of domains is discussed by **Fedi** *et al.* They identify a number of aeromagnetic anomalies in the region with peak-to-trough axes at angles to the local magnetic meridian, and suggest that these can only be effectively explained by the presence of significant components of remanent magnetization. By using an inversion method to estimate the magnetization direction associated with each anomaly, **Fedi et al.** have been able to document the first-order distribution of rotated domains throughout southern Italy. A close agreement is observed with the rotation pattern shown by traditional palaeomagnetic studies. Unlike palaeomagnetic analyses, however, this alternative technique gives no information on the timing of tectonic rotation.

To the NE of Adria, the Carpathians represent a highly arcuate eastward-vergent Alpine fold and thrust belt (Fig. 8), bearing striking structural similarities to the westward-vergent Betic–Rif arc of the western Mediterranean. Recent geological models suggest that the Western Carpathians were influenced by eastward tectonic escape of the Transdanubian Range domain away from a collision zone between the Northern Calcareous Alps and the Southern Alps (Mauritsch & Márton 1995). The onset of escape is variously dated to Late Eocene–Early Oligocene or Early Miocene times, while final emplacement is dated at either the end of the Palaeogene or the Early Miocene (references in Mauritsch & Márton 1995). Data from the North Hungarian Central Range, which forms part of the Transdanubian Range domain, are directly relevant to these models and are presented by **Márton & Márton**. The data document a two-phase history of tectonic rotation. The first stage involved a large-scale northward drift and an accompanying 40–60° anticlockwise rotation during the Burdigalian. This was followed by a second phase of deformation which produced an additional 30° anticlockwise rotation of pre-early Serravalian age, presumably related to final collision with the Northern Calcareous Alps and resultant tectonic escape. The full palaeomagnetic dataset from the Transdanubian Central Range, Southern and Northern Calcareous Alps is sufficiently detailed to resolve the relative motion history between the three terranes, and is discussed in detail by Mauritsch & Márton (1995). Available data from the Western Carpathians to the ENE of the Northern Calcareous Alps are summarized and reviewed by **Krs et al.** Seemingly disparate pole positions obtained from sites throughout the region are observed to lie on small circles centred on the study area. This is taken by **Krs et al.** to reflect the influence of vertical axis rotations on the pole distribution.

Anticlockwise rotations are shown to predominate in this sector of the Carpathian arc and are observed in all three tectonostratigraphic zones, namely the Inner Carpathians, Klippen Belt and Outer Carpathians. The inclination data indicate that the Western Carpathians underwent a rapid northward drift away from initial southern equatorial latitudes during the Permo-Triassic.

A range of magnetostratigraphic applications are presented in the second half of this section. These are arranged by age rather than by geological setting. The first paper, by **Houša et al.**, addresses the long-standing problem of the correlation of the Jurassic–Cretaceous boundary between the Tethyan and Boreal realms. This important boundary is well-defined by biostratigraphy, but exact biostratigraphic correlation between realms is impossible due to an absence of common index species and only poorly defined records in transitional areas. Magnetostratigraphic correlation offers an obvious solution to this problem since polarity reversals are globally synchronous events. **Houša et al.** show, however, that no consensus has been reached on which polarity chron contains the Jurassic-Cretaceous boundary in the Tethyan realm. They discuss new palaeomagnetic data from two sections in the western Carpathians where biostratigraphic control is provided by abundant *caplionellid* faunas. The combined magnetostratigraphic and biostratigraphic datasets locate the boundary within chron M19n (at approximately 149 Ma). Moving up the geological column, **Iorio et al.** (*b*) have erected a detailed polarity chronology for the Hauterivian and Barremian (Early Cretaceous) by analysis of bore cores of shallow-water carbonates from Monte Raggeto in the Southern Apennines. The ultra-fine sequence of reversals recorded by these sediments enables directional changes in the geomagnetic field to be studied at a scale comparable to that of secular variation. Palaeomagnetic parameters do not appear to be related to textural and lithological variations down core, but are instead controlled by geomagnetic factors. Periodicities within the records are found to correlate with predicted Milankovitch periodicities for the Cretaceous. This study serves to highlight the potential of painstaking, detailed palaeomagnetic analyses of suitable lithologies for improving understanding of geomagnetic field variations over geological timescales.

The magnetostratigraphy of Late Miocene sections within two sub-basins of the Caltanisetta basin of Sicily is used by **McClelland et al.** to determine the onset of the Messinian

Salinity Crisis in southern Italy. A commonly encountered problem in palaeomagnetism is the distinction between primary remanences and secondary Recent magnetic components within relatively young geological units. Despite extensive Recent overprinting, **McClelland et al.** have been able to isolate primary or early diagenetic remanences in the studied sections because Pliocene tilting has rotated the pre-tilt remanence away from the overprint direction. The resulting magnetostratigraphies have been correlated with the geomagnetic polarity timescale using additional constraints provided by observed variations in sedimentation rate. The evaporite phase marking the onset of the salinity crisis is found to occur within opposite polarity chrons in the two sections, giving independent evidence for tectonic control on the timing of evaporite formation within the basin. The potential for magnetostratigraphic errors introduced by secondary magnetic components in sections where field tests can not be performed is exemplified by the case study of **Florindo & Sagnotti**. They present results of re-sampling of the Pliocene Valle Ricca section near Rome, which show that a previously identified normal polarity zone within the reversely magnetized section arises from a strong, secondary component of magnetization in sediments close to an ash bed. The normal component is carried by the iron sulphide greigite and is either of late diagenetic origin or is associated with self-reversal behaviour. In either case, correlation of the normal polarity zone with the Reunion sub-chron is demonstrated to be invalid. This example illustrates the dangers of blanket treatment of samples at a single demagnetization level and emphasises the need for thorough demagnetization and rock magnetic experiments of the type carried out by **Florindo & Sagnotti**.

The final paper in this section does not strictly concern the Central Mediterranean region, but instead describes procedures which may be followed when using the Global Palaeomagnetic Database (McElhinny & Lock 1990). **Scalera et al.** point out the need for some form of quality filtering to be applied when selecting data from the 7000+ palaeopoles included in the database. They discuss several interesting effects of data filtering with examples drawn from the Tethyan Belt, and Italy in particular.

Eastern Mediterranean

To the east of Adria lies the complex assemblage or microcontinental and ophilitic terranes of the eastern Mediterranean Tethyan belt (Fig. 9). Late Tertiary to Quaternary reconstructions of this region are relatively well-constrained. In contrast, no overall consensus has been reached on the Mesozoic to Early Tertiary regional palaeogeography. **Robertson et al.** summarize and discuss the range of alternative reconstructions which have been proposed. These can be ascribed to three main variants which differ both in the number of inferred microcontinental units and intervening ocean basins present in the region during the Mesozoic, and in the direction of subduction of Tethyan crust (northwards versus southwards) during basin closure. The preferred model of **Robertson et al.** is one in which continental fragments rifted away from Gondwana during the Late Permian, Early–mid Triassic and Early Cretaceous, and then drifted across Tethys to become amalgamated with Eurasia. The large ophiolites throughout the eastern Mediterranean (e.g. Troodos, Pindos) are considered to have formed in above subduction zone settings, and were rooted in a number of northerly and southerly oceanic strands. Final closure of Tethys in the eastern Mediterranean occurred in the Early Tertiary, leaving only a small remnant in the present easternmost Mediterranean Sea.

Regardless of the overall framework adopted, it is clear that large-scale tectonic rotations were associated with the progressive closure of the eastern Mediterranean Tethys, especially during the later stages of (micro)continental collision. A prerequisite for successful interpretation of palaeomagnetic data from Mesozoic units is adequate coverage from Cenozoic (especially Neogene) sequences. Much of the key work has been undertaken by Kissel, Laj and various co-authors at Gif-sur-Yvette, Paris, who have obtained palaeomagnetic data from some 750 sites in Albania, Greece and Turkey (Kissel & Laj 1988; Kissel et al. 1993; Speranza et al. 1995). These data, and those obtained by other workers (e.g. Horner & Freeman 1983), provide a framework of Cenozoic rotational deformation in the eastern Mediterranean. The curvature of the Hellenic arc is shown to have been acquired tectonically by opposing rotations of its western and eastern terminations. In the west, the data suggest that the external zones of the Albanide–Hellenide belt (Fig. 9) rotated clockwise during the Neogene as a coherent unit from northern Albania to the Peloponnesos (Speranza et al. 1995). **Mauritsch et al.** present results from the Scutari–Pec transverse zone (Fig. 9), which is the inferred northern boundary of this major rotated unit. Both clockwise and anticlockwise rotations are documented close to the Scutari–Pec line. To the north, westerly declinations are consistent with those observed in the Northern

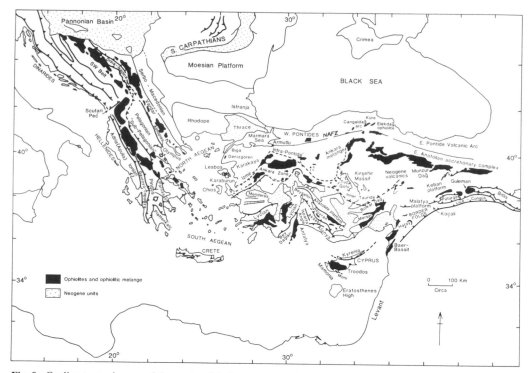

Fig. 9. Outline tectonic map of the eastern Mediterranean region showing the location of major units (from Robertson *et al*. this volume). Not all the units named are discussed by authors in the 'Eastern Mediterranean' section of the volume.

Dinarides (Márton *et al*. 1990) and are not significantly different from predicted African palaeodirections. Clockwise rotations observed to the south of the Scutari–Pec line are consistent with those observed throughout Albania. The data clearly support the idea that the Scutari-Pec line accommodates important relative crustal rotations. However, **Mauritsch et al**. conclude that the external zones to the south did not rotate as a single unit during the Neogene, but instead were divided into a series of fault-bounded blocks which rotated independently in the same sense in response to a common driving force.

Both **Kondopoulou et al**. and **Feinberg et al**. (*b*) examine the more internal zones of the Hellenides in Northern Greece. **Kondopoulou et al**. summarize evidence for systematic Tertiary clockwise rotations extending from southern Albania to the western Greek Rhodope (Fig. 9). In contrast, the eastern Greek Rhodope appears to be unrotated. Clearly the Rhodope can no longer be considered as a relatively stable part of Eurasia, but instead represents a mobile part of the orogenic belt. By integrating other geophysical and geological data, **Kondopoulou et al**.

compare the North Aegean region during the Early Tertiary to the present day active South Aegean subduction zone. They suggest that the observed clockwise rotations in the North Aegean were associated with deformation of a palaeosubduction zone, in a manner similar to that proposed for the Neogene Hellenic arc by Kissel & Laj (1988). The complex tectonic history of this region is further illustrated by **Feinberg et al**. (*b*), who identify multi-component remanences within ophiolitic units which are used to document two separate rotation phases. Two distinct magnetization directions are interpreted as magnetic overprints. The most recent component can be correlated with an Eocene metamorphic event produced by emplacement of the Sithonia granite of the Chalkidiki, and suggests a post-Eocene clockwise rotation of *c*.45°. This is consistent with the sense of rotation observed in other studies. The older remagnetized component appears to result from the Late Jurassic – Early Cretaceous phase of ophiolite emplacement and granite intrusion, and indicates an earlier 70–80° *anticlockwise* rotation of Northern Greece (after removing the effects of the later post-Eocene clockwise rotation). **Feinberg et al.** (*b*) partition this post-Early Cretaceous to pre-Eocene rotation into a component due to movement of the African plate as a whole, and an additional component related to opening of a small ocean basin to the south of the Pelagonian zone (Fig. 9) during the Cretaceous. Alternative explanations are possible, however, and include the effects of indentation by micro-continental units (Olympos, Parnassos; Fig. 9) during closure of the Pindos ocean to the west (as suggested for the southern Pelagonian zone by Morris (1995)).

A controlling influence on Neogene to Recent deformation in the Eastern Mediterranean has been the westward tectonic escape of the Anatolian block away from a zone of incipient continental collision between Arabia and Eurasia in eastern Turkey (Fig. 9). Movement of Anatolia is accommodated by two strike-slip fault zones. The Bitlis suture zone in southeast Turkey (Fig. 9) is predominantly a compressional feature with a small component of sinistral strike-slip. In contrast, the North Anatolian Fault zone (NAFZ) is a major dextrally-slipping intracontinental transform fault, forming the boundary between the Anatolian block and Eurasia to the north (Fig. 9). Dextral movement along the North Aegean Trough (NAT), which forms the lateral continuation of the NAFZ in the North Aegean, has resulted in clockwise rotation of Lemnos island (Westphal & Kondopoulou 1993; **Kondopoulou et al.**). Other clockwise rotations in Evvia further to the west (Kissel *et al.* 1989) have been attributed to an inferred distributed shear zone linking the NAFZ/NAT to the Hellenic trench (McKenzie & Jackson 1989). Against this background, **Piper et al.** present a case study of the central part of the NAFZ itself. Recent theoretical models for distributed rotational deformation across intracontinental transforms would predict a broad zone of clockwise rotations either side of the main North Anatolian fault trace. **Piper et al.** find no evidence for such deformation, however, with the exception of a narrow zone of clockwise rotation where the transform breaks the surface. Instead, variable *anticlockwise* rotations to the north of the NAFZ are attributed by **Piper** *et al.* to back-arc opening of the Black Sea and/or deformation of the Pontide orogenic belt prior to initiation of the NAFZ. Anticlockwise rotations to the south possibly result from lateral translation of blocks along side-splay faults. These results confirm those obtained in a recent study of volcanic rocks along the NAFZ by Platzman *et al.* (1994), who observed no clockwise rotations but 30° of

post-Miocene anticlockwise rotation across the fault zone. Platzman *et al.* (1994) suggest that the entire Anatolian block may be undergoing anticlockwise rotation as a consequence of the collision of Arabia along the Bitlis suture.

One of the key elements in the tectonic mosaic of the eastern Mediterranean is the Troodos ophiolite of Cyprus (Fig. 9). In contrast to other major Tethyan ophiolites in the Mediterranean (e.g. Pindos, Baër Bassit), the Troodos massif never experienced large-scale overthrusting and is essentially *in situ* (Robertson & Xenophontos 1993). The Troodos has been the subject of a large number of palaeomagnetic investigations by various authors, which are summarized by **Morris**. These have addressed both the role of the ophiolite in the regional palaeogeography and the interpretation of the primary sea-floor structures preserved within the Troodos massif. The timing of the well-known 90° anticlockwise rotation of the ophiolite has now been documented in detail. Rotation began soon after eruption of the Troodos extrusives and was complete by the Early Eocene, with at least 30° and possibly up to 45° of rotation taking place within 15 Ma of formation. No single rotation mechanism has been accepted, but a popular idea is that rotation resulted from an anticlockwise torque produced by collision of a trench with the Arabian continental margin to the east. Other palaeomagnetic studies have documented dextral shear along the fossil Arakapas transform fault (now known as the Southern Troodos Transform Fault) and demonstrated the importance of extensional faulting and block tilting in the spreading process.

The last two papers in this section present data from the Sinai peninsula, which is considered to represent a microplate which separated from Africa as a result of Tertiary rifting in the Gulf of Suez. Both **Kafafy et al.** and **Abdeldayem & Tarling** show that the Sinai has not rotated significantly with respect to the main African plate. Important deformation has, however, taken place. **Abdeldayem & Tarling** describe results from Palaeocene, Lower Eocene and Miocene sediments which demonstrate that southwest Sinai is characterized by local tectonic tilting of normal fault blocks. They conclude that tilting is associated with the opening of the Gulf of Suez and that the Sinai region can not be considered as a single rigid plate. The Cretaceous Nubian Sandstones studied by **Kafafy** *et al.* were subjected to an earlier Cenomanian to Maastrichtian phase of deformation, which produced a NE–SW-trending series of folds. An extensive remagnetization identified by **Kafafy** *et al.* is probably of Miocene age, however, and is not related to the main deformation phase. Instead they suggest that the characteristic magnetization of the Nubian Sandstones in the studied region is a chemical remanence acquired by fluid migration during arid Miocene climatic phases.

Applications in volcanology

The remaining papers in the volume highlight the flexibility of the palaeomagnetic method in other fields of research. Palaeomagnetic studies can provide useful insights into volcanic processes. For example, magnetic methods are now widely used to determine the emplacement temperature of pyroclastic flows through analyses of lithic clasts. These acquire a partial thermoremanence (PTRM) during *in situ* cooling to ambient temperature after deposition in a flow. Providing that the original, pre-eruption high blocking temperature remanence is not completely reset during eruption, then an estimate of the emplacement temperature can be determined by thermal demagnetization of the lithic clasts, and is given by the temperature at which the lower blocking temperature PTRM is destroyed. **Bardot *et al.*** apply this method to provide constraints on eruption mechanisms for the volcano of Santorini in the Aegean, through analysis of lithic clasts from the Thera Pyroclastic Formation. **Bardot *et al.*** stress the need for careful checks to ensure that clasts were not altered during eruption, which would produce chemical remanence overprints and result in erroneous emplacement temperature determinations. The authors carried out complementary palaeointensity experiments to test the thermal origin of magnetization. This is a time-consuming procedure but must be viewed as essential if reliable constraints are to be obtained. **De'Gennaro *et al.*** review results of a similar study of the Neapolitan Yellow Tuff around Naples, where studies of zeolitization of trachytic pumices have been used to support the magnetically determined emplacement temperatures in the absence of palaeointensity analyses. They show that scatter of the low blocking temperature PTRM directions and essentially random magnetic fabrics within some of these deposits reflect post-emplacement dislocations of lithic clasts, providing further information on the lithification process. In a related case study, **Incoronato** has been able to distinguish and magnetically date historical lava flows produced during various eruptions of Vesuvius. The method employed is essentially that used in archaeomagnetic dating, whereby

well-defined magnetization directions are compared with records of secular variation of the geomagnetic field obtained from observatory records, archaeological artefacts and other dated lava flows. **Incoronato** stresses the importance of correct identification of characteristic remanent magnetizations by detailed demagnetization experiments.

Archaeomagnetism

The final two papers in the volume give a taste of the broad and varied subject of archaeomagnetism. In addition to magnetic dating of historical artefacts such as oven walls and fired pottery (which acquire a thermoremanent magnetization during cooling from elevated temperatures), archaeomagnetic studies have provided a wealth of geomagnetic information. Archaeological records of the direction and strength of the geomagnetic field complement and extend observatory records. The resultant wider coverage of the spectrum of geomagnetic variations is a prerequisite for improved models of geomagnetic field behaviour. **Evans** gives an overview of both archaeological and geomagnetic results obtained to date throughout the Mediterranean realm, while **Márton** focuses on the detailed records which are now available from Hungary.

Conclusions

This volume only gives a taste of the variety of tectonic, volcanological and archaeological research in the Mediterranean region using magnetic methods. The geographical and geological coverage of the papers makes the volume an ideal entry point to the growing amount of literature on Mediterranean palaeomagnetism. Several overall conclusions can be drawn from the case studies presented herein. Firstly, rotational deformation must be considered as an integral part of the overall deformation in orogenic belts and can not be ignored in palaeogeographic reconstructions. Tectonic rotations on a range of scales are shown to be ubiquitous throughout the mobile belts of the Mediterranean region, with non-rotation being the exception rather than the rule. This should of course be immediately apparent given the highly arcuate geometries of many of the mountain chains. Secondly, the importance of widespread remagnetization events at various stages in the evolution of the Mediterranean orogenic belts is becoming increasingly apparent. A positive note is sounded by the papers in this volume, however, which show that the resulting secondary magnetizations can be as useful as primary directions in providing valuable tectonic constraints. The exceptions are those units which record only present field components of magnetization related to recent weathering. Finally, the range of problems tackled using palaeomagnetism is increasing. This is partly related to refinements in data analysis techniques, but must also be due to a raised awareness of the inherent flexibility of the palaeomagnetic method.

We would like to thank all who participated in the London meeting, and the friendly and ever helpful staff of the Geological Society. We are grateful to the Joint Association for Geophysics for financial support which enabled Eastern European colleagues to attend and make valuable contributions to the London meeting. Our special thanks to the contributors to this volume, including all who kindly provided reviews, for their prompt responses to our many requests. We are grateful to Angharad Hills at the Geological Society Publishing House for her guidance and assistance in producing this volume. Finally, A. M. wishes to thank Fotini and Alice for encouragement and support throughout this project.

References

ALLERTON, S., LONERGAN, L., PLATT, J. P., PLATZMAN, E. S. & McCLELLAND, E. 1993. Paleomagnetic rotations in the eastern Betic Cordillera, southern Spain. *Earth and Planetary Science Letters*, **119**, 225–241.

BANDA, E. & SANTANACH, P. 1992. The Valencia Trough (western Mediterranean): an overview. *Tectonophysics*, **208**, 183–202.

BARTRINA, M. T., CABRERA, L., JURADO, M. J., GUIMERA, J. & ROCA, E. 1992. Evolution of the central Catalan margin of the Valencia Trough (western Mediterranean). *Tectonophysics*, **203**, 219–247.

BECK, M. E. & SCHERMER, E. R. 1994. Aegean palaeomagnetic inclination anomalies. Is there a tectonic explanation ? *Tectonophysics*, **231**, 281–292.

BESSE, J. & COURTILLOT, V. 1991. Revised and synthetic apparent polar wander paths of the African, Eurasian, North American and Indian plates and true polar wander since 200 Ma. *Journal of Geophysical Research*, **96**, 4029–4050.

DEWEY, J. F., HELMAN, M. L., TURCO, E., HUTTON, D. H. W. & KNOTT, S. D. 1989. Kinematics of the western Mediterranean. *In*: COWARD, M. P., DIETRICH, D. & PARK, R. G. (eds) *Alpine Tectonics*. Geological Society, London, Special Publications, **45**, 265–283.

DINARÈS, J., McCLELLAND, E. & SANTANACH, P. 1992. Contrasting rotations within thrust sheets and kinematics of thrust tectonics as derived from palaeomagnetic data: an example from the Southern Pyrenees. *In*: McCLAY, K. (ed.) *Thrust tectonics*, Chapman & Hall, London, 265–275.

HORNER, F. & FREEMAN, R. 1983. Palaeomagnetic

evidence from pelagic limestones for clockwise rotation of the Ionian zone, western Greece. *Tectonophysics*, **98**, 11–27.

KISSEL, C. & LAJ, C. 1988. The Tertiary geodynamical evolution of the Aegean arc: a palaeomagnetic reconstruction. *Tectonophysics*, **146**, 183–201.

——, AVERBUCH, O., DE LAMOTTE, D. F., MONDO, O. & ALLERTON, S. 1993. First palaeomagnetic evidence for a post-Eocene clockwise rotation of the western Taurides Thrust Belt east of the Isparta re-entrant (southwest Turkey). *Earth and Planetary Science Letters*, **117**, 1–14.

——, LAJ, C., POISSON, A. & SIMEAKIS, K. 1989. A pattern of block rotations in central Aegea. *In*: KISSEL, C. & LAJ, C. (eds) *Palaeomagnetic rotations and continental deformation*. NATO ASI Series C, **254**, 115–129.

LAJ, C., JAMET, M., SOREL, D. & VALENTE, J. P. 1982. First palaeomagnetic results from Mio-Pliocene series of the Hellenic sedimentary arc. *Tectonophysics*, **86**, 45–67.

LIVERMORE, R. A. & SMITH, A. G. 1983. Relative motions of Africa and Europe in the vicinity of Turkey. *In*: TEKELI, O. & GÖNCÜOGLU, M. C. (eds) *Geology of the Taurus Belt*. Proceedings of the International Symposium on Geology of Taurus Belt, Ankara, Turkey.

MÁRTON, E., MILICEVIC, V. & VELJOVIC, D. 1990. Paleomagnetism of the Kvarner islands, Yugoslavia. *Physics of the Earth and Planetary Interiors*, **62**, 70–81.

MAURITSCH, H. J. & MÁRTON, E. 1995. Escape models of the Alpine-Carpathian-Pannonian region in the light of palaeomagnetic observations. *Terra Nova*, **7**, 44–50.

McCLELLAND, E. & McCAIG, A. M. 1989. Palaeomagnetic estimates of rotations in compressional regimes and potential discrimination between thin-skinned and deep crustal deformation. *In*: KISSEL, C., & LAJ, C. (eds) *Palaeomagnetic rotations and continental deformation*. NATO ASI Series C, **254**, 365–379.

McELHINNY, M. W. & LOCK, J. 1990. IAGA Global Palaeomagnetic Data Base. *Geophysical Journal International*, **101**, 763–766.

McKENZIE, D. & JACKSON, J. A. 1989. The kinematics and dynamics of distributed deformation. *In*: KISSEL, C. & LAJ, C. (eds) *Palaeomagnetic rotations and continental deformation*. NATO ASI Series C, **254**, 17–31.

MOORES, E. M. & VINE, F. J. 1971. The Troodos massif, Cyprus and other ophiolites as oceanic crust: evaluation and implications. *Philosophical Transactions of the Royal Society, London.* A**268**, 443–466.

MORRIS, A. 1995. Rotational deformation during Palaeogene thrusting and basin closure in eastern central Greece: palaeomagnetic evidence from Mesozoic carbonates. *Geophysical Journal International*, **121**, 827–847.

PITMAN, W. C. 3RD & TALWANI, M. 1972. Sea-floor spreading in the North Atlantic. *Geological Society of America Bulletin*, **83**, 619–646.

PLATZMAN, E. & LOWRIE, W. 1992. Paleomagnetic evidence for rotation of the Iberian Peninsula and the external Betic Cordillera, southern Spain. *Earth and Planetary Science Letters*, **108**, 45–60.

——, PLATT, J. P., TAPIRDAMAZ, C., SANVER, M. & RUNDLE, C. C. 1994. Why are there no clockwise rotations along the North Anatolian Fault Zone? *Journal of Geophysical Research*, **99**, 21705–21716.

ROBERTSON, A. H. F. & DIXON, J. E. 1984. Introduction: aspects of the geological evolution of the Eastern Mediterranean. *In*: DIXON, J. E. & ROBERTSON, A. H. F. (eds) *The geological evolution of the Eastern Mediterranean*. Geological Society, London, Special Publications, **17**, 1–74.

—— & GRASSO, M. 1995. Overview of the Late Tertiary-Recent tectonic and palaeo-environmental development of the Mediterranean region. *Terra Nova*, **7**, 114–127.

—— & XENOPHONTOS, C. 1993. Development of concepts concerning the Troodos ophiolite and adjacent units in Cyprus. *In*: PRICHARD, H. M., ALABASTER, T., HARRIS, N. B. W. & NEARY, C. R. (eds) *Magmatic processes and plate tectonics*. Geological Society, London, Special Publications, **76**, 85–119.

SAVOSTIN, L. A., SIBUET, J.-C., ZONENSHAIN, L. P., LE PICHON, X. & ROULET, M.-J. 1986. Kinematic evolution of the Tethys Belt from the Atlantic Ocean to the Pamirs since the Triassic. *Tectonophysics*, **123**, 1–35.

SPERANZA, F., ISLAMI, I., KISSEL, C. & LAJ, C. 1995. Paleomagnetic evidence for Cenozoic clockwise rotation of the external Albanides. *Earth and Planetary Science Letters*, **129**, 121–134.

VANDENBERG, J., KLOOTWIJK, C. T. & WONDERS, A. H. H. 1978. The late Mesozoic and Cenozoic movements of the Italian peninsula: further palaeomagnetic data from the Umbrian sequence. *Geological Society of America Bulletin*, **89**, 133–150.

VAN DER VOO, R. 1969. Paleomagnetic evidence for the rotation of the Iberian Peninsula. *Tectonophysics*, **7**, 5–56.

—— 1993. *Paleomagnetism of the Atlantic, Tethys and Iapetus oceans*. Cambridge University Press.

WESTPHAL, M. 1977. Comments on: 'Postulated rotation of Corsica not confirmed by new palaeomagnetic data', by K. M. Storetvedt & N. Peterson. *Journal of Geophysics*, **42**, 399–401.

—— 1993. Did a large departure from the geocentric axial dipole hypothesis occur during the Eocene? Evidence from the magnetic polar wander path of Eurasia. *Earth and Planetary Science Letters*, **117**, 15–28.

—— & KONDOPOULOU, D. 1993. Paleomagnetism of Miocene volcanics from Lemnos Island (Northern Aegean): implications for block rotations in the vicinity of the North Aegean Trough. *Annales Tectonicae*, **7**, 142–149.

——, BAZHENOV, M. L., LAUER, J. P., PECHERSKY, D. M. & SIBUET, J.-C. 1986. Palaeomagnetic implications on the evolution of the Tethys Belt from the Atlantic Ocean to the Pamirs since the Triassic. *Tectonophysics*, **123**, 37–82.

Application of net tectonic rotations and inclination analysis to a high-resolution palaeomagnetic study in the Betic Cordillera

A. KIRKER & E. McCLELLAND

Department of Earth Sciences, University of Oxford, Parks Road, Oxford, OX1 3PR, UK.

Abstract: Our initial structural investigations of the Sierra Endrinal thrust system in the western External Betic Cordillera suggested a differential block rotation associated with ramp and tipline structures in the thrust system. However, we present an analysis of new, high-resolution palaeomagnetic sampling which refutes this hypothesis and reveals a multicomponent remanence, including a syn-deformational component of magnetization. Due to the structural complications of the thrust system, our palaeomagnetic analysis required only the use of inclination data to determine the timing of magnetization, and comparison with expected inclinations to determine the degree of tilting prior to magnetization. In addition, a finite rotation study involving the combination of inclined axis rotations was preferred to simple bedding tilt corrections. This study emphasizes the usefulness of these techniques in structurally-complex regions, where there may be no valid basis for implementing a fold test, or applying the simple correction of untilting about strike. The study also demonstrates that parts of a thrust system may develop trends that are highly oblique to the regional norm without involving substantial differential rotation about a vertical axis.

Palaeomagnetism provides the structural geologist with an invaluable tool for studying the rotational component of the kinematics of a deformed region. Most commonly the emphasis has been on determining the component of vertical axis rotation, which is generally undetectable using conventional structural analysis. Thus, regions yielding particularly anomalous declinations have inspired models for rotating crustal blocks, such as western North America (Beck 1976; Luyendyk *et al.* 1980, 1985), the Aegean (Kissel & Laj 1988), and New Zealand (Lamb 1988). Studies such as these rely on the ability to apply an appropriate structural correction to the palaeomagnetic data. This is usually performed by restoring the local bedding to horizontal by untilting about strike. However, in complexly deformed regions this procedure may introduce errors in the estimate of palaeodeclination (MacDonald 1980). For example, structural corrections must allow for any significant plunge of a fold, or for the various episodes of tilting in regions of multiple folding or faulting phases. Use of a valid structural correction has also become important with the recognition of syn-deformational remanences (e.g. McClelland Brown 1983; Courtillot *et al.* 1986; Hudson *et al.* 1989; Villalaín *et al.* 1994), which require partial correction of the bedding tilt for their interpretation. The presence of both syn- and pre-deformational remanences may provide additional constraints on the rotation history of a complexly-deformed region. Bates (1989), using multicomponent remanences, was able to reconstruct a history of finite rotations about inclined axes for a laterally-variable thrust system.

Although detailed studies such as Bates (1989) are essential for the valid integration of palaeomagnetic and structural data in complexly-deformed regions, many previous palaeodeclination studies in orogenic belts have been either, of necessity, for reconnaisance purposes or for investigating large-scale trends (e.g. Van der Voo & Channell 1980; Eldredge *et al.* 1985; Lowrie & Hirt 1986). Thus, the relationship between the apparent vertical axis rotation obtained and the true rotation history, including the detailed mechanisms for accommodating rotation, often has not been determined. The aim of this study was to combine a local structural and kinematic study with high-resolution palaeomagnetic sampling to investigate these issues. This paper compares conventional techniques for structural correction of palaeomagnetic data and for assessing timing of magnetization with some techniques more appropriate to the integration of palaeomagnetic and structural information in complexly-deformed areas. The study forms part of an investigation into the rotational deformation of

From Morris, A. & Tarling, D. H. (eds), 1996, *Palaeomagnetism and Tectonics of the Mediterranean Region*, Geological Society Special Publication No. 105, pp. 19–32.

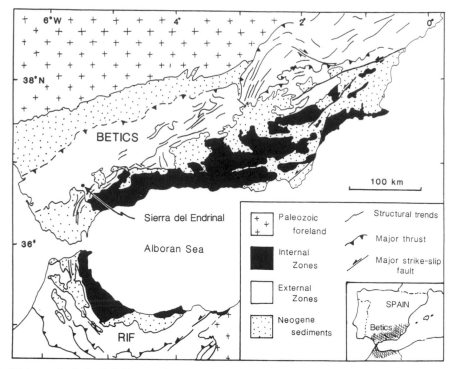

Fig. 1. Major geological subdivisions and structural trends of the Betic Cordillera and Rif Mountains. Dot shows location of the study area, the Sierra del Endrinal.

the western Betic Cordillera of southern Spain (Kirker 1994).

Palaeomagnetism in the External Betic Cordillera

The Betic Cordillera is part of the Alpine chain of mountain belts formed during Tertiary convergence between Africa and Eurasia. It consists of a dominantly metamorphosed interior, known as the Internal Zones, and an unmetamorphosed fold and thrust belt, known as the External Zones (Fig. 1). The External Zones are the remnants of the Mesozoic carbonate platform that formed the southern continental margin of Iberia (García-Hernandez *et al*. 1980), and which were deformed during the Early and Mid-Miocene due to convergence with the Internal Zones. The resulting thrust belt forms a thin-skinned crustal wedge, detached on Upper Triassic evaporitic horizons (Banks & Warburton 1991; Blankenship 1991), and generally trends NE, oblique to the margins of both the Internal Zones and Iberia (Fig. 1).

Palaeomagnetic studies in the External Betic Cordillera have established the presence of clockwise rotations in almost every case (Steiner

et al. 1987; Osete *et al*. 1988, 1989; Platzman & Lowrie 1992; Allerton *et al*. 1993; Villalaín *et al*. 1994). Many of the present declinations obtained are at least 60° in excess of their original Jurassic or Cretaceous azimuths. Although early models for crustal block rotations were dominated by strike-slip fault systems (e.g. Luyendyk *et al*. 1980; Ron *et al*. 1984), many recent studies have demonstrated the important role of differential displacements on thrust faults (Schwartz & Van der Voo 1984; Bates 1989; Channell *et al*. 1990; Potts 1990; Jolly & Sheriff 1993; Thomas *et al*. 1993). Similarly, a dextral shear zone was proposed for the External Betics by Osete *et al*. (1988), but there is recent evidence that thrust faults are largely responsible for accommodating the rotations (Allerton 1994; Kirker 1994).

The study area lies within the thrust system of the Sierra del Endrinal in the western External Betics, near the villages of Grazalema and Villaluenga del Rosario (Figs 1 and 2). An anomalously high clockwise rotation within this region, that of site RJ04 of Platzman & Lowrie (1992), and anomalous structural trends and kinematic data, all consistently clockwise rotated relative to the regional norm by around 40°, indicated a possible differential block

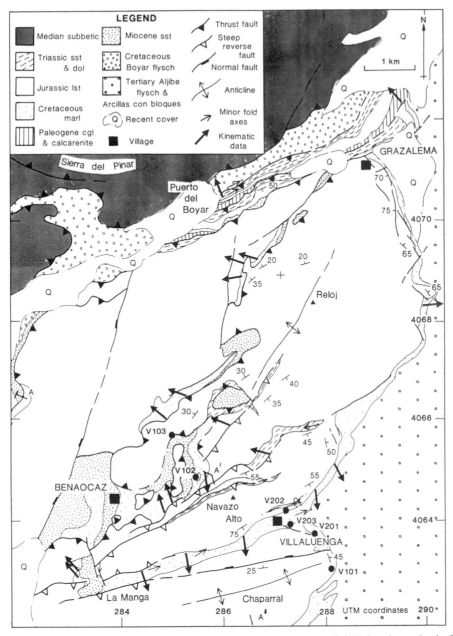

Fig. 2. Geological map of the Sierra del Endrinal, after Moreno Serrano *et al.* (1991) and mapping by Kirker. Kinematic data indicating transport directions of major faults are shown as solid arrows. Dots show the localities of palaeomagnetic sampling sites in this study. Grid spacings on the map are in UTM coordinates.

rotation within the thrust system. Note that reference directions for Iberia used in this study are 342/35 for the Late Jurassic and 002/52 for the Miocene, calculated from Westphal *et al.* (1986). A detailed structural and palaeomagnetic investigation was initiated to test the bounds of the rotated zone, and in particular whether the major thrust faults bounding the zone were responsible for accommodating the differential rotation.

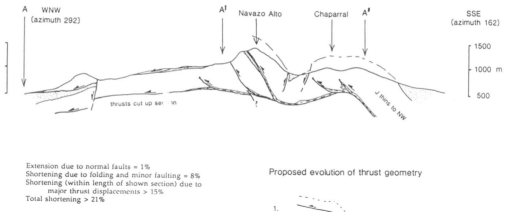

Extension due to normal faults = 1%
Shortening due to folding and minor faulting = 8%
Shortening (within length of shown section) due to
major thrust displacements > 15%
Total shortening > 21%

Proposed evolution of thrust geometry

Cretaceous-Tertiary

Jurassic

Triassic

Fig. 3. Cross-section through the thrust system at Navazo Alto, and cartoon showing its possible evolution. No vertical exaggeration. Section line is shown in Fig. 2.

Local structure of the study area

A series of at least four repetitions of the External Zone stratigraphic sequence crops out between the Puerto del Boyar and the village of Grazalema (Fig. 2). Known as the 'Boyar corridor', it is interpreted by Hoeppener *et al.* (1964), Chauve (1968), Hoppe (1968), and Moreno Serrano *et al.* (1991) as an over-steepened imbricate stack, tilted and disrupted by the later backthrusting of the Sierra del Pinar block (Fig. 2). The highest thrust of the Boyar imbricates can be traced from Grazalema south to the village of Benaocaz, where it bounds a number of small klippen, dipping gently 5–10° to the WNW. A kilometre-scale anticline seen at the peak of Reloj refolds the thrust, and is responsible for its present geometry (Fig. 2). Adjacent calcite fibre lineations and shear fabrics record a west-northwest transport direction (Fig. 2).

Clearly postdating the flat thrust are a series of steep reverse faults that form the cliffs on the north flanks of the peak of Navazo Alto (Fig. 2). The traces of these end abruptly to the west at a steep cross-fault, and to the east their displacement dies out or is partially transferred to minor NE-trending faults, which continue into the core of the Reloj anticline. Kinematic indicators

adjacent to the reverse faults show a N to NNW direction of motion, significantly different to that of the flat thrust (Fig. 2). Since the reverse fault system reaches a series of terminations to the northeast, it probably branches from the flat surface at depth, perhaps as an early ramp structure which has subsequently broken through the hangingwall. Figure 3 shows the envisaged evolution of the fault system in cross-sectional view.

South of Navazo Alto, the hinge area of a tight syncline, at times overturned with reverse faults breaking through the hinge area, is exposed in the valley of Villaluenga del Rosario (Fig. 2). Kinematic data from the sheared Cretaceous marls in the valley show a very uniform, SSE convergence direction. Minor folds reflect this pattern (Fig. 2). Further south, a major anticline plunges ENE, its hinge located along the crest of Chaparral, and its southern flank moderately dipping and apparently unfaulted.

There are several reasons for believing that the Navazo Alto range might have been rotated relative to its surrounds. The trends of structures around Navazo Alto and in the valley of Villaluenga del Rosario are east-northeasterly in too systematic and continuous a fashion to be a random variation of the regional northeasterly trend. The regional trend is resumed abruptly at the northernmost late reverse fault and the Reloj anticline, and north among the isolated klippen, suggesting that the anomalous trends may form a discrete 'block'. In addition, kinematic data are systematically different to the regional NW–SE direction by the same angular difference as the structures producing them, suggesting that they also have been rotated, or that they represent a late stage local strain produced by structures well-advanced toward their present positions. The change to NW-directed kinematic data is also abrupt, coinciding with the regionally trending structures. Lastly, an exploratory palaeomagnetic site, sampled by Platzman & Lowrie (1992) behind Villaluenga del Rosario, produced a declination 25–40° clockwise of the regional average, again a similar angular difference to that suggested by the structure and kinematics.

The reverse faults bounding Navazo Alto to the north provide a simple explanation for the accommodation of such a rotation since they appear to bound the rotated region to the north, and reach a tipline at their easternmost extent. Provided that not all of their displacement is transferred to NE-trending reverse faults, which anyway die out into the core of the Reloj anticline, these faults could accommodate clockwise rotation of their hanging-wall block relative to their footwall. If the reverse faults are connected with a continuation of the flat thrust at depth, the main thrust would also bound the rotated region from below.

Therefore, we could describe the structural history of the Villaluenga region in this way:
(a) initial thrusting towards the W to NW on a gently-dipping fault system, including the Boyar corridor imbricates;
(b) breaking through of ramps as new steep reverse faults, folding of the thrust system, and differential rotation of the Navazo Alto block due to the tip-line geometry of the new faults.

The palaeomagnetic study was designed to test the above hypothesized history by providing an independent and more direct evaluation of the rotational aspects of the deformation.

Palaeomagnetic methods and rock-magnetic properties

All new sites for this study were sampled from Upper Jurassic grey micrites and oolites. This lithology was chosen for its abundance of coherent well-bedded sequences and its successful use in previous studies (Platzman & Lowrie 1992; Villalaín et al. 1994). Sites were chosen purely on the basis of testing particular structures of interest, and as a consequence it was often necessary to collect samples that were affected by re-solution, secondary cleavages, or preferential weathering. However, as the dominant forms of secondary alteration are pressure solution cleavages, it was usually possible to preferentially sample less-affected parts of the rock. A sampling site consisted of between eight and twelve oriented cores, spread sufficiently to minimize the biases due to secular variation and other local weathering effects. The cores were oriented with both sun and magnetic compasses. On average, two standard 2.5 cm diameter samples were obtained from each core, and between eight and fifteen such samples were demagnetized from each site. Sampling sites were positioned in a transect designed to detect a differential rotation of the Navazo Alto block. Sites V101 and V201 were drilled as control points to the south of the Navazo Alto block on the plunging nose of the Chaparral anticline, and sites V202 and V203 within the Navazo Alto block near the village of Villaluenga del Rosario (Fig. 2). Two other sites, V102 and V103, were placed to the north within the flat thrust sheet, with V102 adjacent to the major reverse faults bounding the region of anomalous trends and kinematics (Fig. 2).

Intensity of NRM for most samples was between 0.03 and 1.5 mA m^{-1}, over 50% of which was lost in the first step of thermal demagnetization. Samples were progressively demagnetized using the stepwise thermal method, and their magnetizations were measured on a CCL cryogenic magnetometer. Demagnetization was conducted in steps of 50° up to 300°C, then in steps of 30° or less for higher peak temperatures. This was continued for 10 to 20 steps, until the intensity of the samples had decreased beyond the measurement limit of the cryogenic magnetometer (considered to be 0.01 mA m^{-1}), or until successive measurements became random. After each thermal step, a measurement was made of the bulk susceptibility for each sample, to monitor possible changes in magnetic mineralogy during heating, which may add new, irresolvable components of magnetization to the remanence. IRM acquisition experiments performed on representative samples from each site confirm the findings of Platzman & Lowrie (1992) and Villalaín et al. (1994) for the remanence carriers in the Upper Jurassic limestones: the majority of samples reached within 3% of their peak intensity prior to an applied field of 300 mT, indicating the presence of mainly magnetite, but also minor hematite (Fig. 4a).

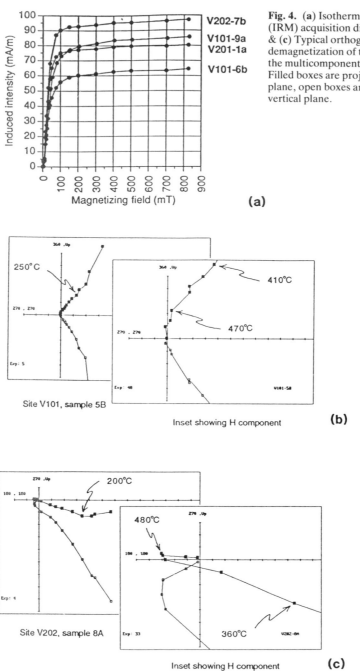

Fig. 4. (a) Isothermal remanent magnetization (IRM) acquisition diagrams for selected samples. (b) & (c) Typical orthogonal vector plots of thermal demagnetization of the Villaluenga samples, showing the multicomponent structure of the remanence. Filled boxes are projections onto the horizontal plane, open boxes are projections onto E–W or N–S vertical plane.

Remanence components and timing of magnetization

Components of remanent magnetization were successfully isolated from all sites (Fig. 4b & c) except V102, whose samples possessed very low intensities and unstable demagnetization during thermal treatment. With few exceptions, the components are characterized by one of three blocking temperature ranges. The first corresponds to a viscous remanence which is completely removed by 200°C, and which is always oriented near the present-day geomagnetic field in southern Spain. The second,

Table 1. *Components of remanence from the Villaluenga sites.*

Site	Component	N	R	k	α_{95}	In situ mean	Tilt corrected	Bedding
V101	up to 200°C	10	10.0	206	3.4	025/66		
	I 250–460°C	13	12.9	166	3.2	048/51	050/43	065/09
	H 460–520°C	5	4.9	77	8.8	011/34	016/29	
V103	up to 200°C	12	11.9	107	4.2	010/60		
	I 200–390°C	8	7.8	37	9.3	044/43	015/42	301/30
V201	up to 150°C	10	9.7	33	8.5	012/57		
	I 200–300°C	9	8.9	55	7.0	036/66	036/25	
	I 330–420°C	10	10.0	203	3.4	067/73	046/33	035/42
	H 420–540°C	7	6.9	78	6.9	004/66	021/26	
V202	up to 150°C	10	9.6	25	9.8	001/48		
	I 200–450°C	9	9.0	167	4.0	026/49	094/43	140/57
	H 480–540°C	8	7.9	48	8.0	193/18	205/−31	
V203	up to 200°C	8	7.4	11	17.4	038/79		
	I 200–330°C	7	6.9	76	7.0	091/56	004/00	154/73
	? 330–480°C	8	7.8	42	8.7	111/27	018/30	(o'turned)

Components are stated in terms of blocking temperature ranges; I & H refer to intermediate and high temperature respectively. Statistical parameters are: number of components used (*N*), normalized vector magnitude (*R*), dispersion parameter (*k*), and 95% confidence limits of the Gaussian mean (α_{95}). Site means are given in declination/inclination format, and bedding orientations as dip direction/dip. Reference directions for Iberia: 342/35 (Late Jurassic); 002/52 (Miocene).

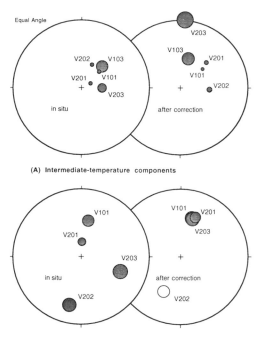

(A) Intermediate-temperature components

(B) High-temperature components

Fig. 5. Equal-angle stereoplots of components of remanence from the Villaluenga sites, *in situ* and after bedding tilt correction. Circles represent the 95% confidence regions for the mean direction of each site. The unshaded circle represents an upper hemisphere, reversed direction.

intermediate unblocking range is between 200°C and 460°C, in which every site has a component with a declination that differs significantly from north (Fig. 5a and Table 1). For sites V201 and V203, more than one component was isolated in this range, those of V201 being coplanar, and therefore probably not independently resolved. The final unblocking range is from 420°C to 540°C. These high-temperature components have lower in both declinations and inclinations than the intermediate-*T* components (Fig. 5b and Table 1). Sites V101, V201, and V202 produced well-defined high-*T* components, and samples from other sites often showed an indication of its existence, in the failure of their intermediate-*T* component to demagnetize directly to the origin of the Zijderveld plot. Thus, the characteristic components isolated from two-component sites probably correspond to intermediates from the three-component sites. Samples with resolved high-*T* components always also exhibited intermediate temperature components, and it is therefore likely that the intermediate-*T* component represents a magnetization that overprints the earlier high-*T* component. In support of this suggestion, high-*T* components have both normal and reverse polarities, indicating acquisition of remanence during relatively slow sedimentation, while intermediate-*T* components are always normal, indicating relatively rapid magnetization. However, confirmation of this may only be obtained using field tests.

Table 2. *Inclination statistics for Villaluenga sites.*

Site	Intermediate temp.			High temp.		
	In situ	Corrected	Mismatch	In situ	Corrected	Mismatch
V101	51	43	9 ± 3	34	29	7 ± 9
V103	43	42	12 ± 9			
V201	73	33	19 ± 3	66	26	8 ± 7
V202	49	43	12 ± 4	18	31	−2 ± 8
V203	56	00	52 ± 7	27	30	6 ± 9
Mean	54	33		36	29	
k	25	10		8	704	
α_{95}	19	30		47	5	

The top part of the table lists the inclination data for both intermediate- and high-temperature components of each site, *in situ* and after correction for bedding tilt. The 'mismatch' columns refer to the discrepancies between the measured bedding/remanence angles and that expected from the inclination of the reference direction, and incorporate the analytical error for each component mean as stated in Table 1. Reference directions for Iberia: 342/35 (Late Jurassic); 002/52 (Miocene). A range of mismatches that straddles zero indicates that magnetization effectively predates any tilting. The lower part of the table gives the mean, dispersion parameter (k), and 95% confidence limits (α_{95}) of the data using inclination-only statistics (McFadden & Reid 1982). The decrease in dispersion for the high-temperature components after correction exceeds the 95% confidence level of an F-distribution with 4 and 4 degrees of freedom. However, the change in dispersion for the intermediate-temperature components is not significant at this level.

The relationship between the intermediate- and high-T components in this study is similar to that described by Villalaín *et al.* (1992, 1994), who also document a multiple magnetization history for Upper Jurassic rocks in the Serranía de Ronda. Using field tests, they show that the high-T component always predates tilting of the beds, but that the intermediate-T component sometimes does not, suggesting that this component represents a syn-deformation remagnetization. Furthermore, the declination of the high-T component is always less than that of the intermediate-T component, and by a constant angular difference. They explain these features by proposing a two-phase history of magnetization:

(i)   The high-T component was locked in during the Late Jurassic, and was followed by a regional anticlockwise rotation, presumably the rotation of Iberia during the Cretaceous (Van der Voo 1969; Galdeano *et al.* 1989);

(ii) the intermediate-T component was locked in during deformation in the Early Miocene, followed by various clockwise block rotations, greater in magnitude than the regional anticlockwise rotation.

Normal field procedure for assessing timing of magnetization is to sample from outcrops of widely-varying bedding orientation for use in a standard fold test, such as those of McElhinny (1964) or McFadden & Jones (1981). However,

the application of a fold test that compared the discontinuous outcrops for each site would invalidate the basic premise for the study by assuming that there was no differential rotation between the sites. Therefore, we have applied an alternative test that does not depend on the structural history. The angle between bedding and remanence must remain constant during any combination of rigid body rotations, and is therefore an appropriate parameter for comparison between sites which may have had varying rotation histories. It follows that a series of sites with a similar inclination after bedding tilt correction, and which possess different structural orientations, are strong evidence for a pre-tilt magnetization.

Table 2 shows the inclination statistics for the Villaluenga sites, before and after tilt correction, calculated using the method of McFadden & Reid (1982). The inclinations of the high-T components for all sites are significantly better clustered after tilt correction, indicating that these components predate tilting. The inclinations of the intermediate-T components cluster neither significantly better nor worse after tilt correction. This suggests that some sites have a pre-tilt magnetization and should be corrected, and others have a syn- or post-tilting magnetization. Although this general result is of little practical use in assessing the timing of magnetization for each site, some things are intuitively clear on a site-by-site basis. For example, the inclination of site V201 changes from an unusually high value to a Mesozoic one after correction. This suggests that this site should be corrected for bedding tilt. Site V203 changes from a clearly Miocene inclination to an unusually low inclination, which suggests that its magnetization postdates some or all of the tilting.

A more quantitative way of estimating the timing of any syn- or post-tilt magnetization relative to the local tilting history is to compare the after-correction inclinations with their expected reference inclination, which is that of the Late Jurassic reference vector for Iberia. Since the angle between bedding and remanence direction should remain constant during rigid-body rotations, the tilt-corrected inclination should be the same as the reference inclination, unless the magnetization post-dates tilting of the beds. Furthermore, any mismatch in the bedding/remanence angles should be equivalent to the minimum amount of tilting that has occurred prior to magnetization. The bedding/remanence angles of high-T components from the Villaluenga sites are the same as the Late Jurassic reference inclination, within analytical errors, confirming their pre-deformation age of magnetization (Table 2). Bedding/remanence angles of the intermediate-T components are generally significantly different to the Miocene reference inclination. The differences, including analytical error, are between 3° and 59°, and on a site-by-site basis represent variable but substantial proportions of the present bedding dips (Table 2). This is strong evidence for magnetization at various stages during the tilting history of those sites. In particular, the inclination mismatch for site V101 indicates a magnetization that completely post-dates tilting at that locality.

In summary, the analysis of before- and after-tilting inclinations supports the interpretation by Villalaín et al. (1992, 1994) of the intermediate-T component as a syn-deformational remagnetization. The high-T component is clearly pre-tilting in age, and its bedding/remanence angles match that of the Late Jurassic reference vector for Iberia, indicating a primary magnetization.

Palaeodeclination results

The in situ declinations of the high-T components are widely variable and do not correspond to previous estimates for the region (Platzman & Lowrie 1992), but after correction for bedding tilt, the four site means are inseparable from each other within analytical

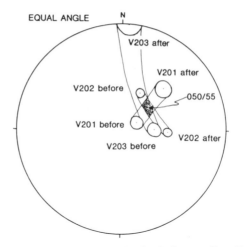

Fig. 6. Untilting trajectories for the intermediate-T components of sites V201, V202, and V203. The coincidence of crossing points indicates that these components were all magnetized part-way through their tilting histories and that there is no differential rotation between them. The uncertainty region for the crossing point is defined by declinations between 041° and 069°, and inclinations of 49° to 59°.

error (Fig. 5b and Table 1). Since these components have been shown to predate deformation, this unexpectedly indicates no differential rotation history between the Navazo Alto block and the Chaparral anticline (see Fig. 2). Analysis of the intermediate-T components is more difficult, since the inclination tests applied above show that Miocene remagnetization has occurred during different amounts of tilting of the Villaluenga sites. However, a stereoplot showing the progressive untilting of the intermediate-T components suggests a possible solution. Fig. 6 shows the paths that the intermediate-T components of V201, V202, and V203 make during the process of tilting about the strike of their beds. These paths cross at the same point within errors, at approximately 050/55, a coincidence that suggests that each component was magnetized in this orientation, part-way through their tilting history. This solution is one in which there is no differential rotation between sites, as suggested by the high-T components. Furthermore, the inclination of 55° is appropriate for a Miocene magnetization, and the implied vertical axis rotation of 36–64° overlaps with the 25–51° implied by the high-T component.

Syn-tilting magnetization is a likely solution for the timing of the intermediate-T component, and is a natural consequence of the Miocene remagnetization shown by Villalaín et al. (1994).

However, there is a fundamental flaw in the precise solution provided by Fig. 6. There is no independent reason for believing that a simple tilting of the beds about their present strike represents the deformation history of each site, particularly in a region where multiple episodes of folding and thrusting have occurred. Some of the bedding/remanence angle mismatches (Table 2) cannot be explained by the scenario illustrated in Figure 6. For example, the intermediate-T component from site V203 has a bedding/remanence angle that deviates from the expected reference angle by 45–59°, but according to the simple tilt model (Fig. 6) its magnetization does not take place until after 76–101° of tilting have taken place. Therefore, a single phase of tilting does not adequately explain the details of the syn-tilting magnetization history indicated by the bedding/remanence angle data. This result is not unexpected, since some sites are located in the hinge area of the Chaparral anticline (Fig. 2), and probably have less-complicated deformation paths than those on the limbs. Since the age of formation of the fold plunge is not known, it is impossible to correct the tilting paths by unplunging the fold, as is sometimes performed.

These points highlight a common source of error in the assumption that rotation has occurred about a vertical axis, and that a bedding tilt correction accurately reconstructs other parts of the tilting history of a site. Even if this assumption were acceptable, considerable uncertainties are also introduced for corrections of high bedding dips (Demarest 1983), which may be common in deformed localities, for example sites RJ04, V202, and V203. The following section applies a technique for structural analysis of palaeomagnetic data that utilises a realistic combination of tilts and rotations, which are compatible with the uncertainties inherent in the palaeomagnetic data.

Finite rotation study

Palaeomagnetic studies in deformed areas are faced with the likelihood that the deformation history of the sampling site cannot be adequately described by tilts about a vertical axis and a single horizontal axis. During general progressive heterogeneous strain, any linear feature (for example a fold axis, remanence vector, or bedding pole) will trace out a continuous curved path in space which is not necessarily confined to a single plane. Linear features within a deforming fold with a changing plunge, or a fault block adjacent to a fault with varying slip, will follow such a path. Thus, the assumption of

discrete tilting/rotation events can only be an approximation.

Even in more simple cases, where the deformation may be described as a series of discrete rotations about inclined axes, the standard correction by rotating about the strike of the beds may produce considerable error in estimating the declination of the remanence (MacDonald 1980). This is because only rotations about orthogonal axes are commutative, and the order of tilt reconstruction is often important. Implicit in the application of a bedding tilt correction is the assumption that the deformation history simply has involved vertical axis rotations and a single tilt. In some cases, where the deformation history has not included rotation about a vertical axis, there may be no reason why an untilting about the final strike reveals any useful information about the rotation history.

One potentially useful approach to such complex situations is the determination of the unique axis and amount of finite rotation that is required to rotate the original bedding and remanence to their present orientations. This construction is known as the *net tectonic rotation* (MacDonald 1980; Fig. 7a). This rotation may have some meaningful interpretation which is immediately obvious. Alternatively, it may be possible to deconstruct this rotation into a meaningful sequence of rotations, perhaps including inclined axis rotations, based on independent information about the structural history.

Net tectonic rotations have been calculated for the intermediate-T and high-T components of the Villaluenga sites, and these are shown in Fig. 7b. Error bounds include uncertainties in the mean directions (α_{95}), and the mismatch between the angle separating the original bed pole and palaeomagnetic reference and that of the present day bedding pole and remanence (given in Table 2). Thus, the error bounds of the net tectonic rotations take into account the effect of syn-tilting magnetization of the intermediate-T components, and assumes a pre-tilting magnetization. This assumption is appropriate in cases where magnetization has taken place after only a few degrees of net tilting, since the objective of this analysis is to retrieve a very general tilting history only. Only the intermediate-T component of site V203 is not shown because of its particularly large bed/remanence angle mismatch. Note that, for V201 and V202, the net tectonic rotations for the intermediate- and high-T components do not precisely coincide, but that the error bounds for the respective components overlap (Fig. 7b). Since

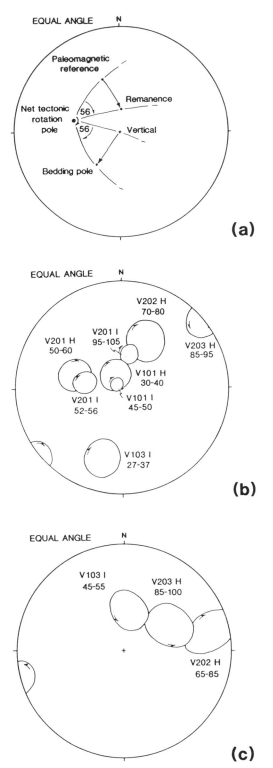

(a)

(b)

(c)

the error bounds are relatively large, the two components may have undergone a slightly different tilt/rotation history.

In order to analyse the possible rotation histories for the Villaluenga sites, we must recall the structural events proposed for the region. These are:

(a) initial thrusting on a fault system, gently-dipping to the ESE, possibly contemporaneous with a regional clockwise rotation of 55–70°;

(b) breakthrough of new steep reverse faults and folding of the thrust system, and the possible differential rotation of the Navazo Alto block due to the tip-line geometry of the new reverse faults.

The net tectonic rotations for the control sites of V101 and V201, previously presumed to be outside the 'rotated zone', suggest that their rotation histories can both be described by a 40–60° rotation about an axis, steeply- to moderately-inclined to the WNW. This is what would be expected if rotation had occurred on gently-dipping thrusts with a regional NNE strike, as proposed in part (a) of the structural history above. The difference between the control sites and those within the Navazo Alto block can therefore be calculated by subtracting the effect of such a rotation from their own net tectonic rotations, assuming that all the sites have undergone step (a). Figure 7c shows the results of this subtraction, which should correspond to the structural history described in part (b) above. The increased uncertainty in Fig. 7c is due to the use of a range of rotation axis inclinations (from 60–90°) and of rotation amounts (from 40–60°) in step (a).

Subtracted resultants from sites V202 and

Fig. 7. (a) The net tectonic rotation is the unique rotation axis and amount that rotates both the palaeomagnetic remanence from its reference vector and the bedding pole from its original vertical position. In the example above, the net tectonic rotation is given by a 56° clockwise rotation about the pole 42/285. (b) Stereoplot of net tectonic rotations for Villaluenga components. The rotation axes lie within the error circles shown. Errors are due to analytical 95% confidence limits on site means, and mismatches between bedding and remanences, before and after correction. Arrows show the sense of rotation and the numbers show the range in amount of rotation. (c) Resultant finite rotations following subtraction of the net tectonic rotations of V101 and V201 from V103, V202, and V203. A region on the stereonet represents the locus of resultants due to a range of subtracted rotation axes (see text), and the errors from (b).

V203 dip moderately to the ENE. The orientations of their axes coincide, yet the amount and sense of rotation differs between the two sites. The details of these rotations may be readily described by folding about a ENE-plunging fold axis, which is compatible with the plunge and trend of folds adjacent to V202 and V203 and the trend of the Navazo Alto block (see Fig. 2). Such a phase of folding is as expected for step (b) of the structural history above. The resultant rotation for site V103, the northerly control point, indicates a different rotation history. Subtraction of step (a) leaves a finite rotation of 45–55° anticlockwise about a NNE-dipping axis. This may reflect formation of the Reloj antiform, whose entire western flank dips WNW (Fig. 2).

Thus, the rotation history of the Villaluenga sites can be adequately described in terms of two components which parallel the proposed structural history above:

(i)   an initial widespread 40–60° clockwise rotation on thrust faults, gently-dipping to the ESE;
(ii) tight folding of the beds at sites V202 and V203 about an ENE-trending fold axis, and folding of the beds at V103 due to formation of the Reloj antiform.

This rotation history requires no differential rotation across faults, as suggested by the initial high-T component declinations. However, it includes neither vertical axis rotations nor tilts about strikelines. Consequently, it predicts greater clockwise rotations (40–60°) in inclined thrust planes than indicated by the straightforward bedding tilt-corrected data (25–50°).

Discussion

One of the consequences of sampling rocks from deformed areas is that they preserve complex remagnetization histories, some components of which may be locked in part-way through tilting of the beds. Such components are largely uninterpretable because of a lack of structural reference horizons. The techniques used in this study were attempts to address uncertainties of this kind. An inclination study, and in particular the comparison of bedding/remanence angles with the reference inclination, is not only capable of establishing the timing of magnetization, but also can provide a quantative estimate of the degree of pre-magnetization tilting in cases of a syn-deformational remanence. In the case of the Villaluenga sites, an inclination study clearly demonstrates each high-T component to be pre-tilting, and each intermediate-T component to be syn- or post-tilting, in a locality where there was no valid basis for implementing a regional fold test.

The fact that such field tests are inappropriate in deformed regions cannot be over-emphasised. There is no reason to suppose that the total kinematic history of the Villaluenga sites may be represented by simple tilting about their present bedding strike. In fact, such an assumption is structurally unrealistic, given the relatively complex structural history of the region. For the Villaluenga sites, we therefore consider the bedding/remanence angle mismatches to be the only valid test of timing of magnetization, even for the high-T components, despite the remarkable coincidence of their orientations after simple tilt correction.

Finite rotation analysis based on structural and kinematic information can provide more realistic structural corrections for palaeomagnetic data. The net tectonic rotations for the Villaluenga sites are varied, and mainly involve inclined axis rotations, suggesting rotation histories that are more complicated than those represented in Fig. 6. Thus, the finite rotation study also suggests that the bedding tilt correction in palaeomagnetism can induce errors in interpretation of the data. Since the rotation axes involved are only slightly inclined from the vertical, the errors in this study mainly concern the magnitude of rotation. However, in other cases that involve a succession of inclined axis rotations, the potential for error is great. The fact that the proposed rotation history contains no rotations about vertical axes poses the question as to whether the simple bedding correction yields any meaningful data at all in most cases. Clearly, it is only appropriate to compare rotations using a consistent reference that is based on the actual structural and rotational history.

For the Villaluenga study, this reference is an axis steeply- to moderately-inclined to the WNW, corresponding to rotations in actual thrust planes. When comparing the proposed histories of sites V202 and V203 to those of sites V101 and V201, no differential rotation of the Navazo Alto block is required using this reference. In addition, the total rotation implied by the high-T components, using a Late Jurassic reference direction, is 40–60°. This is less than the average for the Serranía de Ronda. We submit that the site of Platzman & Lowrie (1992) showing an anomalously large declination in this region has recorded a syn-deformational remanence only, corresponding to the intermediate-T components of this study. Thus, it is likely that it is inappropriate either to compare it with a Jurassic reference or to correct the *in situ* data with a simple bedding tilt.

The palaeomagnetic reconstruction is not the result expected from the structural study. The anomalous kinematic and structural trends of Navazo Alto and the valley of Villaluenga del Rosario must have formed obliquely to the regional transport direction without involving substantial differential rotation. This obliquity may have been controlled by the changes in detachment rheology and the cross-fault geometries that are to be found at the lateral extremities of the Navazo Alto thrusts (Fig. 2), and which may have constrained the mechanically-favourable orientations for fault formation. Pre-existing oblique trends, which are manifest in dome-like interference structures typical of the vicinity (Moreno Serrano et al. 1991), may also have guided formation of the oblique thrusts and folds. Some small differential rotation may also have enhanced the apparent obliquity. The finite rotations for V202 and V203 in Fig. 7c are not precisely horizontal, and may be decomposed to allow more complicated rotation histories than presented here. The errors on their loci were produced by incorporating a 20° uncertainty in rotation.

Although the possibilities for deconstructing the Villaluenga net tectonic rotations are theoretically infinite, the solution posed here is by no means arbitrary. It is based on input from observed structural features, and a large pre-existing body of rotation data. It was originally envisaged that a palaeomagnetic study would provide an independent test of the rotational component of the deformation at Villaluenga. Clearly this cannot be the case, since the problems posed during the palaeomagnetic analysis require substantial input from the structural data for their solution. We would argue that there can never be an independent comparison between structural and palaeomagnetic data in strongly deformed areas. Nevertheless, interactively they provide a real addition to the understanding of the deformation history.

References

ALLERTON, S. 1994. Vertical axis rotation associated with folding and thrusting: an example from the eastern Subbetic of southern Spain. *Geology*, **22**, 1039–1042.

——, LONERGAN, L., PLATT, J. P., PLATZMAN, E. S. & MCCLELLAND, E. 1993. Paleomagnetic rotations in the eastern Betic Cordillera, southern Spain. *Earth and Planetary Science Letters*, **119**, 225–241.

BANKS, C. J. & WARBURTON, J. 1991. Mid-crustal detachment in the Betic system of southeast Spain. *Tectonophysics*, **191**, 275–289.

BATES, M. P. 1989. Palaeomagnetic evidence for rotations and deformation in the Nogueras Zone, Central Southern Pyrenees, Spain. *Journal of the Geological Society, London*, **146**, 459–476.

BECK, M. E. 1976. Discordant paleomagnetic pole positions as evidence of regional shear in the western Cordillera of North America. *American Journal of Science*, **276**, 694–712.

BLANKENSHIP, C. L. 1991. Structure and paleogeography of the External Betic Cordillera, southern Spain. *Marine and Petroleum Geology*, **9**, 256–264.

CHANNELL, J. E. T., OLDOW, J. S., CATALANO, R. & D'ARGENIO, B. 1990. Palaeomagnetically determined rotations in the western Sicilian fold and thrust belt. *Tectonics*, **9**, 641–660.

CHAUVE, P. 1968. *Etude géologique du Nord de la province de Cadix (Espagne méridionale)*. Memorias de Instituto Geológico y Minero de España, **69**.

COURTILLOT, V., CHAMBON, P., BRUN, J. P., ROCHETTE, P. & MATTE, P. 1986. A magnetotectonic study of the Hercynian Montagne Noire (France). *Tectonics*, **5**, 733–751.

DEMAREST, H. H. 1983. Error analysis for the determination of tectonic rotation from paleomagnetic data. *Journal of Geophysical Research*, **88**, 4321–4328.

ELDREDGE, S., BACHTADSE, V. & VAN DER VOO, R. 1985. Paleomagnetism and the orocline hypothesis. *Tectonophysics*, **119**, 153–180.

GALDEANO, A., MOREAU, M. G., POZZI, J. P., BERTHOU, P. Y. & MALOD, J. A. 1989. New paleomagnetic results from Cretaceous sediments near Lisboa (Portugal) and implications for the rotation of Iberia. *Earth and Planetary Science Letters*, **92**, 95–106.

GARCÍA-HERNANDEZ, M., LOPEZ-GARRIDO, A. C., RIVAS, P., SANZ DE GALDEANO, C. & VERA, J. A. 1980. Mesozoic paleogeographic evolution of the external zones of the Betic Cordillera. *Geologie en Mijnbouw*, **59**, 155–168.

HOEPPENER, R., HOPPE, P., DÜRR, S. & MOLLAT, H. 1964. Ein Querschnitt durch die Betischen Kordilleren bei Ronda (SW Spanien). *Geologie en Mijnbouw*, **43**, 282–298.

HOPPE, P. 1968. Stratigraphie und Tektonik der Berge um Grazalema (SW Spanien). *Geologische Jahrbuch*, **86**, 267–338.

HUDSON, M. R., REYNOLDS, R. L. & FISHMAN, N. S. 1989. Synfolding magnetization in the Jurassic Preuss Sandstone, Wyoming-Idaho-Utah thrust belt. *Journal of Geophysical Research*, **94**, 13681–13705.

JOLLY, A. D. & SHERIFF, S. D. 1993. Paleomagnetic study of thrust sheet motion along the Rocky Mountain front in Montana. *Geological Society of America Bulletin*, **104**, 779–785.

KIRKER, A. I. 1994. *Kinematics and rotational deformation of the Betic-Rif arc in southwestern Spain*. DPhil thesis, Oxford.

KISSEL, C. & LAJ, C. 1988. The Tertiary geodynamic evolution of the Aegean arc, a paleomagnetic reconstruction. *Tectonophysics*, **146**, 183–201.

LAMB, S. H. 1988. Tectonic rotations about vertical axes during the last 4 Ma in part of the New

Zealand plate-boundary zone. *Journal of Structural Geology*, **10**, 875–893.

LOWRIE, W. & HIRT, A. M. 1986. Paleomagnetism in arcuate mountain belts. *In*: WEZEL, F.-C. (ed.) *The Origin of Arcs*. Developments in Geotectonics, **21**, Elsevier, 141–158.

LUYENDYK, B. P., KAMERLING, M. J. & TERRES, R. R. 1980. Geometric model for Neogene crustal rotations in southern California. *Geological Society of America Bulletin*, **91**, 211–217.

——, KAMERLING, M. J., TERRES, R. R. & HORNAFIUS, J. S. 1985. Simple shear of southern California during Neogene time suggested by paleomagnetic declinations. *Journal of Geophysical Research*, **90**, 12454–12466.

MACDONALD, W. D. 1980. Net tectonic rotation, apparent tectonic rotation, and the structural tilt correction in paleomagnetic studies. *Journal of Geophysical Research*, **85**, 3659–3669.

McCLELLAND-BROWN, E. 1983. Palaeomagnetic studies of fold development and propagation in the Pembrokeshire Old Red Sandstone. *Tectonophysics*, **98**, 131–149.

McELHINNY, M. W. 1964. Statistical significance of the fold test in paleomagnetism. *Geophysical Journal of the Royal Astronomical Society*, **8**, 338–340.

McFADDEN, P. L. & JONES, D. L. 1981. The fold test in paleomagnetism. *Geophysical Journal of the Royal Astronomical Society*, **67**, 53–58.

——, & REID, A. P. 1982. Analysis of palaeomagnetic inclination data. *Geophysical Journal of the Royal Astronomical Society*, **69**, 307–319.

MORENO SERRANO, F. and other contributors for the Instituto Technológico GeoMinero de España. 1991. *Mapa geológico de España*, 1:50,000, Ubrique (1050).

OSETE, M. L., FREEMAN, R. & VEGAS, R. 1988. Preliminary paleomagnetic results from the Subbetic Zone (Betic Cordilleras, southern Spain): kinematic and structural implications. *Physics of the Earth and Planetary Interiors*, **52**, 283–300.

——, FREEMAN, R. & VEGAS, R. 1989. Paleomagnetic evidence for block rotations and distributed deformation of the Iberian-African plate boundary. *In*: KISSEL, C. & LAJ, C. (eds), *Paleomagnetic rotations and continental deformation*. Kluwer, London. 381–385.

PLATZMAN, E. & LOWRIE, W. 1992. Paleomagnetic evidence for rotation of the Iberian Peninsular and the external Betic Cordillera, southern Spain. *Earth and Planetary Science Letters*, **108**, 45–60.

POTTS, G. J. 1990. A paleomagnetic study of recumbently folded and thermally metamorphosed Torridon Group sediments, Eishort anticline, Skye, Scotland. *Journal of the Geological Society, London*, **147**, 999–1007.

RON, H., FREUND, R. & GARFUNKEL, Z. 1984. Block rotation by strike-slip faulting: structural and paleomagnetic evidence. *Journal of Geophysical Research*, **89**, 6256–6270.

SCHWARTZ, S. Y. & VAN DER VOO, R. 1984. Paleomagnetic study of thrust sheet rotation during foreland impingement in the Wyoming-Idaho overthrust belt. *Journal of Geophysical Research*, **89**, 10 077–10 086.

STEINER, M., OGG, J. & SANDOVAL, J. 1987. Jurassic magnetostratigraphy, 3. Bathonian-Bajocian of Carcabuey, Sierra Harana and Campillo de Arenas (Subbetic Cordillera, southern Spain). *Earth and Planetary Science Letters*, **2**, 357–372.

THOMAS, J.-C., PERROUD, H., COBBOLD, P. R., BAZHENOV, M. L., BURTMAN, V. S., CHAUVIN, A. & SADYBAKASOV, E. 1993. A paleomagnetic study of Tertiary formations from the Kyrgyz Tien-Shan and its tectonic implications. *Journal of Geophysical Research*, **98**, 9571–9589.

VAN DER VOO, R. 1969. Paleomagnetic evidence for the rotation of the Iberian Peninsula. *Tectonophysics*, **7**, 5–56.

——, & CHANNELL, J. E. T. 1980. Palaeomagnetism in orogenic belts. *Reviews of Geophysics and Space Physics*, **18**, 455–481.

VILLALAÍN, J. J., OSETE, M. L., VEGAS, R. & GARCÍA-DUEÑAS, V. 1992. Nuevos resultados paleomagneticos en el Subbetico Interno, implicaciones tectonicas. *In*: *Actas de las sesiones científicas III Congreso Geológico de España, vol. I*. Salamanca, 308–312.

——, OSETE, M. L., VEGAS, R., GARCÍA-DUEÑAS, V. & HELLER, F. 1994. Widespread Neogene remagnetization in Jurassic limestones of the South-Iberian palaeomargin (Western Betics, Gibraltar Arc). *Physics of the Earth and Planetary Interiors*, **85**, 15–33.

WESTPHAL, M., BAZHENOV, M. L., LAUER, J. P., PERCHERSKY, D. M. & SIBUET, J. C. 1986. Paleomagnetic implications on the evolution of the Tethys belt from the Atlantic Ocean to the Pamirs since the Triassic. *Tectonophysics*, **123**, 37–82.

The Neogene remagnetization in the western Betics: a brief comment on the reliability of palaeomagnetic directions

J. J. VILLALAÍN[1], M. L. OSETE[1], R. VEGAS[2], V. GARCÍA-DUEÑAS[3] & F. HELLER[4]

[1] *Departamento de Física de la Tierra, Astronomía y Astrofísica I, Universidad Complutense, Madrid 28040, Spain*
[2] *Departamento de Geodinámica, Universidad Complutense, Madrid 28040, Spain*
[3] *Departamento de Geodinámica, Instituto Andaluz de Geología Mediterránea, CSIC and Universidad de Granada, 18071 Granada, Spain*
[4] *Institut für Geophysik, ETH-Hönggerberg, CH-8093 Zürich, Switzerland*

Abstract: Recent palaeomagnetic investigations in the western Subbetics have shown the existence of a widespread pervasive magnetic overprint of Neogene age in Upper Jurassic rocks. This remagnetization is coeval with the folding deformation in the area. We describe three examples in which the fold test result indicates that the remagnetization is pre-folding, post-folding and synfolding. An evaluation of the consequences of an incorrect interpretation of the remagnetization is presented. Heterogeneous rotational patterns can be explained as a consequence of incorrect evaluation of the timing between folding and overprint acquisition.

In recent years the Betic Cordillera has been the subject of several palaeomagnetic studies carried out for structural purposes. These studies have produced important results which can help in the understanding of the complex tectonic evolution of this area. Palaeomagnetism has demonstrated that the Betic Cordillera has been affected by large rotations since the Mesozoic (Osete *et al.* 1988, 1989; Platzman & Lowrie 1992; Allerton *et al.* 1993; Villalaín *et al.* 1994). These rotations are systematically clockwise (although the values are heterogeneous) in the central and the western regions of the chain (Osete *et al.* 1988, 1989; Platzman & Lowrie 1992; Villalaín *et al.* 1994) and show a much more complex pattern in its eastern zone (Allerton *et al.* 1993; Calvo *et al.* 1994).

A study recently carried out in Upper Jurassic limestones from 20 sites from the northern Gibraltar Arc (Western Subbetics) has revealed that the major part of the NRM of these rocks is dominated by a widespread and pervasive remagnetization of Neogene age (Villalaín *et al.* 1994). This study also demonstrates that the remagnetization is coeval with the Neogene deformation by folding in the Betics.

In this study three examples are described in which the results of the fold test reflect three clearly different situations: Los Canutos (JCA) in which the acquisition of the remagnetization is post-folding, El Torcal (JTO) in which it is

pre-folding and El Valle de Abdalajís (JAB) in which it is synfolding. A total of eight sites have been considered grouped in three localities, in order to perform a fold test in each locality. The results of the localities JCA (sites JCA2 and JCA3) and JTO (sites JTO1, JTO2, JTO3 and JTO4) are described in Villalaín *et al.* (1994). A new locality JAB (sites JAB1 and JAB2) is also included. The palaeomagnetic samples at these three localities in the Internal Subbetics were obtained from grey oolitic limestones and grey and red nodular limestones of Upper Jurassic age. Figure 1 shows the location of the localities.

The three chosen localities are examples of how spurious apparent rotations can be obtained if the fold test is not systematically applied and if the secondary component is not identified and properly corrected.

Laboratory treatment and magnetic analyses of specimens were carried out at the palaeomagnetic laboratory of the ETH, Zurich. The natural remanent magnetization (NRM) of pilot samples from each site was progressively demagnetized by alternating field (AF) and thermal methods. However, thermal treatment was used as a systematic cleaning technique because it was more useful in isolating the magnetic components of the NRM. To detect the possible creation of new magnetic minerals during heating, the low field susceptibility was monitored at each step of thermal treatment. In order to

From Morris, A. & Tarling, D. H. (eds), 1996, *Palaeomagnetism and Tectonics of the Mediterranean Region*, Geological Society Special Publication No. 105, pp. 33–41.

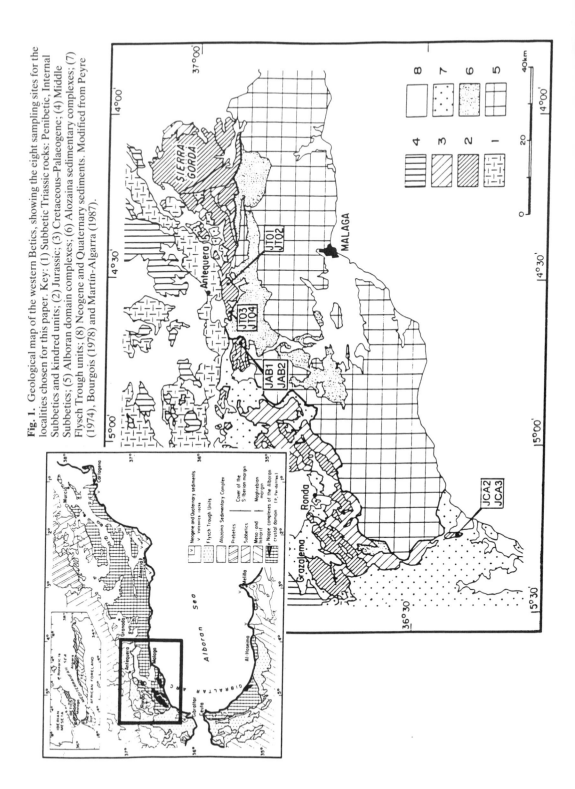

Fig. 1. Geological map of the western Betics, showing the eight sampling sites for the localities chosen for this paper. Key: (1) Subbetic Triassic rocks: Penibetic, Internal Subbetics and kindred units; (2) Jurassic; (3) Cretaceous–Palaeogene; (4) Middle Subbetics; (5) Alboran domain complexes; (6) Alozaina sedimentary complexes; (7) Flysch Trough units; (8) Neogene and Quaternary sediments. Modified from Peyre (1974), Bourgois (1978) and Martín-Algarra (1987).

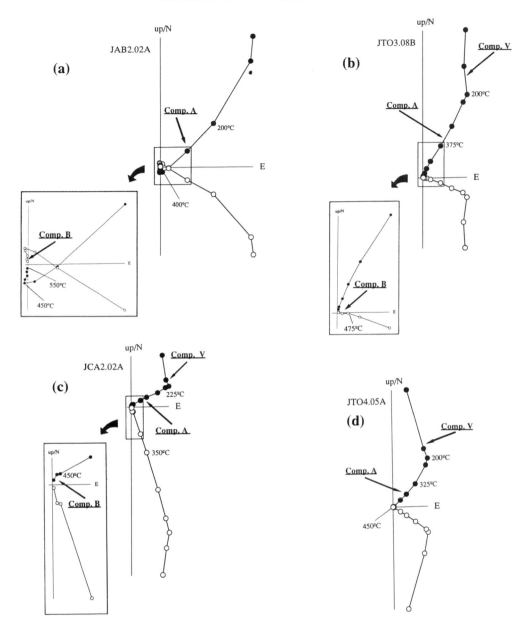

Fig. 2. Thermal demagnetization of four specimens from the three localities documented in this work. All directions have been plotted in geographic coordinates. Solid circles represent projections on the horizontal plane and open circles represent those on the E–W vertical plane.

identify the magnetic mineral responsible for the palaeomagnetic component of the NRM, several experiments were carried out: acquisition of isothermal remanent magnetization (IRM), thermal demagnetization of two IRM components (high and low field), IRM at low temperatures, etc. These procedures and their results are described in Villalaín et al. (1994).

The statistical significance of each local fold test was determined by the McFadden & Jones (1981) method. The hypothesis that the directions from both limbs had originated from a

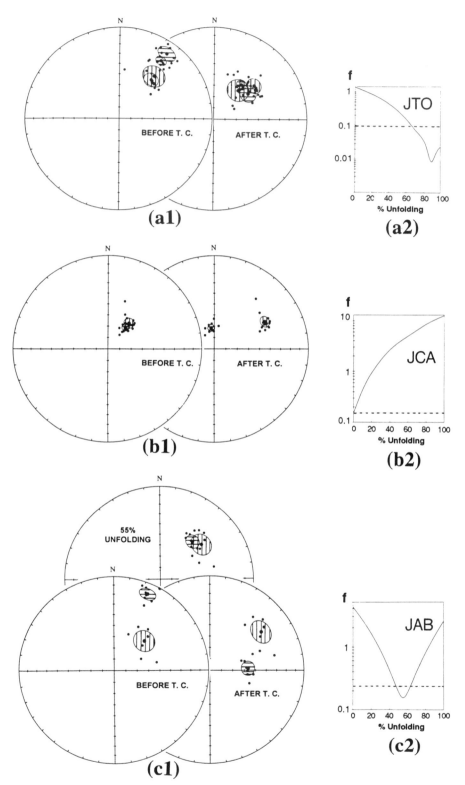

single population was tested by comparing the precision parameters on each limb (McFadden & Lowes 1981). The population considered on each limb is formed by the directions of the particular specimens. In order to detect possible synfolding magnetization, incremental fold tests were performed by rotating the magnetic directions around the strike in progressive increments of bedding dip until the beds were fully restored to horizontal.

Magnetic behaviour

The initial NRM intensities range between 0.1 and 5 mA m^{-1}. A common pattern during thermal demagnetizations has been observed. Figure 2 shows Zijderveld plots during thermal demagnetization from the localities chosen in this work. Three magnetic components could be isolated in most samples.

(1) Component V. This has a low unblocking temperature (maximum between 200–275°C). It is present in all the specimens and gives a direction close to the present day field. It is interpreted as a viscous component.

(2) Component A. This component has been observed in all samples as well. It is responsible for the major part of the non-viscous fraction of the NRM. Its maximum unblocking temperature is 450°C and is always of normal polarity. This component has been interpreted as a secondary Neogene component (Villalaín et al. 1994)

(3) Component B. In many specimens, a third component could be isolated over 450°C. This component is very weak and represents a small part of the NRM. Its maximum unblocking temperature ranges between 525 and 575°C. In contrast to component A, this component exhibits both reversed (Fig. 2a) and normal (Fig. 2b) polarities. The response to the fold test is positive and it has been interpreted as a primary magnetization (Villalaín et al. 1994). In some samples this dual polarity component has an extremely weak intensity. Therefore in these samples it is not possible to calculate its direction, although it is feasible to suggest its presence (Fig. 2c). In other cases it is absent (Fig. 2d), or it is not possible to observe it.

Due to the following reasons, components A and B could be easily mistaken: (a) they have very similar high unblocking temperatures; (b) both components have overlapping coercivity spectra; (c) component B frequently has very low intensity; (d) in some of the samples component B is absent; and (e) both components are carried by magnetite (Villalaín et al. 1994). Therefore a very careful analysis is necessary to differentiate them.

Remagnetization directions

Figure 3 shows fold test and incremental fold test results for the studied region. In Table 1 directions of component A are given for all the sites. Fold test results are given in Table 2. The f parameter (McFadden & Jones 1981) is indicated for each locality before and after tectonic correction, as well as the statistical significance limit F_{95}. Below this value, directions on both limbs are statistically the same at the 95% level of confidence.

In the Los Canutos (JCA) area, component A presents a distribution statistically different, at the 95% level of confidence, after tectonic correction ($f > F$; Table 2). In contrast, it is statistically similar before tectonic correction ($f < F$). This clearly indicates that the remagnetization is post-folding in age. Component A from the El Torcal (JTO) area is clearly pre-folding in age at the 95% level of confidence ($f > F$ before tectonic correction and $f < F$ after tectonic correction; Table 2). In the Valle de Abdalajís (JAB) area component A presents a distribution which is statistically different at the 95% level of confidence in both pre-folding and post-folding configurations (Table 2). In this locality a maximum clustering (minimum in the curve) is found for an intermediate correction (55% of unfolding). Therefore the remagnetization was acquired after 55% of the tilting.

Component A presents a similar magnetic behaviour at all localities and the polarity is always normal which suggests that the remagnetization took place in a relatively short time span (the Neogene is a period of mixed polarity). The computed inclination for the remagnetization after applying the proper tectonic correction (on the basis of an incremental fold test) is 54.5 ± 6.6°, which is consistent with the

Fig. 3. Equal-area projections showing remagnetization directions (component A) after and before tectonic correction at each locality (a1, b1 and c1). Incremental fold tests are also shown (a2, b2 and c2). Mean directions and 95% confidence circles are given for each site in the stereoplots. The incremental fold tests show the McFadden & Jones (1981) parameter f of the component A as a function of the percentage unfolding of bedding tilt for each of the seven localities. Horizontal lines represent the critical value of f at the 95% confidence level. In JTO the acquisition is pre-folding, in JCA it is post-folding and in JAB it is synfolding.

Table 1. *Remanent magnetization parameters for component A*

Site	Lat.(°N)	Long.(°W)	DD/D	N/N_0	k	α_{95}	In situ		Tilt corrected	
							D	I	D	I
JTO1	36.955	4.537	125/12	10/10	60.5	6.3	38.9	45.3	50.2	42.9
JTO2	36.957	4.532	215/16	6/6	49.8	9.6	38.7	40.8	39.9	56.5
JTO3	36.959	4.619	190/45	8/8	60.9	7.2	35.6	14.3	53.3	51.8
JTO4	36.958	4.601	207/27	8/8	163.6	4.3	35.2	24.0	38.6	50.6
JCA2	36.393	5.279	295/20	11/12	242.6	2.9	55.7	72.2	350.2	71.6
JCA3	36.405	5.263	75/30	12/12	98.7	4.4	51.7	67.3	67.0	39.0
JAB1	36.933	4.692	170/80	8/12	89.2	5.9	23.9	12.2	87.1	56.7
JAB2	36.940	4.683	72/21	8/9	37.8	9.1	45.3	52.7	53.2	32.3

DD/D, Dip direction and dip; N/N_0, number of sample directions used in the analysis v. number of samples demagnetized; k and α_{95}, statistical parameters (Fisher 1953); D and I, declination and inclination.

Table 2. *Results of the fold tests for the A component*

Locality	N/N_0	F_{95}	Before TC					After TC					After 55% unfolding				
			f	k	α_{95}	D	I	f	k	α_{95}	D	I	f	k	α_{95}	D	I
JTO	32/32	0.1050	1.536	26.0	5.1	36.9	31.4	0.0243	52.6	3.6	46.4	49.7					
JCA	23/24	0.1533	0.1512	125.9	2.7	38.4	63.4	12.60	10.2	10.0	44.3	58.2					
JAB	16/21	0.2386	4.192	10.7	11.8	32.0	32.8	2.515	16.0	9.5	66.6	45.8	0.1443	49.9	5.3	46.1	43.8

Symbols are the same as in Table 1; f, McFadden & Jones (1981) fold test statistical parameter; F_{95}, 95% significance level value of f. Site JCA was previously corrected for fold plunge.

expected Miocene inclination in this region (Dijksman 1977).

Tectonic implications

In the following, an evaluation of the consequences of an incorrect interpretation of the remagnetization is presented. As has been mentioned before, the secondary component is sometimes the only component that can be detected in the rocks. This may lead to false interpretations if the secondary component were taken as primary. Two mistakes may result: (1) an incorrect age assigned to the palaeomagnetic component (Jurassic instead of Neogene); or (2) an inappropriate tectonic correction applied to this component (a full tectonic correction overestimates the necessary restoration to the palaeohorizontal at the time of the remagnetization). Therefore incorrect identification of this secondary component could introduce more scatter of the data and apparent rotations which are erroneous.

When an *in situ* rotation ($R = D_O - D_E$) is obtained it is computed by the difference between the observed palaeodeclination (D_O)

and the expected palaeodeclination (D_E) calculated from the apparent polar wander path of the continent or tectonic block to which the rock unit is attached (Beck 1989). A computation of the block rotation (in situ rotation) is shown for two cases: (1) the A component has been considered erroneously as the primary Upper Jurassic component; (2) the A component has been identified as a secondary component and the tectonic correction has been established on the basis of the incremental fold test.

Case (1). If the remagnetization is considered to be of Jurassic age, the expected normal polarity palaeodeclination is approximately $D_{E,J}$ = 325° (computed from Steiner et al. 1985; Galbrun et al. 1990; Juárez et al. 1994). The observed declination is obtained after full tectonic correction ($D_{O,TC}$). The calculated rotation is presented in Fig. 4a and in Table 3. The apparent rotations are clockwise but large variations are observed, ranging from 30° to 130°. The rotational scenario is complex and therefore a heterogeneous pattern of rotational deformation is inferred for this region.

Case (2). If the A component has been identified as a Neogene remagnetization, the

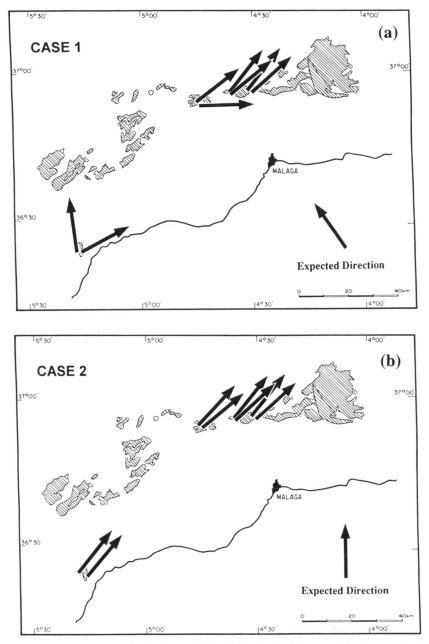

Fig. 4. Maps showing the palaeodeclinations obtained following the Case 1 and Case 2 scenarios. Case 1: Reference declination for Jurassic calculated from Steiner *et al.* (1985), Galbrun *et al.* (1990) and Juárez *et al.* (1994). Case 2: reference declination for Neogene (Dijksman 1977). See explanation in text.

expected declination is $D_{E,N} = 0°$ (obtained from Dijksman 1977). The observed declination ($D_{E,X\%}$) is then computed in each locality using the restoration to the palaeohorizontal given by the incremental fold test. In our cases it is 100%

of the tilting for the four sites of El Torcal ($D_{E,100\%}$), 0% for the two sites from Los Canutos ($D_{E,0\%}$) and 55% for the two localities from El Valle de Abdalajís ($D_{E,55\%}$). The rotational view of Western Betics has changed. Now, a more

Table 3. *Palaeodeclinations and apparent rotations for Case 1 and Case 2*

Site	Case 1 ($D_{E,J}=-35°$)		Case 2 ($D_{E,N}=0°$)	
	$D_{O,TC}$	$R_J=D_{O,TC}-D_{E,J}$	$D_{O,X\%}$	$R_N=D_{O,X\%}-D_{E,N}$
JTO1	50.2	85.2	50.2	50.2
JTO2	39.9	74.9	39.9	39.9
JTO3	53.3	88.3	53.3	53.3
JTO4	38.6	73.6	38.6	38.6
JCA2	351.9	26.9	38.3	38.3
JCA3	61.5	96.5	38.5	38.5
JAB1	87.1	122.1	50.6	50.6
JAB2	53.2	88.2	41.3	41.3

See text for notation. The palaeodeclinations computed for sites JCA1 and JCA2 are in both cases previously corrected for fold plunge.

systematic clockwise rotational pattern of about 45° is shown (Fig. 4b and Table 3).

These results give rise to doubts about previous palaeomagnetic investigations carried out in the Betics in which clockwise but heterogeneous rotations have been found (Osete *et al.* 1989; Platzman & Lowrie 1992; Allerton *et al.* 1993), and in northern Africa where a complex pattern has been evidenced (Platzman 1992). At present the Neogene remagnetization has only been well documented in the western Subbetics. Although some evidence indicates that remagnetization is present in the central Betics (Osete 1988), it is not possible to extrapolate to the rest of the Betics and to northern Africa, but future palaeomagnetic studies should consider this problem.

One of the localities presented here, Los Canutos, was previously investigated by Platzman & Lowrie (1992), who obtained a rotation of around 90°. They also studied Cretaceous rocks in this area which showed a rotation of around 40°. The authors point out that they can not explain the anomalous rotation found between the Jurassic and the Cretaceous in this region with regard to the rest of the Betic Cordillera. Finally they argue that the structure of the Los Canutos anticline is complex so that the tectonic correction possibly could not be applied properly. On the basis of our results we suggest another explanation: Component A has probably been interpreted as a Jurassic direction by these authors. Our JCA3 site at this locality also gives a rotation of 97° if the remagnetization is not positively identified (Table 3). On the other hand, if it is considered as a secondary component and the tectonic correction is properly applied, then a rotation of 39° is obtained which is consistent with the Cretaceous data of Platzman & Lowrie (1992). What about

the rest of the Jurassic localities studied by these authors? Our incremental fold test results show that the magnetization was nearly pre-folding, ranging from 75% to 100% of unfolding (Villalaín *et al.* 1994) in most of the regions they investigated. Therefore, the age of magnetization might have been assigned mistakenly, but the correct tectonic correction applied. We admit that some of the Jurassic directions found by Platzman & Lowrie (1992) correspond to the original magnetization, especially when both Jurassic and Cretaceous rocks were sampled and the angular difference between them is consistent with the counterclockwise rotation of the Iberian plate (Villalaín *et al.* in prep.).

Finally, we would like to point out that the remagnetization can be very useful in constraining the rotational motions in the Betics. In contrast to the disappointing idea that a widespread synfolding remagnetization would impede palaeomagnetic investigations for tectonic purposes, we have demonstrated that the remagnetization allows us to quantify rotations. To be able to use this secondary component, the fold test should be systematically applied. Previous palaeomagnetic results in the western and central Subbetics focused on Jurassic and Cretaceous rocks because Tertiary sediments had too weak magnetizations or strong viscous components (Platzman & Lowrie 1992). Therefore, the observed clockwise rotation could have occurred at any time since the Mesozoic up to the present. In the areas we have investigated clockwise rotations are constrained to the Neogene to present period.

Conclusions

A widespread Neogene synfolding remagnetization has affected the Upper Jurassic limestones from the western Subbetics. The maximum unblocking temperature of the secondary component is 450°C, while the original component exhibits maximum unblocking temperatures between 525°C and 575°C. In addition both components have overlapping coercivity spectra. The original Jurassic component is frequently of very low intensity and is absent in some samples. Therefore, only a very detailed thermal demagnetization procedure can isolate the original Jurassic component. This magnetic behaviour could easily cause errors in the interpretation of the two components. The secondary component could be interpreted erroneously as primary. Enlarged scatter of the data and apparent rotations could be produced if the components are not identified correctly. The systematic application of fold tests and incremental fold tests in this region allows identification of the remagnetization which can be utilized for the reconstruction of tectonic rotations. Systematic rotations have affected the investigated region from the Neogene up to the present.

This work has been supported by the Dirección General de Investigación Científica y Tecnológica DGICYT (Project PB92-0193) and by the European Community (Projet No. 935018MX. This paper is Publication No. 372, Departamento de Geofísica, Universidad Complutense de Madrid.

References

ALLERTON, S., LONERGAN, L., PLATT, J. P., PLATZMAN, E. S. & McCLELLAND, E. 1993. Palaeomagnetic rotations in the eastern Betic Cordillera, southern Spain. *Earth and Planetary Science Letters*, **119**, 225–241.

BOURGOIS, J. 1978. La transversale de Ronda (Cordillères Bétiques, Espagne). Données géologiques pour un modèle d'evolution de l'Arc de Gibraltar. *Annales Scientifique de l'Université de Besançon*, **30**, 1–453.

BECK, M. E. 1989. Block rotations in continental crust: Examples from western North America. *In*: KISSEL, C. & LAJ, C. (eds) *Paleomagnetic Rotations and Continental Deformation*. NATO ASI Series. Series C, Mathematical and Physical Sciences, **254**, 381–385.

CALVO, M., OSETE, M. L. & VEGAS, R. 1994. Paleomagnetic rotations in opposite senses in southeastern Spain. *Geophysical Research Letters*, **21**, 761–764.

DIJKSMAN, A. A. 1977. *Geomagnetic reversals as recorded in the Miocene redbeds of the Calatayud-Teruel basin (Central Spain)*. Thesis, Utrecht.

FISHER, R. A. 1953. Dispersion on a sphere. *Proceedings of the Royal Society of London*, **217**, 295–305.

GALBRUN, B., BERTHOU, P. Y., MOUSSIN, C. & AZÉMA, J. 1990. Magnétostratigraphie de la limite Jurassique–Crétacé en faciès de plateforme carbonatée: la coupe de Bias do Norte (Algarve, Portugal). *Bulletin de la Société géologique de France*, **8**, VI, 1, 133–143.

JUÁREZ, M. T., OSETE, M. L., MELÉNDEZ, G., LANGEREIS, C. G. & ZIJDERVELD, J. D. A. 1994. Oxfordian magnetostratigraphy of the Aguilón and Tosos sections (Iberian Range, Spain) and evidence of a pre-Oligocene overprint. *Physics of the Earth and Planetary Interiors*, **85**, 195–211.

MARTÍN-ALGARRA, A. 1987. *Evolución geológica alpina del contacto entre las Zonas Internas y las Zonas Externas de la Cordillera Bética*. PhD Thesis, University of Granada.

McFADDEN, P. L. & JONES, D. L. 1981. The fold test in palaeomagnetism. *Geophysical Journal of the Royal Astronomical Society*, **67**, 53–58.

—— & LOWES, F. J. 1981. The discrimination of mean directions drawn from Fisher distributions. *Geophysical Journal of the Royal Astronomical Society*, **67**, 19–33.

OSETE, M. L. 1988. *Estudio del magnetismo de las rocas de interés paleomagnético en España*. PhD Thesis. Madrid.

——, FREEMAN, R. & VEGAS, R. 1988. Preliminary palaeomagnetic results from the Subbetic Zone (Betic Cordillera, southern Spain): kinematic and structural implication. *Physics of the Earth and Planetary Interiors*, **52**, 283–300.

——, —— & —— 1989. Palaeomagnetic evidence for block rotations and distributed deformation of the Iberian-African plate boundary. *In*: KISSEL, C. & LAJ, C. (eds) *Paleomagnetic Rotations and Continental Deformation*. NATO ASI Series. Series C, Mathematical and Physical Sciences, **254**, 381–385.

PEYRE, Y. 1974. *Géologie d'Antequera te de sa région (Cordillères Bétiques-Espagne)*. Université de Paris, Publ. Inst. Agron., Paris.

PLATZMAN, E. S. 1992. Palaeomagnetic rotations and the kinematics of the Gibraltar arc. *Geology*, **20**, 311–314.

—— & LOWRIE, W. 1992. Paleomagnetic evidence for rotation of the Iberian Peninsula and external Betic Cordillera, Southern Spain. *Earth Planetary Science Letters*, **108**, 45–60.

STEINER, M. B., OGG, J. G., MELÉNDEZ, G. & SEQUEIROS, L. 1985. Jurassic magnetostratigraphy, 2. Middle-Late Oxfordian of Aguilón, Iberian Cordillera, northern Spain. *Earth and Planetary Science Letters*, **76**, 151–166.

VILLALAÍN, J. J., OSETE, M. L., VEGAS, R., GARCÍA-DUEÑAS, V. & HELLER, F. 1994. Widespread Neogene remagnetization in Jurassic limestones of the South-Iberian palaeomargin (Western Betics, Gibraltar Arc). *Physics of the Earth and Planetary Interiors*, **85**, 15–33.

New constraints on the bending of the Gibraltar Arc from palaeomagnetism of the Ronda peridotites (Betic Cordilleras, Spain)

H. FEINBERG[1], O. SADDIQI[2] & A. MICHARD[1]

[1] *Laboratoire de Géologie, UA 1316 CNRS, Ecole Normale Supérieure, 24 rue Lhomond, 75231 Paris Cedex 05, France*

[2] *Département de Géologie, Faculté des Sciences I, BP 5366, Maârif-Casablanca, Morocco*

Abstract: The study of 210 cores (15 sites) from the Ronda peridotites (Sierra Bermeja and Sierra Alpujata, Alpujarride nappe complex, Betic zone), and from the granites intruding these peridotites and their country rocks shows the occurrence of two stable antipodal directions of magnetization ($D = 46°$, $I = 47°$, $\alpha_{95} = 6.6$). The reverse polarity high-temperature component, only found in peridotites, is carried by hematite, while the normal polarity intermediate-temperature component is carried by magnetite in the peridotites, and by sulphides in the granites. Negative fold tests point to a late magnetization. The acquisition of remanence is attributed to the post-metamorphic cooling of the Alpujarrides, bracketed between 23 and 18 Ma by isotopic and stratigraphic data. Structural data and the homogeneity of the *in situ* mean palaeomagnetic directions preclude significant tilting of the massifs after their magnetization. The observed declination is interpreted as the result of a post-metamorphic, $46° \pm 8°$ clockwise rotation of the Ronda massifs around a vertical axis. These results are compared with those from the Beni Bousera peridotites (southern branch of the Gibraltar Arc). In the latter massif, a *c.* $74° \pm 11°$ anticlockwise rotation has been documented, and dated from the time of cooling of the peridotite unit. Therefore the opposite rotations of the Spanish and Moroccan massifs occurred rapidly during Early Miocene. A tectonic model involving extensional collapse with preferential displacement towards the Atlantic free-margin is favoured.

On the southern margin of Iberia, the Betic Cordilleras constitute one of the major elements of the Western Mediterranean Alpine chain. They are connected to the Morrocan Rif belt through the Gibraltar Arc. The whole chain developed in response to the collision between Africa and Eurasia during late Mesozoic and early Cenozoic times before being dramatically dissected by the late Oligocene–Miocene opening of the Mediterranean basins. The interpretation of the Gibraltar Arc, i.e. of the apparent bending of the Betic-Rifian tectonic units and structural directions around the Alboran sea, was first documented through geological and structural analyses (Andrieux *et al.* 1971; Platt & Vissers 1989; Balanya & Garcia-Dueñas 1987; Frizon de Lamotte *et al.* 1991). Recently, palaeomagnetic studies were performed on sediments belonging to the External Sub-Betic zone (Osete *et al.* 1988; Platzman & Lowrie 1992; Villalaín *et al.* 1994, this volume), and to the Rifian Dorsale Calcaire (Platzman 1992; Platzman *et al.* 1993). These studies showed a clockwise rotation of the Sub-Betic units, and an anticlockwise rotation of the Rifian Dorsale, but ran into difficulties in dating the magnetization

components (primary Mesozoic or secondary Miocene components). In contrast, palaeomagnetic studies of the ultrabasic massifs included either in the External Rif (Beni Malek massif; Elazzab & Feinberg 1994) or in the metamorphic, Internal Rif (Beni Bousera massif; Saddiqi *et al.* 1995) demonstrate the usefulness of these rocks for dating magnetizations and for further documentation of tectonic rotations in the region. In this paper, we focus on the Internal, metamorphic units of the northern branch of the Gibraltar Arc, studying both the ultrabasic massifs (Ronda peridotites) and the associated, syn-orogenic granites. The results from the Betic ultrabasites can be compared with those from the Rifian ones, and from the sedimentary units of the Dorsale and Sub-Betic zones, in order to constrain the tectonic interpretation of the Gibraltar Arc.

Geological setting

The Ronda peridotites *sensu lato* include essentially the ultrabasic massifs of Sierra Bermeja and Sierra Alpujata (Fig. 1). They belong to the Alpujarride nappe complex of the Internal, or

From Morris, A. & Tarling, D. H. (eds), 1996, *Palaeomagnetism and Tectonics of the Mediterranean Region*, Geological Society Special Publication No. 105, pp. 43–52.

43

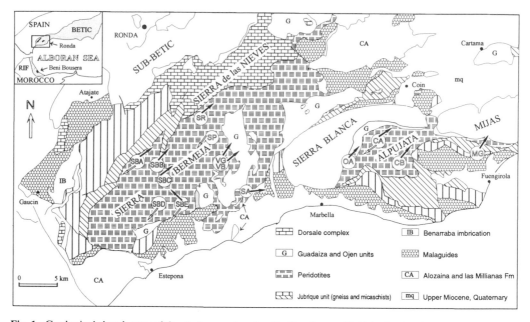

Fig. 1. Geological sketch map of the study area showing the Ronda massifs (Alpujarrides) and location of sampling sites. Arrows indicate *in situ* mean declinations for each site, solid (open) symbols correspond to direct (reverse) inclinations.

Betic zone of the Cordilleras, and constitute the lowest part of the Los Reales nappe (Lundeen 1978; Tubia & Cuevas 1987; Balanya & Garcia-Dueñas 1991). The Los Reales tectonic unit comprises high-grade metamorphic material, namely from top to base Triassic metacarbonates, quartzites and metapelites, schists and gneisses (Jubrique unit; Balanya *et al.* 1993), which are separated from the underlying ultrabasites by a high-temperature shear zone. The Los Reales (Ronda) ultrabasites in turn overthrust the migmatites, schists and marbles of the Ojen–Blanca Alpujarride nappe through another high-temperature shear zone. The Alpujarride nappe complex overlies the Nevado–Filabrides high-grade nappes in the Eastern Betics, and is tectonically overlain by the virtually unmetamorphosed Malaguide complex. Finally, the whole Betic zone is thrust over the External, Sub-Betic zone.

Within the southern branch of the Gibraltar Arc, the Beni Bousera peridotites, included in the Sebtide nappe complex, are the strict equivalent of the Ronda peridotites. All of these ultrabasic massifs are dominated by spinel lherzolite with associated garnet-bearing pyroxenite layers. They are usually regarded as infracontinental mantle rock slivers sampled at the top of a syn-orogenic mantle diapir during the early Miocene (Loomis 1975; Obata 1980; Tubia & Cuevas 1987), or at the top of a pre-

orogenic (Mesozoic), extensional mantle uplift (Reuber *et al.* 1982; Saddiqi *et al.* 1988; Michard *et al.* 1991, 1992; Van der Wal 1993). Regardless of their origin, the Ronda-Beni Bousera peridotites, together with the juxtaposed crustal units, suffered a high-temperature, high to medium-pressure metamorphic evolution (Goffé *et al.* 1989; Bouybaouene 1993; Azanon 1994). The high-temperature evolution ended under low-pressure conditions with the emplacement of leucogranitic dykes rooted in the Ojen migmatites which intrude the Los Reales peridotites and schists (Muñoz 1991). Multi-method isotopic dating of these granites, of the ultrabasites (garnet-bearing pyroxenites), and of the juxtaposed schists indicate that the maximum temperature was reached at 30–25 Ma, and was followed by a rapid cooling up to 18 Ma ago (Priem *et al.* 1979; Polvé & Allègre 1980; Michard *et al.* 1983; Zeck *et al.* 1989, 1992; Monié *et al.* 1991; De Jong *et al.* 1992). The oldest, unconformable deposits with high-grade Alpujarride clasts and peridotites pebbles are Early Burdigalian (19 ± 1 Ma) in age (Olivier 1984; Aguado *et al.* 1990; Durand-Delga *et al.* 1993; Feinberg *et al.* 1990).

Methodology

Measurements were performed on 210 cored samples from 15 sites distributed through the

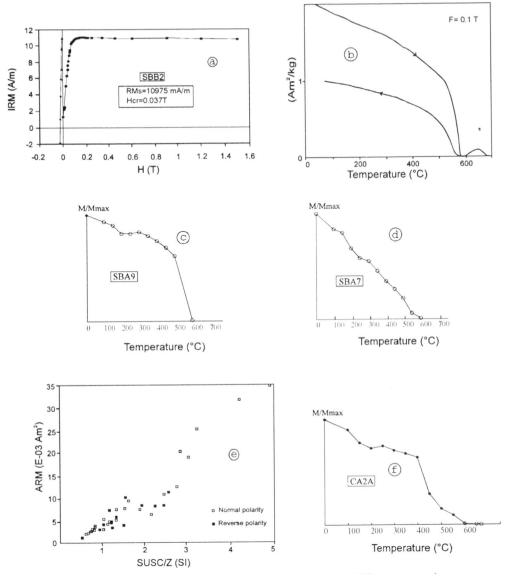

Fig. 2. Magnetic mineralogy of the peridotites; (**a**) IRM acquisition curve; (**b**) Thermomagnetic measurement under vacuum; (**c**, **d**, **f**) decay of NRM during thermal demagnetization; (**e**) diagram of ARM; symbols indicate the NRM polarity of samples.

two main peridotite massifs between Ronda and Malaga (Sierra Bermeja and Sierra Alpujata), in the underlying Ojen nappe (Guadaiza window; sites VG, VB), and in the overlying Jubrique unit (site MG). Most sites consist of ultrabasic samples, while sites SBG, VG,VB and MG consist of granitic samples (SBG being located in leucogranitic dykes intruding the lherzolite of site SBD). Characteristic magnetic directions were isolated using either thermal or alternating field (AF) demagnetization, with very similar results. The directions were selected using a least-square routine (Kirschvink 1980). Site mean directions and associated parameters were calculated using Fisher's statistics (1953). Measurements of IRM (isothermal remanent magnetization), and of ARM (anhysteretic remanent magnetization) were performed to determine the magnetic mineralogy. Natural remanent magnetizations (NRM) were

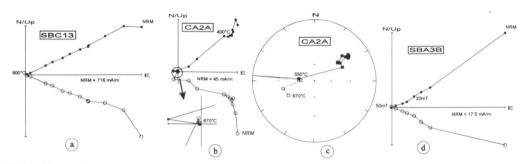

Fig. 3. Vector diagrams and stereographic projections of peridotite samples during stepwise demagnetization. Open (solid) circles indicate projection onto the vertical (horizontal) plane; open (solid) squares correspond to the upper (lower) hemisphere. (**a–c**) Thermal demagnetization; (d) alternating field demagnetization.

measured with a spinner magnetometer for peridotites and a cryogenic magnetometer for the weakly magnetized granites.

Magnetic mineralogy

The NRM intensity of the peridotites varies from site to site with a mean value of 0.25 A m^{-1}, except at the Alijar site (SA) where the values are ten times larger. The variation of NRM intensity is related to the degree of serpentinization (Moody 1976; Lienert & Wasilievski 1979). The magnetic susceptibility (measured with Molspin equipment) has a mean value of $0.8–1 \times 10^{-2}$ SI. The steep gradient of the IRM acquisition curve and the weak value of coercive field indicate that the main magnetic carrier is magnetite or titanomagnetite (Fig. 2a). This is supported by thermomagnetic measurements performed under vacuum which show (Fig. 2b) a Curie temperature close to that of magnetite (580°C). This mineral is also indicated by most thermal demagnetization diagrams, where Curie temperatures between 550°C and 600°C are visible (Fig. 2c & d). These demagnetization curves show either a regular decay (SBA7) or a collapse around 550°C (SBA9). The thermomagnetic analyses also show a small fraction of hematite in these peridotites as shown in Fig. 2b, where a second Curie temperature, typical of that of hematite, is present around 670–680°C. Some thermal demagnetization curves also show the occurrence of hematite (Fig. 2f). Additionally, a study of the magnetic grain size was conducted on 40 samples. The specimens were submitted to an alternating field (peak value = 100 mT) combined with a 0.1 mT continuous field (Tauxe & Wu 1990). The resulting ARM plot (Fig. 2e) shows a good homogeneity of grain size which is apparently independent of specimen magnetization.

In contrast, the NRM of the granites is weak

with a mean value of 2×10^{-4} A m^{-1}. The magnetic susceptibility is also very low ranging from $0.5–5 \times 10^{-5}$ SI. The magnetic mineralogy of the granites consists predominantly of low unblocking temperature minerals (350–400°C) probably corresponding to sulphides associated with a small fraction of magnetite.

In conclusion the magnetic mineralogy of the Ronda peridotites is dominated by rather coarse grains of magnetite associated with a fraction of hematite, whereas magnetic minerals in the granites appear to be dominated by sulphides.

Analysis of palaeomagnetic directions

Nearly all sampled sites in peridotites show the occurrence of a NE direction with the following calculated mean direction: $D = 49.1°$, $I = 46°$, $\alpha_{95} = 12.8°$, $k = 13$ (Table 1). This direction is either of normal polarity and stable up to 600°C (Fig. 3a & b) or of normal polarity up to 550°C and then reversed at higher temperatures (Fig. 3c & d). The same result is obtained with alternating field demagnetization (Fig. 3e & f). A similar NE direction is also observed in the granites with a calculated mean direction of: $D = 40.8°$, $I = 48.3°$, $\alpha_{95} = 6°$, $k = 132$ (Table 1, Fig. 4c & d).

In some cases a reversed direction of magnetization is observed throughout demagnetization (Fig. 4a). This direction, which is only present in 5 sites (Table 2), has the following mean direction: $D = 225°$, $I = -49.4°$, $\alpha_{95} = 6°$, $k = 106$. This reversed polarity direction corresponds to that of the stable, high temperature component observed in samples showing a reversal of polarity during the last demagnetization steps.

A fold test using the foliation planes was performed on all the peridotite sites bearing the NE direction. The mean values calculated before tectonic correction are: $D = 49.1°$, $I =$

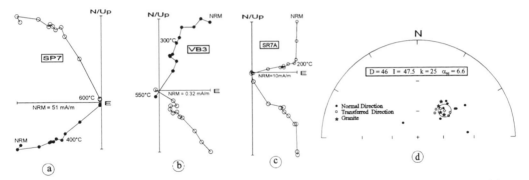

Fig. 4. Same conventions as for Fig. 3. (**a**) A completely reversed peridotite sample; (**b**) granite sample; (**c**) peridotite sample with a low unblocking temperature component; (**d**) stereonet of all characteristic sites mean directions.

Table 1. *Palaeomagnetic results from sites in the Ronda massifs which contain the NE-directed normal polarity component*

Sites	N/N_u	D_b	I_b	k_b	α_{95b}	D_a	I_a	k_a	α_{95a}	Long	Lat
CA	06/11	56.3	29.0	36	09.5	139.0	53.7			4 43.3	36 37.1
CB	11/11	38.7	39.2	28	08.1	023.7	−16.6			4 42.3	36 35.3
OA	11/11	42.1	46.4	34	07.2	256.3	41.0			4 50.3	36 35.3
SA	10/11	84.6	35.4	11	13.5	096.7	12.0			5	36 30.2
SBA	07/11	53.8	37.2	16	13.3	085.8	16.2			5 10.8	36 33.5
SBB	09/11	42.0	40.3	38	07.6	014.8	22.5			5 10.4	36 33.3
SBC	08/11	52.9	38.5	15	12.7	028.0	−3.7			5 10.1	36 32.5
SBE	07/11	314.5	65.9	21	11.5	323.7	−8.2			5 10.5	26 29.3
SR	04/11	51.5	56.3	13	08.0	340.0	36.5			5 4.9	36 38.6
Total		49.1	46.0	13	12.8	21.4	29.2	2	32.5		
VG	7/12	36.2	50	11	16.1					5 3.2	36 33.8
VB	5/5	50.7	50	22	12.2					5 3.2	36 33.8
MG	5/12	33.0	41	13	14.5					4 37.9	36 34.4
SBG	5/12	44.6	51	09	21.2					5 11.3	36 30.4
Total		40.8	48.3	132	6.1						

Upper part: ultrabasic rocks; lower part: leucogranite. N, N_u = number of samples showing the NE component over number of demagnetized and measured samples; D_b, I_b = *in situ* directions, D_a, I_a = directions after tectonic correction; k and α_{95} are Fisher's statistical parameters; Lat., latitude and Long., longitude of sites, in degrees and minutes.

46°, $\alpha_{95} = 12.8°$, $k = 13$, $N = 9$. After tectonic correction the values are: $D = 21.4°$, $I = 29.2°$, $\alpha_{95} = 32.5°$, $k = 2$, $N = 9$. Following McElhinny (1964) this test is negative at the 95% confidence level ($N = 9$, $K_1/K_2 = 0.15$, $F(16,16)\ 5\% = 2.32$). The test is also negative for the SW direction.

In conclusion, the magnetic directions found in the Ronda peridotites and in the late-orogenic granites show a remarkable clustering (except for sites SA and SBE). After taking the antipode of the reverse directions, the mean for all the sites becomes: $D = 46°$, $I = 47.5°$, $\alpha_{95} = 6.6°$, $N = 19$ (Fig. 4f).

Discussion

Age of magnetization and rotation

The Ronda peridotites were affected, together with the juxtaposed crustal units of the Alpujarride nappes, by a late Oligocene–Early Miocene metamorphic event, during which the temperature of the ultrabasites reached at least 750°C (Van der Wal 1993). Isotopic and stratigraphic data document a rapid cooling of the tectonic pile between 23 and 18 Ma (Zeck *et al.* 1989, 1992; Durand-Delga *et al.* 1993). The high- and

Table 2. *Palaeomagnetic results from sites in the Ronda massif which contain the SW-directed reversed polarity component*

Sites	N/N_u	D_b	I_b	k_b	α_{95b}	D_a	I_a	k_a	α_{95a}	Long	Lat
CA	06/11	233.7	−44	20	12	222.0	05.9			4 43.3	36 37.1
SP	10/11	215.5	−55	52	06	265.2	−60.7			5 4.1	36 37.8
SBA	03/11	236.7	−46	20	18	191.3	−24.2			5 10.8	36 33.5
SBD	03/11	219.4	−49	17	19	259.2	−32.7			5 11.3	36 30.4
SBE	03/11	216.7	−51	32	14	185.3	02.2			5 10.5	36 29.3
Total		225.0	−49.4	106	6	217.9	−25	31.5	44		

Same conventions as Table 1.

medium-temperature magnetization components in the peridotites were, therefore, acquired at that time, during the Aquitanian and early Burdigalian. This conclusion is clearly supported by the fact that the 22 ± 4 Ma aged granitic dykes which intrude the peridotites and juxtaposed schists yield the same palaeomagnetic components as the peridotites.

The palaeomagnetic study demonstrates that the Ronda massifs registered a magnetic field reversal during their progressive cooling. The change from reverse to normal field was followed by a $46 \pm 8°$ clockwise rotation of the massifs. This rotation was probably completed before final cooling of the massifs, since some of the samples yield a magnetic component close to recent field direction and stable up to 250°C. This suggestion is supported by the palaeomagnetic data for the Beni-Bousera massif to the South of the Alboran Sea (Saddiqi *et al.* 1995). In this peridotite massif, which is the equivalent of the Ronda massifs to the north, a $74 \pm 11°$ anticlockwise rotation is documented, which occurred early during the cooling evolution of the massif i.e. before 18 Ma, as evidenced by the presence of a magnetic direction close to the recent field direction and which is stable up to 450°C. We conclude that not only the magnetization, but also the rotations of the Gibraltar Arc peridotite massifs occurred during Aquitanian–Burdigalian time.

Vertical axis rotations versus tilting

The magnetization of the Ronda (and Beni Bousera) peridotites postdates the folding of their internal foliation (negative fold test). The absence of a defined paleohorizontal surface linked to the rocks at the time of their magnetization requires a careful consideration of the possible role of tilting in the reorientation of the units after magnetization. There are two direct arguments against an important component of post-magnetization tilting: (i) the

homogeneity of natural remanent magnetization over a large region; and (ii) the shallow dips observed at a regional scale in the Aquitanian–Burdigalian clastic layers, which progressively overlapped the Alpujarride units during their exhumation (Feinberg *et al.* 1990; Durand-Delga *et al.* 1993). The exhumation and consequent cooling of the Alpujarride nappe complex occurred through extensional thinning, involving the operation of low-angle, shallow-dipping normal faults within a previously thickened, ductile pile of nappes (Garcia-Dueñas *et al.* 1992; Balanya *et al.* 1993; Bouybaouene 1993). Under such conditions, block tilting in the extending plates is neither expected theoretically (except at the shallowest, brittle levels of the upper plate), nor actually observed in well documented cases such as the Aegean metamorphic belt (Avigad & Garfunkel 1991). Note that the earliest part of the exhumation and cooling of the Ronda peridotites occurred without any significant tilting or vertical axis rotation, since the high- and medium-temperature components of magnetization are strictly antipodal. Even if we imagine moderate (10–20°) tilting during the cooling and exhumation of the peridotites, this cannot completely explain the observed reorientation of the remanent magnetization, regardless of the orientation of the axis of tilting (Fig. 5). A major component of vertical axis rotation is therefore suggested. This conclusion is also supported by palaeomagnetic results obtained in the adjoining, Sub-Betic zone by Villalaín *et al.* (1994) (see below).

Tectonic implications

Various models are currently suggested to explain the geometry and structure of the Gibraltar Arc: (i) westward indentation of an Alboran microplate (Leblanc & Olivier 1984; Bouillin *et al.* 1986); (ii) oroclinal bending of a previously rectilinear belt (Tubia & Cuevas

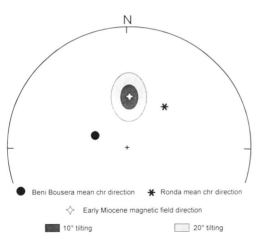

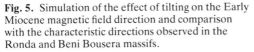

Fig. 5. Simulation of the effect of tilting on the Early Miocene magnetic field direction and comparison with the characteristic directions observed in the Ronda and Beni Bousera massifs.

1987); (iii) distributed dextral strike-slip between Africa and Iberia (Osete *et al.* 1988); and (iv) late orogenic, radial collapse of the Alboran thickened crust (Platt & Vissers 1989). Platzman (1992) suggested that the latter model could be improved by incorporating two relative motions, namely the Africa-Iberia convergence and the westward motion of the centre of the collapsing ridge with respect to its margins (due to its east–west elongation). Platzman (1992) argues that such a model may account for the opposite palaeomagnetic rotations documented in the Sub-Betic and Rifian Dorsale sediments (clockwise and anticlockwise respectively) (Osete *et al.* 1988; Platzman & Lowrie 1992; Platzman 1992; Platzman *et al.* 1993).

Our results are consistent with those of the latter authors, as far as the orientation and magnitude of the rotations observed in the northern and southern branches of the Gibraltar Arc are concerned, but we reach a different conclusion for the age of the rotation. For example, Platzman (1992) suggests that the Sub-Betic and Rifian sedimentary units have probably rotated during the Miocene main orogenic event (dated from uppermost Mid-Miocene to Late Miocene). This cannot apply to the rotation of the peridotite massifs and associated metamorphic units (Alpujarride–Sebtide nappe complex), which rotated during the Early Miocene. Villalaín *et al.* (1994, this volume) documented a widespread remagnetization of the Sub-Betic Jurassic and Cretaceous layers, corresponding in different localities to a pre-, syn- or post-folding, low- to medium-

temperature normal component of magnetization. This component was acquired over a short time span, probably during the thrusting of the Alboran units (Betic zone) onto the Iberian foreland (Burdigalian to Middle Miocene). Since the observed clockwise rotations of the Ronda and Sub-Betic units are equivalent, we suspect that they occurred mostly at the same time, e.g. during Burdigalian times, immediately after the impingement of the Betic units onto the Sub-Betic domain.

Further tectonic implications can be drawn with regard to the initial orientation of the Ronda and Beni Bousera massifs, i.e. of the Alpujarride belt prior to the bending of the Gibraltar Arc. In the Internal Rifian–Betic zones, this bending is reflected by curvature of the well-developed stretching lineation (direction of relative transport of the upper plate), from NNW in the Beni Bousera (Michard *et al.* 1983; Saddiqi *et al.* 1988) to NE–ENE in the Ronda area (Tubia & Cuevas 1987; Balanya & Garcia-Dueñas 1991). It is tempting to associate this structural bending to that of the characteristic palaeomagnetic components since: (i) both these components and the stretching lineations developed during the unloading and cooling evolution of the Alpujarride–Sebtide complex; and (ii) the present-day orientation of the characteristic declinations and of the stretching lineations are virtually parallel in each peridotite massif (Fig. 6). Therefore we suggest that the Ronda and Beni Bousera massifs were both elongated in a N–S direction and were probably in line at the beginning of the tectonic unloading. On Fig. 6, the initial position of these massifs on a meridian some 50 km east of Malaga has been arbitrarily chosen. However, this position (relative to the Iberian and African forelands) seems realistic, in view of the important NNW displacement of the detached Sub-Betic cover (e.g. Frizon de Lamotte *et al.* 1991). The diverging pattern of displacement would reflect gravity-driven tectonics with preferential displacement of the collapsing units toward the Atlantic free-margin.

Conclusions

The peridotites of the Betic–Rifian Internal Zones yield palaeomagnetic data which place additional constraints on the regional deformation. Samples from the Betic Ronda massifs, including those from the leucogranites which intrude the ultrabasites, contain high- and intermediate-temperature characteristic components of magnetization of either reverse or normal polarity, which were acquired during the

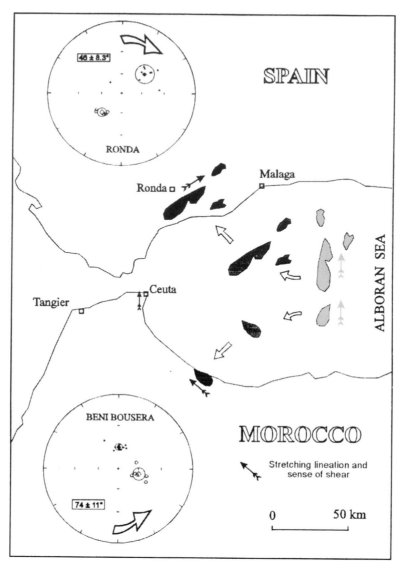

Fig. 6. Tectonic cartoon illustrating the evolution of palaeomagnetic rotations of the Gibraltar Arc peridotites during Early–Mid-Miocene. Stereonets show the site mean directions of the normal and reversed polarity components in each study area.

Early Miocene cooling of the Alpujarride–Sebtide nappe complex. A $46 \pm 8°$ clockwise rotation is documented in the Ronda sites, in contrast to the $74 \pm 11°$ anticlockwise rotation of the Beni Bousera massif (Saddiqi *et al.* 1995). Our results are consistent with those previously obtained from the palaeomagnetic study of Mesozoic sedimentary units adjacent to the Alpujarride–Sebtide nappes. However, the new measurements on the internal, high-temperature rocks constrain the rotation to the Early

Miocene, between 23 and 18 Ma. The Ronda rotation is believed to be slightly younger than that of Beni Bousera, and was probably concentrated in the 20–19 Ma period of dominantly normal polarity magnetic field. This suggests a particularly high rotation rate (*c.* $30°\,Ma^{-1}$) for the Ronda massifs, at about the time of the impingement of the Internal Betic Zones onto the Sub-Betic foreland. Additionally, the correlation of the Early Miocene declinations with the directions of the stretching lineations observed in the peridotites and their country-rocks allows the reconstruction of the geometry of the massifs to a N–S orientation at about 22 Ma, before subsequent extensional bending of the Gibraltar Arc.

Thanks to J. C. Balanya for providing field assistance. Comments and grammatical help from A. Morris and comments from S. Allerton considerably improved the manuscript.

References

AGUADO, R FEINBERG, H., DIDON, J., DURAND-DELGA, M., ESTERAS, M. & MARTIN-ALGARRA, A. 1990. Nuevos datos sobre la edad de las formaciones del Mioceno inferior transgresivas sobre las zonas internas béticas: la formacion de San Pedro de Alcantara (prov. di Malaga). *Revista de la Sociedad Geologica de Espana*, **3**, 79–85.

ANDRIEUX, J., FONTBOTE, J. M. & MATTAUER, M. 1971. Sur un modèle explicatif de l'Arc de Gibraltar. *Earth and Planetary Science Letters*, **12**, 191–198.

AVIGAD, D. & GARFUNKEL, Z. 1991. Uplift and exhumation of high-pressure metamorphic terrains: the example of the Cycladic blueschist belt (Aegean sea). *Tectonophysics*, **187**, 1–15.

AZANON, J. M. 1994. *Metamorfismo di alta presion/baja temperatura, baja presion/alta temperatura y tectonica del Complejo Alpujarride (Cordilleras Bético-Rifenas)*. PhD Thesis, Univ. Granada.

BALANYA, J. C. & GARCIA-DUEÑAS, V. 1987. Les directions structurales du domaine d'Alboran de part et d'autre du détroit de Gibraltar. *Comptes Rendus de l'Acadamie des Sciences, Paris*, **304**, II, 929–934.

—— & —— 1991. Estructuracion de los mantos Alpujarrides al W de Malaga Béticas, Andalucia). *Geogaceta*, **9**, 30–33.

——, AZANON, J. M., SANCHEZ-GOMEZ, M. & GARCIAS-DUEÑAS, V. 1993. Pervasive ductile extension, isothermal decompression and thinning of the Jubrique unit in the Paleogene (Alpujarrides Complex, western Betic Spain). *Comptes Rendus de l'Acadamie des Sciences, Paris*, **316**, 1595–1601.

BOUILLIN, J.P., DURAND-DELGA, M. & OLIVIER PH. 1986. Betic-Rifian and Tyrrhenian Arcs: Distinctive features, genesis, and development stages. *In*: WEZEL, F. C. (ed.) *The origin of Arcs*. Elseviers Science Publishers, Amsterdam, **21**, 281–304.

BOUYBAOUENE, M. 1993. *Etude pétrologique des métapelites des Sebtides supérieures, Rif interne, Maroc*. PhD Thesis, Univ. Rabat.

DE JONG, K., WIJBRANS, J.R. & FERAUD, G. 1992. Repeated thermal resetting of phengites in the Mulhacen Complex (Betic Zone, southeastern Spain) shown by $^{40}Ar/^{39}Ar$ step heating and single grain laser probe dating. *Earth and Planetary Science Letters*, **110**, 173–192.

DURAND-DELGA, M., FEINBERG, H., MAGNE, J., OLIVIER, P. & ANGLADA, R. 1993. Les formations oligo-miocènes discordantes sur les Malaguides et les Alpujarrides et leur implications dans l'évolution géodynamique des Cordillères bétiques (Espagne) et de la Méditerranée d'Alboran. *Comptes Rendus de l'Academie des Sciences, Paris*, **317**, II, p. 679–687.

ELAZZAB, D. & FEINBERG, H. 1994. Paléomagnétisme des roches ultrabasiques du Rif externe (Maroc). *Comptes Rendus de l'Academie des Sciences, Paris*, **318**, 351–357.

FEINBERG, H., MAATE, A., BOUHDADI, S., DURAND-DELGA, M., MAGNÉ, J. & OLIVIER, PH. 1990. Signification des dépôts de l'Oligocène supérieur-Miocène inférieur du Rif interne (Maroc) dans l'évolution géodynamique de l'Arc de Gibraltar. *Comptes Rendus de l'Academie des Sciences, Paris*, **310**, 1487–1495.

FISHER, R. A. 1953. Dispersion on a sphere. *Proceedings of the Royal Society of London*, **A217**, 295–305.

FRIZON DE LAMOTTE, D., ANDRIEUX, J. & GUEZOU, J. 1991. Cinématique des chevauchements néogènes dans l'Arc bético-rifain: discussion sur les modèles géodynamiques. *Bulletin de la Société Géologique de France*, **162**, 611–626.

GARCIA-DUEÑAS, V., BALANYA, J. C. & MARTINEZ-MARTINEZ, J. M. 1992. Miocene extensional detachments in the outcropping basement of the Northern Alboran Basin (Betics) and their tectonic implications. *Geo-Marine Letters*, **12**, 88–95.

GOFFE, B., MICHARD, A., GARCIAS-DUEÑAS, V., GONZALZ LODEIRO, F., MONIE, P., CAMPOS, P., GALINDO-ZALDIVAR, F., JABALOY, A., MARTINEZ, J. M. & SIMANCAS, J. F. 1989. First evidence of high-pressure, low- temperature metamorphism in the Alpujarride nappes, Betic Cordilleras (SE Spain). *European Journal of Mineralogy*, **1**, 139–142.

KIRSCHVINK, J. L. 1980. The least squares line and plane and the analysis of paleomagnetic data. *Geophysical Journal of the Royal Astronomical Society*, **62**, 699–718.

LEBLANC, D. & OLIVIER, PH. 1984. Role of strike -slip faults in the Betic Rifean orogeny. *Tectonophysics*, **101**, 344–355.

LIENERT, B. R. & WASILIEVSKI, P. J. 1979. A magnetic study of the serpentinisation process at Burro Mountain, California. *Earth and Planetary Science Letters*, **43**, 406–416.

LOOMIS, T. P. 1975. Tertiary mantle diapirism, orogeny and plate tectonics east of the Strait of Gibraltar. *American Journal of Science*, **275**, 1–30.

LUNDEEN, M. T. 1978. Emplacement of the Ronda peridotite, Sierra Bermeja, Spain. *Geological Society of America Bulletin*, **89**, 172–180.

McELHINNY, M. W. 1964. Statistical significance of the fold test in paleomagnetism. *Geophysical Journal of the Royal Astronomical Society*, **8**, 338–340.

MICHARD, A., CHALOUAN, A., MONTIGNY, R. & OUAZZANI-TOUHAMI, M. 1983. Les nappes cristallophylliennes du Rif (Sebtides-Maroc), témoins d'un édifice alpin de type pennique incluant le manteau supérieur, *Comptes Rendus de l'Academie des Sciences, Paris*, **296**, 1337–1340.

MICHARD, A., GOFFÉ, B., CHALOUAN, A. & SADDIQI, O. 1991. Les corrélations entre les chaînes bético-rifaines et les Alpes et leurs conséquences. *Bulletin de la Societé Géologique de France*, **162**, 1151–1160.

MICHARD, A., FEINBERG, H., ELAZZAB, D., BOUYBAOUENE, M. & SADDIQUI, O. 1992. A serpentinite ridge in a collision paleomargin setting; the Beni Malek massif, External Rif, Morocco. *Earth and Planetary Science Letters*, **113**, 435–442.

MONIÉ, P., GALINDO-ZALDIVAR, J., GONZALEZ LODEIRO, F., GOFFÉ, B. & JABALOY, A. 1991. $^{40}Ar/^{39}Ar$ geochronology of Alpine tectonism in the Betic Cordilleras (Spain). *Journal of the Geological Society, London*, **148**, 289–297.

MOODY, J. B. 1976. Serpentinization: a review, *Lithos*, **9**, 125–138.

MUÑOZ, M. 1991. Significado de los cuerpos de leucogranitos y de los "gneiss cordieriticos con litoclastos" asociados en la unidad de Guadaiza (Alpujarrides occidentales, Béticas). *Geogaceta*, **9**, 10–13.

OBATA, M. 1980. The Ronda peridotite: Garnet-Spinel-, and Plagioclase-Lherzolites facies and the P-T trajectories of a high-temperature mantle intrusion. *Journal of Petrology*, **21**, 533–572.

OLIVIER, PH. 1984. *Evolution de la limite entre Zones internes et Zones externes dans l'Arc de Gibraltar (Maroc, Espagne)*. PhD Thesis, Univ. Toulouse III.

OSETE, M. L., FREEMAN, R. & VEGAS, R. 1988. Preliminary paleomagnetic results from the Subbetic zone (Betic Cordilleras, Southern Spain): Kinematic and structural implications. *Physics of the Earth and Planetary Interiors*, **52**, 283–300.

PLATT, J. P. & VISSERS, R. L. M. 1989. Extensional collapse of thickened continental lithosphere: A working hypothesis for the Alboran Sea and Gibraltar Arc. *Geology*, **17**, 540–543.

PLATZMAN, E. S. 1992. Paleomagnetic rotation and the kinematics of the Gibraltar arc. *Geology*, **20**, 311–314.

—— & LOWRIE, W. 1992. Paleomagnetic evidence for rotation of the Iberian Peninsula and the external Betic Cordillera, Southern Spain. *Earth and Planetary Science Letters*, **108**, 45–60.

——, PLATT, J. P. & OLIVIER, P. 1993. Paleomagnetic rotations and fault kinematics in the Rif Arc of Morocco. *Journal of the Geological Society, London*, **150**, 707–718.

POLVÉ, M. & ALLEGRE, C. J. 1980. Orogenic lherzolite complexes studied by Rb^{87}-Sr^{87}: a clue to understand the mantle convection processes. *Earth and Planetary Science Letters*, **51**, 71–93.

PRIEM, H. N. A. K., BOELRIJK, N. A. I. M., HEBEDA,

E. H., OEN, I. S., VERDURMEN, E. A.TH. & VERSHURE, R. H. 1979. Isotopic dating of the emplacement of the ultramafic masses in the Serrania de Ronda, Southern Spain. *Contributions to Mineralogy and Petrology*, **70**, 103–109.

REUBER, I., MICHARD, A., CHALOUAN, A., JUTEAU, T. & JERMOUMI, B. 1982. Structure and emplacement of the alpine-type peridotites from Beni Bousera, Rif, Morocco: a polyphase tectonic interpretation. *Tectonophysics*, **82**, 231–251.

SADDIQI, O., REUBER, I. & MICHARD, A. 1988. Sur la tectonique de dénudation du manteau infracontinental dans les Beni Bousera, Rif septentrional, Maroc. *Comptes Rendus de l'Academie des Sciences, Paris*, **307**, 657–662.

SADDIQI, O., FEINBERG, H., ELAZZAB, D. & MICHARD, A. 1995. Paléomagnétisme des roches ultrabasiques des Beni Bousera (Rif Interne, Maroc): conséquences pour l'évolution miocène de l'Arc de Gibraltar. *Comptes Rendus de l'Academie des Sciences, Paris*, **321**, 361–368.

TAUXE, L. & WU, G. 1990. Normalized remanence in sediment of the western equatorial Pacific: relative paleointensity of the geomagnetic field. *Journal of Geophysical Research*, **95**, 12337–12350.

TUBIA, J. M. & CUEVAS, J. 1987. Structures et cinématiques liées à la mise en place des péridotites de Ronda (Cordillères Bétiques, Espagne). *Geodinamica Acta*, **1**, 59–69.

VAN DER WAL, D. 1993. *Deformation processes in mantle peridotites with emphasis on the Ronda peridotite of SW Spain*. PhD Thesis, Utrecht.

VILLALAÍN, J. J., OSETE, M. L., VEGAS, R., GARCIA-DUEÑAS, V. & HELLER, F. 1994. Widespread Neogene remagnetization in Jurassic limestones of the South-Iberian paleomargin (Western Betics, Gibraltar Arc). *Physics of the Earth and Planetary Interiors*, **85**, 15–33.

ZECK, H. P., ALBAT, F., HANSEN, B. T., TORRES-ROLDAN, R. L., GARCIA-CASCO, A. & MARTIN-ALGARA, A. 1989. A 21 ± 2 Ma age for the termination of the ductile Alpine deformation in the internal zone of the Betic Cordilleras, South Spain. *Tectonophysics*, **169**, 215–220.

——, ——, ——, —— & —— 1996. The Neogene remagnetization in the Western Betics: a brief comment on the reliability of Palaeomagnetic directions. *This volume*.

——, MONIÉ, P., VILLA, I. M. & HANSEN, B. T. 1992. Very high rates of cooling and uplift in the Alpine belt of the Betic Cordilleras, Southern Spain. *Geology*, **20**, 79–82.

Palaeomagnetic studies in Morocco: tectonic implications for the Meseta and Anti-Atlas since the Permian

D. KHATTACH[1], D. NAJID[2], N. HAMOUMI[3] & D. H. TARLING[4]

1 Département de Géologie, Faculté des Sciences, Oujda, Morocco

2 Institut Scientifique, Université Mohammed V, Rabat, Morocco

3 Département de Géologie, Faculté des Sciences, Rabat, Morocco

4 Department of Geological Sciences, The University, Plymouth PL4 8AA, UK

Abstract: Most Palaeozoic rocks studied in Morocco show evidence of partial or complete remagnetization that occurred towards the end of Hercynian times (late Carboniferous and Permian). The age of remagnetization can be established by direct comparisons with other Permian palaeomagnetic directions. The late Carboniferous and Permian rocks, and the remagnetizations of this age, show essentially identical directions throughout the Meseta and Anti-Atlas regions of Morocco and suggest that there have been few significant rotations of blocks within the Meseta domain since Permian times. This implies that most later motions in this region have not involved significant rotations about vertical axes or large scale changes in the latitude.

Various authors have attempted to use palaeomagnetism to assess the problem of the extent to which Morocco, or parts of it, have undergone rotations relative to the main African cratonic blocks. Some have proposed post Triassic–Jurassic rotations within the Moroccan Meseta relative to the Anti-Atlas region (e.g. Michard *et al.* 1975; Feinberg *et al.* 1989), while others have argued that this region has been stable relative to Africa since Permian times (e.g. Morel *et al.* 1981; Najid 1986; Najid & Tarling 1989). Palaeomagnetic studies during the last two decades, concentrated on Palaeozoic rocks, have now provided a substantial database that can be used to re-evaluate the possible existence of such Mesozoic or later rotations.

The palaeomagnetic evidence

Morocco can be divided into five structural domains (Fig. 1; Piqué *et al.* 1987, 1991): the Meseta, the Anti-Atlas, the High Atlas, the Rif and cratonic Africa. Palaeozoic data up to the late Carboniferous–Permian (290–300 Ma) are available from the western part of the Meseta and from the Anti-Atlas (Table 1). Many of these results, although from rocks much older than the Permian, are characterized by a clear, distinct remagnetization which occurred in late Hercynian times (Najid 1986; Salmon *et al.* 1988; Khattach *et al.* 1989). This phenomenon is common in many other parts of the world (Creer 1968), including other parts of NW Africa (Abou-deeb & Tarling 1984; Aïfa *et al.* 1990) as well as within the Hercynian chains of Europe (Edel & Coulon 1987; Courtillot *et al.* 1986) and North America (Kent 1985; McCabe *et al.* 1984). In Morocco, Huon (1985) has undertaken microstructural, granulometric and isotopic studies of different Palaeozoic schists and related these properties to different phases of metamorphism of the Hercynian Chain. In the western Meseta, the K/Ar ages indicate that most of the metamorphism was around 300 Ma ago. In a similar study of the coastal Meseta, Rais (1992) determined similar events in the range 290–300 Ma in the Rehamna region. In the Anti-Atlas, radiometric dates of Precambrian and Palaeozoic rocks have commonly been reset to around 290–300 Ma (Choubert *et al.* 1965; Charlot *et al.* 1970; Ducrot & Lancelot 1977; Benziane & Yazidi 1982; Bonhommet & Hassenforder 1985), which also corresponds to the age at which the magnetic remanences appear to have been reset. The remanence directions isolated in these areas of Morocco are clearly similar to each other and also similar to Permian remagnetised directions isolated in Algeria (Table 1 & Fig. 1).

Comparisons of remanence directions of the same age from different domains enables the detection of their relative tectonic motions since the magnetization was acquired. In particular, palaeomagnetic data enable the quantification of different rotations about vertical axes and differential changes in palaeolatitude. However, for the reliable detection of such motions, the precision of the individual measurements must

From Morris, A. & Tarling, D. H. (eds), 1996, *Palaeomagnetism and Tectonics of the Mediterranean Region*, Geological Society Special Publication No. 105, pp. 53–57.

53

Table 1. *Mean site directions and poles*

Formation	Age	D/I	N(n)	k	α95	Lat.	Long	Reference
Anti-Atlas								
Tafroute Dolerites	PreC/uCamb	137/11	2(11)	120	11*	36	48	Khattach et al. (1989)
Ouarzaazte Andesites	Adoud	129/05	3	25	25*	31	58	Khattach et al. (1989)
Anti-Atlas Sediments	1Cam	134/02	7	74	17*	36	55	Khattach et al. (1989)
Bou-Azzer Andesites	578 Ma	129/10	8(43)	11	7	31	57	Hailwood & Tarling (1973)
Cambrian Formations	Camb	142/09	5(17)	–	21*	40	40	Najid (1986)
Lie de vin Series	1Camb	124/–2	8(8)	8	5	29	63	Martin et al. (1987)
Erfoud Limestones	Dev.	139/–2	9	105	4	41	56	Salmon et al. (1988)
Hmar Lakhdad Basalts	Dev.	142/00	1(8)	172	9	43	60	Najid (1986)
A. Torkoz Limestones	Sil/Dev	136/–5	1(7)	51	8	41	57	Khattach & Najid (1994)
Coastal Meseta								
O. Rhebar Volcanics	mCamb	138/13	5	29	14*	33	46	Khattach et al. (1989)
M.B. Rhyodacites	Carb/Perm	146/–6	1(6)	116	6	47	46	Westphal et al. (1979)
Central Meseta								
B. Acila Volcanics	Camb.	138/3	11	66	6	38	52	Khattach & Najid (in press)
Bou Regreg Spilites	Ord.	129/00	9	96	5	31	59	Khattach et al. (1989)
Bou Regreg Lmsts.	Sil/Dev	142/02	10	795	2	40	47	Khattach et al. (1989)
Meseta Sediments	Carb	143/04	4	103	9	40	45	Salmon et al. (1988)
Chougrane Trachyandesites	Carb/Perm	137/–5	1(6)	37	4	40	56	Westphal et al. (1979)
Chougrane Trachyandesites	Carb/Perm	141/–5	1(6)	113	6	43	52	Westphal et al. (1979)
Chougrane Red Ssts.	1Perm	127/–6	45(45)	20	5	32	64	Daly & Pozzi (1976)
Taztot Trachyandesites	1Perm	136/–5	1(12)	80	5	39	57	Daly & Pozzi (1976)

* Too imprecise for tectonic interpretation.

Camb, Cambrian; Adoud, Adoudounian; Sil, Silurian; Dev, Devonian; Carb, Carboniferous; Perm, Permian; l, m and u are lower, middle and upper, respectively. D/I, mean declination/inclination.

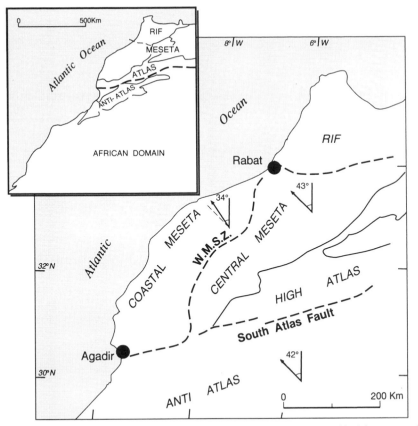

Fig. 1. Declinations for Moroccan blocks and cratonic Africa. The Coastal Meseta block is separated from the Central Meseta block by the Western Meseta Shear Zone (W.M.S.Z.). The mean declination differences are shown for each block. Inset: The main structural domains of Morocco.

exceed the magnitude of the differences. In this review, the limit on precision has been taken as $\alpha_{95} < 10°$ although a more restrictive criterion would have been desirable. The selected data (Table 1) have been used to calculate the late Carboniferous–Permian mean directions (primary or secondary) for the coastal Meseta, central Meseta and stable African blocks. The palaeomagnetic pole positions have then been used to determine the mean pole position for each block (Table 2; Figs 1 & 2).

Tectonic implications

It is clear that there are strong similarities between the pole positions for each block and that there is no evidence for significant post-Permian relative motion about either a vertical axis or in relative palaeolatitude. In particular, there is no detectable difference between the mean directions for the Central Meseta, Anti-Atlas and cratonic Africa. There is only weak

evidence for a possible difference in the palaeomagnetic declinations between the Coastal Meseta and the Central Meseta, Anti-Atlas and cratonic Africa. However, the difference is small, equivalent to 7° clockwise rotation, which is comparable to the precision for the only site accepted for this domain (Table 1). At this stage, it would seem that the available data constrain any such differential motion to less than 10°. This suggests that the West Meseta Shear Zone, which separates the Coastal and Central Meseta (Fig. 1) and is known to have some Hercynian dextral displacement (Piqué *et al.* 1983), may only have had a small post Permian horizontal displacement but this agreement between the palaeomagnetic poles from all of these areas does not exclude other tectonic motions. These could include dip-slip faulting and linear strike-slip motions, but it does exclude any motions that involve significant rotations about vertical axes (>5°). It is also evident that latitudinal displacement between these three blocks, along

56 D. KHATTACH *ET AL.*

Table 2. *North West African Carboniferous to Permian Mean Pole Positions*

Formation	Age	Lat.(°S)	Long(°E)	k	α_{95}	Reference
Abadla Red Beds	Perm.	29	60	59	6	Morel *et al.* (1981)
Adeh Larach Series	Carb/Perm	39	58	1459	2	Henry *et al.* (1992)
Seds & Volcs	Carb.	28	58	39	13*	Abou-deeb & Tarling (1984)
Hassi-Bachir	Carb.	36	59	–	5	Daly & Irving (1983)
B. Abbès Lst	Dev.	46	45	–	7	Aifa *et al.* (1990)

The mean poles are calculated on the data in Table 1, using only sites where the mean direction is defined with α_{95} < 10°. Age abbreviations as Table 1.

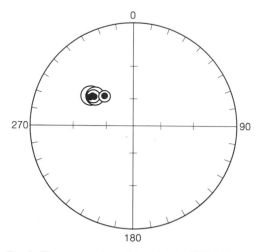

Fig. 2. The mean pole positions for the NW African structural domains (Stereographic polar projection).

the Permian meridian, cannot have exceeded 500 km. The post Permian tectonic regime within the Meseta has predominantly been one of re-activation of pre-existing, largely rectilinear fault systems of predominantly Pre-Cambrian or Hercynian age. Vertical displacement of these terranes has been responsible for most of the post Hercynian sedimentation patterns as well as controlling the response to Atlas deformation (Piqué *et al.* 1983, 1987; Hamoumi 1988). All such deformations clearly occurred with little or no rotation about vertical axes in any of these regions.

Conclusions

The available late Carboniferous–Permian palaeomagnetic results from Morocco do not indicate any significant tectonic rotations between the major terranes south of the Rif Atlas since the Hercynian orogeny. Clearly much more extensive studies are still required to extend the areas of investigation and to substantiate the

present observations (the evidence for the coastal Meseta rests entirely on one studied area). Nonetheless, it is also clear that rotations about vertical axes by greater than some 5–10°, and differential palaeolatitudinal movements on a similar scale are both unlikely. The study also illustrates how palaeomagnetic directions of remagnetized rocks can still be of considerable value in constraining the tectonic evolution of a region. Morocco provides an ideal region for extending such studies as its tectonic blocks have been subjected to both extensional regimes, during the opening of the Central Atlantic, and compressional regimes associated with the formation of the Rif Atlas.

References

ABOU-DEEB, J. M. & TARLING, D. H. 1984. Upper Paleozoic paleomagnetic results from Algeria and Tunisia. *Tectonophysics,* **101**, 143–158.

AÏFA, T., FEINBERG, H. & POZZI, J. P. 1990. Devono-Carboniferous paleopoles for Africa: consequences for Hercynian geodynamics. *Tectonophysics,* **179**, 287–304.

BENZIANE, F. & YAZIDI, A. 1982. Géologie de la boutonnière précambrienne d'Ifni (Anti-Atlas occidental). Notes et Mémoires du Service Géologique, Maroc.

BONHOMMET, M. & HASSENFORDER, B. 1985. Le métamorphisme hercyniens dans les formations tardi- et post-panafricaines de l'Anti-Atlas occidental (Maroc). Données isotopiques Rb-Sr et K-Ar des fractions fines. *Science Géologique Bulletin, Strasbourg,* **38**, 175–183.

CHARLOT, R., CHOUBERT, G., FAURE-MURET, A. & TISSIRANT, D. 1970. Etude géochronologique du Précambrien de l'Anti-Atlas (Maroc). *Notes et Mémoires du Service Géologique, Maroc,* **30**, 99–134.

CHOUBERT, G., DIOURI, M. & FAURE-MURET, A. 1965. Mesures géochronologiques récentes par la méthode A^{40}/K^{40} au Maroc. *Notes et Mémoires du Service Géologique, Maroc,* **24** 53–62.

COURTILLOT, V., CHAMBON, P., BRUN, J. P. & ROCHETTE, P. 1986. A magnetotectonic study of the Hercynian Montagne Noire (France). *Tectonics,* **5**, 733–751.

CREER, K. M. 1968. Palaeozoic palaeomagnetism *Nature*, **219**, 246–250.

DALY, L. & IRVING, E. 1983. Paléomagnétisme des roches carbonifères du Sahara centrale; analyse des aimantations juxtapopsées; configuration de la pangée. *Annales Geophysicae*, **1**, 207–216.

—— & POZZI, J. P. 1976. Résultats paléomagnétiques du Permien inférieur et du Trias marocain; comparaison avec les données africaines et sud américaines. *Earth and Planetary Science Letters*, **29**, 71–80.

DUCROT, J. & LANCELOT, J. R. 1977. Problème de la limite Cambrien–Précambrien: étude radiochronologique par la méthode U-P sur zircons du volcan de Jbel Boho (Anti-Atlas, Maroc). *Canadian Journal of Earth Sciences*, **14**, 12,2771–2777.

EDEL, J. B. & COULON, M. 1987. A paleomagnetic cross-section through the Ardennes and Brabant massifs (France-Belgium). *Journal of Geophysics*, **61**, 21–29.

FEINBERG, H., TAJEDDINE, K. & BOUTAKIOUT, M. 1989. Recherches paléomagnétiques sur le Toarcien et la limite Jurassique/Crétacé au Maroc. *Résumé colloque de Géologie Franco-Marocain, Strasbourg*, 113.

HAILWOOD, E. A. & TARLING, D. H. 1973. Paleomagnetic evidence for a proto-Atlantic Ocean. *In*: TARLING, D. H. & RUNCORN, S. K. (eds) *Implication of Continental Drift to Earth Sciences* NATO Advanced Study Institute, Academic Press London & New York, 37–46.

HAMOUMI, N. 1988. *La plateforme ordovicienne du Maroc: Dynamique des ensembles sédimentaires*. Thèse Doctorat es-Sciences, U.L.P. Strasbourg, France.

HENRY, B., MERABET, N., YELLES, A., DERDER, M. M. & DALY, L. 1992. Geodynamical implications of new paleomagnetic poles from the Permo-Carboniferous of the Sahara. *Bulletin de la Societé Géologique, Strasbourg*, **163**, 403–406.

HUON, S. 1985. *Clivage ardoiser et réhomogénésation isotopique K-Ar dans des schistes paléozoiques du Maroc*. Thèse Université de Strasbourg.

KENT, D. V. 1985. Thermoviscous remagnetization in some Appalachian limestones. *Geophysical Research Letters*, **12**, 805–808.

KHATTACH, D. & NAJID, D. 1994. Mise en évidence de deux réaimantations dévono-carbonifères dans des calcaires siluriens et dévoniens d'Aouinet Torkoz (Anti-Atlas occidental, Maroc. *Comptes Rendus l'Academie des Sciences, Paris*, **318**, II, 487–492.

—— & NAJID, D. Palémagnétisme du volcanisme cambrien de Bou-Acila (Méséta centrale, Maroc). *Journal of African Earth Sciences*, in press.

——, PERROUD, H. & ROBARDET, M. 1989. Etude paléomagnétique de formations paléozoiques du Maroc. *Science Géologique Mémoirs, Strasbourg*, **83**, 97–113.

MARTIN, D. L., NAIRN, A. E. M., NOLTIMER, H. J., PETTY, M. H. & SCHMITT, T. J. 1987. Paleozoic and Mesozoic paleomagnetic results from Morocco. *Tectonophysics*, **44**, 91–114.

MCCABE, C., VAN DER VOO, R. & BALLARD, M. M. 1984. Late Paleozoic remagnetization of the Trenton Limestone. *Geophysical Research Letters*, **11**, 979–982.

MICHARD, A., WESTPHAL, M., BOSSERT, A. & HAMZEH, R. 1975. Tectonique de blocs dans le socle Atlaso-Mesetien du Maroc: une nouvelle interprétation des données géologiques et paléomagnétiques. *Earth and Planetary Science Letters*, **24**, 363–368.

MOREL, P., IRVING, E., DALY, L. & MOUSSINE-POUCHKINE, A. 1981. Paleomagnetic results from Permian rocks of the northern Saharan craton and motions of the Moroccan Meseta and Pangea. *Earth and Planetary Science Letters*, **55**, 65–74.

NAJID, D. 1986. *Palaeomagnetism of Morocco*. PhD Thesis, University of Newcastle-upon-Tyne.

—— & TARLING, D. H. 1989. Permian remagnetizations in Morocco: tectonic implications. *Résumé colloque Géologie Franco-Marocain, Strasbourg*, 159.

PIQUÉ, A., JEANNETTE, D. & MICHARD, A. 1983. The western Meseta shear zone, a major and permanent feature of the Hercynian belt in Morocco. *Journal of Structural Geology*, **2**, 55–61.

——, DAHMANI, M., JEANNETTE, D. & BAHI, L. 1987. Permanence of structural lines in Morocco from Pre-Cambrian to Present. *Journal of African Earth Sciences*, **6**, 247–256.

——, CORNÉ J. J., MULLER, J. & ROUSSEL, J. 1991. The Moroccan Hercynides. *In*: DALLMEYER, R. D. & LÉCORCHÉ, J. P. (eds) *The West African orogens and Circum-Atlantic correlatives*. Springer-Verlag, Berlin, 229–263.

RAIS, N. 1992. *Caractéristiques minéralogique cristallochimique et isotopique (K-Ar) d'un métamorphisme polyphasé de faible intensité. Exemple: les grauwackes cambriennes du Maroc occidental*. Thèse d'Université de Bretagne occidentale, Brest.

SALMON, E., EDEL, J. B. A., PIQUÉ, A. & WESTPHAL, M. 1988. Possible origins of Permian remagnetizations in Devonian and Carboniferous limestones from the Moroccan Anti-Atlas (Tafilelt) and Meseta, *Physics of the Earth and Planetary Interiors*, **52**, 339–351.

WESTPHAL, M., MONTIGNY, R., THUIZAT, R., BARDON, C., BOSSERT, A., HAMZEH, R. & ROLLEY, J. P. 1979. Paléomagnétisme et datation du volcanisme permien, triassique et crétacé du Maroc. *Canadian Journal of Earth Sciences*, **16**, 2150–2164.

Palaeomagnetism and magnetostratigraphy of the Middle Triassic in the Iberian Ranges (Central Spain)

D. REY[1], P. TURNER[2] & A. RAMOS[3]

[1] *Departamento de Recursos Naturais e Medio Ambiente, Universidade de Vigo, 23000 Vigo, Spain*

[2] *School of Earth Sciences, The University of Birmingham, Birmingham B15 2TT, UK*

[3] *Departamento de Estratigrafía, Universidad Complutense, 28040 Madrid, Spain*

Abstract: The study of the structure and origin of remanence in five sections comprising the Rillo Mudstones and Sandstones and Torete Multicoloured Mudstone and sandstone Formations (Middle Triassic, Central Spain) reveals the presence of several components of magnetization of different ages. Detrital hematite and ilmenohaematite carry a primary magnetization of Triassic origin in the form of a DRM or PDRM. In addition, it is likely that an early CRM also contributes to the primary remanence. Goethite and several textural phases of authigenic hematite are responsible for a recent overprint in the form of a late CRM. Thermal demagnetization allowed the isolation of a characteristic Triassic remanence on which a reliable magnetostratigraphy for Anisian–Ladinian times could be based, and correlated with sections of similar age in North America and Asia. However, it did not succeed in entirely removing the CRM associated with the latest component of the overprint which resulted in the primary remanence being biased by residual components of Recent origin.

Remanence acquisition was controlled by the sedimentary environment and the diagenetic processes that took place during burial and uplift. The sedimentary environment controlled the conditions of deposition and facies distribution which limited the occurrence of a particular diagenetic phase within the rock volume. Acquisition of secondary components of magnetization, and the consequent pervasive Holocene magnetization (in the form of a late CRM) is primarily related to coarser grained horizons. Sand body interconnectivity has also played an important role in the process. Primary remanence is better preserved in the finer-grained horizons. A PTRM associated with the uplift is easily removed during thermal demagnetization. Calculation of mean vectors by integrating data from adjacent magnetozones allows correction for the overprint bias. The corrected directions can then be closely compared with results from stable Europe.

The Iberian Ranges constitute an important morphostructural feature of Central Iberia which extends in a NNW–SSE trend between the Cantabrian Mountains and the Levante depression (Fig. 1a). They are interpreted as an intermediate type of chain formed by tectonic inversion of an Early Mesozoic aulacogen (Alvaro *et al.* 1979; Vegas 1989). The formation of the Iberian Ranges is generally explained as the result of concentration of intraplate strain derived from the convergence of the European and African plates in a zone of lithospheric weakness since Mid–Late Mesozoic times (Vegas 1989). Permo-Triassic rocks were deposited in a complex half-graben fault-system where the sedimentation was controlled by reactivation of a pre-existing NW–SE and NE–SW wrench-fault system. This fault system

is a Late Hercynian (Late Carboniferous–Early Permian) feature, which developed as a result of megashear stress associated with relative dextral motion between the pre-Atlantic African and Euro-American plates (Arthaud & Matte 1975). These structures were reactivated as normal faults during the subsequent extensional stages and as reverse faults during compression (i.e. Alpine orogeny), and controlled the structural style and geographic distribution of Permo-Triassic basins along the Iberian Ranges (Arthaud & Matte 1975; Alvaro *et al.* 1979; Ziegler 1982; Sopeña *et al.* 1988).

The stratigraphy and sedimentology of the Permo-Triassic continental sequences of the Iberian Cordillera have been extensively described by Virgili *et al.* (1983); Ramos *et al.* (1986) and Sopeña *et al.* (1988). These authors describe a number of stratigraphic units which are associated with the three discrete lithological facies characteristic of the so-called German Triassic, namely Buntsandstein (siliciclastic),

---

In memoriam Amparo Ramos, who died in late August 1995. A great loss to all who have known her.

From Morris, A. & Tarling, D. H. (eds), 1996, *Palaeomagnetism and Tectonics of the Mediterranean Region*, Geological Society Special Publication No. 105, pp. 59–82.

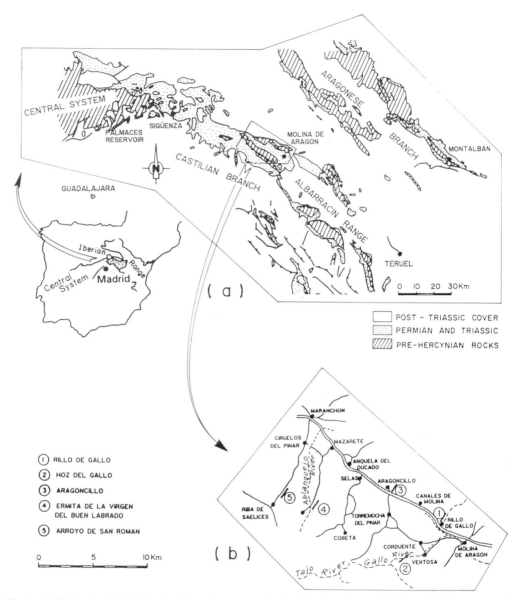

Fig. 1. (a) Geological sketch showing the main structural units (after Sopeña *et al.* 1988) and (b) the geographic location of the sampling sites in the studied area.

Muschelkalk (carbonate) and Keuper (evaporitic). The study area lies in the western part (Castilian Branch) of the Iberian Cordillera in Central Spain (Fig. 1a). In these regions, the Buntsandstein sediments unconformably overlie all previous rocks. They represent a period of major subsidence with accumulation of some 550 m of clastic red bed sequences. The two units studied here, the Rillo Mudstone and Sandstone Formation (RMS Fm) and the Torete Multi-

coloured Mudstone and Sandstone Formation (TMMS Fm), represent the evolution from a distal fluvial environment to the supratidal sedimentation which preceded the Tethys transgression, which in this area is marked by the overlying Tramacastilla Dolomites Formation.

The RMS Fm comprises up to 120 m of red mudstones and sandstones with local basal conglomerates deposited by low sinuosity streams with mixed gravel-sand bed load

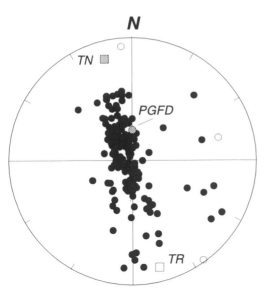

Fig. 2. Stereographic projection showing the distribution of the NRM directions for Rillo Sandstones and Mudstones Formation after bedding correction (Rillo de Gallo Section). PGFD, present geomagnetic field direction; TN, TR, expected normal and reverse Triassic directions in the area.

(Muñoz *et al.* 1992). The sequence evolves upwards to a distal sand-dominated braidplain type of environment where evidence of subaerial exposure indicates more ephemeral sedimentation. A palynological assemblage in the upper part of the unit indicates an Anissian age for the Formation (Ramos 1979). The overlying TMMS Fm consists of thin bedded mudstones and sandstones which often show evidence of evaporite precipitation and occasional interbedding of fine dolomitic horizons showing algal lamination. This unit is up to 40 m thick, and represents a marginal marine environment dominated by supratidal sedimentation (Ramos 1979; Muñoz *et al.* 1992). It shows several palynological assemblages of Ladinian age.

Five sections were sampled for palaeomagnetic and magnetostratigraphic study at 1–2 m intervals. The RMS Fm was sampled in the Rillo de Gallo section (Fig. 1b), whilst the TMMS Fm was sampled in four separate sections; namely: La Hoz del Gallo; Aragoncillo; Ermita de la Virgen del Buen Labrado; and Arroyo de San Roman (Fig. 1b).

Magnetic properties and NRM distribution

Initial NRM intensities for the studied samples range from 0.05 mA m^{-1} to 19.11 mA m^{-1}. The values of initial susceptibility range from 2.38 × 10^{-6} to 34.32 × 10^{-6} SI showing positive correlation with the intensities. The main fluctuations in both parameters were related to colour and grain size. In fine-grained, red-coloured sandstones and mudstones the intensity was higher than in coarser, drab or orange-coloured sandstones. The distribution of NRM directions for the RMS Fm is shown in Fig. 2. Directions are spread along a NNW–SSE-trending great circle, which passes close to both the present geomagnetic field direction (PGFD) and the expected normal and reverse Triassic directions for the area. Most directions are relatively steep and positive. A number of specimens show shallow negative inclinations (less than 30°). It is noticeable that directions are significantly shallower and more scattered in the SE sector of the distribution. NRM vectors in the TMMS Fm generally show steeper directions. A few differences are observed between the sampled sections. In La Hoz del Gallo section, directions are scattered on the NW sector with steep positive inclinations. Shallower directions more rare and show a more scattered distribution. A similar pattern is observed in the Arroyo de San Roman section. In this case directions are located along a NW–SE trend but lie preferentially in the SE sector with generally negative inclinations. In the Ermita de La Virgen del Buen Labrado section, NRM directions appear tightly clustered near the vertical. The Arangoncillo section has a similar distribution to that in Rillo de Gallo, with directions spread along a great circle containing the PGFD in the area. In the following sections we will attempt to show that the magnetizations were acquired at different times and that this accounts for the observed distributions of NRM directions.

Thermal demagnetization results: structure of remanence

Despite a few calcrete horizons, the great majority of samples were red sandstones and mudstones. Thus, thermal demagnetization was considered to be the best possible technique to study the structure of the remanence. A minimum number of two specimens from each sample were demagnetized in 100°C steps up to 400°C and in 50°C steps onwards for the RMS Fm. About 70 specimens from 42 samples were subjected to detailed thermal demagnetization. Similar procedures were applied to 48 samples from four in sections the TMMS Fm. A minimum of one specimen per sample was fully step-wise demagnetized (24 from La Hoz del Gallo, 8 from La Ermita de la Virgen del Buen

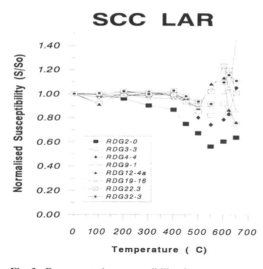

SCC LAR

Fig. 3. Representative susceptibility decay curves (low-field susceptibility).

Labrado, 8 in Aragoncillo section and 7 from the Arroyo de San Roman section). Nine samples were accidentally destroyed later, resulting in a total 38 fully step-wise demagnetized specimens. Another 21 specimens from the La Hoz del Gallo section were subjected to detailed thermal demagnetization up to 400°C. Bulk demagnetization in three steps (400, 500 and 620°C) was accomplished for another 61 specimens from all four sections. The final number of specimens fully demagnetized was approximately 150.

Susceptibility and intensity changes

Changes in low-field susceptibility are not significant up to 400°C (Fig. 3). After that temperature the susceptibility decreases rapidly up to 600°C suggesting dehydration of oxyhydroxides. The increase after 600°C may indicate formation of new finer-grained hematite as the result of heating (Dekkers 1989 a, b). The changes in intensity during thermal demagnetization show progressive removal of the magnetic remanence up to 690°C but a more detailed investigation reveals significant differences in the behaviour of individual specimens. Most of these can be reduced to the two main unblocking temperatures that are present in most samples. The first is represented by a sharp drop in the intensity between room temperature and 100°C. The amount of remanence lost in this interval varies between 20 and 80% (Figs 4 to 6). The second inflexion point is at higher temperatures. Above 600°C the intensity decays very quickly up to 700°C, at which the sample is virtually demagnetized.

Most of the directional changes during thermal demagnetization, occur between 100 and 600°C. The shape of the intensity decay curve (IDC) is then controlled by the number and nature of components present in the specimen. Samples showing a single component of magnetization exhibit a very slow decrease in intensity i.e. dominantly concave-down IDC (Fig. 5b). If two components of magnetization are present the behaviour depends very much on their polarity. If both components are of the same polarity the IDC shows an almost linear decrease of intensity up to 400°C. After that the rate slows down significantly and shows a concave-up IDC (RDG17-6 in Fig. 4b). When a normal component is superimposed on a reverse component of magnetization, the 400 to 600°C segment shows a characteristic concave-up appearance (RDG32-2 in Fig. 4b). Although this type of behaviour can be recognized as a general pattern, many samples behave slightly differently. It will be demonstrated later the reason for this is the different proportion that each component represents in the overall remanence.

Directional changes

During the study three types of demagnetizations were found, which are independent of both stratigraphy and of sampling sites.

Type I magnetizations are characterized by a low unblocking temperature component with steep positive inclination (component A) superimposed on a shallower higher unblocking temperature component with shallow normal or reverse inclinations (component B). During thermal demagnetization the steeper component (close to the PGFD) shows significant directional movement towards the NNW or SSE quadrants. In the NNW, directions are shallow and positive while in the SSE sector the inclinations are shallower and negative. The directional behaviour of type I magnetizations is illustrated in Fig. 4.

Type II magnetizations are characterized by a single component of magnetization. During thermal demagnetization they show little directional change. They show two main unblocking temperatures at 100–200°C and above 500°C. The single component present can be one of the types described above. Type II magnetizations represent a situation in which one of the components completely dominates the remanence (Fig. 5)

Type III magnetizations. Magnetizations comprising two nearly antiparallel components in

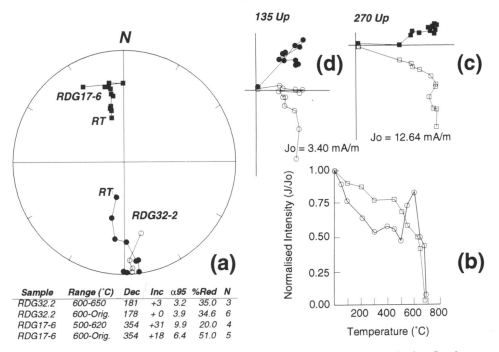

Sample	Range (°C)	Dec	Inc	α95	%Red	N
RDG32.2	600-650	181	+3	3.2	35.0	3
RDG32.2	600-Orig.	178	+0	3.9	34.6	6
RDG17-6	500-620	354	+31	9.9	20.0	4
RDG17-6	600-Orig.	354	+18	6.4	51.0	5

Fig. 4. Thermal demagnetization diagrams for a representative sample of Type I magnetization. Specimens RDG32-2 and RDG17-6, Rillo Mudstone and Sandstone Fm, Rillo de Gallo section. The directional and intensity changes are shown by the (**a**) stereographic projection, (**b**) normalized intensity decay curves (IDC) and (**c** & **d**) orthogonal projection diagrams. The table shows the principal component analysis.

the NNW and SSE quadrants with shallow inclination and overlapping temperature spectra. Quite often antiparallel components are isolated in adjacent temperature intervals (300–500°C and 550°C origin) (Fig. 6).

Type IV magnetizations. Specimens showing directional changes in a NNE–SSW trend at intermediate and high unblocking temperatures. The isolated components of magnetization are intermediate between those described above.

The stable components present in Type I, II and III magnetizations fall into two main categories.

(a) Steep components (components A), which are always positive and close to the PGFD. They are related to intermediate unblocking temperatures and interpreted as Tertiary to Recent in age. Some of them may have been acquired during the present polarity interval. It is also likely that those closer to the vertical may have a magnetization induced during drilling in the laboratory.

(b) Shallow components of magnetization that are nearly antiparallel. These are associated with high blocking temperatures. They represent the characteristic remanence of the rock.

By comparison with other Triassic data of the Iberian Ranges (Turner *et al.* 1989), other areas of Spain (Van der Voo 1969; Parés *et al.* 1988) and Stable Europe (Jowett *et al.* 1987; Piper 1988) they are considered as Triassic magnetizations.

This also indicates that type A components are superimposed on type B components of magnetization.

Whilst stereographic projection of type A components in geographic coordinates shows clustering around the PGFD (Fig. 7a), projection of type B shows after the tectonic correction two discrete groups (Fig. 7b). The first group clusters in the NNW quadrant with inclinations of about 35°. The second one concentrates in the SSE quadrant with inclinations of about −5°. Although both components are virtually antiparallel in declination there are striking differences in the inclinations. The fact that type A components are generally superimposed on type B components indicates that thermal demagnetization did not completely succeed in removing the secondary magnetization. Further evidence of this is presented in Fig. 4.

Thermal demagnetization of specimen RDG32-2 involves considerable directional

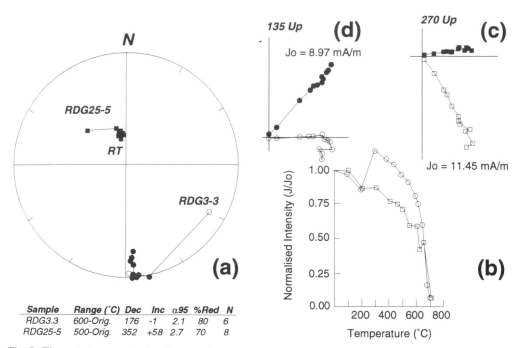

Fig. 5. Thermal demagnetization diagrams for a representative sample of Type II magnetization. Specimens RDG3-3 and RDG25-5, Rillo Mudstone and Sandstone, Rillo de Gallo section.

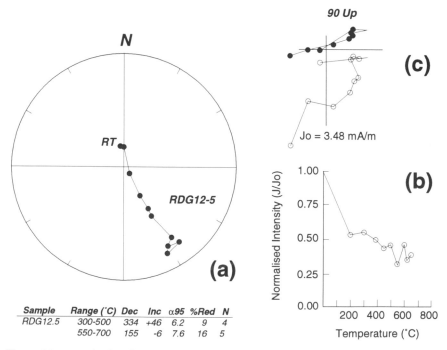

Fig. 6. Thermal demagnetization diagrams for a representative sample of Type III magnetization. Specimen RDG12-5, Rillo Mudstone and Sandstone, Rillo de Gallo section.

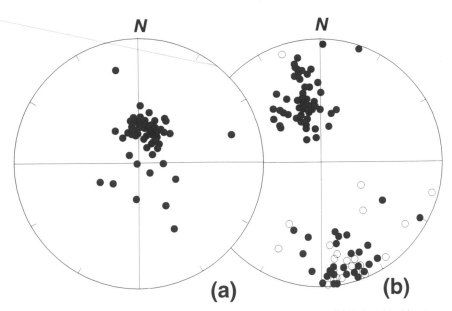

(a) (b)

Fig. 7. (a) Stereographic projection in geographical coordinates of low and (b) high unblocking temperature components of magnetization after tectonic correction (bedding).

migration from a normal steep component to a reverse shallow direction in the SSE quadrant (Fig. 4). Two components can be isolated that are statistically almost indistinguishable. They represent 84 and 35% of the total remanence. Significantly, the highest temperature component is virtually negative, while the other remains positive. As components are distributed in more discrete temperature spectra the differences between them are greater. Sample RDG17-6 (Fig. 4) complements the previous example. The two isolated components are more discretely isolated in two adjacent temperature intervals, but still at very high unblocking temperatures.

Whilst Type I and II magnetizations are randomly distributed in the magnetostratigraphical sequence, Type III and IV magnetizations appear intimately associated to the stratigraphic boundaries in which the characteristic remanence exhibit opposite polarity. Although the latter are relatively rare and infrequent, they provide important clues to the mechanisms by which these rocks acquired their magnetization. In Type III magnetizations the normal polarity Triassic component is superimposed on a reversed polarity Triassic component. The fact that this only occurs close to magnetozone boundaries implies that acquisition of characteristic remanence took place for a period of time that lasted longer than the time taken for the field to reverse its polarity and less

than the time spent in each interval. This is likely to be caused by short-term authigenesis of a magnetic carrier close to the time of deposition. Further consideration of mechanisms of remanence acquisition and magnetic carriers will be discussed in detail later.

Type IV magnetizations also occur close to polarity transition boundaries. Different from type A magnetizations, their directional behaviour is more complex. Intermediate components must have their origin in processes associated with polarity changes. The time of remanence acquisition in the studied rocks seems to be long enough to discard the possibility of remanence acquired during polarity transitions. This type of magnetizations are interpreted as 'condensed' magnetizations. They occur at magnetozone boundaries associated with very fine grained sediments. Very possibly the sedimentation rate was so slow that part of two successive magnetozones are present in the same specimen.

Isothermal remanent magnetization

IRM experiments for representative specimens of the RMS and TMMS Fms show fairly similar results (Fig. 8), with a continuous increase in intensity up to fields of 1.3 Tesla, field at which a few samples started to show signs of saturation. During the back field experiment the samples generally showed relatively high coercivities (H_{cr} between 0.13 and 0.18) and medium

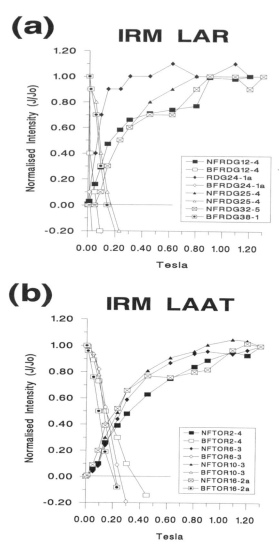

Fig. 8. Curves showing the acquisition of isothermal remanent magnetization for selected specimens of the (a) RMS Fm and (b) TMMS Fm. Prefix NF indicates normal field, BF indicates back field.

destructive fields between 0.2 and 0.38 Tesla. This behaviour is typical of fine-grained hematite (Dunlop 1971; Dekkers 1989*a*). However, one sample from the Rillo Fm (sample RDG24-1a, Fig. 8a) showed substantially lower coercivity and a slight tendency to saturation at intermediate fields. This possibly indicates the presence of a small amount of magnetite.

Iron oxides and the origin of remanence

The petrography and the diagenesis of the Upper Buntsandstein in the Molina de Aragon

area has been a subject of great economic interest since the early seventies. The area was extensively prospected for uranium and oil. Early studies were directed at the composition and provenance (Sanchez *et al.* 1971; Marfil *et al.* 1977). Compositionally the investigated rocks are arkoses and subarkoses with variable proportions of argillaceous matrix cemented by carbonates, silica, iron oxides, and sporadically by barite (Marfil *et al.* 1977). The relative importances of these components and their geochemical characteristics vary greatly along the Iberian Ranges (Garcia-Palacios & Lucas 1977). Provenance, initially interpreted as the re-working of Permian sediments (Virgili 1977), is better understood today as derived from a wide spectrum of metamorphic rocks (low, middle and high rank) (Senchordi & Marfil 1983). These provide a rich source of ferromagnesian minerals whose diagenetic transformations have created the complex mineralogy associated with the ferromagnetic phases.

The investigation of the evolution of the water chemistry during burial has been based on studies of chemical composition and the stable isotopes $\delta^{13}O$ and $\delta^{13}C$ of the carbonates (dolomite, Fe-dolomite, ankerite and calcite constitute major cementing phases in the rock). During deeper burial, carbonate cements have been subjected to various degrees of recrystallization (dissolution-precipitation) and show variable amounts of Fe and Mn (Morad *et al.* 1990). It is thought that temperatures during burial never exceeded 100°C. It has also been suggested, that the structural inversion during Tertiary times allowed the entrance of meteoric water from the present phreatic zone producing the latest diagenetic changes (Turner *et al.* 1989). The sequence of major diagenetic events is outlined in Fig. 9. In this context, the petrographic study presented here is intended to relate the occurrence of the magnetic mineralogical phases to the diagenetic processes that took place during the formation of these rocks and hence to the mechanisms of acquisition of remanence.

Description

The petrological study of the Rillo Mudstone and Sandstone and Torete Multicoloured Mudstone and Sandstone Formations reveals that the magnetic mineralogy (opaques) is not simple. These are present as part of the framework, matrix and cement in the rock, showing a great variety of textural habits. However, their proportion is relatively small and normally constitutes less than 4% of the total rock volume. Their main constituents are hematite, ilmeno-

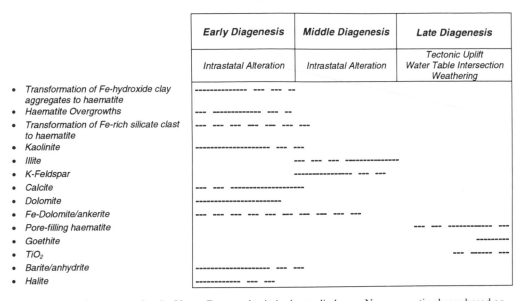

	Early Diagenesis	Middle Diagenesis	Late Diagenesis
	Intrastatal Alteration	Intrastatal Alteration	Tectonic Uplift Water Table Intersection Weathering
• Transformation of Fe-hydroxide clay aggregates to haematite	--------------- --- --- --		
• Haematite Overgrowths	--- -------------- --- --		
• Transformation of Fe-rich silicate clast to haematite	--- --- --- --- --- --- ---		
• Kaolinite	--------------------- --- ---		
• Illite		--- --- --- --------------	
• K-Feldspar		------------------ --- ---	
• Calcite	--- --- ---------------------		
• Dolomite	-------------------------		
• Fe-Dolomite/ankerite	--- --- --- --- --- --- --- --- --- ---		
• Pore-filling haematite			--- --- -------------- ---
• Goethite			----------
• TiO₂			--- ------ ---
• Barite/anhydrite	--------------------- --- ---		
• Halite	-------------- --- ---		

Fig. 9. Diagenetic sequence for the Upper Buntsandstein in the studied area. Non-magnetic phases based on Morad *et al.* (1990).

hematite, goethite and Ti-oxides. The coarse magnetic material occurs as discrete grains, relatively well-rounded, and apparently forming part of the framework. They can appear disseminated throughout the rock , or concentrated in heavy mineral-rich laminae (Fig. 10a) with an accompanying suite of apatite, zircon and more rarely tourmaline (Fig. 10b). Their grain-size is generally smaller than that of other framework grains, very possibly because of their hydrodynamic equivalence. This observation provides the first evidence of their detrital origin. In more detail, the appearance and degree of alteration of the grains is very variable. They can be very well preserved, or show a fairly ragged, embayed or even skeletal appearance (Fig. 10c,d & f). The systematic inspection of well rounded grains shows hematite and ilmeno-hematite as the major constituents (Fig. 10c). The IRM experiments (Fig. 8) suggested the presence of small amounts of magnetite in some of the samples but this mineral has not been positively identified under the microscope. Ilmenite–hematite intergrowths can only be formed during cooling of an igneous rock at high temperature. Their occurrence in some of the well rounded-grains is a critical evidence of their detrital origin.

Fine-grained hematite appears in a great variety of textural phases of which microcrystalline hematite is by far the most common and abundant. As such, it forms part of the matrix in intimate association with clay minerals and other hydroxides and is responsible for most of the red

colouration in the rock (Fig. 10d). Its importance is greater in the finer-grained siltstones and mudstones that dominate the TMMS Fm. Pigmentary hematite is very common in coarser sandstones, forming a thick pore-lining of oversized secondary pores created by framework and/or cement dissolution (Fig. 10e). Other textural phases of microcrystalline hematite include the replacement of rock fragments (of possible metamorphic origin, Fig. 10f), cutans in framework grains, and micas (Fig. 10g). Hematite also occurs as small authigenic overgrowths attached to detrital grains (Fig. 10c), and as needle-like crystals replacing ilmenite or forming aggregates in pore spaces. Sometimes it may constitute a well developed pore-filling cement (Fig. 10h). Complex authigenic textures are also found replacing peloidal concretions of siderite and preserving their characteristic concentric curved fissures and associated to pedogenic horizons. Other authigenic phases relevant to this study, although present in small quantities, are goethite and Ti-oxides. Goethite mainly occurs as secondary pore-coatings. Ti-oxides occur as detrital grains or as an authigenic phase. The latest exist as a pore-filling cement that eventually may grow on the margin of certain grains, or completely replace them.

Interpretation

The complexity of Fe-oxide textural phases detailed above explains the complicated structure

of the remanence described in the previous sections. Their intricate relationship with the other authigenic phases in the rock has provided some elements from which the acquisition of remanence can be modelled. Coarse detrital Fe-oxides are of great importance in establishing the origin of remanence. They must be considered as the potential carriers of the primary Triassic remanence in the form of a DRM or a PDRM. As such, they are a major contributor in the components, isolated during thermal demagnetization, and associated to the highest unblocking temperatures. Acquisition of secondary components of magnetization in the RMS and TMMS Fm is a process associated with burial and uplift. The rich textural variety in which the authigenic iron oxides occur indicates their complex origin and a number of processes. As they are very fine-grained and often appear as replacement textures, their position in the diagenetic sequence is difficult to establish. Traditionally red pigmentation on Buntsandstein rocks of the Iberian Cordillera has been interpreted as of exclusively intrastratal origin (Marfil et al. 1977; Senchordi & Marfil 1983) although a late phase associated with the Holocene inversion has been described (Turner 1988; Turner et al. 1989). The following model is proposed for the diagenetic evolution of iron oxides.

During early diagenesis detrital iron hydroxides and mechanically infiltrated clays acted as precursors of the microcrystalline hematite that primarily constitutes the pigment. Iron hydroxide inversion to hematite is a fairly well documented process (Berner 1969; Walker et al. 1978, 1981). This process has been particularly important in very fine-grained sandstones and mudstones, in which the red colouration is more intense. In coarser sands the clay-iron hydroxides aggregates would be more scarce, or

restricted to the mica rich laminae and the percolation cutans in individual grains. Evidence of alteration and or replacement by hematite in detrital iron-titanium oxides, biotites and ferromagnesian minerals (present mainly as biotites and in the metamorphic rock fragments) indicate additional source of iron during early diagenesis (shallow burial). At this stage periodic mixing between meteoric and marine waters may induce significant changes in the Eh and possibly, to a lesser extent, in the pH controlling the precipitation of iron at different stages. Differences in Mn and Fe content among dolomite and calcite zones in the dolocretes studied by Morad et al. (1990) in this area are interpreted in that way by these authors. If the process is to be completed rapidly (within a few thousand years) much of the remanence acquired would be approximately of the same age as the sediment. There is also evidence (as abundance of replacement textures in authigenic phases, fine grained acicular hematite in primary pores) which indicates that precipitation and dissolution of iron bearing minerals took place during deeper burial thus forming other textural phases of hematite which acquired a post-Triassic remanence. Maximum burial temperatures reported in these rocks are less than 100°C. Most of the microcrystalline hematite that constitutes the pigment must have reached the superparamagnetic stage at some point during burial, being remagnetized during uplift. Secondary components appear thermally distributed on the lower unblocking temperature spectrum.

It is difficult to demonstrate that under such conditions the fine-grained hematite that forms the acicular crystals reached the superparamagnetic stage. Their grain size was possibly too coarse to exceed the relaxation time at

Fig. 10. Photomicrographs illustrating different textural phases of magnetic minerals in the studied rocks. (**a**) In medium to coarse sandstones hematite occurs concentrated in heavy mineral-rich laminae which correspond to sedimentation planes. Note replaced biotite (arrowed). Sample RDG4, Rillo de Gallo section. Reflected light, FWV = 1125 μm. (**b**) Typical association of detrital hematite (light-grey) with apatite (elongated crystals) and titanium oxide (anatase). Sample TOR7, Hoz del Gallo section. Reflected light (oil), FWV = 225 μm. (**c**) Detrital grain of hematite with exsolution lamelae of ilmenite (ilmeno-hematite). The ilmenite is the darker mineral. Note a small syntaxial overgrowth of hematite bounding the grain edge. Sample RDG4, Rillo de Gallo section. Reflected light (oil), FWV = 225 μm. (**d**) Fractured rounded grain of hematite (specularite) within a very fine matrix of microcrystalline hematite (dark background). Sample RDG7. Reflected light (oil), FWV = 225 μm. (**e**) Late pigmentary hematite (black) coating secondary pore spaces (grey). Note the occurrence of an earlier phase outlining rounded detrital grains (arrowed) which pre dates k-feldspar overgrowths. Sample RDG2, Rillo de Gallo section. Transmitted plane polar light (PPL), FWV = 1125 μm. (**f**) Lithic detrital clast extensively replaced by hematite. Note that hematization occurs preferentially after ferromagnesian minerals. Sample RDG33, Rillo de Gallo section. Reflected light (oil), FWV = 180 μm. (**g**) Biotite completely replaced by hematite along cleavage planes. Sample RDG4, Rillo de Gallo section. Reflected light (oil), FWV = 180 μm. (**h**) Embayed detrital and authigenic quartz and feldespar (grey) floating on a late pore-filling hematite cement. Secondary hematite like this is thought to precipitate very late associated to the intersection of phreatic table water during terminal uplift. Reflected light (oil). Sample RDG7, FWV = 725 μm.

temperatures predicted by the Néel equation. Paradoxically their contribution to remanence seems to be very small if any, as the structure of remanence is satisfactorily explained without the complication of long-term authigenesis of hematite. It is understood that the acicular shape and random orientation of these crystals have played an important role in the way in which these phases have been formed, and ultimately in the type of remanence that they carry. Stokking & Tauxe (1990) experimentally demonstrated the preferred growth direction of hematite in the presence of an applied field. Their results indicate that the induced CRM is parallel to the field. These authors also found that precipitation of two generations of hematite under orthogonal field directions results in acquisition of parallel or antiparallel CRM to both fields directions as well as intermediate between the two fields. This depended on grain size and the exchange energy between the phases involved. In a similar fashion, changing field directions and periodic precipitation of hematite in the rock must have resulted in the growth of acicular crystals and be the cause of their random distribution. The CRM acquired in this fashion may result in a sort of random magnetic direction (if we consider the random orientation of the crystals), rather than a clear magnetic signal with a preferred orientation (as the individual crystals cancel each other out). During the terminal phase of uplift, the intro-duction of phreatic water starts to dominate the chemistry of pore fluids, changing the pH and Eh conditions. This results in a flush of the formation and generation of secondary porosity by dissolution. It is likely that the exhumation of originally red sands eroded in nearby areas acted as a source of extraformational iron, also dissolution of iron-rich carbonates and the most unstable ferromagnesian grains inputs iron into the system. The semi-arid climatic conditions in the area resulted in extensive precipitation of hematite, mainly as a pore-filling cement. This process is of great importance in coarse sand-stones, in which the generation of secondary porosity has also being more extensive. Eventu-ally this secondary phase postdates the primary replacements. It is thought that this process occurred very quickly, as it is controlled by the equilibrium of iron in the groundwater. In a later stage, exhumation of the rock produced the alteration of hematite to goethite by weathering. Both mechanisms will generate a CRM of Recent age. Preservation of the original rem-anence, would depend on the amount of damage (dissolution) caused to the original detrital grains.

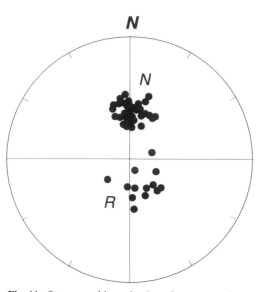

Fig. 11. Stereographic projection of uncorrected NRM directions for normal (N) and reverse (R) magnetozones.

Separation of stable components of magnetization

The stable directions of magnetization for each individual specimen were calculated using the principal components analysis (Kirschvink 1980). As-Zijderveld projection of thermal demagnetization data gave a minimum of three points in which definition of linear segments could be calculated for nearly 100% of speci-mens. Type I, III and IV magnetizations showed two or more stable components during thermal demagnetization, while Type II magnetization showed only one. The directional analysis was based on isolated components that showed maximum angular deviations values smaller than 10°. These stable components were gener-ally isolated in thermally discrete intervals associated with unblocking temperatures be-tween 100 and 600°C and above 600°C.

Secondary component analysis: the overprint

Group N in Fig. 11 shows the NRM directions (before bedding-correction) of specimens show-ing normal type B components. Directions are well clustered and close to the present geomag-netic field direction. The NRM from samples showing reversed type B components (group R, Fig. 11), shows slightly greater dispersion and a more southern direction. The increase in disper-sion is due to two main causes: greater angular differences between the recent and ancient components, thus inducing a more viscous behaviour and variation in the relative proportion of magnetic carriers within each specimen. The results of the fold test for the NRM directions was inconclusive (*K* improved by 0.1). Tilt correction resulted in significant steepening of inclinations and shift to the west. The increasing discrepancy between the PGFD and NRM directions suggest that one of the main components contributing to the NRM postdates folding. Investigation of directional changes in the 100–600°C interval shows migration along a great circle which passes through the PGFD and the expected Triassic magnetization. Principal component analysis allows isolation of stable component in the interval (100–600°C). These directions cluster around the PGFD (Fig. 7b). It has been demonstrated during the discussion of Type III magnetizations that type B components are thermally distributed. This results in unblocking temperatures overlap of type A and type B components in the 100–600°C interval and explains the presence of stable intermediate components. It is also possible the acquisition of a PTRM during Tertiary times by burial-uplift re-setting of the finest carriers. Its contribution to the remanence seems to be very small. In general terms, the main differences between NRM directions are related to the presence of normal and reverse components above 600°C (type B components). All the samples showed similar distributions regardless of sampling site or stratigraphic position. Based on this and considering the proximity of the NRM directions to the PGFD, it can be interpreted that an important event of remanence acquisition seems to have taken place between Tertiary and Recent times during a period of normal polarity. This is consistent with authigenesis of hematite during the terminal phase of uplift and may be related to the intersection of the water table at the time. It is likely that the penetration of oxidizing waters induced changes in the pore-fluids leading to precipitation of hematite. In coarse and well interconnected sand bodies, good porosity and permeability led to an almost pervasive magnetization. Some of the original hematite has been inverted to goethite by present-day weathering, giving a characteristic orange-yellowish colouration to the rock. In finer sands and mudstones, penetration of those fluids was restricted to dissolution-enlarged fractures. Mica rich horizons acted as permeability barriers, thus preserving the original magnetic carriers and very often the pigment. This favoured the preservation of the remanence and the red colouration in the rock.

Characteristic magnetization

Stable components of magnetization were isolated by principal component analysis of Kirschvink (1980), generally with a 5° region limit. Two groups of directions of characteristic remanence of opposite polarity were isolated (Fig. 7b). The means of these two groups are 180° apart with respect to their declination values but significant differences are found between the absolute values of their inclinations. The great circle between them passes in the vicinity of the present geomagnetic field direction in the area. This indicates that thermal demagnetization did not completely remove the secondary component associated with the late authigenic phases of hematite. It can be interpreted that directions of characteristic magnetization in the higher unblocking temperature reflect the directions of the magnetic field during the sedimentation of the Rillo Mudstones and Sandstones and Torete Multicoloured Mudstones and Sandstones Formations but that they are slightly biased by residual components of recent origin.

As described above the results of the fold test were inconclusive. The main reason is the very small differences in the strike and dip observed in the sampling sites (130/15, 144/16, 104/20 and 106/19). However, the age of the characteristic remanence has been interpreted as Triassic (primary), based on three types of evidence.

(a) The petrological study shows very well preserved grains of detrital hematite and ilmeno-hematite. Burial temperatures are too low to produce thermal re-set of the primary remanence.

(b) Thermal demagnetization achieves an important reduction of the scatter. Directions are segregated into two discrete groups of opposite polarity.

(c) These components can be used to construct a coherent magnetostratigraphy that can be correlated between sections and compared with other sections of similar age in North America and Eurasia, as will be shown later in this paper.

The discrepancy in inclination values between normal and reverse polarity Type B components can be explained by the presence of a recent CRM component (type A component). Thermal demagnetization did not succeed in removing the overprint entirely. Mainly because the large crystals of authigenic hematite (specularite) must have unblocking temperatures close to those of the detrital grains. Isolated components in the 600 to 690°C interval are consequently slightly biased towards the PGFD. This results in shallowing and weakening of the reverse

RILLO DE GALLO SECTION

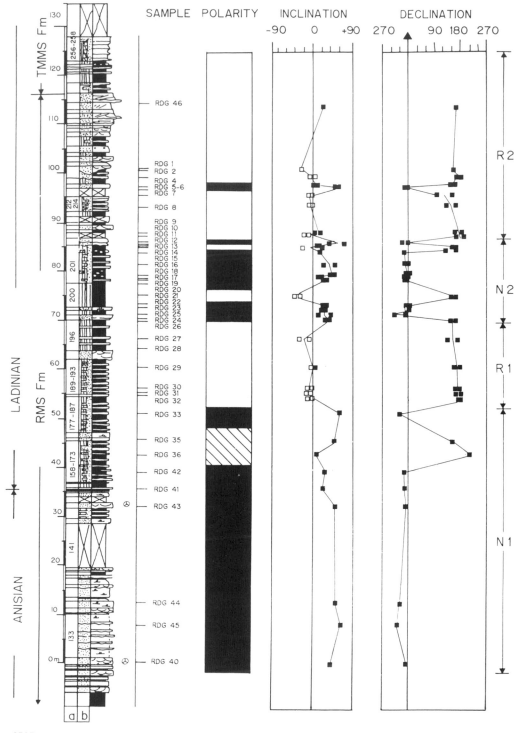

Fig. 12. Magnetostratigraphy of Rillo Mudstones and sandstones at the Rillo de Gallo section.

components (as they are the resultant of almost antiparallel vectors) and in steepening of the normal components. To maintain the observed antiparallelism, the ancient Triassic declination must have been very close to the PGFD at the time in which the overprint was acquired. Otherwise both normal and reverse directions would have been shifted towards the PGFD, thus resulting in larger declination discrepancies. This is consistent with the rapid acquisition of recent components supported by the petrographic study.

Magnetostratigraphy

To avoid the most strongly remagnetized horizons sampling was restricted to the finer-grained sandstones and mudstones. Thermal demagnetization experiments were then performed in the laboratory in order to systematically remove secondary components of magnetization. At least the components associated with the pigment and microcrystalline hematite (associated with the thermally distributed unblocking temperatures) and with goethite (physical destruction by dehydration occurs around 300°C) were successfully removed during the process. The magnetic stratigraphy of the section was then constructed based on components in the higher unblocking temperature range. To construct the magnetic stratigraphy the declination and inclination of a minimum of two specimens from each sample are plotted against their stratigraphical level. It needs to be noted that the directions used correspond to extracted components isolated during thermal demagnetization by means of the principal component analysis of Kirschvink (1980). Eventually some samples (5 or 6) did not pass the test and single components isolated above 550°C were used.

The magnetostratigraphy of the RMS Fm at the Rillo de Gallo section (Fig. 12) shows horizons of normal and reverse polarities clearly segregated into discrete stratigraphic intervals. Each of them represents a magnetozone. In the lowermost part of the section a long period of normal polarity is identified and named as N1. Fairly coarse sand bodies evolving to finer-grained sandstones and mudstones constitute the dominant lithology of the interval. They are interpreted as fluvial distal deposits with some degree of marine influence towards the top (Ramos 1979; Muñoz et al. 1992). Sample spacing was severely constrained by a continuous succession of coarse sandstone beds. Only finer horizons interbedded were sampled. Some of them may show slightly steeper than expected inclinations (RDG 41) indicating that the recent

components have not been entirely removed during thermal demagnetization. Samples RDG35 and RDG36 show intermediate components of magnetizations (cross hatched area in Fig. 12). The directional trend followed by these two specimens may well indicate a very short lived interval of reverse polarity that was missed by the sample spacing. The interpretation of data collected in these magnetozones agrees quite well with the magnetostratigraphy previously proposed by Turner et al. (1989). However, a significant improvement in precision has been achieved by means of selective sampling of the fine-grained horizons and the use of the principal component analysis to determine the component associated to the higher unblocking temperature. A palynological assemblage at the stratigraphic horizon 149 indicates Anisian-Ladinian age. The upper boundary N1 and the lower limit of the overlying R1 magnetozone is marked by specimen RDG35 showing intermediate components of magnetization. This sample was collected from a muddy very fine-grained horizon. The slow sedimentation rate associated with this type of horizon (Muñoz et al. 1992) probably resulted in the contraction of the two magnetic episodes to such an extent that both are present within the specimen volume. At this stratigraphic level the sand bodies are generally thinner, finer-grained and laterally less continuous. The abundance of interbedded mudstones is greater than in N1 evidencing the vertical evolution to a more distal fluvial environment. Interconnectivity between sand bodies is poorer, and permeability much less (due to their finer grain size) resulting in minor diagenetic damage of the primary remanence. R1 magnetozone spans samples RDG32 and RDG27 representing a period of reverse polarity. Its upper boundary is in sample RDG26. The bottom specimen of the sample (RDG26-3) shows reverse polarity and the top specimen (RDG26-4b) indicates normal polarity showing a sharp change of polarity between R1 and N2 magnetozones. The N2 magnetozone is characterized by being of mainly normal polarity with a short-lived reversal of the field at the upper part of the interval. The lithology is dominated by mudstones and very fine-grained sandstones with slightly coarser sands and dolocretes interbedded. Nodular structures (possibly of siderite) related to pedogenesis may indicate the beginning of some degree of marine influence for which there is clearer evidence higher in the section (Ramos 1979; Muñoz et al. 1992). Coarser-grained horizons (RDG16 and RDG12) inclinations are slightly steeper, showing once more the link between sedimentary facies, diagenetic processes and quality of

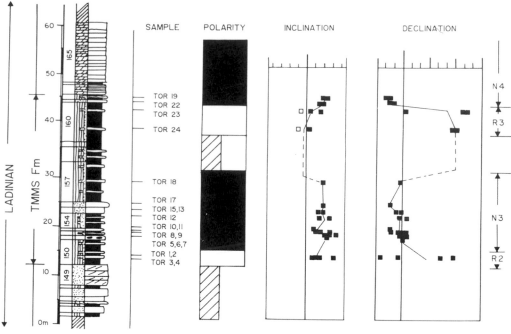

Fig. 13. Magnetostratigraphy for the Torete Mudstones and Sandstones Formation at the La Hoz del Gallo.

remanence preservation. Two short-lived reversals occurred at the base (RDG21) and near the top of the magnetozone (RDG13 and RDG14). The uppermost part of the RMS Fm comprises a period of mainly reverse polarity (R2). A brief interval of normal polarity occurs in the magnetozone. The occurrence of this short event is very gradational, indicating that a higher sedimentation rate allows a fuller record of the polarity reversal. Lithologically, R2 magnetozone consists of very fine sandstones and mudstones interbedded with slightly coarser sands. Some of the beds show structures of current bipolarity as clear indicators of marine influence (Ramos 1979). The uppermost part of the section is fairly coarse, as the result of eustatic changes (Muñoz *et al.* 1992), and consequently only one horizon (RDG46) met the sampling conditions. The isolated direction during thermal demagnetization shows an intermediate component of magnetization. However, because of its shallow inclination and its declination just west of south it has been interpreted as a reverse component. The thermal demagnetization shows the consistent removal of a normal, steep component of recent age but did not succeed in its complete removal, causing the shallowing effect observed.

We attributed reverse polarity to the interval despite the gap between samples RDG1 and RDG46 by incorporating the data of Turner *et al.* (1989) for the same stratigraphic interval.

The overlying TMMS Fm in the La Hoz del Gallo comprises mudstones with interbedded fine grained sandstones whose occurrence is less frequent towards the top of the unit (Fig. 13). The unit has been dated as Upper Ladinian on the base of pollen assemblages. The presence of algal lamination and evaporite precipitation (most of the gypsum crystals are replaced by late calcite) has been interpreted as clear evidences of supratidal influence. Quite often the sediment was poorly consolidated making sampling very difficult. The material was so feeble that some of the samples did not survive transportation to the laboratory. Sampling was unavoidably confined to the best cemented beds. At the base of the section samples TOR1 to 4 define a period of reverse polarity. The magnetostratigraphy seems to be condensed (Type III magnetization). This period is interpreted as part of the underlying R2 magnetozone and therefore named as R2. Its upper boundary lies between TOR1 and TOR 5. This magnetozone is poorly defined at the La Hoz del Gallo Section but more

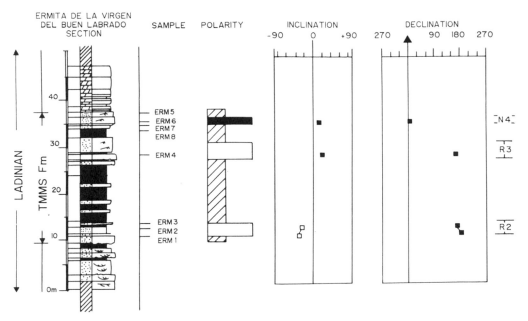

Fig. 14. Magnetostratigraphy for the Torete Mudstones and Sandstones Formation at La Ermita del Buen Labrado sections.

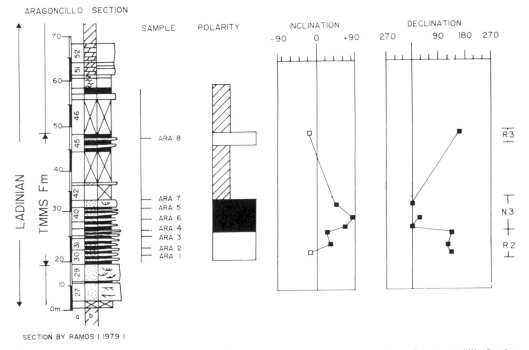

Fig. 15. Magnetostratigraphy for the Torete Mudstones and Sandstones Formation at the Aragoncillo Section.

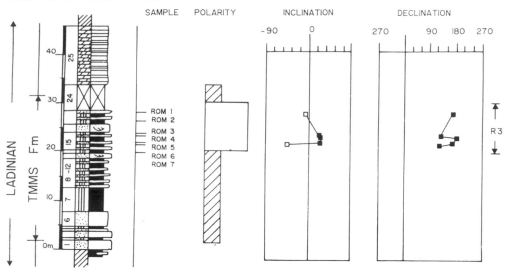

Fig. 16. Magnetostratigraphy for the Torete Mudstones and Sandstones Formation at the Arroyo de San Roman section.

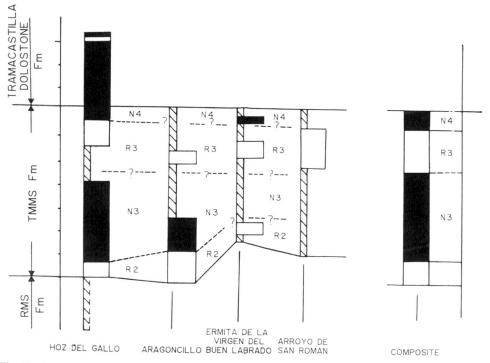

Fig. 17. Correlation and composite magnetostratigraphic sequence for the Torete Mudstones and Sandstones Formation.

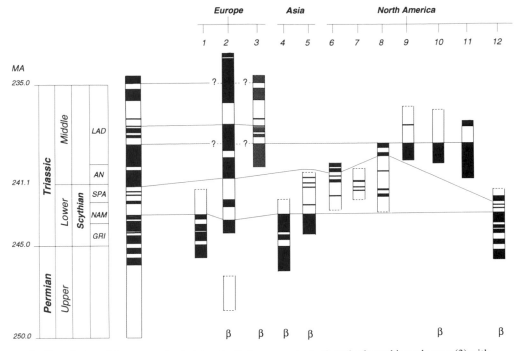

Fig. 18. Correlation of the composite Anisian–Ladinian magnetostratigraphy from this study area (3) with Turner *et al.* (1989) for the area (2). A Lower to Middle Triassic magnetostratigraphy is also proposed based in correlations with other sections from West Cumbria, UK (St Bees Sandstone; Rey 1992), the Sichuan Province in China, (4, Heller *et al.* 1988; 5, Steiner *et al.* 1989), North America (6, Shive *et al.* 1984, Chugwater Fm, WY, USA; 7, Lienert & Helsley 1980, Moenkopi Fm, UT, USA; 8, Helsley & Steiner 1974, Moenkopi, CO, USA; 9, Molina-Garza *et al.* 1991, Moenkopi, NM, USA; 10, Molina-Garza *et al.* 1991, Anton Chico, Yucatan, Mexico; 11, Elston & Purucker 1979, Moenkopi, AZ, USA; 12, Ogg & Steiner 1991, Canadian Arctic). β indicates biotratisgraphically constrained sections.

clearly existent at La Ermita del Buen Labrado Section (Fig. 14) by samples ERM1 and ERM2. It is also depicted at the Aragoncillo Section by samples ARA1 and ARA2 (Fig 15). The sandier character of the latter gives poor resolution. N3 overlays R2 magnetozone. N3 is defined by samples TOR5 to TOR18 at the La Hoz del Gallo Section (Fig. 13). N3 constitutes a relatively long and quiet normal event, also depicted at the Aragoncillo section (Fig. 15). The upper limit of the magnetozone is difficult to establish as it comprises a very friable pack of mudstones that very rarely crops out in the field. N3 magnetozone is followed by a relatively short period of reverse polarity (R3). R3 is defined by samples TOR23 and TOR24 at la Hoz del Gallo Section (Fig. 13), sample ERM4 at La Ermita del Buen Labrado Section (Fig. 14), and can also be identified in sample ARA8 at the Aragoncillo Section (Fig. 15). At the Arroyo de San Roman Section (Fig. 16) samples ROM1 to ROM7 seem to suggest its occurrence. Again the quality of the data is heavily constrained by grain size and

sandbody interconnectivity. The upper limit of R3 with the overlying N4 magnetozone is clearly defined at La Hoz del Gallo section by samples TOR19 to TOR23 (Fig. 13). It is also present al La Ermita del Buen Labrado Section (Fig. 14), depicted by sample ERM6.

The data for the Torete Formation allows the establishment of a preliminary magnetostratigraphy (Fig. 17) for this stratigraphic unit. It is based upon a fairly complete section (La Hoz del Gallo) and partial data collected from another three sections (Aragoncillo, Ermita del Buen Labrado and Arroyo San Roman). The composite magnetostratigraphy for the Rillo and Torete Formations is given in Fig. 18 (column 3). Figure 18 establishes a correlation with Turner *et al.* (1989) for the area. Their version has been improved by means of the principal component analysis and reduction of sampling spacing. In this figure a proposed Mid-Triassic magnetostratigraphy for the area is correlated with other published magnetostratigraphic sections for North America and Eurasia. These are mainly

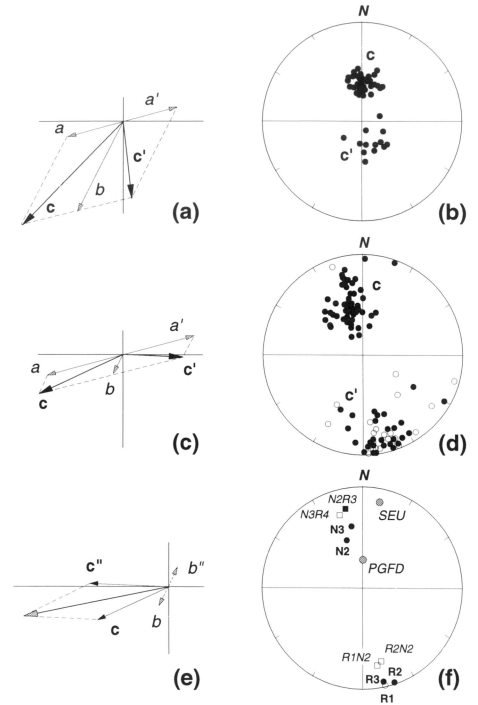

Fig. 19. (**a**) Graphical representation of the contribution of the primary and secondary components of magnetization to the NRM (prior to thermal demagnetization). *a* & *a'* = expected normal and reverse Triassic directions; *b* = present geomagnetic field direction; **c** & **c'** = resultant. (**b**) Distribution of NRM directions in a normal (N3) and reversed (R1) magnetozones. (**c**) The contribution of primary (*a* & *a'*) and secondary (*b*) components after thermal demagnetization. (**d**) Stereographic projection of components isolated above 600°C. (**e**) Method of correction for the isolated component bias due to unremoved overprint. **c"** = inverted **c'** and *b"* = inverted *b'*. (**f**) Compensated mean directions. SEU, Stable Europe Middle Triassic Reference Direction.

based on revised data of Molina-Garza *et al.* (1991) for North America (also based in red beds), Ogg & Steiner (1991), in the Canadian Arctic, Heller *et al.* (1988), Haag & Heller (1991) and Steiner *et al.* (1989) from Sichuan province in North China. The four later are based on well-dated carbonate sections. Based on these correlations an Early to Mid-Triassic magnetostratigraphy is proposed (Fig. 18).

Sedimentological controls on the preservation of primary remanence and modelling of the structure of remanence

The degree of directional stability and intensity behaviour during thermal demagnetization is consistent with petrographic evidence indicating several mineralogical phases contributing to the remanence. Evidence of preservation of primary remanence comes from the presence of coarse detrital hematite and ilmeno-hematite. During thermal demagnetization the components associated with the high unblocking temperature are segregated into two near antiparallel groups which are stratigraphically distributed in the sections. The processes of acquisition of primary remanence were not spatially variable as it is the quality of preservation of the ancient Triassic magnetization. The latter is widely controlled by facies distribution and diagenetic response of the different lithologies during burial and uplift. The variations in demagnetization behaviour can be attributed to a variability of magnetic remanence carriers (petrological inhomogenities within the specimen). Steep components of magnetization produce a recent overprint on the much shallower ancient magnetization on which the Triassic magnetostratigraphy was based. The petrological data indicates that authigenesis of hematite took place at two particular moments during the geological evolution of the rock: at very early stages of burial, precipitating microcrystalline hematite (pigment) replacing iron-rich minerals (e.g. lithic clast and biotites), and a much later stage, during the retrodiagenesis associated with the terminal uplift and weathering of the exhumed rock. It is also thought that each mineralogical phase acted as a magnetic carrier which reflects the direction of field at the time of their occurrence in the rock thus resulting in the characteristic multicomponential magnetization observed. Thermal demagnetization did not remove completely the overprint and most of the isolated directions at high temperature appear polluted by a small residual overprint.

The contribution of type A and type B components to the remanence can be theoretically modelled. A simplified 2D version of the vectorial balance before thermal demagnetization is presented in Fig. 19a. *a* and *a'* represent the theoretical Triassic normal and reverse components, *b* represents the PGFD, **c** and **c'** are the resultant between them. The scalar magnitude of the vectors is estimated from known elements in the construction (dips of *b*, **c** and **c'** are known, a is estimated from mid-Triassic directions for stable Europe at the same hypothetical latitude) after attributing an arbitrary value to *a*. The construction is consistent with the distribution of the NRM direction in normal and reverse magnetozones (Fig. 19b). Figure 19c shows the same type of construction after thermal demagnetization with a small proportion of the overprint still remaining. The diagram explains the discrepancy in inclination between normal and reverse isolated components, and is consistent with the assumption that the Triassic direction is the major contributor to the isolated component. This fact allowed the construction of the magnetostratigraphy, despite the fact that the overprint was not completely removed during thermal demagnetization. A modified version of McFadden & McElhinny's (1981) reversal test can be used to correct for bias caused polluted B components. Under this construction it is possible to subtract *b* by inversion of **c'** and consequent addition to **c** (Fig. 19e). In such configuration, directional inversion of **c'** to **c''** and consequent addition of **c** to **c''** has an important property: **d** = **c**+**c''**= *a*!, being *a* the *pure* Triassic component. Inversion of a leads to inversion of the *b* component of *a* (*b''*). Therefore addition of **c** to **c''** results in *b* and *b''* cancelling each other out, as they are exactly of the same magnitude and antiparallel.

This model can be exploited to correct the bias caused by the recent overprint. The proposed method consist in recalculating the means of contiguous magnetozones of opposite polarity once the directions of one of them has been inverted. As magnetozones do not contain the same number of samples, to achieve statistical significance compensated means have to be calculated. Means were then calculated combining consecutive magnetozones (R1-N2, N2-R3, R3-N3 and R3-N4) to minimize age differences between directions. From above, mean directions for the studied area are given in Table 1 and the overall mean recalculated from the complete collection of data is given in Table 2.

The overall mean direction for the Molina de Aragon area compares quite well with the Triassic expected direction in the area, despite significant discrepancies that can be *a priori* attributed to the rotation of Iberia consequent with the Bay of Biscay opening during middle Mesozoic times. However confirmation of the

Table 1. *Mean directions for combined magnetozones in studied area*

Combined magnetozones	Dec	Inc	N	α_{95}	R	K
R1N2	169	−15	24	5.8	23.2	27.2
N2R2	349	+13	28	6.1	26.7	21.0
R2N3	166	−17	28	7.5	26.1	14.3
N3R3	344	+17	14	11.2	13.0	13.5

Table 2. *Overall mean directions for the complete collection of data*

	Dec	Inc	N	α_{95}	K	VGP Lat P	VGP Long P	VGP P Plat
N	346.4	32.0	46	4.0	29.3			
R	167.2	02.9	46	5.3	16.9			
All	347.2	14.9	89	4.4	12.40	49.6	204.60	3.80

results implies the acceptance of the following assumptions.

(a) The age differences between adjacent magnetozones is sufficiently small to discard any significant directional differences as the result of continental drift during Triassic times in this area.

(b) The overprint remaining at high temperature can be considered as a single component of magnetization.

(c) The number of samples used in the calculation averages the proportion of the magnetic phases in the rock.

The data presented indicate that the compensated palaeomagnetic directions represent quite accurately the direction of the palaeomagnetic field during Mid-Triassic times for the area, as it has been deduced from the directions extracted during component characterization of the formation.

Conclusions

(a) The study of the structure and origin of remanence in the Rillo and Torete Fms reveals the presence of several components of magnetization of different ages. Detrital hematite and ilmeno-hematite carry a primary magnetization of Triassic origin in the form of a DRM or PDRM. It is also likely that an early CRM also contributed to the primary remanence. Goethite and several textural phases of authigenic hematite are responsible for a recent overprint in the form of a CRM. Acquisition of a PTRM during the uplift also contributes to the overprint.

(b) Thermal demagnetization allows the iso-lation of a characteristic Triassic remanence but did not succeed in removing entirely the CRM associated with the youngest component of the overprint. Despite the primary remanence being slightly biased by residual components of recent origin, a reliable Triassic magnetostratigraphy for Anissian–Ladinian times can be accurately established and correlated, within the basin, and with other sequences in North America and Asia.

This work was funded by project PB88-0070 of the CAYCIT. DR was supported by an FPU grant of the Ministry of Education of Spain. Completion of the paper was carried out thanks to an Integrated action between Spain (CAYCIT) and the UK (British Council). The authors are very grateful to David Collinson for providing the Nuffield Laboratory facilities of Newcastle University. Figures modified from Torsvik (1992) IAPD program.

References

ALVARO, M., CAPOTE, R. & VEGAS, R. 1979. Un modelo de evolucion geotecténica para la Cadena Celtiberica. *Acta Geologica Hispanica*, **14**, 172–177.

ARTHAUD, F. & MATTE, P. 1975. Les decrochements tardy-herciniques du sud-Ouest de l'Europe. Geometrie et essai de reconstruction des conditions de deformation. *Tectonophysics*, **25**, 139–171.

BERNER, R. A. 1969. Goethite stability and the origin of red beds. *Geochimica et Cosmochimica Acta*, **33**, 267–273.

DEKKERS, M. J. 1989a. Magnetic properties of natural goethite-I. Grain-size dependence of some low- and high-field related rockmagnetic parameters measured at room temperature. *Geohysical Journal*, **97**, 323–339.

—— 1989b. Magnetic properties of natural goethite-II. TRM behaviour during thremal and alternating field demagnetization and low temperature treatment. *Geophysical Journal*, **97**, 341–355.

DUNLOP, D. J. 1971. Magnetic properties of fine particles hematite. *Annales of Geophysics*, **27**, 269–263.

ELSTON, D. P. & PURUCKER, M. 1979. Detrital magnetization in red beds of the Moenkopi Formation. *Journal of Geophysical Research*, **84**, 1653–1665.

GARCIA-PALACIOS, M. & LUCAS, J. 1977. Le Bassin Triassique de la Branche Castellane de la Chaine Iberique. II. Geochemique. *Cuadernos de Geologia Iberica*, **4**, 355–368.

HAAG, M. & HELLER, F. 1991. Late Permian to Early Triassic magnetostratigraphy. *Earth and Planetary Science Letters*, **107**, 42–54.

HELLER, F., LOWRIE, W., HUAMEI, L. & JUNDA, W. 1988. Magnetostratigraphy of the Permo-Triassic boundary section at Shangsi (Guangyuan, Sichuan Province, China). *Earth and Planetary Science Letters*, **88**, 348–356.

HELSEY, C. E. & STEINER, M. B. 1974. Paleomagnetism of the Lower Triassic Moenkopi Formation. *Geological Society of America Bulletin*, **85**, 457–464.

JOWETT E. C., PEARCE, G. W. & RYDZEWSKY, A. 1987. A Mid-Triassic paleomagnetic age of the Kupferschiefer mineralization in Poland. Based on a revised Apparent Polar Wander Path for Europe and Russia. *Journal of Geophysical Research*, **92**, B1, 581–598.

KIRSCHVINK, J. L. 1980. The Least-squares line and plane and the analysis of palaeomagnetic data. *Geophysical Journal of the Royal Astronomical Society*, **62**, 699–718.

LIENERT, B. R. & HELSLEY, C. E. 1980. Magnetostratigraphy of the Moenkopi Formation at Bear Ears, Utah. *Journal of Geophysical Research*, **85**, 1475–1480.

MARFIL, R., DE LA CRUZ, B. & DE LA PEÑA, J. A. 1977. Procesos diagenéticos en las areniscas del Buntsandstein de la Cordillera Ibérica. *Cuadernos de Geologiía Ibérica*, **4**, 411–422.

McFADDEN, P. L. & McELHINNY, M. W. 1981. Classification of the reversal test in paleomagnetism. *Geophysical Journal International*, **103**, 725–729.

MOLINA-GARZA, R. S., GEISSMAN, J. W., VAN DER VOO, R., LUCAS, S. G. & HAYDEN, S. N. 1991. Paleomagnetism of the Moenkopi and Chinle formations, in Central New Mexico: implications for the North American Apparent Polar Wander Path and Triassic magnetostratigraphy. *Journal of Geophysical Research*, **96**, B9, 14239–14262.

MORAD, S., AL-AASM, I. S., RAMSEYER, K., MARFIL, R. & ALDAHAM, A. A. 1990. Diagenesis of carbonate cements in Permo-Triassic sandstones from the Iberian Range, Spain: evidence from chemical composition and stable isotopes. *Sedimentary Geology*, **67**, 281–295.

MUÑOZ, A., RAMOS, A., SANCHEZ-MOYA, Y. & SOPEÑA, A. 1992. Evolving fluvial architecture during a marine transgression: Upper Buntsandstein, Triassic, central Spain. *Sedimentary Geology*, **75**, 257–281.

OGG, J. G. & STEINER, M. B. 1991. Early Triassic magnetic polarity time scale-integration of magnetostratigraphy, ammonite zonation and sequence stratigraphy from stratotype sections (Canadian Artic Archipelago). *Earth and Planetary Science Letters*, **107**, 69–87.

PARES, J. M., BANDA, E. & SANTANACH, P. 1988. Palaeomagnetic results from the southeastern margin of the Ebro Basin (northeastern Spain). Evidence of Tertiary clockwise rotation. *Physics of the Earth and Planetary Interiors*, **52**, 267–282.

PIPER, J. D. A. 1988. *Palaeomagnetic database*. Open University Press, Milton Keynes.

RAMOS, A. 1979. *Estratigrafía y paleogeografía del Pérmico y Triásico al Oeste de Molina de Aragon (Provincia de Guadalajara)*. Seminarios de Estratigrafía, Serie Monografías, 6, Facultad de Ciencias Geológicas de la Universidad Complutense de Madrid.

——, SOPEÑA, A. & PEREZ-ARLUCEA, M. 1986. Evolution of Buntsandsteinfluvial sedimentation in north-west Iberian Ranges (Central Spain). *Journal of Sedimentary Petrology*, **56**, 862–875.

REY, D. 1992. *Palaeomagnetism and Magnetostratigraphy of continental Red Bed Sequences of Permian and Triassic Age from Western Europe*. Unpublished PhD Thesis. The University of Birmingham, UK.

SANCHEZ, V., MARFIL, R., DE LA CRUZ, B. & DE LA PEÑA, J. A. 1971. *Abstracts VII International Sedimentological Congress*, Heildelberg, 85–86.

SENCHORDI, E. & MARFIL, R. 1983. Estudio petrológico de las facies Saxonienses y Buntsandstein de la zona del Pobo de Dueñas (Cordillera Ibérica). *Boletín Geológico y Minero*, **94–95**, 448–471.

SHIVE, P. N., STEINER, M. B. & HUYCKE, D. T. 1984. Magnetostratigraphy, paleomagnetism and remanence acquisition in the Triassic Chugwater Formation of Wyoming. *Journal of Geophysical Research*, **89**, 1801–1815.

SOPEÑA, A., LOPEZ, J., ARCHE, A., PEREZ-ARLUCEA, M., RAMOS, A., VIRGILI, C. & HERNANDO, S. 1988. Permian and Triassic rift basins of the Iberian Peninsula. *In*: MANSPEIZER, W. (ed.) *Triassic–Jurassic Rifting: Continental Breakup and the origin of the Atlantic Ocean and passive margins*, part B. Developments in Geotectonics, **22**, Elsevier, 757–786.

STEINER, M. B., OGG, J., ZHANG, Z. & SUN, S. 1989. The Late Permian/Early Triassic magnetic polarity time scale and plate motions of South China. *Journal of Geophysical Research*, **94**, B6, 7343–7363.

STOKKING, L. B. & TAUXE, L. 1990. Multicomponent magnetization in synthetic hematite. *Physics of the Earth and Planetary Interiors*, **65**, 109–124.

TORSVIK, T. H. 1992. *Interactive Analysis of Palaeomagnetic Data (IAPD)*. Computer Software package, University of Bergen, Norway.

TURNER, A. 1988. *The Diagenesis and Palaeomagnetism of Permian and Triassic Sediments from*

Central Spain. PhD Thesis, The University of Aston in Birmingham, UK.

TURNER, P., TURNER, A., RAMOS , A. & SOPEÑA, A. 1989. Palaeomagnetism of Permo-Triassic Rocks in the Iberian Cordillera, Spain: Acquisition of Secondary and Characteristic Remanence. *Journal of the Geological Society, London*, **146**, 61–76.

VAN DER VOO, R. 1969. Palaeomagnetic evidence for the rotation of the Iberian Peninsula. *Tectonophysics,* **7**, 1, 5–56.

VEGAS, R. 1989. Palaeomagnetismo y modelos geodinámicos en la Peninsula Ibérica. *Cuadernos de Geologia Ibérica*, **12**, 75–82.

—— & BANDA, E. 1982. Tectonic framework and Alpine evolution of the Iberian Peninsula. *Earth Evolution Sciences*, **4**, 320–343.

VIRGILI, C. 1977 Le Trias du Nord de l'Espagne. *Bulletin due Bureau de Recherches Geologiques et Minieres*, **12**, IV, 3, 205–213.

VIRGILI, C., SOPEÑA, A., RAMOS, A. ARCHE, A. & HERNANDO, S. 1983. El relleno posthercinico y el comienzo de la sedimentacion mesozoica. *In*: FONTBOTE, J. M. (ed.) *Geologia de España (Libro Jubilar de J. M. Rios)*. Instituto Geologico y Miñero de España, Madrid, **2**, 25–36.

WALKER, T. R., WAUGH, B. & CRONE, A. J. 1978. Diagenesis in first-cycle desert alluvium of Cenozoic age, southwestern United States and northwestern Mexico. *Geological Society of America Bulletin*, **89**, 19–32.

WALKER, T. R., LARSON, E. C. & HOBLITT, R. P. 1981. Nature and Origin of Hematite in the Moenkopi Formation (Triassic), Colorado Plateau: A Contribution to the origin of Magnetism in Red Beds. *Journal of Geophysical Research*, **86**, 317–333.

ZIEGLER, P. A. 1982. *Geological atlas of Western and Central Europe*. Shell International Petroleum. Maaschappi, s. v.

ZIJDERVELD, J. D. A. 1967. A. C. demagnetization of rocks: analisis of results. *In*: COLLINSON, D. W., CREER, K. M. & RUNCORN, S. K. (eds) *Methods in Palaeomagnetism*. Developments in Solid Earth Physics 3, Elsevier, Amsterdam, 254–258.

Palaeomagnetic study of Jurassic limestones from the Iberian Range (Spain): tectonic implications

M. T. JUÁREZ[1], M. L. OSETE[1], R. VEGAS[2], C. G. LANGEREIS[3] & G. MELÉNDEZ[4]

[1]Department of Geophysics, Faculty of Physics, University Complutense of Madrid, Madrid 28040, Spain
[2]Department of Geodynamics, Faculty of Geology, Complutense University of Madrid, Spain
[3]Paleomagnetic laboratories, 'Fort Hoofdijk', Institute of Earth Sciences, Utrecht University, The Netherlands
[4]Department of Paleontology, Faculty of Sciences, University of Zaragoza, Spain

Abstract: A palaeomagnetic investigation has been carried out in the Iberian Range (Spain). Seven localities have been sampled. Two stable magnetization components have been found in all investigated sections. A high-temperature primary component of Oxfordian age shows alternatively normal and reversed polarities. A consistent low-temperature component appears in all the studied sites; it has normal polarity and passes the fold test indicating its pre-Oligocene–Miocene age. Six localities show counterclockwise declinations for both components, the high-temperature component having a Jurassic direction and the low-temperature component an early Cretaceous direction. Only the southernmost section shows a clockwise rotation for both components. This indicates that the whole Iberian Range can not be considered as a part of Stable Iberia. The angular difference (c. 15°) between both components is the same in all investigated sites. This implies that: (1) based on the palaeomagnetic direction of both components, the low-temperature remagnetization was probably acquired in the Cretaceous during the rotation of the Iberian Plate; (2) the clockwise rotation observed in the southernmost section took place after the acquisition of the remagnetization.

The rotation of the Iberian plate is still an open problem despite many palaeomagnetic investigations (e.g. Van der Voo 1969; Van der Voo & Zijderveld 1971; Vanderberg 1980; Schott et al. 1981; Galdeano et al. 1989). This is mainly due to lack of an adequate Mesozoic APWP for Iberia. This is particularly caused by difficulties in determining which zones in the Iberian Peninsula should be considered as representative of the Iberian Plate. It is therefore important to define the so-called 'Stable Iberia' in order to obtain an APWP for Iberia during the Mesozoic.

The Iberian Range has been traditionally considered as one of the most important units in the Iberian Peninsula and as a part of Stable Iberia. Palaeomagnetic directions obtained from studies in rocks from the Iberian Range have, therefore, been often used to define the Iberian APWP (e.g. Vanderberg 1980; Schott & Peres 1987; Moreau et al. 1992). It is still a matter of debate, however, whether the Iberian Range can be considered as Stable Iberia or not (Osete 1988; Van der Voo 1993).

In addition, recent palaeomagnetic investi- gations have demonstrated that remagneti- zations are frequently important in orogenic systems (e.g. McCabe & Elmore 1989; Kent et al. 1987; Villalaín et al. 1994), thereby bringing into doubt many previous palaeomagnetic re- sults. Palaeomagnetic remagnetizations have sometimes been erroneously interpreted as primary remanences, leading to incorrect defi- nitions of apparent pole positions. In particular, in the Iberian Range a very important wide- spread remagnetization of probably Cretaceous age has been suggested by Moreau et al. (1992) and Juárez et al. (1994). We have carried out, therefore, a palaeomagnetic study in the Iberian Range in order to determine: (1) which zones of this mountain system can be used to define the Iberian APWP; and (2) which zones have suffered this important reported remagneti- zation.

Geological setting and sampling

The Iberian Range is a large mountain chain located in eastern Iberia which extends in a

From Morris, A. & Tarling, D. H. (eds), 1996, *Palaeomagnetism and Tectonics of the Mediterranean Region*, Geological Society Special Publication No. 105, pp. 83–90.

83

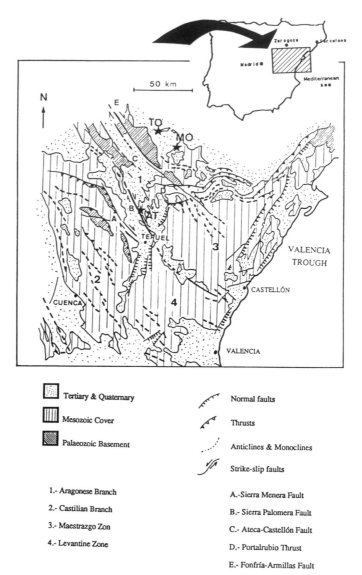

Fig. 1. Geological sketch map showing the location of the sampling sites. TO, Tosos region; MO, Moneva region; AT, Aguatón Region.

general NW–SE direction from the Cantabrian Range, in the north, to the Betic Range, in the south (Fig. 1). Two main mountain alignments or 'branches' can be recognized within the Iberian Range: the so-called northeastern or Aragonian Branch, and the southwestern or Castilian Branch. Both branches are separated by Tertiary grabens. Two additional regions can be also distinguished in the eastern part of the Range: the Maestrazgo Zone and the Levantine Zone (Fig. 1). Some notable differences with true alpine ranges led some authors to classify

the Iberian Range as an 'intermediate alpine range', between true alpine and platform structures (Julivert *et al.* 1972).

This intraplate mountain belt formed by tectonic inversion of a rifted zone formed during the Mesozoic extensional tectonics of the Iberian Peninsula (e.g. Alvaro *et al.* 1979; Vegas & Banda 1982). This inversion took place during the Palaeogene (Oligocene to lowermost Miocene) and caused crustal shortening that was accommodated in the Palaeozoic basement by thrusts which uplifted Hercynian rocks, and by

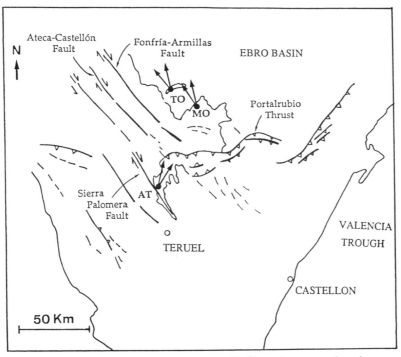

Fig. 2. Sketch map showing the structural setting for the studied area. Arrows are the palaeomagnetic directions obtained for P and S components. TO, Tosos region; MO, Moneva region; AT, Aguatón region.

the reactivation of pre-orogenic, NW–SE-trending faults as right-lateral fracture zones. This crustal shortening was accommodated in the Mesozoic cover by means of folds and monoclines, as well as by low-angle thrusts (Fig. 1).

In the Aragonian Branch, in its boundary with the Maestrazgo Zone, lateral motion along the main NW–SE-trending strike-slip faults has been related to the formation of E–W, winding cover thrusts (Calvo Hernandez 1993). Moreover, the activity of these main faults (Sierra Menera, Sierra Palomera, Ateca–Castellón and Fonfría–Armillas) caused the rupture of the cover in many places giving rise to flower structures, as well as to transgressive duplexes, as described by Calvo Hernandez (1993). This compressive stage is attributed to the Pyrenean Ibero-Europe collision. A subsequent extensional phase, from lower Miocene up to present, was established in connection with the opening of the Valencia trough (Vegas *et al.* 1979). The main NNE–SSW-trending extensional faults and the related Neogene intramontane basins formed during this extensional phase (Fig. 1).

A total of 450 cores were collected from seven localities, distributed in three different regions

(Tosos, Moneva and Aguatón regions; Fig. 2) covering a large area in the Iberian Range. Four sites have been sampled in the Tosos region, two sites in the Moneva region and one site in the Aguatón region. The sampled lithologies are grey pelagic limestones of Oxfordian (late Jurassic) age.

A previous detailed magnetostratigraphic study has been carried out in the Tosos region (Juárez *et al.* 1994). The outcrops in this region are situated on both limbs of a kilometric wavelength anticline with a NE–SW direction. Hence a regional fold test could be performed.

Laboratory procedures

Thermal and AF demagnetization techniques have been applied to a selection of about 60 pilot samples. Thermal treatment was found to be more effective in isolating the different magnetic components, and was therefore systematically applied to the remaining samples. All the samples were heated from room temperature up to 600°C in temperature intervals varying between 15°C and 100°C. Special care was necessary at high temperatures (over 400°C) to better constrain one of the magnetic components

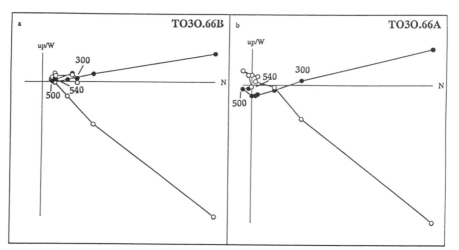

Fig. 3. Demagnetization diagrams of two samples from the same core, showing the influence of the viscous component at high temperatures. In (**a**) the viscous component has not been removed, whereas in (**b**) the procedure described in the text has been used to eliminate this viscous component before measurement of the remanence.

because of the acquisition of a viscous component of the magnetization, probably carried by new fine-grained magnetite created during heating. This viscous component was removed by leaving the specimen at rest within the zero field of the cryogenic magnetometer before making the measurement. The resting time varied between 2 and 15 minutes depending on the specimen and the temperature step. It is important to point out that no high temperature components could have been isolated without removal of this viscous component due to its extremely high magnetization. Figure 3 shows an example of the influence of this viscous magnetization.

Palaeomagnetic directions have been determined by principal component analysis (Kirschvink 1980). Magnetic susceptibility was monitored after each step of the thermal treatment to detect any change in the mineralogy occurring during heating. In addition, isothermal remanent magnetization (IRM) experiments (acquisition and subsequent thermal demagnetization) were carried out.

Results

Thermal demagnetization shows a similar pattern in most specimens: two stable components could be identified, both different from the present-day magnetic field (Fig. 4). A low temperature component (S-component) is isolated after heating to 200°C, and has a maximum

unblocking temperature of 350–450°C. A subsequent high temperature component (P-component) is removed at 540–580°C. The S-component accounts for most of the NRM intensity, whereas the P-component has a very low intensity (approximately 10% of the initial NRM intensity).

The IRM demagnetization experiments indicate that both components of the magnetization are probably carried by magnetic minerals with maximum unblocking temperatures ranging from 350°C up to 580°C. This is consistent with a magnetite carrier with varying grain size and/or titanium content.

The main difference observed between P- and S-components is in their palaeomagnetic polarity. The P-component shows alternatively normal and reversed polarities, and it has been used to construct six magnetostratigraphic sequences which are well-correlated between sections (Juárez 1994; Juárez *et al.* 1994). In contrast the S-component exhibits only normal polarities suggesting that it is of secondary origin.

A fold test could be performed in the Tosos region (Juárez *et al.* 1994), which shows that both components have a pre-folding origin, indicating a pre-Oligocene–Miocene age of acquisition of both magnetization components.

The mean directions for both components in each region are given in Table 1, together with the overall mean directions combining all regions except the Aguatón area.

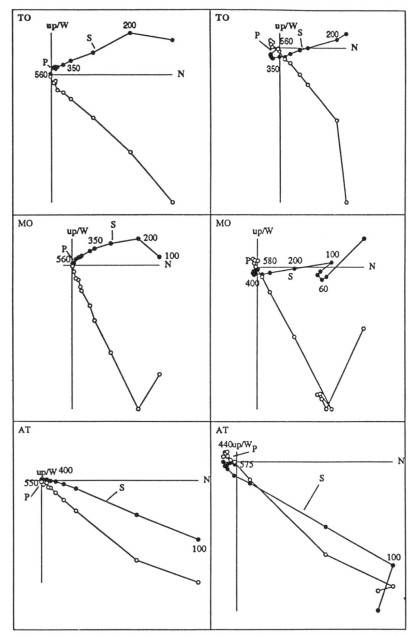

Fig. 4. Zijverveld plots showing demagnetisation behaviour of two typical specimens from each studied area, with normal and reversed polarities of the P-component. Closed symbols represent the horizontal projection and open symbols are the vertical projection. (TO, Tosos region; MO, Moneva region, AT, Aguatón region).

Discussion and conclusions

Palaeomagnetic results

We interpret the P-component as the primary magnetization component acquired during the Oxfordian, and the S-component as a remag- netization acquired during a normal polarity period of the Earth's magnetic field. Taking into account the extremely similar magnetic behavior observed in all the studied regions, we conclude that the S-component was acquired at the same time in all these areas.

Table 1 *Palaeomagnetic directions (Dec. and Inc.) and Fisherian statistics parameters (K and α_{95}) obtained in each region, together with the average calculated for the locality data.*

Region	Lat, Long	Component	n	Geographical corodinates				Bedding tilt corrected			
				Dec	Inc	K	α_{95}	Dec	Inc	K	α_{95}
MO	41.1°N, 0.7°W	P	83	332.1	41.5	12.6	4.6	329.1	48.4	12.3	4.6
		S	76	344.7	48.7	52.1	2.3	343.2	56.1	48	2.4
TO	41.3°N, 1.4°W	P	268	309.3	56.9	1.9	9.2	324.1	40.6	9.9	2.9
		S	231	355.6	62.0	2.1	8.9	340.9	44.9	31.0	1.7
AT	40.7°N, 0.7°W	P	69	285.9	45.5	11.3	5.3	14.9	37.0	11.3	5.3
		S	69	282.5	55.5	46.6	2.5	27.1	34.2	46.6	2.5
All		P	351	319.9	52.4	2.4	6.3	325.3	42.4	10.6	2.4
		S	307	349.9	57.7	2.7	6.2	341.4	47.7	30.3	1.5

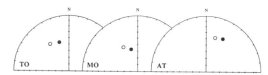

Fig. 5. Equal-area projection of the average palaeomagnetic directions obtained in each region. Open symbols correspond to the P-component and closed symbols to the S-component.

At all the investigated sites, except the Aguatón site, the average palaeomagnetic direction obtained for the P-component is rotated to the west. This is consistent with the expected Jurassic direction because it shows the 35° counterclockwise rotation expected for the Iberian Peninsula with respect to 'Stable Europe' (Van der Voo 1967, 1969; Van der Voo & Zijderveld 1971; Galdeano et al. 1989; Van der Voo 1993). The S-component also exhibits a westerly declination showing an average 15° clockwise rotation with respect to the P-component (Fig. 5). In the Aguatón region both components are rotated to the east, but the relative rotation between P and S components remains constant. This indicates that an apparent rotation of 45°–50° of this site with respect to the Tosos and Moneva sites took place after the acquisition of the secondary S-component.

Considering the results from some previous palaeomagnetic studies in the Iberian Peninsula (Galdeano et al. 1989; Moreau et al. 1992) together with the palaeomagnetic direction obtained for the S-component we can conclude that this remagnetization was probably acquired during the lower Cretaceous.

The existence of a remagnetization affecting the Jurassic outcrops in the Iberian Range was previously suggested by Osete (1988) and has been confirmed in the Maestrazgo zone (southeastern Iberian Range) by Moreau et al. (1992). Our results demonstrate that this remagnetization has also affected the Jurassic limestones in our studied area.

The palaeomagnetic direction obtained for the S-component is also consistent with the palaeomagnetic direction that Vanderberg (1980) obtained for red siltstones of Jurassic age close to Soria (northwestern Iberian Range). This result, originally interpreted as a primary direction, has already been reinterpreted as a late Cretaceous remagnetization by Schott & Peres (1987), in agreement with our results. Therefore a very important remagnetization, of probable Cretaceous age appears to have affected a very large area of or perhaps even the whole Iberian Range.

It is also interesting to point out that a remagnetization affecting Jurassic sediments has been previously reported in other areas of the Iberian peninsula: Galdeano et al. (1989) suggested that a thermochemical remagnetization affected late Jurassic to Valangian sediments located close to Lisbon (Portugal).

Finally, our results can also be compared with those obtained by Schott et al. (1981) for the mid-Jurassic in the Messejana dyke. The pole position obtained by these authors is very close to that calculated from our S-component, although Schott et al. (1981) consider it to be of primary, mid-Jurassic age. The similarity with our S-component pole position suggests that the Messejana results could also represent a Cretaceous remagnetization.

Complementary studies and revisions of previous palaeomagnetic investigations are needed to constrain the origin, extent and timing of this apparently widespread and important remagnetization.

Table 2 *Pole positions (latitude and longitude) and confidence circles (α₉₅) obtained by other authors from mid-Jurassic to late Cretaceous units compared with the pole positions calculated from the P and S components of the present study*

Age	Plong	Plat	α₉₅	Locality	Reference
Mid-Jurassic	236	71	8	Messejana dyke	Schott *et al.* (1981)
Oxfordian	255	55	6	Iberian Range	Steiner *et al.* (1985)
Oxfordian*	214	70		Lisbon	Galdeano *et al.* (1989)
Kimmeridgian*	204	70	4	Soria (Iberian Range)	Vanderberg (1980)
Kimmeridgian*	188	67		Lisbon	Galdeano *et al.* (1989)
Berriasian	252	61	3	Algarve	Galbrun *et al.* (1990)
Hauterivian–Barremian	275	57	4	Lisbon	Galdeano *et al.* (1989)
Barremian–Aptian	226	74	11	Iberian Range/Lisbon	Moreau *et al.* (1992)/ Galdeano *et al.* (1989)
Late Cretaceous	197	66	3	Lisbon	Van der Voo & Zijderveld (1971)
P-component	252	58	3	Iberian Range	This study
S-component	238	71	2	Iberian Range	This study

* Remagnetized directions (see text).

Tectonic implications

The primary (Jurassic) directions obtained for the Tosos and Moneva regions, showing a 35° counterclockwise rotation, are consistent with the expected direction for Stable Iberia. The relative rotation implied by the difference between the primary and secondary components of magnetization reflects the movement of the Iberian plate between the times of acquisition. Therefore a counterclockwise rotation of 15° would have occurred, as a consequence of the opening of the Bay of Biscay, between the Late Jurassic (Oxfordian) and the acquisition time of the remagnetization (probably early Cretaceous). These data are compatible with the tectonic history of the so-called 'Stable Iberia'.

An average clockwise rotation of 45–50° can be observed in both components of the magnetization in the Aguatón region with respect to those in the Tosos and Moneva regions. This rotation must have taken place after the remagnetization event.

The clockwise rotation of the P and S components in the Aguatón area means that some form of rotational deformation must have accompanied the compressive Palaeogene deformation of the Mesozoic cover. Hence the zone of Aguatón does not correspond to Stable Iberia although it is part of the Iberian Range tectonic unit.

How can this different tectonic behavior be explained? One possibility lies in the treatment of the data: the sampled localities could have a more complex geological structure than that considered here, involving two or more move-

ments of the sediments. Therefore an inappropriate tectonic correction might have been applied giving rise to an apparent rotation. Alternatively, the observed rotated declination of both components of magnetization in the southernmost region reflects a real tectonic rotation about a vertical axis. In this scenario it is important to consider the different response of the Mesozoic cover to the dextral movement along the NW–SE-trending basement faults. In the Tosos and Moneva regions, the Mesozoic cover does not record in situ rotational deformation but instead the dextral motion along the Fonfría–Armillas fault is accommodated by folding (Figs 1 & 2). This motion and that along the Ateca–Castellón fault, is absorbed in the frontal E–W-trending Portalrubio Thrust (Figs 1 & 2) as suggested by Calvo Hernandez (1993).

In the sampled area of the Aguatón region, the dextral motion along the Sierra Palomera fault results in clockwise rotation of the Mesozoic cover as shown by the rotation of the palaeomagnetic vectors. This is due to the vicinity of the area to the rupture zone of the cover, which causes that part of the detached cover to be broken in relatively small panels. This is in accordance with the structures described in the cover and caused by the activity of the basement fault. These structures include, as mentioned before, positive flower structures and duplexes. Moreover in this region no frontal thrust can accommodate the right-lateral displacement of the Sierra Palomera fault.

This tectonic scenario implies that different tectonic behaviors can be found in regions where crustal shortening has an important shear component, as is the case of the Iberian Ranges. This should be taken into consideration when interpreting palaeomagnetic data from rotated and non-rotated sites.

We thank William Lowrie for his constructive criticisms of the manuscript. This work has been supported by the DGICYT (project PB93-0193) and by the European Union (project CT94-0114). Publication number 373, Dpto de Geofísica, Universidad Complutense de Madrid.

References

ALVARO, M. R., CAPOTE, R. & VEGAS, R. 1979. Un modelo de evolución geotectónica para la Cadena Celtibérica. *Acta Geologica Hispania*, **14**, 172–177.

CALVO HERNANDEZ, J. M. 1993. *Cinemática de las fallas discontínuas en el sector central de la Cordillera Ibérica*. Doct. Thesis. University of Zaragoza.

GALBRUN, B., BERTHOU, P. Y., MOUSSIN, C. & AZÉMA, J. 1990. Magnétostratigraphie de la limite Jurassique-Crétacé en faciès de plateforme carbonatée: la coupe de Bias do Norte (Algarve, Portugal). *Bulletin de la Societé Géologique de France*, **8**, VI, 1, 133–144.

GALDEANO, A., MOREAU, M. G., POZZI, J. P., BERTHOU, P. Y. & MALOD, J. A. 1989. New Paleomagnetic results from Creaceous sediments near Lisbon (Portugal) and implications for the rotation of Iberia. *Earth and Planetary Science Letters*, **92**, 95–106.

JUÁREZ, M. T. 1994. *Estudio paleomagnético y magnetoestratigráfico del periodo Jurásico en el Sistema Ibérico*. Doct. Thesis, Univ. Complutense, Madrid.

——, OSETE, M. L., MELÉNDEZ, G., LANGEREIS, C. G. & ZIJDERVELD, J. D. A. 1994. Oxfordian magnetostratigraphy of the Aguilón and Tosos sections (Iberian Range, Spain) and evidence of a pre-Oligocene overprint. *Physics of the Earth and Planetary Interiors*, **85**, 195–211.

JULIVERT, M., FONTBOTÉ, J. M., RIBEIRO, A. & CONDE, R. N. 1972. *Memoria explicativa del mapa tectónico de la Península Ibérica y Baleares*. Instituto Geologia y Minero de España.

KENT, D. V., ZHENG, X., ZHNAG, W. Y. & OPDYKE, N. D. 1987. Widespread late Mesozoic to recent remagnetization of Paleozoic and lower Triassic sedimentary rocks from South China. *Tectonophysics*, **139**, 133–144.

KIRSCHVINK, J. L. 1980. The least-squares line and plane and the analysis of paleomagnetic data. *Geophysical Journal of the Royal Astronomical Society*, **62**, 699–718.

McCABE, C. & ELMORE, R. D. 1989. The occurrence and origin of Late Paleozoic remagnetization in the sedimentary rocks of North America. *Reviews in Geophysics*, **27**, 471–494.

MOREAU, M. G., CANEROT, J. & MALOD, J. A. 1992. Paleomagnetic study of Mesozoic sediments from the Iberian Chain (Spain). Suggestions for Barremian remagnetization and implications for the rotation of Iberia. *Bulletin de la Societé Géologique de France*, **163**, 4, 393–402.

OSETE, M. L. 1988. *Estudio del magnetismo remanente de las rocas de interés paleomagnético en España*. Doct. Thesis, Univ. Complutense, Madrid.

SCHOTT, J. J. & PERES, A. 1987. Paleomagnetism of the lower Cretaceous redbeds from northern Spain. Evidence of a multistage acquisition of magnetization. *Tectonophysics*, **139**, 239–253.

——, MONTIGNY, R. & THUIZAT, R. 1981. Paleomagnetism and potassium argon age of the Messejana dike (Portugal and Spain): angular limitation to the rotation of the Iberian Peninsula since the middle Jurassic. *Earth and Planetary Science Letters*, **53**, 457–470.

STEINER, M. B., OGG, J. G., MELÉNDEZ, G. & SEQUEIROS, L. 1985. Jurassic magnetostratigraphy, 2. Middle-Late Oxfordian of Aguilón, Iberian Cordillera, northern Spain. *Earth and Planetary Science Letters*, **76**, 151–166.

VAN DER VOO, R. 1967. The rotation of Spain: Palaeomagnetic evidence from the Spanish Meseta. *Palaeogeography, Palaeoclimatology, Palaeoecology*, **3**, 393–416.

—— 1969. Paleomagnetic evidence for the rotation of the Iberian Peninsula. *Tectonophysics*, **7**, 1, 5–56.

—— 1993. *Paleomagnetism of the Atlantic, Tethys and Iapetus Oceans*. Cambridge University Press.

—— & ZIJDERVELD, J. D. A. 1971. Renewed paleomagnetic study of the Lisbon volcanics and implications for the rotation of Iberian Peninsula. *Journal of Geophysical Research*, **76**, 3913–3921.

VANDERBERG, J. 1980. New paleomagnetic data from the Iberian Peninsula. *Geologie en Mijnbouw*, **59**, 49–60.

VEGAS, R. & BANDA, E. 1982. Tectonic framework and Alpine evolution of the Iberian Peninsula. *Earth and Evolutionary Science*, **4**, 320–343.

——, FONTBOTÉ, J. M. & BANDA, E. 1979. Widespread, Neogenen rifting superimposed on Alpine regions of the Iberian Peninsula, paleogeographic and morphostructural implications. *Instituto Geografia Nacional (Spain), Special Publications*, **201**, 109–128.

VILLALAÍN, J. J., OSETE, M. L., VEGAS, R., GARCÍA-DUEÑAS, V. & HELLER, F. 1994. Widespread Neogene remagnetization in Jurassic limestones from the south-Iberian palaeomargin (Western Betics, Gibraltar Arc). *Physics of the Earth and Planetary Interiors*, **85**, 15–33.

Inclination error linked to sedimentary facies in Miocene detrital sequences from the Vallès–Penedès Basin (NE Spain)

M. GARCÉS[1], J. M. PARÉS[2] & L. CABRERA[1]

[1] *Departament de Geologia Dinàmica, Geofísica i Paleontologia, Universitat de Barcelona, Zona Universitària de Pedralbes, 08071-Barcelona, Spain*
[2] *Institute of Earth Sciences, CSIC, Martí i Franqués s/n, 08028-Barcelona, Spain*

Abstract: Palaeogeographic reconstructions based on kinematic and other integrated models are rarely in perfect agreement with palaeomagnetic data. Furthermore, the reliability of rocks that record the geomagnetic field is often questioned, particularly for detrital remanent magnetization (DRM). Among the diverse types of magnetization bias, *inclination shallowing* caused by several mechanisms such as gravitational torques upon deposition and post-depositional compaction can be a major source of error in palaeolatitudinal estimations. Inclination shallowing for a variety of fine-grained alluvial sediments from the Neogene Vallès–Penedès Basin (NE Spain) has been observed to be dependent on sedimentary facies. The correlation of inclination shallowing to sediment fabric is investigated in samples from 11 stratigraphic profiles. Each profile has a dominant lithology, from thinly laminated to edafized silts and massive muds and breccias. A measure of particle alignment in sediment can be determined by means of the *degree of anisotropy* (P) and *shape symmetry* (T) AMS parameters. Since each sedimentary facies shows a particular range of P and T, a representative mean AMS ellipsoid can be estimated from each lithologically homogeneous profile. Estimated mean remanence inclination (I_{NRM}) has a strongly consistent relationship to the mean AMS ellipsoid and a positive linear correlation is inferred between ln (P) and ln tan (I_{NRM}). Given the extreme case of no sediment anisotropy, i.e. P = 1, inclination error would be cancelled and I_{NRM} would record the true magnetic field inclination of the sampling site. The regression function predicts a corrected magnetic inclination of 60°, consistent with that of the expected geomagnetic dipole for the studied site and age.

Palaeomagnetism is based on the capability of rocks to record variations of the earth's magnetic field. The reliability of natural remanent magnetization (NRM) of rocks can be questioned, however, especially when carried by detrital particles. Acquisition of detrital remanent magnetization (DRM) can be biassed due to gravitational torques acting upon deposition, resulting in an inclination shallowing of the remanence vector. Palaeogeographic reconstructions based on the shallowed palaeomagnetic data would lead to a false interpretation of the studied sites with a drift towards equatorial latitudes.

The action of gravitational torques on ferromagnetic particles in sediment is influenced by a broad variety of factors, such as grain size and shape, content of clay minerals, water salinity, and flow density among others (Levi & Banerjee 1990; Deamer & Kodama 1990). All these parameters may influence the DRM acquisition at the time of deposition and during subsequent burial and compaction. Several postdepositional physical processes may disrupt the primary DRM, particularly in continental environments

(Tucker 1980, 1987; Tauxe & Badgley 1988). Reworking processes may cause either DRM randomization or postdepositional DRM (PDRM) remagnetization, depending on the sedimentary conditions. Sediment water content is particularly important in order to allow particle movement (Khramov 1968). If a new steep-trending PDRM replaces the primary shallow DRM a progressive correction of the inclination error takes place.

As a result of all these factors, the record of geomagnetic field may be markedly influenced by the early sedimentary evolution. Thus it would be useful to study how DRM may be influenced by sedimentary conditions as well as what sort of errors each type of sediment may produce. This would enable palaeomagnetists to follow better sampling strategies to achieve the goals of their study.

Previous works on natural redeposited sediments have proved a tangential relationship between DRM inclination (I_{DRM}) and ambient field inclination (I_H) (King 1965; Tauxe & Kent 1983; Løvlie & Torsvik 1984):

From Morris, A. & Tarling, D. H. (eds), 1996, *Palaeomagnetism and Tectonics of the Mediterranean Region*, Geological Society Special Publication No. 105, pp. 91–99.

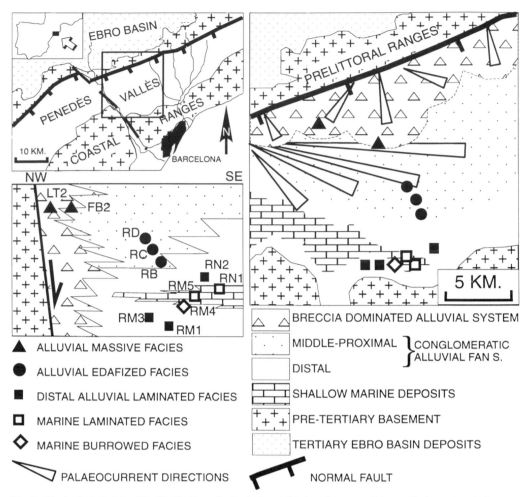

Fig. 1. Geological sketch of the Vallès-Penedès basin (top left); location map of the studied area in relation to the main Middle/Upper Miocene depositional systems (right) and the lithostratigraphic framework (bottom left).

$$\tan I_{DRM} = f \times \tan (I_H) \qquad (1)$$

where f is the so-called flattening factor, which is basically lithology dependent. A study by Jackson et al. (1991) focused on the estimation of f by means of the anisotropy of magnetic remanence. In artificial sediments they found a good agreement with equation (1) by attributing to f the ratio of vertical to horizontal axes of an anisotropic detrital remanence tensor. But the approximation to this DRM ellipsoid is not simple since both sediment and particle remanence anisotropy have to be estimated. Moreover, uncertainities about the angular relationship between particle remanence and shape long axis (Løvlie 1993) make this method essentially non-applicable to natural sediments.

Despite these difficulties, however, Collombat et al. (1993) tested a simplification of the Jackson et al. (1991) model in natural sediments. In recent deep sea deposits they found a positive correlation between I_{DRM} and the anisotropy of anhysteretic remanent magnetization (ARM) where f was the ratio of vertical to horizontal axes of the ARM ellipsoid to the power of a coefficient n: $(ARM_z/ARM_x)^n$. A similar correlation was achieved by using the anisotropy of magnetic susceptibility (AMS) instead of ARM, indicating that both ARM and AMS parameters qualitatively measured the same fabric.

In this study, facies control on the DRM is investigated in Neogene alluvial and shallow marine deposits from the Vallès-Penedès basin. Estimation of the AMS ellipsoid is used to link the sediment fabric to the inclination shallowing. Finally, correlation of AMS to I_{DRM} is tested in order to obtain an estimation of the true magnetic field inclination.

Geological setting and lithologies

The Vallès-Penedès is one of the largest onshore half-grabens integrated within the Neogene extensional structure of the northeastern Spanish continental margin (Roca & Guimerà 1992). Basin geometry and orientation was controlled by the main NE–SW-trending basement fault system. The sedimentary basin infill consists of thick sequences of alluvial deposits. Alluvial systems drained from the NW basin margin and the maximum sediment thickness (up to 3000 m) occurs along the northwestern hanging-wall side. During the Late Burdigalian and Langhian, the half-graben was affected by extensive marine transgressions, resulting in the generation of alluvial fan to fan delta-bay systems and minor carbonate platforms. The transition from continental to marine deposits is recorded both in the western Vallès and the Penedès area (Cabrera et al. 1991; Bartrina et al. 1992; Fig. 1).

In the studied area two main Middle to Late Miocene alluvial systems coexisted and interferred with each other: a large, radially spread conglomeratic fan with WNW–ESE palaeocurrent trends, and a breccia-dominated alluvial system consisting of a set of coalescing small-scale fans against the hanging wall (Fig. 1). The main conglomeratic alluvial fan had a larger source area from where the pre-Tertiary Ebro basin basement and Eocene sedimentary cover were eroded and shed into the Vallès area. The small breccia alluvial fans were directly related to the activity of the main boundary fault. Their source areas were restricted to the nearest reliefs of the Prelittoral Range which consisted of Palaeozoic intrusive, volcanic and metamorphic rocks.

Proximal, middle and distal alluvial fan facies assemblages have been recognized in both alluvial fan system sequences. Moreover, the middle Miocene distal facies of the main conglomeratic fan are laterally related to mudstone- to sandy-dominated transitional facies of Langhian age. Samples were collected from 11 sites, ranging from both proximal to distal alluvial fan and marine transitional facies assemblages. Each site consists of a 10 m to 100 m thick stratigraphic profile of a relatively homogeneous lithology. Depending on the dominant fabric of the sampled fine-grained layers, profiles were classified into three main groups of facies: laminated, massive and bioturbated.

Laminated facies are the most representative of the distal alluvial fan and the interbedded transitional marine facies assemblages. The common feature in both facies is the undisrupted depositional laminated fabric. Bioturbation is rare. Sites RM1, RM3 and RN2 consist of alternating thin layers of flood plain-mud flat red sands and silts. Sites RM5 and RN1 consist of transitional interdistributary bay laminated brown silts. The bottom of the marine facies consists of highly burrowed red and grey silts (site RM4) overlain by transgressive bioclastic sandstones. The burrowed layers area may correspond to former distal alluvial laminated deposits which were later reworked by marine organisms during the marine transgression.

Edafized facies are the most abundant in the middle conglomeratic alluvial fan assemblages (Sites RB, RC and RD). Lamination is rarely preserved due to edafization and burrowing. Palaeosoils are quite diverse, ranging from carbonate to pseudo-hydromorphous soils.

Massive facies (Sites FB2 and LT2) are prevalent along the breccia-dominated alluvial system. They consist of alternating layers of mudstones and breccias that originated by mass flow transport and deposition. Sediment fabric has neither particle sorting nor depositional lamination and mud layers include abundant floating clasts and clast nests.

Independent of their fabric characteristics, fine-grained sediments from both the conglomeratic alluvial fan and the breccia alluvial system have a rather homogeneous mineral composition. Quartz, chlorite, K-feldspar and muscovite are the most common minerals identified by X-ray diffraction analysis.

Measurement procedures

Both AMS and NRM analyses were carried out on the same set of samples. Sample AMS was determined before the NRM thermal demagnetization procedure to avoid noise from the new magnetic minerals precipitated upon heating. Between 10 and 60 samples per site were analysed.

The AMS ellipsoid was estimated using a Kappabridge KLY-2 susceptibility bridge. A good estimation of the AMS principal axes was obtained by measuring induced magnetization in 15 sample positions. AMS parameters such as foliation (F), lineation (L), shape symmetry (T) and degree of anisotropy (P, P') (Jelinek 1981) were calculated from the estimated sample AMS principal axes. Site mean AMS parameters were estimated on every profile (Table 1).

NRM analyses were carried out by standard

Table 1. *Site mean NRM and AMS parameters estimated at 11 sites of diverse lithologies*

Site	NRM					AMS		
	N	Dec	Inc	k	α_{95}	L	F	P
RM5	10	2	1	10	16	1.004	1.161	1.165
RN1	25	2	22	19	7	1.001	1.161	1.161
RM1	17	346	16	11	12	1.006	1.055	1.061
RN2	15	1	32	26	8	1.004	1.053	1.058
RM3	17	18	35	15	11	1.002	1.031	1.033
RM4	13	344	40	11	13	1.006	1.022	1.028
RD	26	12	44	24	6	1.001	1.026	1.027
RB	47	352	43	10	7	1.001	1.022	1.024
RC	59	359	45	12	6	1.001	1.021	1.023
LT2	15	5	42	35	7	1.006	1.021	1.028
FB2	31	10	56	28	5	1.001	1.006	1.007

RM5, RN1, shallow marine laminated facies; RM1, RN2, RM3, distal alluvial laminated facies; RM4, shallow marine-transitional burrowed facies; RD, RB, RC, middle alluvial edafized facies; LT2, FB2, breccia-dominated alluvial massive facies.

stepwise thermal demagnetization procedures. Three samples per stratigraphic level were stepwise demagnetized for magnetostratigraphic purposes. For the goals of this study, only samples with stable and straight high temperature components were considered. Calculation of characteristic directions was made by means of least squares analysis. Directions with overlapping signatures, transitional polarities or those partly disrupted by spurious laboratory magnetization were excluded from the analysis. Mean directions and Fisherian statistics were estimated for each site (Table 1).

Magnetic properties and mineralogy

Magnetic mineralogy

In order to identify the ferromagnetic content, progressive IRM acquisition up to 1 Tesla and thermal demagnetization was carried out following the Lowrie (1990) routine on a representative set of samples. The analysis of all lithologies shows a similar pattern. The magnetic mineralogy consists of two phases, hematite and magnetite, although their relative contents are significantly variable. Hematite is essentially the only ferromagnetic mineral in the distal alluvial and shallow marine laminated facies. Similarly, middle alluvial bioturbated facies are also dominated by hematite, although magnetite is identified in variable amounts. Beside these facies, massive sediments from the breccia alluvial system have a relatively high magnetite content. Sediment ferromagnetic composition may be controlled by diverse factors. The igneous and metamorphic-rich source area of

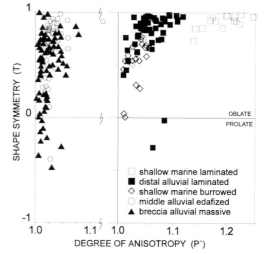

Fig. 2. Hrouda–Jelinek anisotropy diagram. Symbols refer to the different facies types (Note that X-axis is duplicated for the interval from $P' = 1$ to $P' = 1.1$).

the breccia alluvial system is at least partially responsible for the dominance of magnetite in the ferromagnetic fraction. Alternatively, longer duration of particle transport may have facilitated alteration of magnetite grains in sediments from the main conglomeratic alluvial fan.

Anisotropy of magnetic susceptibility

AMS properties of sediments were found to be deeply dependent on the sedimentary fabric (Fig. 2). AMS ellipsoids determined from

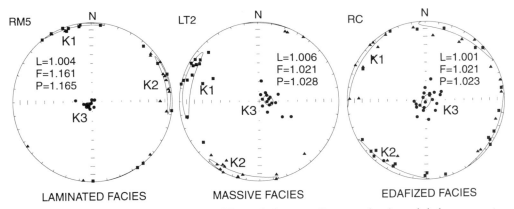

Fig. 3. Representative AMS fabrics of three different facies types. Squares, triangles and circles represent maximum (K_1), intermediate (K_2) and minimum (K_3) AMS principal axis (L: K_1/K_2; F: K_2/K_3; P: K_1/K_3). Ellipses represent the angular error of the site mean principal axis.

laminated facies show a very strong sedimentary fabric with a high magnetic foliation (Fig. 3) and with K_3 perpendicular to bedding. Shallow marine laminated silts show the strongest anisotropy with $1.150 < P < 1.250$, while laminated samples from alluvial flood plain deposits have mean $P \approx 1.07$ (Fig. 2). Massive facies have the weakest anisotropy, as low as $P = 1.007$, and bioturbated facies have intermediate values of $P \approx 1.025$. Despite the higher standard deviation of the principal AMS axes, both massive and bioturbated facies have oblate mean AMS ellipsoids with mean K_3 still perpendicular to bedding (Fig. 3).

As a relation between AMS and sedimentary facies exists (Fig. 3), magnetic fabric must be dependent on both the depositional dynamic and gravitational conditions and the post-depositional reworking processes. The tendency of elongated grains to lie horizontally on the depositional surface causes the common disc-shaped sedimentary fabric, with K_3 perpendicular to bedding. Effective particle sorting and steady deposition driven by aqueous flow results in thinly laminated and strongly anisotropic sedimentary fabric. On the contrary, almost instantaneous sedimentation of mass flow deposits cannot produce such effective particle arrangement. Gravitational torques have a secondary role in particle orientation as density flow increases and matrix supports particles. As a result, massive deposits have a weak anisotropy.

Post-depositional reworking of sediment causes randomization of the primary particle alignment. Disruption of the primary depositional fabric due to bioturbation is followed by a decrease in sediment anisotropy (Fig. 2). The standard deviation of mean AMS principal axes increases as particle statistical randomization progresses. Site mean AMS ellipsoids, however, may still preserve a sedimentary-like principal axis orientation.

Natural remanent magnetization

The analysis of the NRM gives a good insight into the sedimentary control on DRM acquisition, and demonstrates the close link of the remanence inclination with the sedimentary fabric.

Thermal demagnetization of both laminated and massive facies identifies in a well-defined two component NRM (Fig. 4a & b). A low temperature secondary component is removed at 250–300°C. Above this point, progressive demagnetization results in a very straight high temperature characteristic component. Demagnetization of bioturbated facies may follow less linear, more complicated paths (Fig. 4c), depending on the degree and timing of post-depositional reworking and the effectiveness of PDRM acquisition. For these reasons a higher number of unresolved and overlapping multiple component NRMs have been found in these sediments.

The most noticeable feature is the extremely shallow inclination of the characteristic remanence observed in all the profiles consisting of laminated sediments (Fig. 5a). Estimated site mean inclinations are always below 25°. In contrast, steep mean inclinations are found in sites consisting of massive deposits (Fig. 5b). The steepest mean site inclination of 56° estimated in this study better approaches the expected true geomagnetic field inclination,

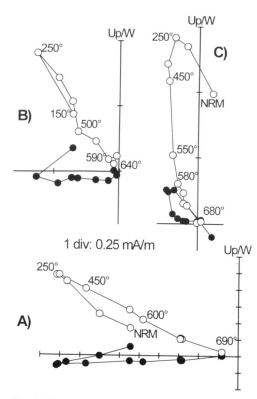

1 div: 0.25 mA/m

Fig. 4. Thermal demagnetization vector end point diagrams of: (**a**) distal alluvial laminated; (**b**) breccia-dominated alluvial massive; and (**c**) middle alluvial edafized facies. Closed symbols, declinations; Open symbols, projection of inclinations onto N–S vertical plane; Temperatures in °C.

which is thought to be about 60°. Mean inclinations obtained from the bioturbated facies (Fig. 5c) are found to have intermediate values around 40°.

Positive results of conglomerate tests carried out on some intraformational mud clasts give evidence of DRM or early PDRM as the main source of sediment magnetization. A detrital origin for the remanence is also supported by the common occurrence of gravity-induced palaeomagnetic inclination errors. This conclusion is particularly valid for sediments displaying extremely shallow inclinations that can not be induced by compaction alone (see next section). The source area control on the magnetic mineralogy points also to a detrital origin for the ferromagnetic particles.

Inclination shallowing and magnetic fabric

DRM inclination shallowing has been widely reported from natural detrital sediments and

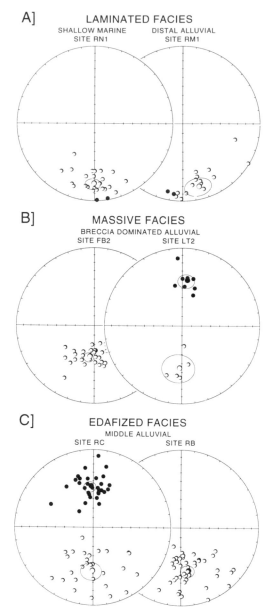

Fig. 5. ChRM directions of: (**a**) shallow marine and distal alluvial laminated facies; (**b**) breccia dominated massive facies; and (**c**) middle alluvial edafized facies. Estimated mean normal and reversed directions are enclosed by α_{95} error circles.

re-deposition experiments in the laboratory. Although compaction may be a source of inclination shallowing (Arason & Levi 1990), gravitational torques acting at the time of deposition may have a more important role in

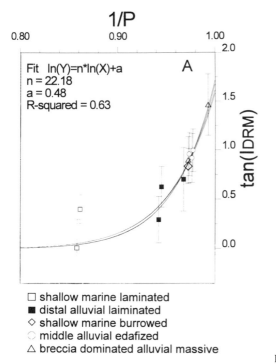

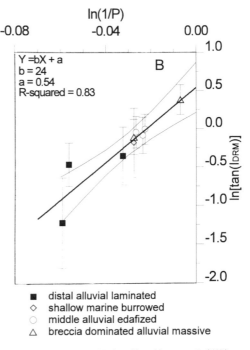

□ shallow marine laminated
■ distal alluvial laiminated
◇ shallow marine burrowed
○ middle alluvial edafized
△ breccia dominated alluvial massive

Fig. 6. Relationship of tangent of DRM inclination versus inverse of P. Points represent site mean data. Y-axis error bars are calculated from site mean ChRM α_{95} angular errors. Each symbol represents a particular facies type. Similar fits result from four different sets of data: (**a**) fitting the whole data set; (**b**) shallow marine laminated facies excluded; (**c**) breccia-dominated massive facies excluded; (**d**) and both (**b**) and (**c**) excluded. Calculated values of a and b for fit (**a**) are shown.

Fig. 7. Relationship of $\ln(\tan(I_{DRM}))$ versus $\ln(1/P)$. Points represent site mean data (strongly anisotropic shallow marine laminated facies are excluded). Error bars indicate ChRM α_{95} angular errors. Values of a, b and R^2 of the best linear fit are indicated. Lines above and below the linear fit represent the limits of the regression function confidence intervals.

this phenomena. Since strongly flattened magnetizations are found in present-day uncompacted sediments, there is no need to call upon lithostatic loading to explain inclination errors. From laboratory evidence (Deamer & Kodama 1990), compaction-induced shallowing is found to be mostly restricted to the early post-depositional stages, when a large volume loss occurs due to dewatering and rearrangement of platy clay particles (Deamer & Kodama 1990). Further lithostatic pressure does not produce more significant shallowing.

The maximum DRM inclination shallowings recorded in sediments in the present study are too large to be caused by lithostatic pressure. Moreover, neither the amount of inclination error nor the degree of AMS foliation correlate with the vertical location of sites in the lithostratigraphic framework (Fig. 1). Both magnetic parameters are linked with sedimentary facies,

which is basically dependent on the depositional and early post-depositional sediment history.

The link between magnetic remanence and AMS fabric has been studied in order to test the validity of using AMS data to correct inclination errors. Defining the flattening factor f from equation (1) as $(AMS_z/AMS_x)^b$ (following Collombat *et al.* 1993), and assuming that for site mean AMS ellipsoids the minimum susceptibility axis K_3 is perpendicular to bedding, then

$$\ln(\tan(I_{DRM})) = -b \times \ln(P) + \ln(\tan(I_H)) \quad (2)$$

where: I_{DRM} = DRM inclination; I_H = ambient field inclination; P is the degree of anisotropy, i.e. the ratio between maximum and minimum AMS principal axis (K_1/K_3); and b is a constant. The same equation can be written as

$$Y = b \times X + a \quad (3)$$

where $Y = \ln(\tan(I_{DRM}))$, $X = -\ln(P)$ and $a = \ln(\tan(I_H))$. Fitting this equation to the estimated site mean I_{DRM} and P of each of the 11 profiles studied here gives the result shown in Fig. 6. Error bars indicate the a_{95} of the estimated site mean directions. A positive correlation between both parameters is deduced with a coefficient of regression $R^2 = 0.63$. The best fit function consistent with equation (3) has $b = 22$ and $a = 0.48$. The regression does not change significantly if we omit data from the extremes of the data distribution (Fig. 6), indicating that no discrete data point has a decisive weight on the result. As we are concerned with the prediction capability towards decreasing anisotropy, we can obtain a better correlation by not considering the rare very strong anisotropic sediments (Fig. 7). The best fit of sites with $P < 1.1$ gives $b = 24$ and $a = 0.54$ with $R^2 = 0.8$, that is, 80% of variability of Y from equation (3) is attributed to variability of X.

If we consider the very extreme case of a perfectly isotropic rock, i.e. $P = 1$, no DRM inclination error should be expected and $a = \ln (\tan (I_H)) = \ln (\tan (I_{DRM}))$. Therefore, the true geomagnetic field inclination I_H may be calculated from the estimated value of a. The fitting function predicts a true field inclination of 60°, with lower to upper 95% confidence limits of 51° to 67°. Similar results can be predicted from correlation of discrete sample data instead of site mean data. The coefficient of regression, however, falls dramatically due to the unbalanced secular variation remaining in each individual sample.

It must be pointed out that the slope b of equation (3) is dependent on several intrinsic properties of sediments. The estimated regression is, therefore, only valid for data from this specific study. It cannot be applied to correct inclination shallowing from other localities without considering all the possible lithologic-dependent variables.

Conclusions

The occurrence of remanence inclination shallowing is not unusual in sediments, especially when dealing with recent alluvial sediments, where most remanence is thought to be carried by detrital grains. Inclination errors for a variety of fine-grained alluvial and shallow marine sediments from the Neogene Vallès–Penedès Basin have been observed to be dependent on their sedimentary facies. Undisrupted thinly laminated distal alluvial facies provide the most biassed inclination data. Post-depositionally reworked facies, such as subaerial edafized silts and underwater burrowed muds, have less clustered palaeomagnetic directions and more reliable steeper mean inclinations. The best estimate of the true geomagnetic field is,

however, recorded in sites consisting of mass flow deposits. Given their very low anisotropy, massive facies give the most reliable data, that almost approximate the expected true geomagnetic field inclination.

The strong facies control on the DRM can be investigated by means of AMS parameters. The degree of anisotropy P correlates to the I_{DRM} in such a way that the true field inclination can be estimated by extrapolation from the fitting function. To achieve a high coefficient of correlation site mean data are required, so that secular variation is averaged out. Since AMS parameters are controlled by several lithology-related variables such as paramagnetic content and anisotropy of magnetic particles, the regression achieved for this study cannot fit data from other sites. The dependence on such factors makes an estimation necessary of co-efficient b from equation (3) for each particular case.

This work was supported by a PhD scholarship from the Spanish Ministry of Education and Science, the CICYT project GEO89-0831, CAICYT project BB90-0575 and DGICYT project PB91-0096.

References

ARASON, P. & LEVI, S. 1990. Compaction and inclination shallowing in deep-sea sediments from the Pacific Ocean. *Journal of Geophysical Research*, **95**, B4, 4501–4510.

BARTRINA, M. T., CABRERA, L., JURADO, M. J., GUIMERÀ & ROCA, E. 1992. Evolution of the Catalan margin of the Valencia trough (western Mediterranean). *Tectonophysics*, **203**, 219–247.

CABRERA, L., CALVET, F., GUIMERÀ, J. & PERMANYER, A. 1991. *El registro sedimentario miocénico en los semigrabens del Vallès-Penedès y de El Camp: Organización secuencial y relaciones tectónica-sedimentación.* I Congreso del Grupo Español del Terciario, Libro-Guía Excursión no 4, Vic.

COLLOMBAT, H., ROCHETTE, P. & KENT, D. V. 1993. Detection and correction of inclination shallowing in deep sea sediments using the anisotropy of anhysteretic remanence. *Bulletin de la Societé Géologique de France*, **164**, 1, 103–111.

DEAMER, G. A. & KODAMA, K. P. 1990. Compaction-induced inclination shallowing in synthetic and natural clay-rich sediments. *Journal of Geophysical Research*, **95**, B4, 4511–4529.

JACKSON, M. J., BANERJEE, S. K., MARVIN, J. A., LU, R. & GRUBER, W. 1991. Detrital remanence, inclination errors, and anhysteretic remanence anisotropy: quantitative model and experimental results. *Geophysical Journal International*, **104**, 95–103.

JELINEK, V. 1981. Characterization of the magnetic fabrics of rocks. *Tectonophysics*, **79**, T63–T67.

KING, R. F. 1955. The remanent magnetism of

artificially deposited sediments. *Monthly Notices of the Royal Astronomical Society, Geophysics Supplement,* **7**, 115–134.

LEVI, S. & BANERJEE, S. 1990. On the origin of inclination shallowing in redeposited sediments. *Journal of Geophysical Research,* **95**, B4, 4383–4389.

LØVLIE, R. 1993. Experimental determination of the relationship between magnetic moment and grain geometry of PSD magnetite grains. *Physics of the Earth and Planetary Interiors,* **76**, 105–112.

—— & TORSVIK, T. 1984. Magnetic remanence and fabric properties of laboratory-deposited hematite-bearing red sandstone. *Geophysical Research Letters,* **11**, 3, 229–232.

LOWRIE, W. 1990. Identification of ferromagnetic minerals in a rock by coercivity and unblocking temperature properties. *Geophysical Research Letters,* **17**, 159–162.

ROCA, E. & GUIMERÀ, J. 1992. The Neogene structure of the eastern Iberian margin: structural constraints on the crustal evolution of the Valencia trough (western Mediterranean). *Tectonophysics,* **203**, 203–218.

TAUXE, L. & BADGLEY, C. 1988. Stratigraphy and remanence acquisition of a palaeomagnetic reversal in alluvial Siwalik rocks of Pakistan. *Sedimentology,* **35**, 697–715.

—— & KENT, D. V. 1983. Properties of a detrital remanence carried by hematite from a study of modern river deposits and laboratory redeposition experiments. *Geophysical Journal Royal Astronomical Society,* **77**, 543–561.

TUCKER, P. 1980. A grain mobility model of post-depositional realignment. *Geophysical Journal Royal Astronomical Society,* **63**, 149–163.

—— 1987. Magnetization of unconsolidated sediments and theories of DRM. *In*: CREER, K. M., TUCHOLKA, P. & BARTON, C. E. (eds) *Geomagnetism of baked clays and recent sediments.* Elsevier, Amsterdam, 9–28.

Consequences of post-collisional deformation on the reconstruction of the East Pyrenees

P. KELLER[1] & A. U. GEHRING[2]

[1] *Departamento de Geofísica y Meteorología, Universidad Complutense, Ciudad Universitaria, 28040 Madrid, Spain*

[2] *Department of Environmental Science, Policy and Management, University of California, Berkeley, CA 94720, USA*

Abstract: Post-collisional structures are generally ignored in tectonic reconstructions of the Pyrenees. The South Pyrenean thrust belt is characterized by different tectonic rotations about vertical axes. A new statistical analysis of palaeomagnetic data shows that the observed rotations are homogeneous within discrete zones which only partly coincide with the tectonic units of the thrust belt. The combination of the palaeomagnetic information with structural and sedimentological data demonstrates that the actual structure of the southeastern Pyrenees is due to interference of Palaeogene (collisional) and Neogene (post-collisional) deformations. Accounting for this structural interference leads to significantly less shortening than calculated by earlier reconstructions of the southeastern Pyrenees. This example clearly shows that the small contribution made by post-collisional tectonics to the overall deformation can strongly influence the structure of orogenic belts.

Studies of different mountain belts have shown that orogenic processes continue for some time after plate collision. A classic example is the Himalaya, where episodic uplift and extension post-date plate collision (e.g., Harrison *et al.* 1992). Most investigations of such post-collisional processes have been concerned with the emplacement of magmatic material (Anovitz & Chase 1990; Turner *et al.* 1992), the collapse of orogens (i.e., post-collisional extension; Dewey 1987; Malavieille & Taboada 1991), and the uplift that follows compressive tectonics (Sengör & Kidd 1979; Molnar *et al.* 1993). Post-collisional compressive or transcurrent tectonics have been relatively neglected. This is partly because the amount of post-collisional deformation is rather small, and also because the younger structures are often difficult to distinguish from the older ones. Therefore post-collisional deformation is often ignored in structural models and palinspastic reconstruction of orogenic belts.

The Pyrenees in southwestern Europe were primarily formed by collision between the European and the Iberian plates during the Cretaceous to Early Neogene (Muñoz 1992). Apart from the predominant thrust tectonics, Neogene pull-apart basins in the eastern Pyrenees indicate that post-collisional deformation has occurred (Cabrera *et al.* 1988). Several palaeomagnetic studies conducted in the southeastern Pyrenees demonstrate that the area has undergone tectonic rotations about vertical axes (Dinarés 1992; Burbank *et al.* 1992). In some areas it has been demonstrated that the rotation was subsequent to thrust tectonics (Keller & Gehring 1992).

In order to reconstruct post-collisional tectonics in the southeastern Pyrenees, structural information is combined with published palaeomagnetic data which were obtained from Permian to Oligocene sediments. A new methodological approach for quantifying tectonic rotation is outlined which compensates the palaeomagnetic data for plate motion during this period. As a result an accurate restoration of the Palaeogene thrust geometry can be described. The consequences of the post-collisional deformation for the palinspastic and structural reconstruction of the Pyrenees are discussed.

Geological setting

The structure of the Pyrenees is characterized by northern and southern foreland belts, which are thrust onto the European and Iberian continental margins, respectively (Roure *et al.* 1989). The Axial Zone, a complex of Palaeozoic basement rocks, separates the two foreland belts (Fig. 1a). The southern foreland belt of the central and eastern Pyrenees is divided into two structurally different parts by the Segre Transform Zone (STZ) (Fig. 1b). The western part, the South Central Unit (SCU; Seguret 1972) consists of

From Morris, A. & Tarling, D. H. (eds), 1996, *Palaeomagnetism and Tectonics of the Mediterranean Region*, Geological Society Special Publication No. 105, pp. 101–109.

101

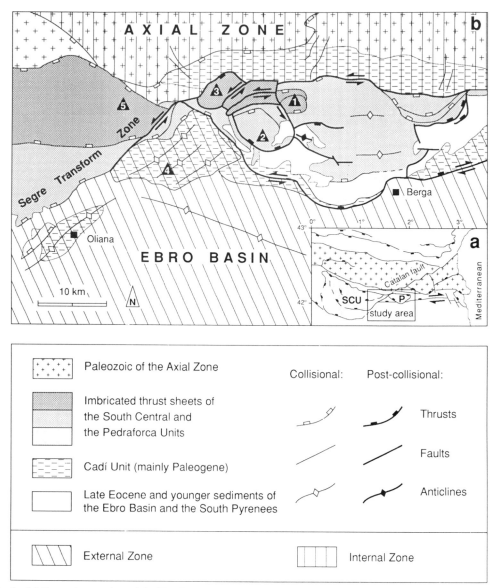

Fig. 1. (a) Tectonic overview of the East Pyrenees and location of study area. SCU, South Central Unit; P, Pedraforca Unit. **(b)** Tectonic map of the South Central Unit and the Pedraforca Unit. The subdivision into the External and the Internal Zone is based on sense of rotation about vertical axes, derived from palaeomagnetic data (Keller *et al.* 1994). Neogene structures are indicated with bold lines; Palaeogene and older structures with fine lines. Triangles signify orientation points for the restoration in Figure 4: 1, Pedraforca; 2, Sierra del Verd; 3, Puig Galliner; 4, Port del Compte; 5, Montan.

Mesozoic sediments that form three major thrust sheets. The eastern part consists of two tectonic units, the Pedraforca and the Cadí Units (Fig.1b; Muñoz *et al.* 1986). The Pedraforca Unit is the structural and stratigraphic equivalent of the SCU (Ullastre *et al.* 1987; Vergés *et al.* 1992), and as with the SCU, it can be

subdivided into three thrust sheets (Vergés & Martínez 1988). The Cadí Unit is the structurally lowermost and youngest thrust sheet of the southeastern foreland belt. This unit is distinguished from the Pedraforca and the SCU by its reduced Mesozoic stratigraphy and the presence of a thick sequence of Palaeogene

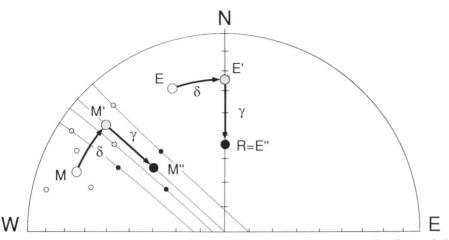

Fig. 2. Example of the compensation of palaeomagnetic data for plate motion. The data (small open circles) are of the same magnetisation age. The compensated directions (small black circles) are referred to as tectonic directions and are used in the structural reconstruction. M, mean direction of magnetic measurements; E, calculated coeval expected direction; R, reference direction.

sediments. The last movements in the southeastern Pyrenees related to Pyrenean collision are documented by early Oligocene folds in the Ebro basin (e.g., Williams 1985; Muñoz 1992). Neogene pull-apart basins in the Axial Zone (Pous *et al.* 1986; Cabrera *et al.* 1988) and deformed Early Miocene conglomerates on the Pedraforca Unit (Durand-Delga *et al.* 1989) indicate post-collisional deformation of the Pyrenees.

Keller *et al.* (1994) used palaeomagnetic data to map zones of tectonic rotation in the southeastern Pyrenees. The Internal Zone combines the northern part of the Cadí Unit and the Axial Zone, and shows no tectonic rotation. The External Zone consists of the southern part of the Cadí Unit and the adjacent folded Ebro basin, and exhibits an anticlockwise rotation (Fig. 1b).

Palaeomagnetic data analysis

Method to compensate for plate motion

Palaeomagnetic studies in the southeastern Pyrenees have been performed on a variety of lithologies of Permian to Oligocene age (Van Dongen 1967; Burbank *et al.* 1992; Dinarès *et al.* 1992; Keller & Gehring 1992). The palaeomagnetic directions may reflect both intra-plate deformation and large-scale plate motion. In order to separate the effect of intra-plate deformation it is necessary to compensate the data for the plate motion that has occurred since the rocks were magnetised. This can be achieved using the following procedure.

First, an arbitrary reference direction R is

chosen. For data with a coeval magnetisation age a mean direction M is calculated, and an expected direction E is determined, based on the appropriate palaeopole (Fig. 2). The deviation of E from R is expressed by two angles, δ for the declination, and γ for the inclination. Rotating about δ and γ transforms E to E" via E', whereby E" = R (Fig. 2). Second, the rotation about δ and γ is applied to the population of directions with the mean M. This transforms M to M" via M', by preserving the angular distribution of the population. M has the same angular relation to E as M" to R. Since R is arbitrary, the directions with M" have no palaeomagnetic significance, and therefore are referred to as *tectonic directions*. The deviation of the tectonic directions from R is a measure of intra-plate deformation.

To quantify intra-plate deformation of an area, tectonic directions of different magnetization ages are combined, and a mean tectonic direction is calculated using Fisher's (1953) statistics. The use of these statistics implies a rotational symmetric Gaussian distribution of directions around their mean (Fisherian distribution). The combination of populations with different Fisherian distributions, does not presuppose a new Fisherian distribution. Therefore the populations of tectonic directions have first been tested to ensure that they have an adequate fit to a Fisherian distribution (Watson & Irving 1957).

Pattern and dating of tectonic rotation

The tectonic directions for the southeastern Pyrenees were calculated for the Pedraforca

Table 1. *Palaeomagnetic data and their compensation for plate motion*

Site mean Dec/Inc	References	Age	Expected Dec/Inc	δ	γ	Tectonic Dec/Inc	χ_A^2/χ_θ^2	Mean	
Internal Zone									
350/03	5	eP	339/−2	21	47	011/50			
342/04	5	lP	350/12	10	33	353/37			
333/33	5	lP	350/12	10	33	341/56		Dec:	357
341/24	5	eT	337/22	23	23	007/47	0.0/0.6	Inc:	48
327/28	3	PT	337/22	23	23	348/51		α_{95}:	7.6
325/23	3	PT	337/22	23	23	346/46		k:	53.5
012/39	2	mE	003/49	−03	−04	009/35			
003/61	3	lE	003/49	−03	−04	001/57			
External Zone									
331/65	3	eO	006/57	−06	−12	325/51			
352/54	3	eO	006/57	−06	−12	341/41			
330/52	3	eO	006/57	−06	−12	324/38			
329/64	3	eO	006/57	−06	−12	324/50			
317/62	3	eO	006/57	−06	−12	316/48		Dec:	336
329/53	3	eO	006/57	−06	−12	324/39	0.2/0.9	Inc:	39
350/32	2	mE	003/49	−03	−04	347/28		α_{95}:	6.0
342/28	2	mE	003/49	−03	−04	339/24		k:	43.3
337/27	2	mE	003/49	−03	−04	334/23			
354/40	1	mE	003/49	−03	−04	351/36			
346/48	1	mE	003/49	−03	−04	343/44			
355/40	1	mE	003/49	−03	−04	352/36			
344/48	1	eE	003/49	−03	−04	341/44			
Pedraforca Unit									
025/51	3	lE	003/49	−03	−04	023/47		Dec:	034
051/40	3	lE	003/49	−03	−04	047/36	0.0/0.8	Inc:	38
027/35	3	lE	003/49	−03	−04	024/31		α_{95}:	7.7
038/40	4	lE	003/49	−03	−04	035/36		k:	83.5

Reference: 1, Burbank *et al.* (1992); 2, Dinarès *et al.* (1992); 3, Keller (1992); 4, Keller & Gehring (1992); 5, Van Dongen (1967).

All NRM directions are converted into lower hemisphere polarity.

The expected directions were calculated for the sampling location using published palaeopoles for the age of magnetization: Early Permian (eP) palaeopole is from Van der Voo (1990), Permo-Triassic (P-T) and Early Triassic (eT) palaeopoles are from VandenBerg & Zijderveld (1982), and Late Permian (lP), Late–Early Eocene (l-eE) and Early Oligocene (eO) from Westphal *et al.* (1986).

The fit of tectonic directions to the Fisherian distribution was tested with χ^2-tests for azimuthal (χ^2_A), and the angular (χ^2_θ) distribution (Watson & Irving 1957).

Unit, the Internal, and the External Zone separately (Table 1). The reference direction used in this study is due north with an inclination of 45° ($R = 0°/45°$). The mean tectonic direction for the Internal Zone shows no significant difference from R. This suggests that no tectonic rotation has occurred since the rocks were magnetised (Fig. 3). In contrast the other two zones show significantly different tectonic mean directions that indicate rotations about vertical axes. The Fisherian distribution of tectonic directions argues in favour of homogeneous rotation within each zone. It also demonstrates that rotation occurred after acquisition of the youngest magnetisation (Table 1). The External Zone reveals an anticlockwise rotation of 24°,

and in the Pedraforca Unit a clockwise rotation of 34° (Fig. 3 & Table 1).

In the External Zone, the palaeomagnetic data were obtained from sediments of Early Eocene to Early Oligocene age (Table 1). The anticlockwise rotation of this zone is thus younger than Early Oligocene, and has been explained by differential compression during the last stage of Pyrenean convergence in the Mid-Oligocene (Vergés & Burbank in press).

The palaeomagnetic data from the Pedraforca Unit were obtained from a weathered Cretaceous limestone bounded by an erosion horizon. This horizon is partially covered by conglomerates which also seal the contact between the Pedraforca and the Cadí Unit (Fig.

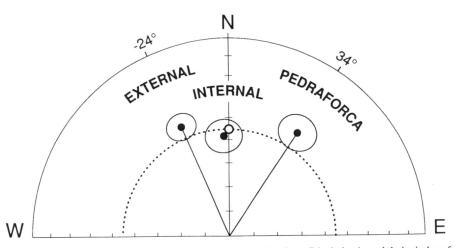

Fig. 3. Equal-area stereographic projection of tectonic mean directions (black dots), and their circles of confidence (α_{95}) that show the rotation pattern of the three tectonic zones in the southeastern Pyrenees. The reference direction (open circle) used for the compensation of the data is $R = 0°/45°$. Data are from Table 1.

1b). Keller & Gehring (1992) demonstrated that the magnetization in this limestone results from chemical weathering. This was achieved by reconstruction of the palaeoweathering horizon using the dip of the overlying conglomerates. Thus, the age of magnetization is younger than the emplacement of the Pedraforca Unit (Mid-Eocene) but older than the deposition of the conglomerates. In the frontal part of the Pedraforca Unit the conglomerates are syn-thrusting in age and strongly deformed (e.g. Riba 1976). Only the uppermost unit of the conglomeratic succession extends above an erosional horizon onto the Pedraforca thrust sheet (Megías & Posadas 1981). The lowest part of these frontal conglomerates is dated as mid-Eocene (Early Lutetian) (Seguret 1972; Martínez et al. 1988; Puigdefàbregas & Souquet 1986). Masriera & Ullastre (1985) obtained a Late Eocene age for the upper conglomeratic unit. Furthermore, the erosional horizon on the Pedraforca Unit is correlated with an erosional horizon in the Ebro Basin (Megías & Posadas 1981). There the horizon is covered by con-glomerates that are dated as Mid-Oligocene from fossil records (Riba 1976). A Mid-Oligocene or younger age of the conglomerates which cover the erosion horizon on the Pedra-forca Unit is also confirmed by palynological data (Durand-Delga et al. 1989). These data imply that the dextral rotation of the Pedraforca Unit is of Mid-Oligocene or younger age, and is therefore post-thrusting (see also Keller et al. 1994). In the Axial Zone a few kilometres north of the Pedraforca Unit, pull-apart basins which

formed during the Neogene as the result of dextral strike-slip faulting (Pous et al. 1986; Cabrera et al. 1988) document post-collisional deformation. Thus, it is very likely that the post-thrusting clockwise rotation in the south-eastern Pyrenees is also of Neogene age.

Post-collisional deformation

Structural evidence in the Pedraforca Unit

Post-collisional tectonics in the Pedraforca Unit are indicated by deformation of the mid-Oligocene and younger conglomerates (Guérin-Desjardins & Latreille 1961; Durand-Delga et al. 1989; Keller & Gehring 1992). Between the Pedraforca Unit and the Internal Zone the conglomerates are truncated by a left-lateral fault, which reactivates the thrust contact (Ull-astre et al. 1987; Durand-Delga et al. 1989). In the northwestern part of the Pedraforca Unit, the conglomerates are offset by NE–SW-trending left-lateral faults that divide the Lower Cretaceous limestone into three blocks (Fig. 1b, location 3). These strike-slip faults are cut by a NNW–SSE-trending fault which also affects the conglomerates (Durand-Delga et al. 1989). Towards the south, this fault ends in a thrust that superimposes Mesozoic sediments on to con-glomerates (Fig. 1b, location 2). East of the NNW–SSE-trending fault the conglomerates are folded and cut by sub-vertical faults (Keller & Gehring 1992). The deformation along these faults decreases towards the central part of the

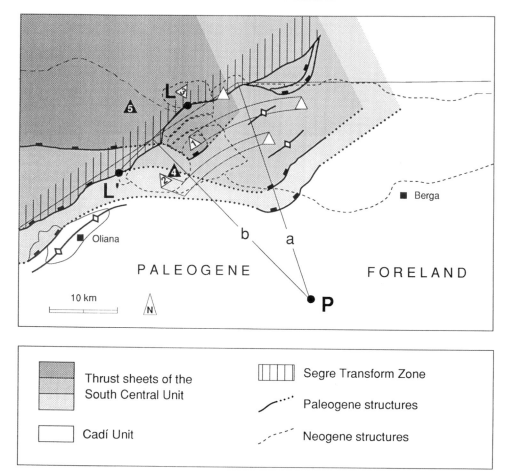

Fig. 4. Reconstruction of the Palaeogene thrust geometry in the southeastern Pyrenees by rotating the Pedraforca Unit 34° anticlockwise, and restoring the post-collisional structures. Locations with white (black) triangles were (not) affected by post-collisional deformation. The position of the black and the unnumbered white triangles corresponds with Fig. 1.

Pedraforca Unit, where no post-thrusting deformation was found. Mesozoic sediments are thrust onto conglomerates in the eastern part of the Pedraforca Unit (Fig. 1b; Solé Sugrañes 1973; Martínez *et al.* 1988). In the south, a steep fault (dip > 50° c.f. Megías & Posadas 1981) forms a complex contact that interferes with the Palaeogene thrust geometry (Fig. 1b). Near Berga this fault separates Mesozoic sediments of the Pedraforca Unit from Oligocene conglomerates of the Ebro Basin (Guérin-Desjardins & Latreille 1961; Losantos *et al.* 1989). West of Berga this fault forms the contact between the Cadí Unit and the conglomerates of the Ebro Basin or forms a structure within the Cadí Unit (Fig. 1).

Palaeomagnetic data from the southeastern

Pyrenees show the Pedraforca Unit as an isolated block with a large post-thrusting rotation compared to the External and the Internal Zones (Figs 1b and 3). The post-collisional structures seen in Fig. 1b can be explained by clockwise rotation of this block. The northern and western margin of the block are left-lateral faults which reactivate the thrust contacts (Ullastre *et al.* 1987), as is expected for clockwise post-thrusting rotation. The southern margin is represented by a steep transpressive fault that interferes with the Palaeogene thrust geometry (Fig. 1b). In the following discussion the rotated block is referred to as the *Pedraforca Block* because it consists mainly of this unit. The internal, post-thrusting deformation of the block shows a dominance of left-lateral strike-slip faults, thrusts and folds. The distribution of these transpressive structures reveals intense deformation at the north-western and south-eastern corners, but no deformation in the central part of the block. Such a structural pattern is characteristic of spatial complications due to clockwise rotation (Terres & Sylvester 1981). For example, the sinistral shear between three blocks of Lower Cretaceous limestone in the north-western part of the Pedraforca Unit (Fig. 1b) can be explained by movement opposing the sense of rotation of the Pedraforca Block. The three-dimensional shape of the Pedraforca Block is a result of a complex interference of post-thrusting wrench tectonics with the Palaeogene thrust geometry.

Reconstruction of the Palaeogene thrust geometry

Reconstruction of the Palaeogene thrust geometry requires back-rotation of the Pedraforca Block and restoration of the post-collisional structures. The Pedraforca Block was restored by a 34° anticlockwise rotation about the pole *P* (Fig. 4). *P* was determined graphically as the intersection of line *a* and *b*. Line *a* is the angular bisector between the lineations of the STZ and the northern border of the Pedraforca Unit. These two left-lateral faults were active during the rotation, and can thus be considered as tangents to the rotational movement of the block. Line *b* is the centre normal between location *L* and its restored position *L'*, based on a Cretaceous facies reconstruction along the STZ (Ullastre *et al.* 1987). In addition the elements of the Pedraforca Unit that show post-collisional deformation were restored by compensating for thrusting (i.e. at location 2), and by moving along strike-slip faults (i.e., at location 3). The resulting pattern in Fig. 4 is area balanced in the map plane.

Consequences for the reconstruction of the Pyrenees

The restoration of the post-collisional deformation rearranges the Palaeogene thrusts and folds in the Pedraforca Unit oblique to the dominant E–W strike of structures in the Pyrenees, but parallel to the NE–SW-striking STZ, as well as to folds and thrusts at Oliana and Port del Compte (Fig. 1). Several authors interpreted the STZ during the Palaeogene as a lateral ramp, and the oblique structures east of it as the result of southward thrusting over the lateral ramp (e.g., Vergés & Muñoz 1990). Accordingly, the restored Pedraforca Unit can

be interpreted as part of the SCU that was thrust during the Palaeogene over the STZ (Fig. 4). This reconstruction agrees well with sedimentological data that emphasize the similarity of the Cretaceous strata from the SCU and the Pedraforca Unit (Souquet 1977; Ullastre *et al.* 1987). Furthermore, the imbrication of both tectonic units into three thrust sheets (Vergés & Burbank, in press; Fig. 1) suggests a structural relationship during Palaeogene thrust tectonics. One consequence of the above reconstruction is that the thick sequence of Cretaceous limestone of the SCU and the Pedraforca Unit had to be deposited to the northwest, and the reduced Cretaceous of the Cadí Unit to the southeast of the STZ. This indicates that the STZ and its prolongation into the Catalan Fault (Fig. 1a) are the palaeogeographic boundaries responsible for the variable stratigraphic development of the Cretaceous in the Pyrenees (Souquet & Mediavilla 1976; Souquet 1977).

The close agreement of structural, palaeomagnetic, and sedimentological data clearly shows that the actual structure of the southeastern Pyrenees is the result of an interference between collisional and post-collisional deformations. Previous reconstructions of the Pyrenees only take into account collisional tectonics (e.g., Williams 1985; Vergés & Martínez 1988; Muñoz 1992), and do not consider rotation about vertical axes, as observed in the Pedraforca Unit (Keller & Gehring 1992). Another consequence of the interference between collisional and post-collisional deformation is a disturbed initial geometry for the calculation of shortening caused by plate collision. In the southeastern Pyrenees a massive reduction of shortening is estimated if the Pedraforca Unit is not restored to the northern end of the Cadí Unit, but re-rotated and structurally combined with the SCU (Fig. 4).

This example from the southeastern Pyrenees demonstrates that post-collisional tectonics can have a strong influence on the geometry of orogenic belts, even though they usually represent only a small part of the deformation.

The authors wish to thank Y. Porchet for field assistance, W. Alvarez, A. Hirt and W. Lowrie for helpful discussions, and M. Ford and J.A. Tait for critical reading of the manuscript and their constructive comments. This research was supported by the Swiss National Science Foundation (grants 2000-5.617 and 8220-028438).

References

ANOVITZ, L. M. & CHASE, C. G. 1990. Implications of post-thrusting extension and underplating for

P-T-t paths in granulite terranes; a Grenville example. *Geology,* **18**, 466–469.

BURBANK, D. W., PUIGDEFÀBREGAS, C. & MUÑOZ, J. A. 1992. The chronology of the Eocene tectonic and stratigraphic development of the eastern Pyrenean foreland basin, northeast Spain. *Geological Society of America Bulletin,* **104**, 1101–1120.

CABRERA, L., ROCA, E. & SANTANACH, P. 1988. Basin formation at the end of a strike-slip fault: the Cerdanya Basin (eastern Pyrenees). *Journal of the Geological Society, London,* **145**, 261–268.

DEWEY, J. F. 1987. Extensional collapse of orogenes. *Tectonics,* **7**, 1123–1139.

DINARÈS, J. 1992. *Paleomagnetisme a les unitats sudpirinenques superiors. Implicacions estructurals.* PhD thesis, University of Barcelona.

——, MCCLELLAND, E. & SANTANACH, P. 1992. Contrasting Rotations within Thrust Sheets and Kinematics of Thrust Tectonics as derived from Palaeomagnetic data; an example from the Southern Pyrenees. *In:* MCCLAY, K. (ed.) *Thrust tectonics.* Chapman and Hall, London, 265–275.

DURAND-DELGA, M., MÉON, H., MASRIERA, A. & ULLASTRE, J. 1989. Effets d'une tectonique compressive, affectant du Miocène supérieur, daté palynologiquement, dans la zone de la Pedraforca (Pyrénées Catalanes, Espagne). *Comptes Rendus de l'Academie des Sciences, Paris,* **308**, série ll, 1091–1098.

FISHER, R. A. 1953. Dispersion on a sphere. *Proceedings of the Royal Society of London,* **A217**, 295–317.

GUÉRIN-DESJARDINS, B. & LATREILLE, M. 1961. Etude géologique dans les Pyrénées espagnoles entre les Fleuves Segre et Llobregat. *Revue de l'Institute Francaise de Pétrolgie,* **16**, 922–940.

HARRISON, T. M., COPELAND, P., KIDD, W. S. F. & YIN, A. 1992. Raising Tibet. *Science,* **225**, 1663–1670.

KELLER, P. 1992. *Paläomagnetische und strukturgeologische Untersuchungen als Beitrag zur Tektogenese der SE-Pyrenäen.* PhD thesis, Zürich, ETH.

—— & GEHRING, A. U. 1992. Different weathering stages indicated by the magnetisation of limestone: An example from the southeast Pyrenees, Spain. *Earth and Planetary Science Letters,* **111**, 49–57.

——, LOWRIE, W. & GEHRING, A. U. 1994. Palaeomagnetic evidence for post-thrusting rotation in the Southeast Pyrenees. *Tectonophysics,* **293**, 29–42.

LOSANTOS, M., ARAGONÈS, E., BELÁSTEGUI, X., PALAU, J. & PUIGDEFÀBREGAS, C. 1989. *Mapa Geològic de Catalunya (1 : 250 000).* Servei Geològic de Catalunya, Barcelona.

MALAVIEILLE, J. & TABOADA, A. 1991, Kinematic model for post-orogenic Basin and Range extension. *Geology,* **19**, 555–558.

MARTÍNEZ, A., VERGÉS, J. & MUÑOZ, J. A. 1988. Secuencias de propagación del sistema de cabalgamientos de la terminación oriental del manto del Pedraforca y relación con los conglomerados sinorogenicos. *Acta Geologica Hispanica,* **23**, 357–396.

MASRIERA, A. & ULLASTRE, J. 1985. Puntualizacion acerca de las relaciones entre el Eoceno marino de Montcalb-La Corriu, el de Sant Llorenç de Monrunys y los conglomerados continentales encajantes (Pirineo Catalan). *Estudios Geologicos,* **41**, 385–390.

MEGÍAS, A. G & POSADAS, M. 1981. Precisiones sobre la colocation del manto de Pedraforca (Pirineo oriental, España). *Estudios Geologicos,* **37**, 221–225.

MOLNAR, P., ENGLAND, P. & MARTINOD, J. 1993. Mantel Dynamics, Uplift of the Tibetian Plateau, and the Indian Monsoon. *Reviews of Geophysics,* **31**, 357–396.

MUÑOZ, J. A. 1992. Evolution of a Continental Collision Belt. ECORS-Pyrenees Crustal Balanced Cross-section. *In:* MCCLAY, K. (ed.) *Thrust tectonics.* Chapman & Hall, London, 235–246.

——, MARTÍNEZ, A. & VERGÉS, J. 1986. Thrust sequences in the eastern Spanish Pyrenees. *Journal of Structural Geology,* **8**, 399–405.

POUS, J., JULIA, R. & SOLÉ-SUGRAÑES, L. 1986. Cerdanya Basin geometry and its implication on the Neogene evolution of the eastern Pyrenees. *Tectonophysics,* **129**, 355–365.

PUIGDEFÀBREGAS, C. & SOUQUET, P. 1986. Tectosedimentary cycles and depositional sequences of the Mesozoic and Tertiary from the Pyrenees. *Tectonophysics,* **129**, 173–203.

RIBA, O. 1976. Syntectonic unconformities of the Alto Cardener, Spanish Pyrenees. *Sedimentary Geology,* **15**, 213–233.

ROURE, F., CHOUKROUNE, P., BERASTEGUI, X., MUÑOZ, J. A., VILLIENA, A., MATHERON, P., BAREYT, M., SEGURET, M., CÁMARA, P. & DERAMOND, J. 1989. ECORS deep seismic data and balanced cross-sections; Geometric constraints on the evolution of the Pyrenees. *Tectonics,* **8**, 41–50.

SEGURET, M. 1972. *Etude tectonique des nappes et séries décollées de la partie central du versant sud des Pyrénées.* Publication US-TELA, série Geologie Structural, 2, Montpellier.

SENGÖR, A. M. C. & KIDD, W. S. F. 1979. Postcollisional tectonics of the Turkish-Iran Plateau and a comparison with Tibet. *Tectonophysics,* **55**, 361–376.

SOLÉ SUGRAÑES, L. 1973. Algunos aspectos de la tectónica del Prepirineo Oriental entre los ríos Segre y Llobregat. *Acta Geologica Hispanica,* **8**, 81–89.

SOUQUET, P. 1977. *Le Crétacé supérieur sud-pyrénéen en Catalogne, Aragon et Navarre.* PhD thesis, Toulouse.

—— & MEDIAVILLA, F. 1976. Nouvelle hypothése sur la formation des Pyrénées. *Comptes Rendus de l'Academie des Sciences, Paris,* **282**, 2139–2142.

TERRES, R. R. & SYLVESTER, A. G. 1981. Kinematic analysis of rotated fractures and blocks in simple shear. *Bulletin of the Seismological Society of America,* **71**, 1593–1601.

TURNER, S., SANDIFORD, M. & FODEN, J. 1992. Some geodynamic and compositional constraints on "postorogenic" magmatism. *Geology,* **20**, 931–934.

ULLASTRE, J., DURAND-DELGA, M. & MASRIERA, A. 1987. Argumentos para establecer la estructura del sector del pico de Pedraforca a partir del análisis comparativo del Cretáceo de este macizo con el de la región de Sallent (Pirineo catalán). *Bolletin Geológico y Minero*, **48**, 3–22.

VAN DER VOO, R. 1990. Phanerozoic palaeomagnetic poles from Europe and North America and comparison with continental reconstructions. *Reviews of Geophysics*, **28**, 167–206.

VAN DONGEN, P. G. 1967. The rotation of Spain; Palaeomagnetic evidence from the eastern Pyrenees. *Palaeogeography, Palaeoclimatology, Palaeoecology*, **3**, 417–432.

VANDENBERG, J. & ZIJDERVELD, J. D. A. 1982. Paleomagnetism in the Mediterranean area. Alpine Mediterranean Geodynamics. *Geodynamic Series, American Geophysical Union*, **7**, 83–112.

VERGÉS, J. & BURBANK, D. in press. Eocene–Oligocene thrusting and basin configuration in the eastern and central Pyrenees (Spain). *In*: FRIEND, P. & DABRIO, C. (eds) *Tertiary Basins in Spain*.

—— & MARTÍNEZ, A. 1988. Corte compensado del Pirineo oriental. Geometría de las cuencas de antepaís y edades de emplazamiento de los mantos de corrimiento. *Acta Geologica Hispanica*, **23**, 95–105.

—— & MUÑOZ, J. A. 1990. Thrust sequences in the southern central Pyrenees. *Bulletin de la societé géologique de la France*, VI, **8**, 265–271.

——, —— & MARTÍNEZ, A. 1992. South Pyrenean fold- and thrust-belt: Role of foreland evaporitic levels in thrust geometry. *In*: McCLAY, K. (ed) *Thrust Tectonics*. Chapman and Hall, London, 255–264.

WATSON, G. S. & IRVING, E. 1957. Statistical methods in rock magnetism. *Monthly Notices of the Royal Astronomical Society, Geophysical Supplement*, **7**, 289–300.

WESTPHAL, M., BAZHENOU, M. L., LAUER, J. P., PECHERSKY, D. M. & SIBUET, J. C. 1986. Palaeomagnetic implications on the evolution of the Tethys belt from the Atlantic Ocean to the Pamirs since the Triassic. *Tectonophysics*, **123**, 37–83.

WILLIAMS, G. D. 1985. Thrust tectonics in the south central Pyrenees. *Journal of Structural Geology*, **7**, 11–17.

Palaeomagnetic dating and determination of tectonic tilting: a study of Mesozoic–Cenozoic igneous rocks in central West Portugal

J. L. PEREIRA[1,2], A. RAPALINI,[1,3] D. H. TARLING[1] & J. FONSECA[4]

[1]*Department of Geological Sciences, University of Plymouth, Plymouth PL4 8AA, UK*

[2]*Present address*: *School of Environmental Sciences, University of East Anglia, Norwich NR4 7TJ, UK*

[3]*Present address*: *Departamento de Ciencias Geologicas, Faculdad de Cencos Exactas y Naturales (UBA), Pabellon 2, Ciudad Universitaria, 1428 Buenos Aires, Argentina*

[4]*Departamento de Fisica, Instituto Superior Ticnico, 1096 Lisboa, Portugal*

Abstract: Samples of igneous rocks (36 sites, 218 samples) associated with two main phases of the evolution of the Lusitanian Basin can be dated magnetically to latest Jurassic–early Cretaceous or late Cretaceous–Eocene. Directions of remanence in rocks belonging to the older phase require a 35° rotation to become consistent with directions from stable Europe, while those of younger age do not require such rotation. On a finer scale, the remanent directions can be used to ascertain the probable tectonic tilt at different times of the igneous activity and hence provide intermediate evidence on the tectonic evolution of this area, including rotations of individual tectonic blocks about both vertical and horizontal axes.

The Iberian microplate is one of the South European micro-continents. It underwent a transpressive phase during the last phases of the Hercynian orogeny, with the formation of NNE–SSW-trending sinistral strike-slip faults and NNW–SSE-trending dextral strike-slip faults. This was followed by an extensional phase in the Jurassic related to the opening of the North Atlantic. The Bay of Biscay opened later as Iberia rotated away from Europe, following the formation of Magnetic Oceanic Anomaly M6 (*c.* 127 Ma) and was completed by Anomaly 25 (*c.* 59 Ma) times, subsequent to which the Iberian microplate has largely moved as part of the European block. During the Late Alpine Orogeny, another extensional phase was followed by a tectonic inversion (Ribeiro *et al.* 1980). Each of these phases imprinted structures on both the on- and off-shore sectors of the Iberian peninsula. The response to these post-Hercynian tectonic phases was commonly the reactivation of pre-existing Hercynian features that still influence the distribution of present-day microseismic activity (Fonseca 1989).

The Lusitanian Basin (Fig. 1) is one of the major tectonic structures that originated during the Late Hercynian and NNE–SSW-trending dextral fault planes of this age were re-activated during the Late Triassic extensional regime to form a succession of westward dipping tilt blocks (Guiry 1984). Evaporites were deposited on these blocks during the Lower Jurassic (Ribeiro *et al.* 1980) and, in the Late Jurassic, these formed diapirs that moved through the overlying thick sedimentary layers, mostly along pre-existing fault planes. (These still form salt walls within fault systems, as at Caldas da Rainha and Rio Maior – site RM2.) This halokinetic activity was associated with the intrusion of some sills and dykes; the first of the two main magmatic stages to affect the Basin. This initial igneous activity extended southwards to at least Sesimbra, where salt diapiric structures are closely related to the dyke swarms. The second phase of magmatic activity was between the Late Cretaceous and Early Tertiary (Eocene) and occurred mostly in the Lisbon region (known as the Lisbon Volcanic Complex) with associated plutonism in the Sintra region (NW of Lisbon). This activity is generally presumed to be related to an Early Alpine stage (Aires-Barros 1979). Both magmatic phases can be distinguished petrologically, but some magmatic activity is thought to have persisted between the two main phases.

This paper presents results of a palaeomagnetic study of these igneous rocks and baked contacts. This is an extension of an initial study (Pereira *et al.* 1992a, b; Tarling *et al.* 1992) which demonstrated a stable remanence in some localities that enabled a palaeomagnetic distinction between the two main igneous episodes.

From Morris, A. & Tarling, D. H. (eds), 1996, *Palaeomagnetism and Tectonics of the Mediterranean Region*, Geological Society Special Publication No. 105, pp. 111–117.

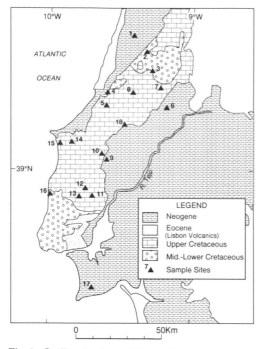

Fig. 1. Outline Geological Map of the Lusitanian Basin and the Sampling Sites. The site numbers are: 1, Mte Redondo; 2, Mte Real; 3, Leiria; 4, Nazare; 5, Famalicao; 6, Cela; 7, Alqueidào Serra; 8, Porto de Mós; 9, Cercal; 10, Serra Todo Mundo; 11, S. Socorro ; 12, Alqueidào; 13, Runa; 14, Rolica; 15, Papõa; 16, Ribamar; 17, Sesimbra; 18, Rio Maior

The igneous rocks which formed during the earlier episode have remanences that showed an anticlockwise rotation by about 35° with respect to stable Europe, identical to that expected as a consequence of the later opening of the Bay of Biscay (Van der Voo & Zijderveld 1971; McElhinny 1983; Galdeano *et al.* 1989; Besse & Courtillot 1991). Igneous rocks of the later phases did not show such declination deviations. In addition, it was clear that the individual site directions appeared to give directions that provided evidence for tectonic tilts that differ from those defined by the present bedding in the carbonate sediments into which the igneous rocks were intruded, i.e. these beds had been tilted at other angles and in other directions during other phases of their tectonic evolution. The present study extends these previous studies, particularly in an attempt to determine the tectonic tilt that had occurred before and after such intrusions. A fuller account, with more examples of vector behaviour, will be published elsewhere, but these preliminary results provide an interesting example of the

complexities of the tectonic record in parts of the Mediterranean Basin that are of general application.

Sampling, methodology and analyses

A total of 218 samples of igneous rocks and baked contacts were obtained by drilling standard core samples at 36 different sites within 18 localities. Many other sites were visited but were commonly deeply weathered and other sites were collected in which the baked contacts, usually in limestones, were too weak to measure. At most sites, it was not possible to use standard procedures to determine the tilt of the beds into which the intrusions had taken place and it would not, in any case, have been possible to determine, on geological grounds, whether the bedding tilt of the carbonate sediments into which they had been intruded had occurred before, during or after the igneous activity. However, if the igneous intrusions could be dated, then it should be possible to estimate the summation of tectonic tilts that occurred before and after the intrusion. The natural remanent magnetization of these samples was measured using Digico and Molspin spinner magnetometers with a noise level of *c.* $0.01 \, \text{mA m}^{-1}$. Standard stepwise thermal and alternating field demagnetization (Tarling 1983) was applied separately to at least two samples from each site and the vector analyses were undertaken both visually and using computer techniques. The computer analyses were primarily principal component analyses (Kirschvink 1980) and directional consistency (previously the stability index of Tarling & Symons 1967).

Magnetic mineralogy

The coercivity and blocking temperature spectra determined during partial demagnetization indicated that, with some exceptions, the remanence is carried by magnetic minerals with a coercivity less than 100 mT and unblocking temperatures around 350°C. This suggests that the predominant magnetic mineral is likely to be a titanium-rich titanomagnetite. There was no evidence for the presence of significant quantities of ilmenohaemite series minerals, suggesting that weathering has not significantly affected the rocks sampled. This confirms the field observations that reddening associated with weathering was absent in the sites sampled. In some localities, particularly at Monte Redondo (E3 & E5) and Sobral (E21–24), the magnetic mineralogy was dominated by minerals with low relaxation times and no useful characteristic component could be identified.

Table 1 *Site mean directions*

Site	R_s	In situ		Tectonic corrected			
		$D°$	$I°$	$D°$	$I°$	k	$\alpha_{95}°$
Late Jurassic, possibly Early Cretaceous							
Porto de Mós							
15	6/5	154	−21	160	−44	51	10.7
Early Cretaceous							
Runa							
25	6/6	154	−63			226	4.5
26	5/3	224	−40			153	10.0
27	4/4	137	−48			246	5.9
Mte Redondo			(sites 2 & 4, no vector consistency)				
1	5/3	266	−22	268	5	269	7.5
3	5/4	357	65			148	7.6
5	6/5	346	67			90	9.7
Mte Real							
7	6/6	28	26	47	31	24	13.9
Leiria							
8	6/5	314	10	323	47	241	4.9
S. Bartolomeu							
9	7/4	44	55	359	45	95	9.5
10	6/6	355	23	24	51	75	7.8
Famalicào							
11	6/5	148	−15	159	−58	63	9.7
12	6/6	272	45			445	3.2
12th	2	92	−53			134	21.7
13	5/3	288	−13			192	8.9
Serra Todo o Mundo							
17	6/6	125	−37			159	5.3
18	6/6	127	−36			608	2.7
19	6/6	121	−42			386	3.4
F. Alqueidào, Sobral							
21	4/4	345	65			925	3.0
22	6/6	356	72			134	5.8
23	5/5	357	53			200	5.4
24	6/6	359	68			331	3.7
Sesimbra							
SS1	6/6	317	9	336	47	109	6.4
SS2	6/6	335	8	346	52	36	11.3
Rio Maior							
RM2	7/6	312	37	323	54	262	4.1
Late Cretaceous							
Rolica							
28	5/5	255	60			185	5.6
Papõa							
29	6/6	172	−43			41	10.5
Ribamar							
30	6/5	180	−46			109	7.3
31	6/6	169	−48			123	6.1
Early Tertiary							
Cascais Sintra							
CS1	6/2	192	−4			261	4.9
CS2	6/4	182	−33			368	4.8
CS3	6/4	345	38			11	28.9
CS6	7/7	352	48			31	11.0
Cascais (Guincho)							
CS4	7/7	343	44			73	7.1
CS5	7/7	12	38			61	7.8
Lisbon Volcanics			(Site CVL2, no vector consistency)				
CVL1	6/5	317	64			94	7.9
CVL3	8/7	355	−7			55	8.2

R_s is the ratio of the number of specimens collected with those used in the calculation. The precision estimates, k, and α_{95}, are based on Fisher (1953). All the vectors were isolated using alternating magnetic fields, apart from the two vectors from site 12. Bedding tilt corrections were only measurable at 10 sites.

Table 2 *Expected geomagnetic field directions within Iberia*

Period	Ma	Euler Pole 40°N 4°W		Euler Pole 38.8°N 9.3°W	
		D	*I*	*D*	*I*
Early Tertiary	37–66	0	47	359	46
Late Cretaceous	67–97	7	40	5	38
Early Cretaceous	98–144	354	38	352	37
Late Jurassic	145–176	354	31	352	30
Hauterivian–Berremian				347	Galdeano *et al.* 1989)
Aptian				348	Galdeano *et al.* 1989)

Based on Van der Voo (1993) for central Iberia (40°N, 4°W) and converted to Lisbon (38.8°N, 9.3°W) using the inclined geocentric dipole model (Tarling 1988).

Palaeomagnetic observations

The palaeomagnetic properties of different sites (Table 1) are considered on the basis of their known or attributed age as summarized by Aires-Barros (1979) and these are compared with the expected directions interpolated from the European apparent polar wander path after correction for the opening of the Bay of Biscay (Table 2). The extant results are only summarized here, with only a few examples of vector behaviour during stepwise demagnetization as this study is ongoing. Nonetheless, the vectors were quite well defined (Fig. 2) and the linear vectors for different samples from the same site were generally consistent (Table 1).

Late Jurassic

Only the *Porto de Mós* site is of almost certain Late Jurassic age on geological and petrological grounds. The mean declination is consistent with a rotated Iberian direction, but somewhat more consistent with the rotation proposed by Galdeano *et al.* (1989) than by Van der Voo (1993). The reversed polarity supports the probable Upper Jurassic age assignment. The *in situ* mean inclination is somewhat too shallow for such an age but the tectonically corrected value is rather too steep, suggesting that the intrusion occurred after some tilting, (c. 10° to the SE) had taken place and a further 10° occurred subsequent to the intrusion.

The Early Cretaceous

The 25 site mean directions form two main groups, northwesterly or southeasterly. The NW group is of positive inclination and the SE group is a negative inclination, indicating the presence of both normal and reversed polarities. The presence of so many reversed remanences for a period when the geomagnetic field is considered

to be predominantly of normal polarity suggests that several assignments are incorrect, but the mean directions of these two groups also require a 35° rotation to make most of them consistent with the first phase of rotation of Iberia proposed by Galdeano *et al.* (1989), suggesting a pre-Tertiary age, but not distinguishing, at this stage, between potentially Late Jurassic or Late Cretaceous ages. Some of the reversed polarities sites, e.g. at Runa, may be indicative of a Late Jurassic age, but subjected to tilting that resulted in their *in situ* directions differing from those expected. However, in many sites (26, 27, 1, 7, 8, 9, 11, 12, 13, 17, 18, 19, SS1, RM2) the mean declinations differ significantly from those expected for any Mesozoic–Tertiary age, with or without the rotation of Iberia, despite high precision and local consistency between site mean directions, e.g. the three Serro Todo o Mundo sites (Table 1). In several instances, the tilt-corrected declinations were more consistent with the expected Mesozoic direction than were the *in situ* declinations, e.g. sites 8, 9, 11, SS1 and SS2. This suggests that, in at least these sites where it could be tested, the declinations have been rotated, at least in part, by localised block motions involving rotations about vertical axes, but this hypothesis cannot be tested in most sites as the local bedding tilt could not be ascertained close enough to the sampling site. Similarly, the mean inclinations for which local bedding tilts were known were more consistent with the expected Early Cretaceous inclination after correction for the observed bedding. The effect was generally to increase the inclination, suggesting a predominantly northerly directed post intrusion tilt. The only exception was for site 1 (Monte Redondo) where only 3 sample vectors were acceptable and the site mean direction was inconsistent with the other sites both before and after bedding correction. It would thus appear that most of the remanences associated with the

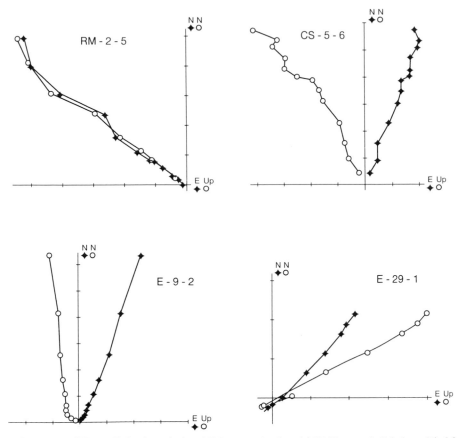

Fig. 2. Examples of Vector Behaviour during AF demagnetization. (**a**) RM 2, sample 5, is from Rio Maior and probably of Early Cretaceous age. (**b**) Lower Tertiary Cascais, site CS5, sample 6. (**c**) Late Cretaceous site E-9, sample 2. (**d**) Late Cretaceous site E-29, sample 1.

igneous intrusions have been titled subsequent to the intrusion, i.e. post Early Cretaceous times. This is particularly well illustrated by the Sesimbra sites (Fig. 3).

The Late Cretaceous

Only four sites from three localities were of this age and none could be corrected for local bedding. Site 28 (Rolica) appears completely anomalous, with the other sites all of reversed polarity and with *in situ* declinations and inclinations consistent with them not having been tilted substantially since they were intruded.

The Early Tertiary

The site directions at the three localities (8 sites) are somewhat scattered, although reasonably well defined (apart from site CS3), and with both normal and reversed polarity. No local bedding tilts could be determined and so only the *in situ*

mean site directions could be considered. The mean site declinations were generally northerly or southerly, consistent with the age assignment, and the mean site inclinations were mostly also consistent with a very late Cretaceous or early Tertiary age. However both sites CS1 and CVL3 had mean inclinations substantially shallower than expected for this age, possibly implying that both may have been subjected to northerly tilting, subsequent to emplacement, by some 40°.

Discussion

It is evident that the igneous intrusions and their baked contacts carry an identifiable component that, in unweathered rocks, seems likely to be associated with the time of intrusion. In most sites, irrespective of age, there is also some evidence for post-emplacement tilting, although this conclusion must be qualified by the observation that there are few sites for which the local

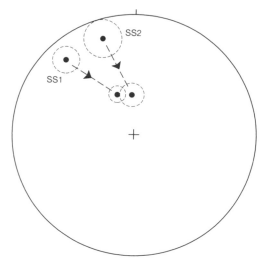
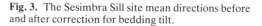

Fig. 3. The Sesimbra Sill site mean directions before and after correction for bedding tilt.

bedding tilt could be determined. It can certainly be concluded that this palaeomagnetic technique has the potential for determining the extent to which the sediments into which they were intruded had been tilted before or after the emplacement. However, it is also clear that such assessments also require good age control on the igneous rocks which, in this example, can be provided, albeit with some uncertainty, by the deflection of declinations by the rotation of the Iberian microplate. The fundamental uncertainty is, nonetheless, the extent to which the directions of remanence within the sites can be considered to represent the geocentric dipole field at that time. The intrusions were likely to have been at shallow depths, but may well have taken some hundreds of years to acquire their remanence and hence secular variations could have been averaged out in obtaining the mean site direction. This is, however, less likely for extrusive or extremely shallow intrusive rocks. This exercise has also illustrated a fundamental problem in all structural palaeomagnetic studies – the scale over which the observations are relevant. In this study, the intrusions were mostly along fractures or faults that were probably in existence prior to the intrusion. Thus the sampled site could, for example, merely be a sliver of rock that may well have acted as a 'roller bearing' during the intrusion. The suggested rotations would then only be applicable over a few cubic metres. Nonetheless, the consistency of the site mean remanences within different locations, particularly Serra Todo o Mundo and Sesimbra, suggest that the

palaeomagnetic observations may well represent the rotations of individual fault blocks, as first identified palaeomagnetically in the Mediterranean region by Freund & Tarling (1979).

References

AIRES-BARROS, L. 1979. Actividade Magmea Pas-Palaeozsica no Continente Portugues – Elementos para una Sintese Critica. *Ciencias da Terra*, **5**, 145–198.

BESSE, J. & COURTILLOT, V. 1991. Revised and synthetic Apparent Polar Wander Paths of the African, Eurasian, North American and Indian Plates and True Polar Wander since 200 Ma. *Journal of Geophysical Research*, **96**, 4029–4050.

FISHER, R. A. 1953. Dispersion on a sphere. *Proceedings of the Royal Society of London*, **A217**, 295–305.

FONSECA, J. 1989. *Seismicity and Regional Tectonics of the Estremadura, SW Portugal*. PhD Thesis, University of Durham, United Kingdom.

FREUND, R. & TARLING, D. H. 1979. Preliminary Mesozoic palaeomagnetic results from Israel and inferences for a microplate structure in the Lebanon. *Tectonophysics*, **60**, 189–205.

GALDEANO, A., MOREAU, M. G., POZZI, G. P., BERTHOU, P. Y. & MALOT, J. A. 1989. New Palaeomagnetic Results from Cretaceous sediments near Lisbon, Portugal and Implications for the Rotation of Iberia. *Earth and Planetary Science Letters*, **92**, 95–106.

GUIRY, F. 1984. *Evolution Sedimentaire et Dynamique du Bassin Marginale Ouest-Portugais au Jurassique, Province d' Estremadure (Secteur de Caldas da Rainha-Montejunto)*. Thèse d'État, Université de Lyon I, France.

KIRSCHVINK., J. L. 1980. The least-squares line and plane and the analysis of palaeomagnetic data. *Geophysical Journal of the Royal Astronomical Society*, **62**, 699–718.

MCELHINNY, M. W. 1973. *Palaeomagnetism and Plate Tectonics*. Cambridge University Press.

PEIRERA, J. L. V., FONSECA, J. F. B. D., RAPALINI, A. & TARLING, D. H. 1992a. Palaeomagnetic study of fault re-activation: a progress report (abs). *23rd General Assembly European Seismological Community*, Prague, 16.

——, ——, —— & —— 1992b. Estudo Paleomagnetico da Reactivacao de Falhas (abs). *8th Conference Nacional Fisica*, Portugal, 429–430.

RIBEIRO, A., ANTUNES, M. T., FERREIRA, M. P., ROCHA, R. B., SOARES, A. F. & ZBYSZESWKI, G. 1980. *Introduction la Geologie Generale du Portugal*. Serviços Geolôgicos de Portugal, Lisboa, Portugal.

TARLING, D. H. 1983. *Palaeomagnetism*. Chapman & Hall, London.

—— 1988. Secular variations of the geomagnetic field – the archaeomagnetic record. *In*: STEPHENSON, F. R. & WOLFENDALE, A. W. (eds) *Secular Solar and Geomagnetic Variations in the last 10 000 years*. Kluwer Acad. Publ., Dordrecht, 349–365.

—— & Symons, D. T. A. 1967. A Stability Index of remanence in palaeomagnetism, *Geophysical Journal of the Royal Astronomical Society*, **12**, 443–448.

——, Fonseca, J., Rapalini, A. & Pereira, J. L. 1992. *Investigative Palaeomagnetic Studies of Rocks in Southern Portugal*. Unpublished British Council report.

Van der Voo, R. 1993. *Palaeomagnetism of the Atlantic, Tethys and Iapetus Ocean*. Cambridge University Press.

—— & Zijderveld, J. D. A. 1970. Renewed Palaeomagnetic Study of the Lisbon Volcanics and Implications for the Rotation of the Iberian Peninsula. *Journal of Geophysical Research*, **76**, 3913–3921.

Palaeomagnetism and palaeogeography of Adria

J. E. T. CHANNELL

Department of Geology, University of Florida, Gainesville, FL 32611, USA

Abstract: An 'African' Mesozoic apparent polar wander path (APWP) based on rotated North American data is compared with palaeomagnetic data from Adria and surrounding regions. The form of the APWPs from the Southern Alps, Umbria-Marche (Italy) and the Transdanubian Central Range (Hungary) is similar to that of the 'African' APWP. The Umbria–Marche and Transdanubian APWPs are rotated relative to the African APWP, and the rotation is best explained by local anticlockwise tectonic rotations of 20° and 30° for the NW Umbria–Marche and Transdanubian regions, respectively. Permian to Cretaceous palaeomagnetic poles from the Southern Alps are not distinguishable from the African APWP. The more recently derived Cretaceous poles from Istria, Cretaceous poles from Iblei (Sicily), Gargano and Apulia are all consistent, within the confidence limits of the data, with the African APWP. Rifting in the Ionian Sea/Eastern Mediterranean either predated the mainly Jurassic-Cretaceous palaeomagnetic dataset or did not produce recognizable rotation of Adria relative to Africa. The Northern Calcareous Alps are often considered to have been contiguous with the Southern Alps during the Mesozoic, however, the NCA (Liassic) poles are closer to coeval poles for Europe, and are rotated in excess of 80° clockwise relative to the African APWP. The Northern Calcareous Alps were probably detached from the Southern Alps during Late Triassic and Jurassic time, and did not undergo the 'African' rotation which affected Adria.

The role of Adria, the continental lithosphere ringed by deformed Periadriatic Mesozoic continental margins, is central to evolutionary models of Alpine orogeny (Fig. 1). The view of Adria as a fixed promontory of the African plate, first proposed by Argand (1924), was advocated by others (e.g. Channell *et al.* 1979) on the basis of the distribution of sedimentary facies, deformation history and palaeomagnetism. There are a number of lines of evidence that indicate that the two plate (Africa/Europe) model for Alpine orogeny is an oversimplification. (1) The Ionian Sea is probably floored by ancient (Triassic?) oceanic crust (De Voogd *et al.* 1992). Deep water rifts are documented from mid-Permian in Sicily (Catalano *et al.* 1991) and Crete (Kozur & Krahl 1987), and from Early Triassic in Southern Apennines (Lagonegro, Marsella *et al.* 1993) implying a Permo-Triassic branch of Tethys beneath the thick Cenozoic sedimentary cover of the Eastern Mediterranean and extending as far west as Sicily/Southern Apennines. Mesozoic basins are not confined to the deformed margins of Adria but cut across the Adriatic Sea, linking the Umbria–Marche and Ionian basins (see Zappaterra 1994) (2) Eo-Alpine deformation at the northern rim of Adria (*c.* 130 Ma) predates the onset of North Atlantic seafloor spreading and resulting N–S relative motion of Africa/Europe

(*c.* 100 Ma). (3) The kinematics of Late Cretaceous deformation in the Alps indicate westward transport, inconsistent with the approximately N–S relative motion of Africa and Europe (Platt *et al.* 1989); however, this argument may be mitigated by supposing that the Northern Calcareous Alps were not part of Adria (Africa) but an independent microplate. (4) Focal mechanisms of large earthquakes in the Periadriatic Belt indicate E–W-oriented extension in peninsular Italy and E–W compression in the Dinarides and Hellenides, implying present-day rotation of Adria relative to Africa (Anderson 1987; Anderson & Jackson 1987). (5) Eocene–Oligocene nappe tectonics in the Hellenides indicate an important phase of shortening, consistent with westward motion of Adria at this time.

Apart from a few early studies, palaeomagnetic data from Adria first became available in the 1970s. The early publications indicated that Permian and Mesozoic palaeomagnetic poles from the Southern Alps lay close to the African apparent polar wander path (APWP) (see Zijderveld *et al.* 1970; Channell & Tarling 1975; VandenBerg & Wonders 1976). Palaeomagnetic data from NW Umbria–Marche (Lowrie & Alvarez 1975; VandenBerg *et al.* 1978), Gargano (VandenBerg 1983) and Istria (Márton & Veljovic 1983) were interpreted to indicate

From Morris, A. & Tarling, D. H. (eds), 1996, *Palaeomagnetism and Tectonics of the Mediterranean Region*, Geological Society Special Publication No. 105, pp. 119–132.

119

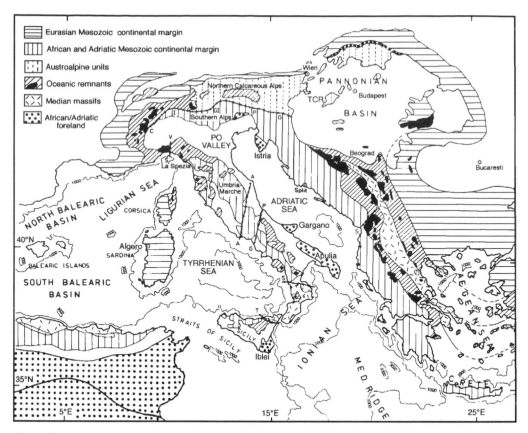

Fig. 1. Structural elements of part of the Mediterranean region. TCR, Transdanubian Central Range of the Pannonian Basin. Tectonic lines: T, Taormina Line; S, Sangineto Line; P–G, Pescara-Gaeta Line; A–A, Ancona–Anzio Line; S–Z, Sestri–Voltaggio Line. Periadriatic Line segments; C, Canavese; I, Insubric; G, Giudicarie; P, Pusteria; G, Gail; S–P, Scutari–Pec Line (modified after Channell *et al.* 1979).

Eocene rotation of peninsular Italy relative to Africa. This interpretation was not unequivocal due to uncertainties in the African APWP and in the role of local thrust sheet rotation, which is largely anticlockwise throughout peninsular Italy.

Reviews of palaeomagnetic data from Adria (Lowrie 1986; Heller *et al.* 1989) stressed these uncertainties, and came to the conclusion that 'declinations of Cretaceous limestones from Istria and Gargano are rotated by about 17° anticlockwise relative to coeval directions from the southern Alps, which may imply that these regions are not autochthonous, or that the Adriatic promontory became decoupled from the African plate in the late Tertiary' (Heller *et al.* 1989). In more recent literature, statements such as 'the evidence for net CCW rotation of the 'hard core' of the Adriatic region with respect to Africa since the late Cretaceous is established' (Márton & Nardi 1994) and 'the

mean direction for Cretaceous rocks from the Apulia–Gargano foreland implies a 31° counter-clockwise rotation during Tertiary' (Scheepers & Langereis 1994) illustrate the widely held view that Tertiary rotation of Adria relative to Africa is indicated by palaeomagnetic data.

Van der Voo (1993) noted that the mean direction of the Early Cretaceous result from Gargano (VandenBerg 1983) was miscalculated. The revised mean direction yields a pole which is less discrepant with the African APWP. Another of the key results favoring rotation of Adria, that from Istrian Cretaceous limestones (Márton & Veljovic 1983), has been superseded by higher quality data from the same region (Márton & Pugliese 1989; Márton *et al.* 1990). Van der Voo (1993) concluded that 'there are no indications that any of the "autochthonous" areas (Iblei, Gargano, Southern Alps, Istria) show systematic differences' (with each other or with the African APWP).

Table 1 *Permo–Triassic palaeomagnetic poles from the Southern Alps (Italy)*

Location	Age	Pole Lat °N	Pole Long °E	A₉₅ (dp/dm)	Reference
1 Dolomites	Ladinian/Carnian	48	240	9	Manzoni (1970)
2 Lombardy	Late Anisian	63	229	8	Muttoni & Kent (1994)
3 Dolomites	Scythian	42	233	3.4/6.8	Channell & Doglioni (1994)
4 Lombardy	Late Permian	47	237	6.3	Kipfer & Heller (1988)
5 Lombardy	Late Permian	48	239	5	Kipfer & Heller (1988)
6 Lombardy	Early Permian	43	243	10	Heiniger (1979)
7 Lombardy	Early Permian	38	245	8	Heiniger (1979)
8 Lombardy	Early Permian	35	248	14	Heiniger (1979)
9 Bolzano	Early Permian	45	239	4	Zijderveld et al. (1970)

Statistical parameters after Fisher (1953)

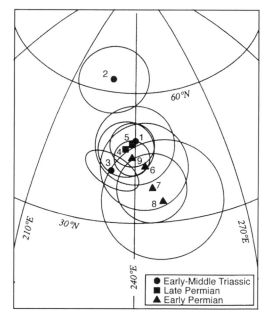

Fig. 2. Permo-Triassic palaeomagnetic poles from the Southern Alps. 95% confidence circles (A₉₅) or ellipses (dp/dm) are shown. For key to poles, see Table 1.

The contrasting interpretations of palaeomagnetic data from Adria are a result of: (1) uncertainties in the African APWP, particularly in the Late Jurassic–Early Cretaceous interval where there is an apparent cusp (hairpin) in the APWP, (2) uncertainties in the distribution in Adria of allochthony (detachment of sedimentary cover from the underlying continental basement) and local thrust sheet rotation. Since the Van der Voo (1993) review, new palaeomagnetic data from Adria have become available. In addition, Mesozoic APWPs (particularly that from North America) have undergone revision, allowing computation of a new African APWP for comparison with the data from Adria.

Permo-Triassic palaeomagnetic data from Adria

Permo-Triassic palaeomagnetic data for Adria are from the Southern Alps (Table 1, Fig. 2). The paucity of Permo-Triassic palaeomagnetic data from Africa and other Gondwana continents results in poor definition of the Late Permian to Late Triassic portion of the African APWP. Estimates of the African APWP in this interval can be provided by rotating North American and European palaeomagnetic data to African coordinates. Figure 3 shows: (1) the Early Permian pole and the Late Triassic–Early Jurassic poles for Africa compiled by Van der Voo (1993) from Gondwanan palaeomagnetic data (intervening poles are too poorly defined); (2) North American mean poles (Van der Voo 1993) rotated to African coordinates using rotation parameters (Latitude = 66.95°, Longitude = −12.02°, Ω = 75.55°) from Klitgord & Schouten (1986); and (3) European mean poles (Van der Voo 1993) rotated to African coordinates by rotation parameters (Latitude = 53.4°, Longitude = 1.5°, Ω = 55.7°) derived from Klitgord & Schouten (1986) and Srivastava & Tapscott (1986).

The majority of the Permo-Triassic poles from the Southern Alps lie close to the Permian and Early Triassic poles on the European APWP rotated to African coordinates (compare Figs 2 and 3). Late Permian poles for Europe lie close to European Early Triassic poles and to an Anisian (Mid-Triassic) pole (Théveniaut et al. 1992) indicating an apparent stasis in the European APWP for Early Permian to Early Triassic time (Fig. 3). The Permian and Triassic

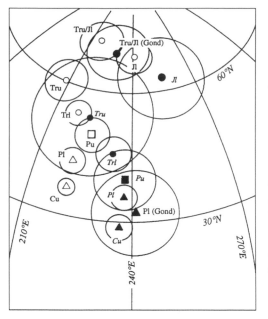

Fig. 3. 'African' palaeomagnetic poles for Upper Carboniferous (Cu) to Lower Jurassic (Jl). Open symbols: North American mean poles (Van der Voo 1993) rotated to African coordinates by rotation parameters (Latitude = 66.95°, Longitude = −12.02°, Ω=75.55°) from Klitgord & Schouten (1986). Closed symbols: European mean poles (Van der Voo 1993) rotated to African coordinates by rotation parameters (Latitude = 53.4°, Longitude = 1.5°, Ω=55.7°) derived from Klitgord & Schouten (1986) and Srivastava & Tapscott (1986). Pl (Gond) and Tru/Jl (Gond) signify Lower Permian and Upper Triassic/Lower Jurassic mean poles for Africa, respectively, compiled by Van der Voo (1993) from Gondwana continents (intervening mean poles are too poorly defined). 95% confidence circles (A₉₅) are shown.

poles on the rotated North American APWP are significantly different to those from the Southern Alps and to the rotated European APWP. There are two explanations for this discrepancy. Firstly, the rotation parameters used to transpose North American data to African coordinates (Klitgord & Schouten 1986), although adequate for latest Triassic time, are not generally applicable to Triassic and Permian time due to the shifting configuration of Pangaea. Secondly, the majority of Early Triassic poles for North America are derived from the Colorado Plateau which has rotated clockwise relative to the North American craton. In the Van der Voo (1993) compilation, Colorado Plateau poles have been rotated counterclockwise by 5.4° to compensate for this tectonic

rotation, however, a larger adjustment (13°–14°) has been suggested (Kent & Witte 1993) although this remains controversial (see Kodama *et al.* 1994).

Jurassic–Cretaceous palaeomagnetic data from Adria

A recent compilation of the Jurassic–Cretaceous palaeomagnetic data from the Southern Alps (Channell *et al.* 1992a) was based on 46 site mean directions divided into four age groups: Senonian, Hauterivian–Barremian, Kimmeridgian–Tithonian and Callovian–Oxfordian. When compared with the Besse & Courtillot (1991) synthetic APWP for Africa or the mean NW African poles compiled by Van der Voo (1993), the Southern Alpine Cretaceous poles are more or less consistent with the African poles, but the Jurassic poles are offset (Fig. 4a,b).

For the Bosso/Sentino/Burano river valleys of NW Umbria–Marche, the declinations are rotated (anticlockwise) by about 20° relative to coeval mean declinations from the Southern Alps (VandenBerg *et al.* 1978; Channell 1992). Further NW in the Umbria–Marche, at Presale/Giordano, the declinations are more westerly, being rotated (anticlockwise) by about 40° relative to the results from the Southern Alps (Márton & D'Andrea 1992; Channell 1992). These rotations are interpreted as local tectonic rotations of thrust sheets in the Umbria–Marche Apennines, the amount of anticlockwise rotation generally increases to the northwest. The orientation of fold axes in Umbria–Marche becomes more westerly towards the north, although there is no consistent relationship between site mean declination and fold axis orientation (Hirt & Lowrie 1988). Thrust sheet rotations do not, therefore, entirely postdate folding, implying rotation pre- and synfolding.

Due to differences in timing of subsidence of Mesozoic carbonate platform in Umbria–Marche and the Southern Alps, Pliensbachian pelagic sediments suitable for palaeomagnetic study are available in Umbria–Marche, whereas pre-Callovian Jurassic palaeomagnetic data are not available from the Southern Alps. Kimmeridgian–Barremian (Late Jurassic–Early Cretaceous) and Pliensbachian palaeomagnetic poles from the Bosso/Sentino/Burano river valleys in NW Umbria–Marche (Channell 1992) can be converted to Southern Alpine coordinates by rotating the mean declinations by 20°. The resulting 'Southern Alpine' APWP for Jurassic–Cretaceous time, comprises seven palaeomagnetic poles for Pliensbachian (Early Jurassic) to Senonian (Late Cretaceous), four

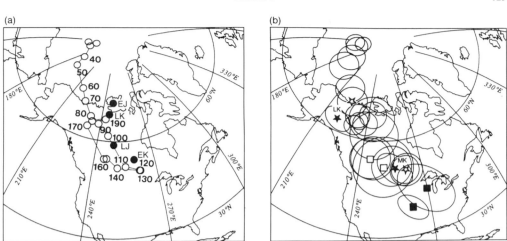

Fig. 4. (a) Open circles: African synthetic polar wander path (APWP) from 190 Ma to 10 Ma (Besse & Courtillot 1991). Closed circles: mean poles for West Gondwana (NW Africa coordinates) for Early Jurassic (EJ), Late Jurassic (LJ), Early Cretaceous (EK) and Late Cretaceous (LK) (Van der Voo 1993), compiled from African and South American palaeomagnetic data. **(b)** The Besse & Courtillot (1991) African APWP (a) with 95% confidence circles (A$_{95}$) highlighting the Late Jurassic poles (open squares) and the Middle (MK) and Late Cretaceous (LK) poles (open stars). Southern Alpine poles for Late Jurassic (closed squares) and Middle (MK) and Late Cretaceous (LK) (closed stars) are indicated with 95% confidence ellipses (dp/dm).

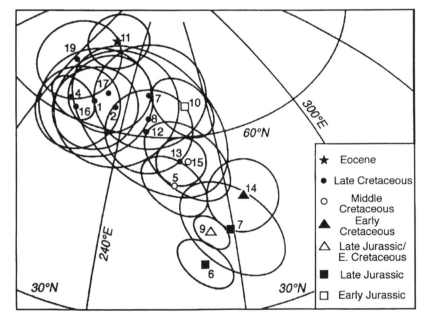

Fig. 5. Palaeomagnetic poles from Adria. For key to poles, see Table 2. The poles from NW Umbria–Marche are compensated for the supposed 20° anticlockwise tectonic rotation. 95% confidence circles (A$_{95}$) or ellipses (dp/dm) are shown.

from the Southern Alps and three from Umbria–Marche rotated to Southern Alpine coordinates (Table 2). These poles can now be compared with Cretaceous poles from Iblei (Sicily), Istria, Apulia and Gargano (Fig. 5, Table 2). Palaeo-

magnetic data from the Southern Apennines affected by anticlockwise thrust sheet rotation (e.g. Scheepers *et al.* 1993; Scheepers & Langereis 1994) are not included in the analysis, nor are the data from southern Apulia (Tozzi *et al.* 1988)

Table 2 *Jurassic–Cretaceous palaeomagnetic poles from Adria*

	Location	Age	Pole Lat °N	Pole Long °E	A$_{95}$ (dp/dm)	Reference
1	Istria	Late Cretaceous	64	228	9.3	Márton & Pugliese (1989)
2	Istria	Late Cretaceous	66	232	6.7	Márton & Pugliese (1989)
3	Istria	Late Cretaceous	67	248	9	Márton et al. (1990)
4	S. Alps	Late Cretaceous	63	220	2.7/4.5	Channell et al. (1992a)
5	S. Alps	Mid-Cretaceous	52	257	3.7/6.2	Channell et al. (1992a)
6	S. Alps	Late Jurassic (Kimm/Tithonian	39	263	3.0/5.4	Channell et al. (1992a)
7	S. Alps	Late Jurassic (Call/Oxfordian	44	270	6.0/10.0	Channell et al. (1992a)
8	NW Umbria	Late Cretaceous (rotation modified)	50 (63)	271 (249)	9.5/15.3	Channell (1992)
9	NW Umbria	Late Jur/Early Cret (rotation modified)	30 (44)	281 (265)	2.0/3.5	Channell (1992)
10	NW Umbria	E. Jur. (Pliensb.) (rotation modified)	51 (65)	279 (261)	3.9/5.9	Channell (1992)
11	Gargano	Eocene	75	224	5	Speranza & Kissel (1993)
12	Gargano	Late Cretaceous	61	249	7	Channell (1977)
13	Gargano	Late Cretaceous	56	259	4	VandenBerg (1983)
14	Gargano	Early Cretaceous	49	274	6	VandenBerg (1983)
15	Apulia	Mid-Cretaceous	56	261	4.9/8.1	Márton & Nardi (1994)
16	Iblei	Late Cretaceous	62	223	7	Schult (1973)
17	Iblei	Late Cretaceous	63	229	7	Barberi et al. (1974)
18	Iblei	Late Cretaceous	64	236	10	Gregor et al. (1975)
19	Iblei	Late Cretaceous	69	212	7	Grasso et al. (1983)

which are affected by local clockwise tectonic rotation (Tozzi 1993).

Due to rapid apparent polar wander in Jurassic and Cretaceous time and the apparent cusp in the Adria APWP (Fig. 5), the sequential age of poles is critical to the interpretation. The one Early Jurassic (Pliensbachian) pole position lies at high latitude (Fig. 5, Table 2). The pole positions appear to shift to low latitudes by latest Jurassic time, and then move NW, to higher latitudes, during the Cretaceous.

Jurassic–Cretaceous APWPs for Africa

The Mesozoic African APWP is not well defined by African palaeomagnetic data. Most African APWPs utilize palaeomagnetic data from other continents rotated into African coordinates (e.g. Westphal et al. 1986; Besse & Courtillot 1991; Van der Voo 1993) (Fig. 4a). For the Jurassic–Cretaceous segment of synthetic APWPs, the North American data are particularly important due to the relative lack of data of this age from other continents. In the last few years the North American Jurassic-Cretaceous APWP has been the subject of debate. The controversy has centred around palaeomagnetic poles derived from the middle Jurassic Corral Canyon sequence and Glance Conglomerate of southern

Arizona (May et al. 1986; Kluth et al. 1982), and the middle Jurassic Summerville Formation of Utah (Bazard & Butler 1992) which define a Jurassic North American APWP at about 60–65°N latitude extending from 110°E longitude at c. 172 Ma to 150°E longitude at c. 149 Ma (Fig. 6). These poles are very different (by about 20° of arc) from higher latitude poles considered to be approximately the same age from the White Mountains in New England (Opdyke & Wensink 1966; Van Fossen & Kent 1990). Although the age of the Moat Volcanics pole (MV in Fig. 6 at 78.7°N, 90.3°E) from the White Mountains (Van Fossen & Kent 1990) is not well constrained (see Butler et al. 1992), the Corral Canyon and Glance Conglomerate results from Arizona (CC and GC in Fig. 6) appear to be affected by remagnetization from neighbouring Cretaceous plutons, and by tectonic rotations (Van Fossen & Kent 1992a; Hagstrum 1994). The high latitude Moat Volcanics pole (Van Fossen & Kent 1990) is supported by c. 175 Ma remagnetizations from sedimentary and igneous rocks in the Newark Basin (Witte & Kent 1989, 1990; Witte et al. 1991), and by a synfolding magnetization in the Sassamansville diabase, Newark Basin (Kodama & Mowery 1994) (Fig. 6). Courtillot et al. (1994) have constructed synthetic Jurassic North American APWPs from

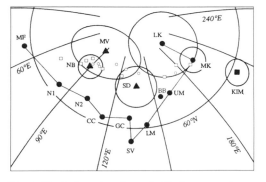

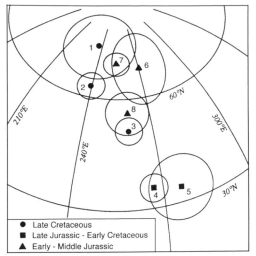

Fig. 6. North American Jurassic–Cretaceous poles.
Open squares and circles: two end member APWPs
for North America in the interval from 190 Ma (left)
to 130 Ma (right) derived from other continents and
various continental reconstructions (after Courtillot
et al. 1994). Solid symbols: poles from North
American palaeomagnetic data. Line approximates
to APWP advocated by May & Butler (1986). Key:
Jurassic poles: MF, Moenave Fm (Ekstrand & Butler
1989); N1 and N2, Newark Supergroup (Smith &
Noltimier 1979); CC, Corral Canyon (May & Butler
1986); GC, Glance Conglomerate (Kluth *et al.* 1982);
SV, Summerville Fm (Bazard & Butler 1992); LM
and UM, Morrison Fm (Steiner & Helsley 1975); BB,
Brushy Basin Member (Morrison Fm) (Bazard &
Butler 1994); MV, Moat Volcanics (Van Fossen &
Kent 1990); NB, Newark B remagnetization (Witte &
Kent 1991); SD, Sassamansville Diabase (Kodama &
Mowery 1994); KIM, New York kimberlite (Van
Fossen & Kent 1993). Cretaceous poles: MK, mid-
Cretaceous pole (Globerman & Irving 1988; Van
Fossen & Kent 1992b); LK, Late Cretaceous pole
from Montana (Diehl 1991). 95% confidence circles
(A_{95}) or ellipses (dp/dm) are shown for some poles.

Eurasian, African and South American palaeomagnetic data and a variety of continental
reconstructions. The two end member synthetic
North American APWPs for 190–130 Ma follow
a narrow latitudinal band between 70°N and
77°N from 82°E to 190°E longitude (open
symbols in Fig. 6), and appear to favor the high
latitude North American APWP. In view of
these recent results, a revision of the May &
Butler (1986) APWP for North America appears
warranted.

For the purposes of this study, we define a
Jurassic–Cretaceous APWP for North America
based on six poles, which are rotated to African
coordinates to give an 'African' APWP for this
time interval (Table 3). A latest Jurassic–earliest
Cretaceous cusp in the resulting 'African'
APWP (Fig. 7) is defined by the 143 Ma pole of
Van Fossen & Kent (1993) from kimberlites in
central New York State (pole 4 in Fig. 7) and by
the 145 Ma African Swartruggens kimberlite
pole of Hargraves (1989) (pole 5 in Fig. 7). The

Fig. 7. Selected North American Jurassic and
Cretaceous poles rotated to African coordinates using
rotation parameters of Klitgord & Schouten (1986).
95% confidence circles (A_{95}) or ellipses (dp/dm) are
shown. For key to poles, see Table 3.

synthetic North American APWP of Courtillot
et al. (1994) does not identify the cusp (Fig. 6),
however, cusps in APWPs are difficult to
identify in time averaged APWPs during intervals of rapid apparent polar wander. The
'African' APWP (Table 3, Fig. 7) is similar to the
Adria APWP (Table 2, Fig. 5).

Márton & Nardi (1994) have pointed out that
Cretaceous poles from Gargano and Apulia
(Fig. 5, Table 2) differ from the Late Cretaceous
poles from Africa, and have interpreted this to
indicate post-Cretaceous relative rotation of
Adria and Africa. The Apulia–Gargano poles lie
along the APWP towards the lower latitude Late
Jurassic poles (Fig. 5). The Apulia/Gargano
poles may, therefore, be older than the bulk of
the Late Cretaceous poles from Adria. The
progression of pole ages (Fig. 5) appears to
support this interpretation. The Apulian pole
(pole 15 in Fig. 5) (Márton & Nardi 1994) is from
the Calcare di Bari Formation which has an age
range from Valanginian to Early Turonian
(Ciaranfi *et al.* 1988). This Mid-Cretaceous pole
lies close to the Mid-Cretaceous pole from the
Southern Alps (pole 5 in Fig. 5).

Transdanubian Central Range (Hungary)

Márton & Márton (1983) pointed out that the
Mesozoic APWP from the Transdanubian Cen-
tral Range (sampled in NE Hungary between
Lake Balaton and the Danube river) has similar

Table 3 *Jurassic–Cretaceous poles rotated to African coordinates*

Age (Ma)	Location	N. Amer. Coords Lat °N	Long °E	A95 (dp/dm)	Rotation Poles Lat °N	Long °E	Ω	Africa Coords Lat °N	Long °E	Reference
1 c. 80	Elkhorn Mts (Montana)	80	190	9.6	76.55	−20.73	29.6	74.7	236.6	Diehl (1991)
2 c. 90	Arkansas/White Mts	71	194	3.7	73.82	−19.48	34.28	63.1	239.4	Globerman & Irving (1988) and
3 c. 120					66.3	−19.9	54.25	51.4	259.7	Van Fossen & Kent (1992b)
4 c. 143	Ithaca (NY)	58	203	3.8	66.4	−18.2	61.0	35.5	267.0	Van Fossen & Kent (1993)
5 c. 145	Swartruggens (S. Africa)			9.0				33.9	276.0	Hargraves (1989)
6 c. 166	White Mts (NY)	79	90	7.1/10.2	66.95	−12.02	75.55	68.3	270.4	Van Fossen & Kent (1990)
7 c. 175	Newark Basin (NJ)	73	90	3.4	66.95	−12.02	75.55	69.9	254.1	Witte & Kent (1991)
8 c. 175	Sassamanville (NJ)	72	137	5.8	66.95	−12.02	75.55	56.2	259.5	Kodama & Mowery (1994)

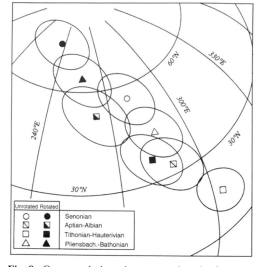

Fig. 8. Open symbols: palaeomagnetic poles from the Transdanubian Central Range (after Márton & Márton 1983). Closed symbols are same poles compensated for a 30° anticlockwise tectonic rotation (see Table 4). 95% confidence ellipses (dp/dm) are shown.

shape to the African APWP. The APWP from this region is, however, rotated away from the African APWP and lies closer to the NW Umbria–Marche APWP. Márton (1984) interpreted the similarity of the Transdanubian APWP and the NW Umbria–Marche APWP, and their divergence from the African APWP, as indicating 35° anticlockwise rotation of a coherent plate ('Unit A') which included the north part of the Pannonian Basin and peninsular Italy south of the Po Plain. This tectonic 'unit' cannot be ratified on geological/tectonic grounds, and the coincidence of the Transdanubian and the NW Umbria–Marche APWPs is fortuitous. As stated above, the NW Umbria–Marche APWP is rotated relative to the Southern Alpine 'African' APWP due, we believe, to local (*c.* 20°) anticlockwise rotation of NW Umbria–Marche thrust sheets. The Mesozoic section in the Transdanubian Central Range of Hungary has affinities with the Southern Alps, although the physical connection with the Southern Alps is obscured by Cenozoic deformation and subsidence of the Pannonian Basin (see Royden & Báldi 1988). From Márton & Márton (1983), we have compiled the site mean directions from the Transdanubian Central Range into four age groups (Table 4). The resulting pole positions for the four age groups (Fig. 8) can be rotated into close concordance with the 'African' APWP (compare Figs 7 and 8)

Table 4 *Jurassic–Cretaceous palaeomagnetic poles from Transdanubian Central Range (after Márton & Márton 1983)*

Age group (N)	Before Tilt correction Dec./Inc. k/α₉₅	After Tilt correction Dec./Inc. k/α₉₅	Pole Lat °N/ Long °E (dp/dm)	Pole Lat °N/ Long °E (30° c/w rotation)
Senonian (4)	323.5/62.4 52/13	318.1/53.2 195/7	56.0/280.3 (6.4/9.2)	74.1/236.4
Aptian–Albian (6)	289.3/41.0 57/9	292.9/42.3 57/9	33.0/289.5 (6.8/11.1)	52.9/263.5
Tithonian–Haut. (7)	287.3/25.6 2/62	270.0/40.0 49/9	16.4/303.7 (6.3/10.5)	36.7/282.6
Pliensbach–Bath. (9)	297.0/52.8 58/7	303.9/49.2 52/7	44.1/286.9 (6.3/9.6)	63.7/257.1

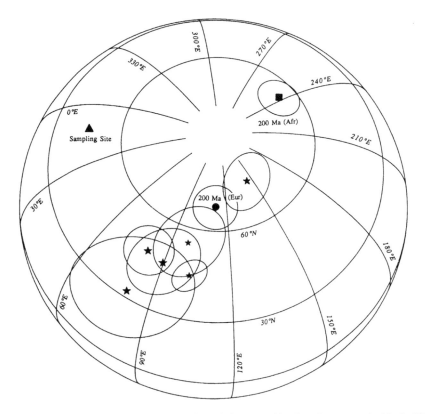

Fig. 9. Liassic palaeomagnetic poles from the Northern Calcareous Alps (stars) compared with the 200 Ma 'African' and 'European' poles of Besse & Courtillot (1991) (after Channell *et al.* 1992b). 95% confidence circles (A₉₅) or ellipses (dp/dm) are shown.

by compensating for a local anticlockwise rotation of 30°. The *shape* of the APWP from the Transdanubian Central Range indicates the African affinity of this region, and the deviation from the African APWP is best explained by

local (*c.* 30°) anticlockwise tectonic rotation. In view of the nature of deformation beneath the Pannonian Basin and in NW Umbria–Marche, the concordance of the NW Umbria–Marche and Transdanubian APWPs is coincidental and

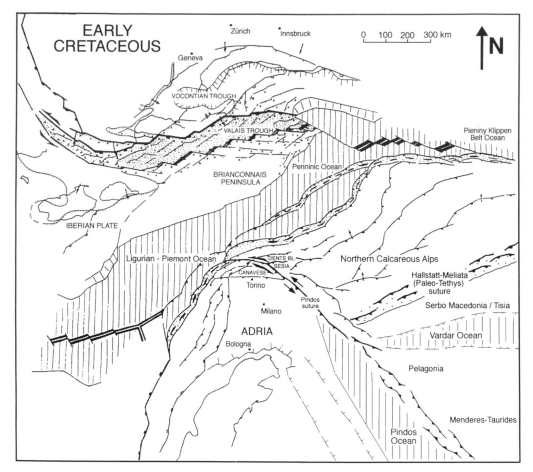

Fig. 10. Schematic Early Cretaceous palaeogeography of the Mediterranean region. Palaeogeography of Vocontian Trough, Valais Trough and Penninic Ocean after Stampfli & Marchant (1995).

is due to anticlockwise rotation of thrust sheets in the two regions.

Northern Calcareous Alps

The Northern Calcareous Alps (NCA) are comprised mainly of Triassic platform limestones, which are generally unsuitable for palaeomagnetic study. At a few locations in the NCA, the Triassic limestones are overlain by Liassic Adnet Limestone, a red nodular cephalopod-rich limestone deposited on swells and submarine highs. The Adnet Limestone usually carries a well-defined hematite magnetization component, and the age of the magnetization is constrained by fold tests and by a conglomerate test, indicating early remanence acquisition at some sites (Channell *et al.* 1992*b*). For sites yielding 'primary' magnetizations, the palaeomagnetic poles are all rotated clockwise

from the coeval African poles (Fig. 9). The magnitudes of the rotations relative to the African pole vary from 36° to 102°. The rotations relative to the European pole are 31–52° and it was suggested that the data signify that the NCA did not undergo the 'African' rotation observed in Southern Alpine palaeomagnetic data (Channell *et al.* 1992*b*).

Palaeogeography of Adria

Palaeomagnetic data and Mesozoic facies analysis from the Northern Calcareous Alps (NCA) imply the existence of a branch of Tethyan ocean between the NCA and the Southern Alps during Jurassic time which was partly closed in the Early Cretaceous (Fig. 10). Basinal Upper Triassic 'Hallstatt facies' sediments indicate a trough on the present-day southern edge of the NCA which was probably continuous with the Meliata trough exposed in Hungary and Slovakia (Kozur 1991). Ophiolitic rocks of Ladinian–Carnian age indicate oceanic rifting at this time in the Meliata trough. Evidence for oceanic rifting in the 'Hallstatt' trough is less clear, however, eclogites which may be metamorphosed remnants of oceanic crust can be traced within the Austroalpine basement from the southern Koralpe, through the Kreuzeck to the Texelgruppe (southern Otztal nappe) (Thöni & Jagoutz 1992; Miller 1970, 1990) to the Sesia–Lanzo zone (Oberhänsli *et al.* 1985). Evidence from the Carpathians implies that the Hallstatt–Meliata Trough was contiguous with the Palaeo-Tethys (*sensu* Şengör *et al.* 1980) and closed in Oxfordian time (Kozur 1991). Eo-Alpine continental collision (and the resulting 130–110 Ma high pressure metamorphism in the Sesia, Dente–Blanche and Canavese regions) postdates the apparent closure of the Hallstatt–Meliata Trough, and may mark subsequent (Early Cretaceous) closure of the northern extension of the Pindos Ocean (Fig. 10).

The Vardar Zone in the Dinarides and Hellenides indicates that the Vardar Ocean in this region was a Jurassic and Cretaceous ocean basin, possibly initiated as a back arc basin during Hallstatt–Meliata subduction (Fig. 10). Older Triassic oceanic remnants are exposed in the Pindos zone of Greece, the Eastern Mediterranean (Cyprus) and Antalya zone of Turkey, where early–mid-Triassic continental rifting was followed by mid–late-Triassic oceanic rifting (see Robertson *et al.* 1991). Mid-Permian deep water sediments and fauna in the Sicanian Zone of Sicily (Catalano *et al.* 1991) and Crete (Kozur & Krahl 1987) may imply rifting from this time along the Eastern Mediterranean. The onset of deep water conditions in the Lagonegro Basin of the Southern Apennines is Early Triassic (Scythian–Anisian) (Marsella *et al.* 1993), although this Triassic to Tertiary basin was probably never floored by oceanic crust.

Conclusions

The close concordance of APWPs from Adria with the African APWP remains enigmatic. As mentioned above, there are sound arguments (based on seismicity, crustal structure and deformation history) against Adria having been a rigid appendage of Africa since Early Mesozoic time. For the Southern Alps, NW Umbria–Marche and the Transdanubian Central Range (TCR), the APWPs are African in *shape*; although for NW Umbria–Marche and the TCR, the APWP is rotated by 20° and 30°, respectively, from the African APWP. The offsets are best interpreted as due to local tectonic rotations of the sedimentary cover. The Cretaceous poles from Istria, Apulia, Gargano and Iblei (Sicily) are not distinguishable from the African APWP. As the Southern Alps, Istria, Apulia, Gargano and Iblei are the regions of Adria which are most likely to be autochthonous, the results imply *either* no rotation of Adria relative to Africa *or* an Adria/Africa rotation pole distant from the Mediterranean and close to the Mesozoic Adria/Africa palaeomagnetic poles.

References

ANDERSON, H. 1987. Is the Adriatic an African promontory? *Geology*, **15**, 212–215.
ANDERSON, H. & JACKSON, J. 1987. Active tectonics of the Adriatic region. *Geophysiscal Journal International*, **91**, 937–983.
ARGAND, E. 1924. La tectonique de l'Asie. *Proceedings of the International Geological Congress.* **XIII**, 171–372.
BARBERI, F., CIVETTA, L., GASPARINI, P., INNOCENTI, F. & SCANDONE, R. 1974. Evolution of a section of the Africa–Europe plate boundary: palaeomagnetic and volcanological evidence from Sicily. *Earth and Planetary Science Letters*, **22**, 123–132.
BAZARD, D. R. & BUTLER, R. F. 1992. Palaeomagnetism of the Middle Jurassic Summerville Formation, east central Utah. *Journal of Geophysical Research*, **97**, 4377–4385.
—— & —— 1994. Palaeomagnetism of the Bushy Basin Member of the Morrison Formation: implications for the Jurassic apparent polar wander. *Journal of Geophysical Research*, **99**, 6695–6710.
BESSE, J. & COURTILLOT, V. 1991. Revised and Synthetic Apparent Polar Wander Paths of the African, Eurasian, North American and Indian Plates, and True Polar Wander since 200 Ma. *Journal of Geophysical Research*, **96**, 4029–4050.
BUTLER, R. F., MAY, S. R. & BAZARD, D. R. 1992. Comment on 'High-latitude palaeomagnetic poles from middle Jurassic plutons and Moat Volcanics in New England and the controversy regarding Jurassic apparent polar wander for North America' by Mickey Van Fossen & Dennis V. Kent, *Journal of Geophysical Research*, **97**, 1801–1805.
CATALANO, R., DI STEFANO, P. & KOZUR, H. 1991. Permian circumpacific deep-water faunas from the western Tethys (Sicily, Italy). *Palaeogeography, Palaeoclimatology, Palaeoecology*, **87**, 75–108.
CHANNELL, J. E. T. 1977. Palaeomagnetism of limestones from the Gargano peninsula, and the implication of these data. *Geophysical Journal of the Royal Astronomical Society*, **51**, 605–616.
—— 1992. Palaeomagnetic data from Umbria (Italy): implications for the rotation of Adria and Mesozoic apparent polar wander paths. *Tectonophysics*, **216**, 365–378.
—— & DOGLIONI, C. 1994. Early Triassic palaeomag-

netic data from the Dolomites (Italy). *Tectonics*, **13**, 157–166.

—— & TARLING, D. H. 1975. Palaeomagnetism and the rotation of Italy. *Earth and Planetary Science Letters*, **25**, 177–188.

——, BRANDNER, R., SPIELER, A. & STONER, J. S. 1992b. Palaeomagnetism and palaeogeography of the Northern Calcareous Alps (Austria). *Tectonics*, **11**, 792–810.

——, D'ARGENIO, B. & HORVÁTH, F. 1979. Adria, the African promontory, in Mesozoic Mediterranean palaeogeography. *Earth Science Reviews*, **15**, 213–292.

——, DOGLIONI, C. & STONER, J. S. 1992a. Jurassic and Cretaceous palaeomagnetic data from the Southern Alps (Italy). *Tectonics*, **11**, 811–812.

CIARANFI, N., PIERI, P. & RICCHETTI, G. 1988. Note alla carta geologica della Murge e del Salento (Puglia centromeridionale). *Memorie della Societa Geologica Italiana*, **41**, 449–460.

COURTILLOT, V., BESSE, J. & THÉVENIAUT, H. 1994. North American Jurassic apparent polar wander: the answer from other continents? *Physics of the Earth Planetary Interiors*, **82**, 87–104.

DE VOOGD, B., TRUFFERT, C., CHAMOT-ROOKE, N., HUCHON, P., LALLEMANT, S. & LE PICHON, X. 1992. Two-ship deep seismic soundings in the basins of the Eastern Mediterranean Sea (Pasiphae Cruise). *Geophysical Journal International*, **109**, 536–552.

DIEHL, J. F. 1991. The Elkhorn Mountains revisited: new data for the Late Cretaceous palaeomagnetic field of north America. *Journal of Geophysical Research*, **96**, 9887–9894.

EKSTRAND, E. J. & BUTLER, R. F. 1989. Palaeomagnetism of the Moenave Formation: implications for the Mesozoic North American apparent polar wander path. *Geology*, **17**, 245–248.

FISHER, R. A. 1953. Dispersion on a sphere. *Proceedings of the Royal Society of London*, **A217**, 295–305.

GLOBERMAN, B. R. & IRVING, E. 1988. Mid-Cretaceous palaeomagnetic reference field for North America: restudy of 100 Ma intrusive rocks from Arkansas. *Journal of Geophysical Research*, **93**, 11,721–11,733.

GRASSO, M., LENTINI, F., NAIRN, A. E. M. & VIGLIOTTI, L. 1983. A geological and palaeomagnetic study of the Hyblean volcanic rocks, Sicily. *Tectonophysics*, **98**, 271–295.

GREGOR, C. B., NAIRN, A. E. M. & NEGENDANK, J. F. W. 1975. Palaeomagnetic investigations of Tertiary and Quaternary rocks. IX: The Pliocene of southeast Sicily and some Cretaceous rocks from Capo Passero. *Geologische Rundschau*, **64**, 948–958.

HAGSTRUM, J. T. 1994. Remagnetization of Jurassic volcanic rocks in the Santa Rita and Patagonia Mountains, Arizona: implications for North American apparent polar wander. *Journal of Geophysical Research*, **99**, 15,103–15,113.

HARGRAVES, R. B. 1989. Palaeomagnetism of Mesozoic kimberlites in southern Africa and the Cretaceous apparent polar wander curve for Africa. *Journal of Geophysical Research*, **94**, 1851–1866.

HEINIGER, C. 1979. Palaeomagnetic and rock magnetic properties of Permian volcanics in the western Southern Alps. *Journal of Geophysics*, **46**, 397–411.

HELLER, F., LOWRIE, W. & HIRT, A. M. 1989. A review of palaeomagnetic and magnetic anisotropy results from the Alps. *In*: COWARD, M. P., DIETRICH, D. & PARK, R. G. (eds). *Alpine Tectonics*. Geological Society, London, Special Publications, **45**, 399–420.

HIRT, A. M. & LOWRIE, W. 1988. Palaeomagnetism of the Umbrian–Marches orogenic belt. *Tectonophysics*, **146**, 91–103.

KENT, D. V. & WITTE, W. K. 1993. Slow apparent polar wander for North America in the Late Triassic and large Colorado Plateau rotation. *Tectonics*, **12**, 291–300.

KIPFER, R. & HELLER, F. 1988. Palaeomagnetism of Permian red beds in the contact aureole of the Tertiary Adamello intrusion (northern Italy). *Physics of the Earth Planetary Interiors*, **52**, 365–375.

KLITGORD, K. D. & SCHOUTEN, H. 1986. Plate kinematics of the central Atlantic. *In*: VOGT, P. R. & TUCHOLKE, B. E. (eds) *The Western North Atlantic Region*. The Geology of North America, **M**. Geological Society of America, 351–378.

KLUTH, C. F., BUTLER, R. F., HARDING, L. E., SHAFIQULLAH, M. & DAMON, P. E. 1982. Palaeomagnetism of Late Jurassic rocks in the northern Canelo Hills, southeastern Arizona. *Journal of Geophysical Research*, **87**, 7079–7086.

KODAMA, K. P. & MOWERY, M. 1994. Palaeomagnetism of the Sassamansville diabase, Newark Basin, southeastern Pennsylvania: support for Middle Jurassic high-latitude palaeopoles for North America. *Geological Society of America Bulletin*, **106**, 952–961.

——, CIOPPA, M. T., SHERWOOD, E. & WARNOCK, A. C. 1994. Palaeomagnetism of baked sedimentary rocks in the Newark and Culpeper basins: evidence for the J1 cusp and significant Late Triassic apparent polar wander from the Mesozoic basins of North America. *Tectonics*, **13**, 917–928.

KOZUR, H. 1991. The evolution of the Meliata-Halstatt ocean and its significance for the early evolution of the Eastern Alps and Western Carpathians. *Palaeogeography, Palaeoclimatology, Palaeoecology*, **87**, 108–135.

—— & KRAHL, J. 1987. Erster Nachweis von Radiolarien im tethyalen Perm Europas. *Neues Jahrbuch für Geologie und Palaeontologie Abhandlungen*, **174**, 357–372.

LOWRIE, W. 1986. Palaeomagnetism and the Adriatic promontory: a reappraisal. *Tectonics*, **5**, 797–807.

—— & ALVAREZ, W. 1975. Palaeomagnetic evidence for the rotation of Italy. *Journal of Geophysical Research*, **80**, 1579–1592.

MANZONI, M. 1970. Palaeomagnetic data of Middle and Upper Triassic age from the Dolomites (eastern Alps, Italy). *Tectonophysics*, **10**, 411–424.

MARSELLA, E., KOZUR, H. & D'ARGENIO, B. 1993. Monte Facito Formation (Scythian – Middle Carnian). A deposit of the ancestral Lagonegro

Basin in Southern Apennines. *Bollettino de Servisio Geologico Italia*, **110**, 225–248.

MÁRTON, E. 1984. Tectonic implications of palaeomagnetic results for the Carpatho-Balkan and adjacent areas. *In*: J. E. DIXON & A. H. F. ROBERTSON (eds) *The Geological Evolution of the eastern Mediterranean*. Geological Society, London, Special Publications, **17**, 645–654.

—— & MÁRTON, P. 1983. A refined apparent polar wander curve for the Transdanubian Central Mountains and its bearing on the Mediterranean tectonic history. *Tectonophysics* **98**, 43–57.

—— & NARDI, G. 1994. Cretaceous palaeomagnetic results from Murge (Apulia, southern Italy): tectonic implications. *Geophysical Journal International*, **119**, 842–856.

—— & VELJOVIC, D. 1983. Palaeomagnetism of the Istria Peninsula, Yugoslavia. *Tectonophysics*, **91**, 73–87.

——, MILICEVIC, V. & VELJOVIC, D. 1990. Palaeomagnetism of the Kvarner islands, Yugoslavia. *Physics of the Earth Planetary Interiors*, **62**, 70–81.

MÁRTON, P. & D'ANDREA, M. 1992. Palaeomagnetically inferred tectonic rotations of the Abruzzi and northwestern Umbria. *Tectonophysics*, **202**, 43–53.

—— & PUGLIESE, N. 1989. Results of a pilot palaeomagnetic study of the Trieste Karst (Italy). *Studi Trentini di Sciense Naturali Geologia Trento*, **65**, 195–208.

MAY, S. R. & BUTLER, R. F. 1986. North American Jurassic apparent polar wander: implications for plate motion, palaeogeography and Cordilleran tectonics. *Journal of Geophysical Research*, **91**, 11519–11544.

——, ——, SHAFIQULLAH, M. & DAMON, P. E. 1986. Palaeomagnetism of Jurassic volcanic rocks in the Patagonia Mountains, southeastern Arizona: implications for the North American 170 Ma reference pole. *Journal of Geophysical Research*, **91**, 11,545–11,555.

MILLER, C. 1970. Petrology of some eclogites and metagabbros of the Oetztal Alps, Tirol, Austria. *Contributions to Mineralogy and Petrology*, **28**, 42–56.

—— 1990. Petrology of the type locality eclogites from the Koralpe and Saualpe (Eastern Alps), Austria. *Schweizerische Mineralogische und Petrographische Mitteilungen*, **70**, 287–300.

MUTTONI, G. & KENT, D. V. 1994. Palaeomagnetism of latest Anisian (Middle Triassic) sections of the Prezzo Limestone and the Buchenstein Formation, Southern Alps, Italy. *Earth and Planetary Science Letters*, **122**, 1–18.

OBERHÄNSLI, R., HUNZIKER, J. C., MARTINOTTI, G. & STERN, W. B. 1985. Geochemistry, geochronology and petrology of Monte Mucrone: an example of eo-Alpine eclogitization of Permian granitoids in the Sesia-Lanzo zone, western Alps. *Chemical Geology*, **52**, 165–184.

OPDYKE, N. D. & WENSINK, H. 1966. Palaeomagnetism of rocks from the White Mountain plutonic-volcanic series in New Hampshire and Vermont. *Journal of Geophysical Research*, **71**, 3045–3051.

PLATT, J. P., BEHRMANN, J. H., CUNNINGHAM, P. C.,

DEWEY, J. F., HELMAN, M., PARISH, M., SHEPLEY, M. G., WALLIS, S. & WESTON, P. J. 1989. Kinematics of the Alpine arc and the motion history of Adria. *Nature*, **337**, 158–161.

ROBERTSON, A. H. F., CLIFT, P. D., DEGNAN, P. & JONES, G. 1991. Palaeogeographic and palaeotectonic evolution of the Eastern Mediterranean Neotethys. *Palaeogeography, Palaeoclimatology, Palaeoecology*, **87**, 289–343.

ROYDEN, L. H. & BÁLDI, T. 1988. Early Cenozoic tectonics and palaeogeography of the Pannonian and surrounding regions. *In*: L. H. ROYDEN & F. HORVATH (eds) *The Pannonian Basin: a study in basin evolution*. American Association of Petroleum Geologists, Memoirs, **45**, 1–16.

SCHEEPERS, P. J. J. & LANGEREIS, C. G. 1994. Palaeomagnetic evidence for counter-clockwise rotations in the southern Apennines fold-and-thrust belt during the Late Pliocene and middle Pleistocene. *Tectonophysics*, **239**, 43–59.

——, —— & HILGEN, F. J. 1993. Counter-clockwise rotations in the southern Apennines during the Pleistocene: palaeomagnetic evidence from the Matera area. *Tectonophysics*, **225**, 379–410.

SCHULT, A. 1973. Palaeomagnetism of Upper Cretaceous volcanic rocks in Sicily. *Earth and Planetary Science Letters*, **19**, 97–100.

ŞENGÖR, A. M. C., YILMAZ, Y. & KETIN, I. 1980. Remnants of a pre-late Jurassic ocean in northern Turkey: fragments of Permian-Triassic Palaeo-Tethys? *Geological Society of America Bulletin*, **91**, 599–609.

SPERANZA, F. & KISSEL, C. 1993. First palaeomagnetism of Eocene rocks from Gargano: widespread overprint or non rotation? *Geophysical Research Letters*, **20**, 2627–2630.

SMITH, T. E. & NOLTIMIER, H. C. 1979. Palaeomagnetism of the Newark trend igneous rocks of the north central Appalachians and the opening of the central Atlantic Ocean. *American Journal of Science*, **279**, 778–807.

SRIVASTAVA, S. P. & TAPSCOTT, C. R. 1986. Plate kinematics of the North Atlantic, *In*: P. R. VOGT & B. E. TUCHOLKE (eds) *The Western North Atlantic Region*. The Geology of North America, **M**. Geological Society of America, 379–404.

STAMPFLI, G. M. & MARCHANT, R. H. 1995. Geodynamic evolution of the Tethyan margins of the Western Alps. *In*: LEHNER, P. *et al.* (eds) *Deep Structure of Switzerland – Results from NFP 20*. Birkhäuser AG, Basel.

STEINER, M. & HELSLEY, C. E. 1975. Reversal pattern and apparent polar wander for the Late Jurassic. *Geological Society of America Bulletin*, **86**, 1537–1543.

THÉVENIAUT, H., BESSE, J., EDEL, J. B., WESTPHAL, M. & DURINGER, P. 1992. A Middle Triassic palaeomagnetic pole for the Eurasian plate from Heming (France). *Geophysical Research Letters*, **19**, 777–780.

THÖNI, M. & JAGOUTZ, E. 1992. Some new aspects of dating eclogites in orogenic belts: Sm-Nd, Rb-Sr, and Pb-Pb isotopic results from the Austroalpine saualpe and Koralpe type-locality (Carinthia/Styria, southeastern Austria). *Geochimica et Cosmocimica Acta*, **56**, 347–368.

TOZZI, M. 1993. Assetto tettonico dell'A-vampaese Apulo meridionale (Murge meridio-nali-Salento) sulla base dei dati strutturali. *Geologica Romana*, **29**, 95–111.

TOZZI, M., KISSEL, C., FUNICELLO, R., LAJ, C. & PAROTTO, M. 1988. A clockwise rotation of Southern Apulia? *Geophysical Research Letters*, **15**, 681–684.

VANDENBERG, J. 1983. Reappraisal of palaeomagnetic data from Gargano (south Italy). *Tectonophysics*, **98**, 29–41.

—— & WONDERS, A. A. H. 1976. Palaeomagnetic evidence for large fault displacement around the Po Basin. *Tectonophysics*, **33**, 301–320.

——, KLOOTWIJK, C. T. & WONDERS, A. H. H. 1978. The late Mesozoic and Cenozoic movements of the Italian peninsula: further palaeomagnetic data from the Umbrian sequence. *Geological Society of America Bulletin*, **89**, 133–150.

VAN DER VOO, R. 1993. *Palaeomagnetism of the Atlantic. Tethys and Iapetus Oceans*. Cambridge University Press, Cambridge, UK.

VAN FOSSEN, M. C. & KENT, D. V. 1990. High-Latitude Palaeomagnetic Poles From Middle Jurassic Plutons and Moat Volcanics in New England and the Controversy Regarding Jurassic Apparent Polar Wander for North America. *Journal of Geophysical Research*, **95**, 17503–17516.

—— & —— 1992a. Reply. *Journal of Geophysical Research*, **97**, 1803–1805.

—— & —— 1992b. Palaeomagnetism of 122 Ma plutons in New England and the mid-Cretaceous palaeomagnetic field in North America: true polar wander or large scale differential mantle motion. *Journal of Geophysical Research*, **97**, 19,651–19,661.

—— & —— 1993. Palaeomagnetic study of 143 Ma kimberlite dikes in central New York State. *Geophysical Journal International*, **113**, 175–185.

WESTPHAL, M., BAZHENOV, M. L., LAUER, J. P., PERCHERSHY, D. M. & SIBUET, J. C. 1986. Palaeomagnetic implications on the evolution of the Tethys belt from the Atlantic Ocean to the Pamirs since the Triassic. *Tectonophysics*, **123**, 37–82.

WITTE, W. K. & KENT, D. V. 1989. A middle Carnian to early Norian (225Ma) palaeopole from sediments of the Newark basin, Pennsylvania. *Geological Society of America Bulletin*, **101**, 1118–1126.

—— & —— 1990. The palaeomagnetism of red beds and basalts of the Hettangian Extrusive Zone, Newark Basin, New Jersey. *Journal of Geophysical Research*, **95**, 17,533–17, 545.

—— & —— 1991. Tectonic implications of a remagneti-zation event in the Newark Basin. *Journal of Geophysical Research*, **96**, 19569–19582.

——, —— & OLSEN, P. E. 1991. Magnetostratigraphy and palaeomagnetic poles from Late Triassic -earliest Jurassic strata of the Newark basin. *Geological Society of America Bullettin*, **103**, 1648–1662.

ZAPPATERRA, E. 1994. Source-rock distribution model of the Periadriatic region. *American Association of Petroleum Geologists*, **78**, 333–354.

ZIJDERVELD, J. D. A., HAZEN, G. J. A., NARDIN, M. & VAN DER VOO, R. 1970. Shear in the Tethys and the Permian palaeomagnetism in the Southern Alps, including new results. *Tectonophysics*, **10**, 639–661.

Palaeomagnetic evidence of block rotations in the Matese Mountains, Southern Apennines, Italy

M. IORIO[1,3], G. NARDI[2], D. PIERATTINI[2] & D. H. TARLING[3]

[1] Geomare Sud, Institute of Marine Science, CNR, Via Vespucci 10, Napoli, Italy

[2] Dipartimento di Scienze della Terra, Università di Napoli, Largo San Marcellino 10, Napoli, Italy

[3] Department of Geological Sciences, University of Plymouth PL4 8AA, UK

Abstract: After rigorous geological and petrological study to exclude all Miocene rock samples that may have been affected by post-deposition disturbances and selecting only sites where the Miocene rocks appear to be rigidly attached to their Mesozoic substrate, over 100 samples were collected from the Monte Matese region. Over 200 oriented specimens, cut from these samples, were subjected to rigorous thermal and alternating field demagnetization. Strict criteria for the definition and within-site consistency of remanent vectors showed that most sites were dominated below 200°C or 20 mT by a magnetization associated with the present geomagnetic field direction. In almost all sites, any higher coercivity or blocking temperature components were inconsistent with the exception of one locality near Pescorosito. Re-sampling of two outcrops at this locality, 100 m apart, provided 26 samples that met all of the geological, petrological and palaeomagnetic criteria. The low temperature/coercivity component is the same as the present geomagnetic field in the locality, but all samples from eight of the nine sites have identical remanence vectors between 200 and 400/450°C. Petrological and tectonic considerations imply that this remanence must be associated with very early diagenesis and is thus probably of Mid-Miocene but possibly Early Miocene age. As these sites are autochthonous relative to their Mesozoic basement, they indicate a 40° anticlockwise rotation of the Monte Matese block, relative to Apulia, since Mid-Miocene times. Thus most of the post-Early Cretaceous rotation of this block has occurred since Mid-Miocene times. It is suggested that similar age constraints may also apply to other rotated blocks within the Southern Apennines.

The Southern Apennine mountain belt is characterized by a large number of nappes of shallow-water carbonates and basinal sediments that were originally deposited on the western margin of the Apulian–Adriatic promontory (D'Argenio et al. 1975) particularly during Jurassic and Cretaceous times, with later clastic deposits of Tertiary age. The Mesozoic carbonate succession includes several stratigraphic gaps (Carannante et al. 1988, 1992; Ruberti 1992), with two main gaps marked by bauxitic levels in the Middle and Upper Cretaceous. This sequence was emplaced as a series of nappes, predominantly during the Mid-Miocene, and the area is still seismically active. The mountain belt (Fig. 1), mostly belonging to the Mesozoic–Early Tertiary palaeogeographic domain (Marsella et al. 1992; D'Argenio et al. 1993), can be subdivided into a number of lithostratigraphic units. These include the Monte Maggiore–Matese unit, within which the Monte Matese massif is of key importance because of its location on the NW of the Southern Apennine

belt. This block (Fig. 2) is bordered by two main regional strike-slip faults (Ortolani et al. 1992) which were active at different times during the Pleistocene, and contains several NW–SE-trending normal faults, some of which are characterized by a vertical displacement of up to 1 km. All previous palaeomagnetic studies of Mesozoic and Tertiary rocks from different blocks within the Southern Apennines have shown anticlockwise rotations relative to the Apulian-Adriatic foreland, which is assumed to be a relatively stable authochthonous block (Channell & Tarling 1975; Incoronato et al. 1988; Iorio & Nardi 1988). In order to establish whether similar rotations have also occurred in the Monte Matese block, sampling had previously been undertaken of Lower Cretaceous carbonates at S. Sbregavitelli (Iorio & Nardi 1992). Subsequently, the Middle Miocene Longano carbonates were investigated in the Monte Matese region to provide a palaeomagnetic control for a different time in order to enable the tectonic evolution of this massif to be quantified.

From Morris, A. & Tarling, D. H. (eds), 1996, *Palaeomagnetism and Tectonics of the Mediterranean Region*, Geological Society Special Publication No. 105, pp. 133–139.

133

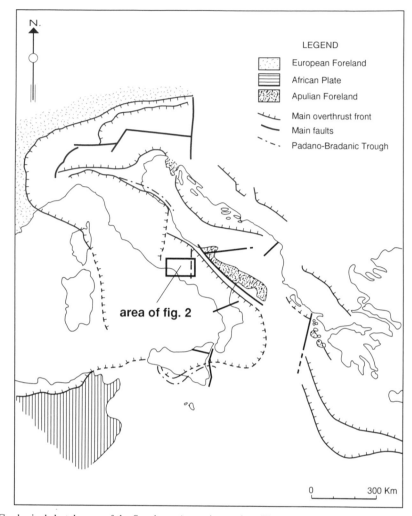

Fig. 1. Geological sketch map of the Southern Apennine region. The location of the Monte Matese sampling location is shown in black.

Sampling and analysis

Sampling was attempted at some 40 localities throughout an area of 150 km² comprising Monte Matese Massif, Monte Venafro and Monte Ausoni. The Miocene deposits included: (i) Burdigalian calcareous rocks containing bryozoans and litothamnion (the Cusano Formation); and (ii) Serravallian marly limestones and marls with orbulinas (the Longano Formation) and sandy clay, mostly turbidity sediments (the Pietraroia Formation). Many of the Cusano and Pietraroia localities proved unsuitable for further study on the basis of evidence for alteration, the degree of fracturing, unsuitable lithologies or lack of adequate tectonic control. However, some of the Longano sediments were

semi-pelagic wackestone and packstone, rich in planktonic foraminifera (Carannante & Simone 1988), and localities could be found where the bedding planes were well defined and consistent. From such localities, hand samples were oriented and bedding attitudes measured using a magnetic compass as the rocks were known to be magnetically very weak. These samples were subsequently drilled in the laboratory to provide an average of three or more cylindrical specimens per sample. At this stage, petrological study of the samples was used to exclude all samples showing evidence of post-deposition disturbances at both site and sample levels, e.g. bioturbation, fractures and evidence for disturbance in the clay fraction. On this basis, some 30 sites were excluded, leaving only ten localities

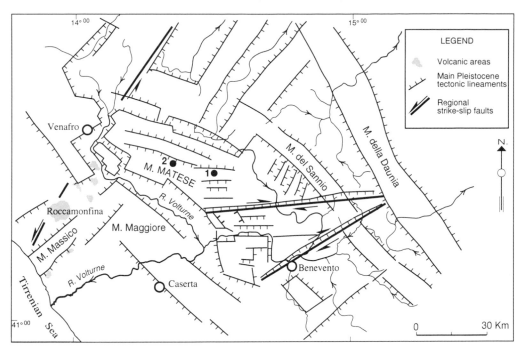

Fig. 2. The regional structure of the Matese Mountain region. The sampling locality for the Middle Miocene at Pescorosito is marked as 1, and the Middle Cretaceous locality (Iorio & Nardi 1992) as at 2 (Serra Sbregavitelli). Simplified from Ortolani *et al.* (1992).

(100 samples) that met these stringent petrological requirements. From these localities, two specimens from each locality (i.e. an uninterrupted area of a few 100 m^2 containing several beds) were selected for measurement of their palaeomagnetic properties. Initially, the natural remanent magnetization (NRM) of these 200 specimens was measured at the Palaeomagnetic Laboratory of Gainesville University (USA) using a cryogenic magnetometer. One specimen from each site was thermally demagnetized in 25°C increments, using a Schoensted furnace, until no longer measurable. On the basis of vector analyses of these samples, the other specimens were demagnetized through the range of 250–450°C. Comparisons were then made between the vectors identified in specimens from the same hand samples and between the vectors isolated in different specimens from different hand samples from the same site (Iorio & Nardi 1988). These showed such poor consistency at nine of the ten remaining localities that the result could not be considered reliable. Vectors isolated by thermal demagnetization of specimens from the remaining locality at Pescorosito showed such a high internal consistency that it was considered that further sampling should be undertaken.

At Pescorosito the Longano Formation forms part of a monoclinal structure where Lower and Middle Miocene deposits transgressively overlie the Coniacian–Lower Senonian limestones. At this locality, the formation is 4–7 m thick and includes well stratified 2–7 cm thick beds of alternating marly limestones and marls. A total of 26 hand samples were taken from eight different beds at two outcrops (sites 1–5 and 6–9) about 100 m apart. The area was also re-examined and it was established that the sampled beds were rigidly attached to the local basement. Each of the samples was also examined petrologically to ensure that there were no indications of post-deposition disturbance. The remanence and susceptibility of cores taken from these samples were then measured at the Palaeomagnetic Laboratory of the Geophysical Institute of Hungary (Budapest) using a JR-4 spinner magnetometer and a KLY-2 susceptibility bridge. Two specimens per site were stepwise demagnetized; one using AF, the other thermally (Fig. 3). The remaining specimens from sites in which the thermally demagnetized specimens showed no susceptibility change by 400–450°C were then thermally demagnetized between 150° and 200°C and then in 50°C increments until the magnetic remanence decreased to the noise-level of the JR-4. For sites where the susceptibility changed, the remaining

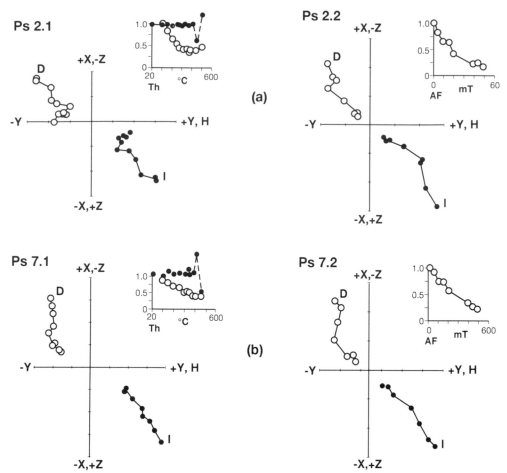

Fig. 3. Examples of demagnetization behaviour. The Cartesian plots illustrate thermal demagnetization on the left and alternating field demagnetization on the right. Open symbols represent the horizontal component and solid symbols show the inclination (horizontal v. vertical). The graphs illustrate the change in intensity (open symbol) and susceptibility (solid symbol – thermal only) as a function of treatment level. The average initial intensities of remanence are given in Table 1; the initial susceptibilities are 49.5 m SI (P2.1) and 50.0 m SI (P7.1).

specimens were AF demagnetized at 10 mT, 20 mT, 30 mT, 40 mT and then in increments of 5 mT until the intensity of magnetic remanence decreased to the noise-level of the magnetometer. In view of the susceptibility changes during thermal demagnetization in Budapest, the results from the thermal analyses at Gainesville (Iorio & Nardi 1988) were discarded because of the lack of control on such thermochemical changes. The isothermal remanent magnetization (IRM) properties of two samples, from sites 5 and 6 (Fig. 4), were also studied in the Palaeomagnetic Laboratory of University of Naples, using a Highmoor magnet (peak field 1 T) and a Molspin magnetometer.

Palaeomagnetic results

At least two components were identified by both thermal and AF treatment in specimens from all sites from Pescorosito, except for site 6 for which the remanences were destroyed by 250°C and AF demagnetization could only identify a northerly and moderately steep direction, close to the present geomagnetic field direction. The same component was also found in all other specimens as the low coercivity/low blocking temperature component, generally removed by 250°C or 20 mT. In these specimens, visual and computer analyses (mainly Kirschvink 1980) isolated a clear component, particularly between 250°C and 350°C, that was identical to the

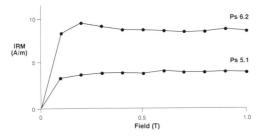

Fig. 4. Isothermal remanent magnetization acquisition curves. Samples are from sites 5 (PS5) and 6 (PS6); both became saturated between 0.1 and 0.2 T.

component isolated during AF demagnetization between 25 and 50 mT. This second component persisted until the remanence either became unmeasurable or thermochemical changes occurred, usually around 400–450°C. Excluding site 6, and including all samples from the remaining sites, the mean vector was well defined, $D = 313°$, $I = 57°$ ($k = 172$, $\alpha_{95} = 4°$; Table 1). While other components could sometimes be identified above 400–450°C, these were mutually inconsistent and were excluded in view of the probability of thermochemical changes at such temperatures. Although only two samples were studied for their IRM remanence properties, both behaved similarly, becoming completely saturated between 0.1 and 0.2 T (Fig. 4). This, combined with the shape of the thermal demagnetization curves (Fig. 3a, b) indicates that the dominant magnetic mineral is titanomagnetite. In these sediments, this mineral is likely to be associated with diagenesis, although neither the identification of magnetite nor its primary nature can be considered established at this stage. Nonetheless, it is reasonable to assume that the magnetization existing in these carbonate rocks is contemporaneous with diagenesis. Furthermore, diagenesis in such rocks usually occurs rapidly after deposition and thus can be considered to be, in practice, contemporaneous with the age of deposition.

Discussion and conclusions

A very stringent set of rejection criteria have been applied to the selection of the sites and samples prior to undertaking the palaeomagnetic analyses. The palaeomagnetic criteria have similarly been stringently applied. Consequently nearly 400 samples have been rejected, primarily on geological and petrological criteria, and only 26 sample observations (c. 5% of the total collection and <1% of the sites visited) were found to fulfil all requirements. Although the

remaining data are few, and must therefore be treated cautiously, this is in part compensated by their quality and by the fact that the same component has been identified in every sample from every site, whether by thermal or AF demagnetization (except site 6 where no high temperature component could be isolated). However, the uniformity of the bedding in the two localities means that no fold test can be undertaken and other methods must be used to assess the age of this component of magnetization. The magnetization appears to be carried by titanomagnetites that formed during diagenesis, which probably occurred almost immediately after deposition in such lithologies. Such early cementation would also inhibit subsequent chemical change which would, in any case, be likely to produce more highly oxidized magnetic minerals than have been observed. The association of the low temperature/coercivity component with the present geomagnetic field also means that the high temperature/coercivity component must have been acquired earlier. The *in situ* inclination is completely inconsistent with any Miocene or later field direction (Table 2). Consequently the magnetization must have been acquired either prior to or during the earliest stages of tilting as it is only in this position that it would conform to a geomagnetic averaged inclination. It is unlikely that it was acquired during the early stages of folding, during which the flushing of fluids or other tectonically induced effects might have occurred, as the carbonate cementation would have isolated the magnetic minerals from such processes. Such remagnetization would, in any case, be discontinuous and variable within the 1000 m^3 of sediment over which such a highly uniform, univectoral magnetization has been identified. The simplest and most likely hypothesis is that the remanence is diagenetic in origin, i.e. acquired in the Mid-Miocene field, and that the sampled beds, together with their Mesozoic basement, have been subsequently tilted to their present orientation, prior to the acquisition of the low temperature/coercivity viscous remanence during the last 1 Ma.

Considering first the Lower Cretaceous data obtained by Iorio & Nardi (1992) from the Monte Matese block, the Serra Sbregavitelli mean direction has an inclination similar to those reported for the Apulian Foreland (Table 2), but shows a 40° difference in declination with respect to the Early Cretaceous Gargano mean direction. Iorio & Nardi (1992) have argued that there has actually been some 66° clockwise rotation of the Matese block relative to Africa at some unspecified time since the Lower Cre-

Table 1. *Pescorosito site mean high temperature vectors*

Site No.	NRM mA m^{-1}	N/n	In situ Decl	In situ Inc	Bedding corrected Decl	Bedding corrected Inc	Palaeopole Lat	Palaeopole Long
1	6.6	6/6	299.7	38.6	306.9	59.7	51	301
2	6.1	6/6	307.0	39.6	318.0	59.5	59	296
3	14.7	2/2	297.2	36.8	302.6	58.2	46	301
4	3.8	2/2	297.8	32.5	302.5	53.9	44	294
5	5.9	2/2	307.0	35.0	316.3	55.1	55	289
6	5.0	2/2	–	–	–	–	–	–
7	19.5	2/2	311.7	35.1	322.8	54.2	59	283
8	8.4	2/2	299.7	33.6	305.5	54.8	47	295
9	6.7	2/2	316.3	36.8	329.7	54.6	65	279
Mean	9.0	8/9 sites	304.5	36.1	313.1		56.6 ($k = 172$, $\alpha_{95} = 4.2°$)	

The mean directions are based on the high coercivity/high blocking temperature components, usually isolated above 350°C or above 45 mT. The initial average intensity of remanence (NRM) is the average value for the specimens from each site prior to demagnetization. The number of samples collected is N and the number accepted is *n*. The precisions for the mean directions of sites 1 and 2 are $k = 124$, $\alpha_{95} = 6.0°$ and $k = 80$, $\alpha_{95} = 7.5°$ respectively, but are indeterminate for the other sites as N is only 2.

Table 2. *Mean directions and poles for Apulia and Monte Matese*

Locality	Age	Dec	Inc	Lat°N	Long°	Reference
Apulia Foreland (41.5°N 16.0°E)						
Gargano	Early Cretaceous	313	38	38	278	Vandenberg (1983)
Gargano	Late Cretaceous	328	38	56	259	Vandenberg (1983)
Murge	Late Cretaceous	327	38	56	261	Márton & Nardi (1994)
Monte Matese (42.0°N 15.0°E)						
S. Sbregavitelli	Early Cretaceous	274	40	18	300	Iorio & Nardi (1992)
Pescorosito	Early Mid-Miocene	313	57	53	293	This paper

taceous, as they consider that Apulia has independently rotated anticlockwise by some 15° (Márton & Nardi 1994). While the problem of the extent to which Apulia can be considered 'African' is of major importance (see Channell, this volume), only the timing of such differential motions will be discussed here. As no other Miocene palaeomagnetic data are available for the Southern Apennines, the Miocene field must be modelled. The Iorio & Nardi (1992) hypothesis means that two models need to be considered: (i) the Miocene average field direction in Apulia can be represented by the present axial dipole field ($D = 0°$, $I = 61°$), hence assuming no rotation of Apulia since Miocene times; and (ii) the Miocene field was that of the axial dipole field but this has been rotated anticlockwise, by some 15° to become $D = 345°$, $I = 61°$, as a consequence of tectonic rotations since and possibly including Miocene times. The new Early–Mid-Miocene Pescorosito mean inclination reported here agrees with that predicted on both models, within the errors, but the mean

declination differs by 47° or by 32° relative to those predicted by models (i) and (ii) respectively. This suggests that most of the tectonic rotation of the Matese Block, relative to either Apulia or Africa, must have taken place since at least Mid-Miocene times and possibly only since the Late Miocene. It is suggested that most other tectonic rotations in the Southern Apennines are also likely to be very young. Clearly such an interpretation needs to be tested by further studies of Miocene rocks within the region, although strict geological, petrological and palaeomagnetic criteria, such as used in the current study, will be fundamental to such tests.

References

CARANNANTE, G. &SIMONE, L. 1988. *In:* Cocco, E. & D'ARGENIO, B. (eds) *L' Appennino Campano–Lucano nel quadro geologico dell' Italia meridionale. Guida all' Escursione.* 74th Congresso della Società Geologica, Italiana, 169–172.

——, D'ARGENIO, B., DELLO IACOVO, B., FERRERI, V.,

MINDSZENTY, A. & SIMONE, L. 1992. Studi sul carsismo cretacico dell' appennino Campano. *Memorie della Societe Geologica Italiana*, **41**, 733–759.

CHANNELL, J. E. T. & TARLING, D. H. 1975. Paleomagnetism and rotation of Italy. *Earth and Planetary Science Letters*, **25**, 177–188.

D'ARGENIO, B., PESCATORE, T. & SCANDONE, P. 1975. Structural pattern of Campania–Lucania Apennines. *In*: OGNIBEN, L. & PRATURLON, A. (eds) *Structural Model of Italy*. Quaderni de la Ricerca Scientifica C.N.R. **90**, 313–328.

——, FERRANTI, L., MARSELLA, E., PAPPONE, G. & SACCHI, M. 1993. From the Lost Lago Negro Basin to the present Tyrrhenian: the Southern Apennines between compression and extension. *In*: *Origin of Sedimentary Basins*. Field Trip Guide Book, 134.

INCORONATO, A., MARSELLA, E. & PAPPONE, G. 1988. Studio paleomagnetico nelle successioni lagonegresi tra Vietri di Potenza e Savoia di Lucania (Appennino campano–lucano). *74th Congresso della Societe Geologia Italiana*. A339–A342.

IORIO, M. & NARDI, G. 1988. Studi paleomagnetici su rocce mesozoiche del Matese Occidentale. *74th Congresso della Societa Geologia Italiana, A343–A345*.

—— & —— 1992. Studi paleomagnetici sul mesozoico del Matese Occidentale. *Memorie della Societa Geologica Italiana,* **41**, 1253–1261

KIRSCHVINK, J. L. 1980. The least-squares line and plane and the analysis of palaeomagnetic data. *Geophysical Journal of the Royal Astronomical Society,* **62**, 669–718.

MARSELLA, E., PAPPONE, G., D'ARGENIO, B., BALLY, A. & CIPPITELLI, G. 1992. L'origine interna dei terreni Lagonegresi e l'assetto tettonicao dell' Appennine Meridionale. *Rendiconti dell' Accadamia Sciente Fisica e Matematica, Napoli.* **59**, 73–101.

MARTON, E. & NARDI, G. 1994 The Apulia region has been part of Africa? New palaeomagnetic results from Murge. *Geophysical Journal International,* **119**, 842–856.

ORTOLANI, F., CAIAZZO, C., PAGLIUCA, S., PALCSCANDOLO, G., SCHIATTARELLA, M. & TOCCACELI, R. M. 1992 Evoluzione geomorfologica e tettonica quaternaria dell' appenino centro-meridionale. *Convegno Excursione, 6–10 Iuglio 1992, Studi Geologici Camerti.*

PRUBERTI, D. 1992 Le lacune stratigrafiche nel Cretacico del Matese centro-settentrionale. *Bolloltino Società Geologica Italiana,* **111**, 283–289.

VANDENBERG, J. 1983. Reappraisal of paleomagnetic data from Gargano (South Italy) *Tectonophysics,* **98**, 29–41.

Lack of Late Miocene to Present rotation in the Northern Tyrrhenian margin (Italy): a constraint on geodynamic evolution

M. MATTEI[1], C. KISSEL[2], L. SAGNOTTI[3], R. FUNICIELLO[1] & C. FACCENNA[1]

[1] Dipartimento di Scienze Geologiche, Terza Università, Via Ostiense, 169-00154 Roma,
Italy
[2] Centre des Faibles Radioactivities CNRS-CEA, 91198 Gif sur Yvette, France
[3] Istituto Nazionale di Geofisica, Via di Vigna Murata, 605-00143 Roma, Italy

This is an extended abstract which discusses palaeomagnetic results from Neogene formations in the Tyrrhenian margin of northern and central Italy. These data have been previously published by Sagnotti et al. (1994) and Mattei et al. (1995b).

Palaeomagnetic data have been widely used in reconstructing the geodynamic evolution of the Central Mediterranean. The pioneering research mainly focused on the tectonic interpretation of palaeomagnetic results from Jurassic to Eocene limestone formations from the Umbria–Marche Apennines (Lowrie & Alvarez 1974; Channell & Tarling 1975; Channell et al. 1978; VandenBerg et al. 1978). These studies identified a counterclockwise rotation which was believed to extend across the entire Italian peninsula. This rotation was subsequently identified in coeval formations from other sectors of the 'Adriatic Promontory' (e.g. Márton & Veljovic 1983; VandenBerg 1983; Lowrie 1986). The age of the rotation was initially defined as Mesozoic-Early Cenozoic. The quality and consistency of these data led to their widespread incorporation into geodynamic models for the Neogene evolution of the peninsula. In many models (e.g. Hill & Hayward 1985; Sartori 1990; Doglioni 1991; Castellarin et al. 1992), the counterclockwise rotation of the Mesozoic and Early Cenozoic formations was assumed to have occurred during the post-Mid-Miocene opening of the Tyrrhenian Sea (to the west) and to have affected all the geodynamic provinces of the Italian peninsula. In the last few years, however, structural and stratigraphic studies (summarized in CNR 1989) have shown that the main tectonic events in the peninsula took place in the Late Miocene to Pleistocene time interval. As a result, the focus of palaeomagnetic research has shifted to Neogene sedimentary sequences (predominantly clays and marls), in order to constrain the rotational deformation accompanying the main phases of the evolution of the Tyrrhenian Sea–Apennine chain–Adriatic foreland system.

Significant Plio-Pleistocene rotations are now recognized as a common feature in southern Italy (Aifa et al. 1988; Sagnotti 1992; Scheepers & Langereis 1993; Scheepers et al. 1993; Scheepers 1994). A varied regional pattern is observed, with rotations in opposite senses between the southern Apennines (CCW) and the Calabrian Arc and Sicily (CW). No tectonic rotation was found, however, in coeval deposits from the Apulian foreland (Scheepers 1992) or the Bradano foredeep (Scheepers 1994).

These studies of Neogene sediments have concentrated mainly in southern Italy and Sicily. On the Tyrrhenian margin to the north, palaeomagnetic data have only previously been obtained from a small outcrop of the Pliocene San Vincenzo rhyolite, where no tectonic rotation was observed (Lowrie & Alvarez 1979). The present contribution summarizes the results we have obtained from Late Miocene to Pleistocene marine clays and marls exposed along the Tyrrhenian coast. These data indicate no significant rotation of the northern Tyrrhenian margin during the later Neogene, and place additional constraints on the geodynamic evolution of the region.

Geological setting

The area of study is termed the Tuscan–Latium Tyrrhenian margin, and forms the western sector of the northern Italian peninsula (Fig.1). The region has a complex geological history, which is mainly related to geodynamic processes associated with the opening of the Tyrrhenian Sea. The regional basement is represented by small outcrops of Palaeozoic metamorphic rocks. The overlying sedimentary sequences are internal to the platform, and were deposited on both oceanic and continental crust. A series of eastward-verging thrusts affects both the sedimentary sequences and their metamorphic basement (Pandeli et al. 1988). Following the compressional phase, a post-orogenic extensional regime developed along the Tuscan–

From Morris, A. & Tarling, D. H. (eds), 1996, Palaeomagnetism and Tectonics of the
Mediterranean Region, Geological Society Special Publication No. 105, pp. 141–146.

141

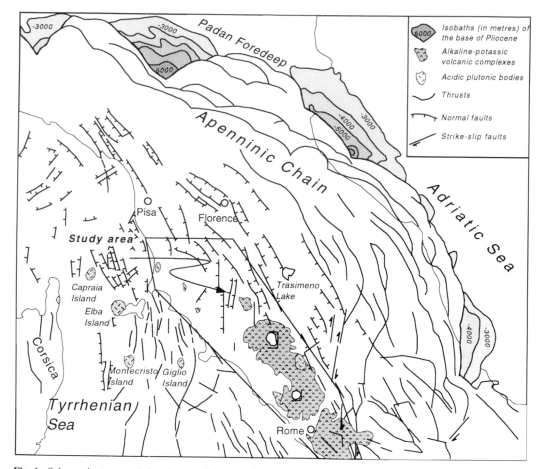

Fig. 1. Schematic structural sketch map of the main Neogene deformation in the northern Italian peninsula (redrawn from CNR 1989).

Latium Tyrrhenian margin during the Late Miocene, with associated normal faulting. This extensional phase was accompanied by the formation of large sedimentary basins, which are mainly oriented NW–SE. Coeval, minor NE–SW-trending basins developed as transverse structures. The sedimentary basins are filled by continental, brackish and marine sequences of the 'Neoautochthonous' sedimentary cycle, dating from the Late Miocene to the Pleistocene. The sediments unconformably cover the compressional Apenninic structures. The extensional phase was also accompanied by diffuse magmatic activity. The extensional stress field migrated progressively eastward from the Tyrrhenian coast to the axis of the Apennines chain, where it is active today.

Sampling, palaeomagnetic methods and results

Palaeomagnetic samples were collected from 60 sites in the grey-blue, marine, marly-clayey units of the Tuscan–Latium Tyrrhenian margin between Pisa and south of Rome (Alban Hills). Sampled units are Upper Miocene (Messinian) to Lower Pleistocene in age, and form part of a sequence of sediments that were deposited in the post-orogenic extensional basins. The availability of fresh, unweathered outcrops of fine-grained, undisturbed units allowed a near-uniform distribution of sampling sites. The sampled units can be considered effectively autochthonous and were unaffected by compressional tectonics or significant tectonic trans-

DECL = 357.5°   INCL = 56.5°   N

k = 84.3 α_{95} = 2.5°   N = 39

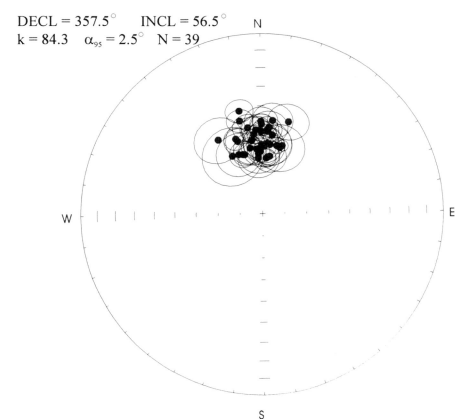

Fig. 2. Palaeomagnetic mean directions for the analysed sites after tectonic correction. Schmidt equal-area projection (lower hemisphere). The ellipses show the α_{95} of each site.

port. Between nine and 16 cores were collected per site, and specimens were stepwise demagnetized using AF or thermal techniques. The magnetic mineralogy was studied using: IRM acquisition curves; thermal demagnetization of three component IRMs; hysteresis behaviour; and thermomagnetic analyses. Full details of laboratory measurements and statistical analyses are given in Sagnotti *et al.* (1994) and Mattei *et al.* (1995*b*).

Useful results were obtained from 39 sites, where the magnetization is carried by magnetite and/or iron sulphides. The magnetization at 17 of the remaining sites was too low to measure, even using a cryogenic magnetometer, and four sites were characterized by coarse-grained magnetite which showed unstable behaviour during demagnetization. The 39 site means have low within-site dispersions. The site mean data pass both fold and reversal tests, suggesting that the primary remanence has been successfully identified. The overall mean direction after tectonic correction is:

$$D = 357.5°; I = 56.5°; k = 84.3;$$
$$\alpha_{95} = 2.5°; N = 39 \text{ (Fig. 2)}.$$

This indicates that there has been no regional tectonic rotation of the Tuscan–Latium Tyrrhenian margin since Late Messinian–Early Pleistocene times, in spite of strong faulting, tilting and local folding. It thus forms a non-rotating hinterland for the well-known, widespread rotational deformation of the Apennines. Local rotations were detected only in the Volterra basin, where the mean of 13 sites is rotated 10° counterclockwise from the expected direction for the geocentric axial dipole field at the locality.

Discussion

Palaeomagnetic research in Plio-Pleistocene sedimentary sequences has been mainly concentrated in southern Italy and Sicily (e.g. Sagnotti

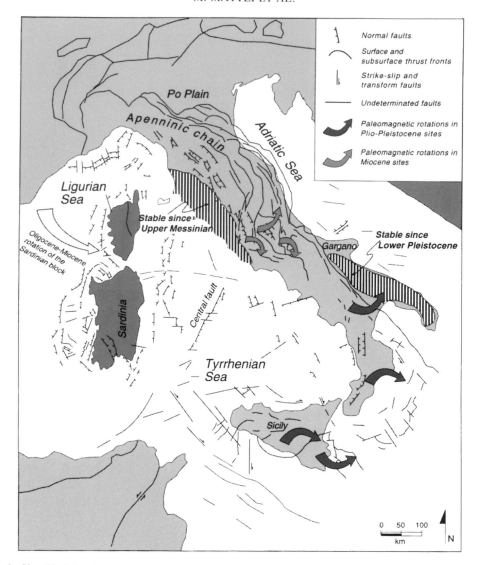

Fig. 3. Simplified sketch of the Neogene rotational patterns deserved in the circum-Tyrrhenian region. The complex pattern of Plio-Pleistocene rotations in southern Italy contrasts with the lack of rotations in the northern Tyrrhenian margin. References are given in the text.

1992; Scheepers 1994), where a complex pattern of tectonic rotations has been recognized (Fig. 3). In particular, the Pleistocene tectonic evolution of the southern Apennines, Calabrian arc and Sicily has been accompanied by very rapid and complex rotations. These rotations have affected both the folded and thrust structures of the external Apenninic front and the extensional basins of the Tyrrhenian margin. The undeformed Pleistocene units of the Bradano region were not rotated, however, indicating that at least during the Pleistocene the Apulian platform behaved as an independent structure with respect to the Apennine chain.

Our results show that the northern Tyrrhenian margin did not undergo significant regional rotation since the Late Messinian. As a consequence, the hypothesis that rotations measured in southern Italy result from geodynamic processes operating over the whole central Mediterranean region can be excluded. The different tectonic history of northern Italy with respect to southern Italy and Sicily has been previously identified from other geological, geophysical and volcanological datasets (e.g. Kastens & Mascle 1990). Serri *et al.* (1991) proposed that the different magmatic evolution of the volcanic provinces of southern Italy with respect to those in the north can be related to subduction of two distinct slabs consisting of oceanic lithosphere in the south and continental lithosphere in the north.

In northern Italy, palaeomagnetic data are mainly concentrated in Mesozoic–Cenozoic sedimentary sequences of the Apennines chain. The data indicate large counterclockwise rotations in the Umbria–Marche Apennines (to the east of our study area) and a complex pattern of rotation in the Central Apennines (Márton & D'Andrea 1992; Dela Pierre *et al.* 1992; Mattei *et al.* 1995*a*). Our new data from the Tyrrhenian margin suggest two possible interpretations of the counterclockwise rotations observed in the Umbria region: (1) the counterclockwise rotations are older than the Late Messinian, and may have affected larger crustal blocks than merely the present-day Umbria region (Adria?); or (2) the counterclockwise rotations are post-Messinian in age, but are confined to the allochthonous structures of the Apennine chain, as previously suggested by Hirt & Lowrie (1986) and Channell (1992) for the Umbria region and by Dela Pierre *et al.* (1992) and Mattei *et al.* (1995*a*) for the Central Apennines. These two possibilities are not mutually exclusive, and two distinct, successive phases of rotation are also possible.

We suggest that the tectonic evolution of the northern Tyrrhenian Sea–Apennine chain involved: (1) thrusting and folding of the external domains, with accompanying rotation of allochthonous units; followed by (2) eastward migration of the orogenic system, and subsequent gravitational collapse of the western sector. This second phase involved no tectonic rotation of the syn-rift and post-rift sequences. In this model, the rotations observed in the Apennine chain are not related to post-Tortonian rifting of the Tyrrhenian Sea, but instead are linked to translation of the allochthonous structures.

This non-rotating hinterland mechanism is not common in the rifted regions of the Mediterranean. In the Balearic Sea, for example, extension was accompanied by counterclockwise rotation of the Corsica–Sardinia block (e.g. Montigny *et al.* 1981; Rehault *et al.* 1982). We suggest that the difference in rifting mechanisms relates to differences in crustal structure: a normal continental crust in the case of the Sardinian block (where rifting was an active geodynamic process which caused drifting and rotations), versus a thickened and weak, tectonized crust in

the northern Apennine chain (where post-orogenic collapse did not cause rotation).

We are grateful to D. Wise for stimulating comments on a previous draft of this paper.

References

AIFA, T., FEINBERG, H. & POZZI, J. P. 1988. Pliocene–Pleistocene evolution of the Tyrrhenian arc: paleomagnetic determination of uplift and rotational deformation. *Earth and Planetary Science Letters*, **87**, 438–452.

CASTELLARIN, A., CANTELLI, L., FESCE, A. M., MERCIER, J. L., PICOTTI, V., PINI, G. A., PROSSER, G. & SELLI, L. 1992. Alpine compressional tectonics in the Southern Alps. Relationships with the N-Apennines. *Annales Tectonicae*, **6**, 62–94.

CHANNELL, J. E. T. 1992. Paleomagnetic data from Umbria (Italy): implications for the rotation of Adria and Mesozoic apparent polar wander paths. *Tectonophysics*, **216**, 365–378.

—— & TARLING, D. H. 1975. Paleomagnetism and rotation of Italy. *Earth and Planetary Science Letters*, **25**, 177–188.

——, LOWRIE, W., MEDIZZA, F. & ALVAREZ, W. 1978. Paleomagnetism and tectonics in Umbria, Italy. *Earth and Planetary Science Letters*, **39**, 199–210.

CNR 1989. *Synthetic structural-kinematic map of Italy.* Scale 1 : 2.000.000. CNR Progetto Finalizzato Geodinamica

DELA PIERRE, F., GHISETTI, F., LANZA, R. & VEZZANI, L. 1992. Paleomagnetic and structural evidence of Neogene tectonic rotation of the Gran Sasso range (central Apennines, Italy). *Tectonophysics*, **215**, 335–348.

DOGLIONI, C. 1991. A proposal for the kinematic modelling of W dipping subduction. Possible applications to the Tyrrhenian–Apennines system. *Terra Nova*, **3**, 423–434.

HILL, C. & HAYWARD, A. 1985. Structural constraints on the Tertiary plate tectonic evolution of Italy. *Marine and Petroleum Geology*, **5**, 2–16.

HIRT, A. & LOWRIE, W. 1986. Paleomagnetism of the Umbrian–Marches orogenic belt. *Tectonophysics*, **146**, 91–103.

KASTENS, K. A. & MASCLE, J. 1990. The geological evolution of the Tyrrhenian sea: an introduction to the scientific results of ODP Leg 107. *In*: KASTENS K. A., MASCLE J. *et al.*, *Proceedings of the Oceanic Drilling Program, Scientific Results*, **107**, 3–26.

LOWRIE, W. 1986. Paleomagnetism and the Adriatic promontory: a reappraisal. *Tectonics*, **5**, 797–807.

—— & ALVAREZ, W. 1974. Rotation of Italian peninsula. *Nature*, **251**, 285–288.

—— —— 1979. Paleomagnetism and rock magnetism of the Pliocene rhyolite at San Vincenzo, Tuscany, Italy. *Journal of Geophysics*, **45**, 417–432.

MÁRTON, E. & VELJOVIC, D. 1983. Paleomagnetic results from the Istria peninsula (Yugoslavia). *Tectonophysics*, **91**, 73–87.

MÁRTON, P. & D'ANDREA, M. 1992. Paleomagnetically inferred rotations of the Abruzzi and northwestern Umbria. *Tectonophysics,* **202,** 43–53.

MATTEI, M., FUNICIELLO, R. & KISSEL, C. 1995*a.* Paleomagnetic and structural evidence of Neogene block rotations in the central Apennines (Italy). *Journal of Geophysical Research,* in press.

——, KISSEL, C. & FUNICIELLO, R. 1995*b.* No tectonic rotation of the Tuscan margin (Italy) since Upper Messinian: structural and geodynamical implications. *Journal of Geophysical Research,* **100,** 17863–17883.

MONTIGNY, R., EDEL, J. B. & THUIZAT, R. 1981. Oligo–Miocene rotation of Sardinia: K-Ar ages and paleomagnetic data of Tertiary volcanics. *Earth and Planetary Science Letters,* **54,** 261–271.

PANDELI, E., PUXEDDU, M., GIANELLI, G., BERTINI, G. & CASTELLUCCI, P. 1988. Paleozoic sequences crossed by deep drillings in the Monte Amiata geothermal region (Italy). *Bollettino della Società Geologica Italiana,* **107,** 593–606.

REHAULT, J. P., BOILLOT, G. & MAUFFRET, A. 1982. The western Mediterranean basin. *In:* STANLEY, D. J. & WEZEL, F. C. (eds) *Geological evolution of the Mediterranean basin.* Springer-Verlag, 101–129.

SAGNOTTI, L. 1992. Paleomagnetic evidence for a Plio-Pleistocene counterclockwise rotation of the Sant'Arcangelo basin, Southern Italy. *Geophysical Research Letters,* **19,** 135–138.

SAGNOTTI, L., MATTEI, M., FUNICIELLO, R. & FACCENNA, C. 1994. Paleomagnetic evidence for no tectonic rotation of the Central Italy Tyrrhenian margin since Upper Pliocene. *Geophysical Research Letters,* **21,** 481–484.

SARTORI, R. 1990. The main results of ODP leg 107 in the frame of Neogene to Recent geology of Perityrrhenian areas. *In:* KASTENS, K. A., MASCLE, J. *et al.* (eds) *Proceedings of the Oceanic Drilling Program, Scientific Results,* **107,** 715–730.

SCHEEPERS, P. J. J. 1992. No tectonic rotation for the Apulia–Gargano foreland in the Pleistocene. *Geophysical Research Letters,* **19,** 2275–2278.

—— 1994. Tectonic rotations in the Tyrrhenian arc system during the Quaternary and late Tertiary. *Geologica Ultratrajectina,* **112,** 1–352.

—— & LANGEREIS, C. G. 1993. Analysis of NRM directions from Rossello composite: implications for tectonic rotations of the Caltanisetta basin, Sicily. *Earth and Planetary Science Letters,* **119,** 243–258.

——, —— & HILGEN, F. 1993. Counterclockwise rotations in the southern Appennines during the Pleistocene: paleomagnetic evidence from the Matera area. *Tectonophysics,* **225,** 379–410.

SERRI, G., INNOCENTI, F., MANETTI, P., TONARINI, S. & FERRARA, G. 1991. Il magmatismo neogenico-quaternario dell'area tosco-laziale-umbra: implicazioni sui modelli di evoluzione geodinamica dell'Appennino settentrionale. *Studi Geologici Camerti,* 1991/1, 429–463.

VANDENBERG, J. 1983. Reappraisal of paleomagnetic data from Gargano (South Italy). *Tectonophysics,* **98,** 29–41.

——, KLOOTWIJK, C. T. & WONDERS, A. A. H. 1978. Late Mesozoic and Cenozoic movements of the Italian peninsula: further paleomagnetic data from the Umbrian sequence. *Geological Society of America Bulletin,* **89,** 133–150.

The pattern of crustal block rotations in the Italian region deduced from aeromagnetic anomalies

M. FEDI[1], G. FLORIO[2] & A. RAPOLLA[2]

[1] *Dipartimento di Scienza dei Materiali, Università di Lecce, Via per Arnesano, Lecce, Italy*

[2] *Dipartimento di Geofisica e Vulcanologia, Università di Napoli 'Federico II', Largo S. Marcellino 10, 80138 Napoli, Italy*

Abstract: In contrast to palaeomagnetic studies that consider outcropping rocks, the analysis of aeromagnetic anomalies can give averaged information about large volumes of deep-seated rocks. The recognition and analysis of anomalies having an 'abnormal shape' in the central Mediterranean area have yielded original and interesting results in recent years. Accurate interpretation of these anomalies gives information on the nature of magnetization of the deep crust and on the geodynamic history of this region. Results suggest that the southern Italian area can be divided into zones which experienced different tectonic rotations in both clockwise and anticlockwise senses. The pattern of rotations deduced in this way agrees well with that determined by standard palaeomagnetic analyses.

The structural complexity of the Mediterranean area and the difficulty in reconstructing its geodynamic history have attracted the attention of many palaeomagnetists. In this paper we describe a different approach to the detection of crustal block rotations, based on the analysis of aeromagnetic anomalies. Such an approach can prove interesting in a regional context because, unlike the palaeomagnetic method, it considers deep-seated sources and gives information on extensive areas. Moreover, it seems opportune to obtain information on block rotations from another independent method, especially when a case as complex as that of the Mediterranean area is considered.

This method of analysis was conceived after the recognition of dipolar magnetic anomalies whose peak-to-trough axes are not parallel to the local magnetic meridian. Such occurrences have been reported in various areas of the world (e.g. Books 1962; Fedi & Rapolla 1988; Alma-Valdivia *et al.* 1991). We refer to these magnetic anomalies as 'abnormal shape magnetic anomalies' (Fig. 1). A number of these anomalies can be noted in the aeromagnetic field of the southern Italian area (Fedi & Rapolla 1988, 1990; Fig. 2). These are characterized by a relatively short wavelength in the Tyrrhenian area and a longer wavelength in the Italian peninsular region. The angle between the peak-to-trough axes and the magnetic meridian (α angle) is never less than about 35–40°. An important feature of these anomalies is that they occur in groups of several anomalies with

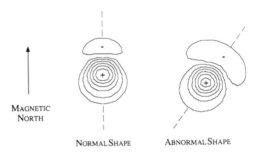

Fig. 1. Example of magnetic anomalies at the latitude of the southern Italian area. If the peak-to-trough axis forms an angle with the direction of magnetic north the anomaly is said to have an 'abnormal shape'.

MAGNETIC NORTH

NORMAL SHAPE   ABNORMAL SHAPE

consistent orientation of their peak-to-trough axes. Another extended and isolated abnormally shaped magnetic anomaly is present in North Sardinia and is characterized by an α angle of 30°.

The sources of these groups of anomalies are quite different as they originate in areas with distinct geological histories: the sources in the Tyrrhenian Sea are mafic volcanic and igneous structures generated during the opening of this basin; in the Italian peninsula area, the magnetic anomalies form a belt coincident with the Apennines compressional front and their sources may be deep, possibly intrusive bodies. Source depth is estimated to be between 15 and 20 km (Florio 1992). The Sardinian anomaly is

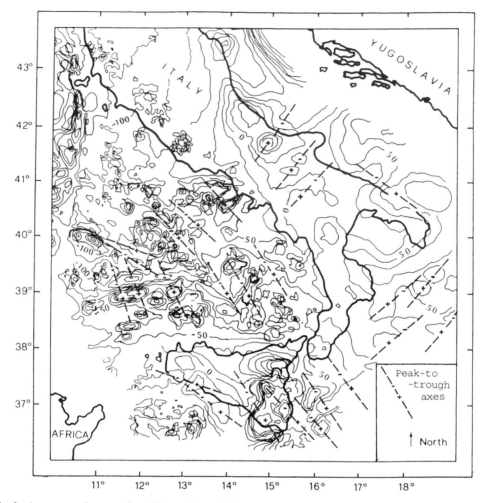

Fig. 2. Aeromagnetic anomalies of the southern Italian area. The abnormally shaped anomalies are indicated with a bar (after Fedi & Rapolla 1990).

probably generated by Oligocene–Miocene calc-alkaline volcanic structures that extend from the surface to a depth of about 10 km (Galdeano & Ciminale 1987). Four possible hypotheses for the shape of these magnetic anomalies are analysed in the following sections.

Interference (coalescence) effects between magnetic anomalies

Coalescence between closely spaced anomalies can cause a deflection of peak-to-trough axes. Simple tests on synthetic sources show, however, that only small α angles can be produced by this effect (Florio 1992). The possibility of interference with other anomalies can be checked by careful study of the aeromagnetic map. In the case of the southern Italian area, the effects of coalescence can be safely excluded for almost all the magnetic anomalies which have abnormal shapes.

Source body shape anisotropy

A magnetic anomaly will have an abnormal shape if the length-to-width ratio of the source body is higher than approximately 10 (e.g. for dykes) and the greater dimension forms an angle with magnetic north. An effective test for the effect of source body shape can be carried out by simply upward continuing the observed magnetic field to a significant altitude (Fedi & Rapolla 1990). This relies on the fact that at a sufficiently great height the magnetic field of any

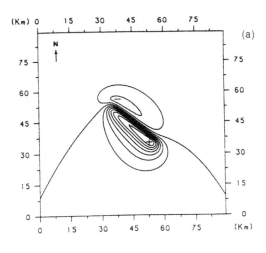

(a)

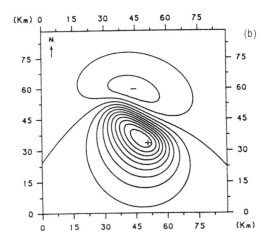

(b)

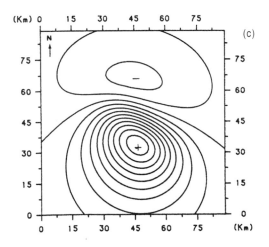

(c)

source becomes dipole-like, and so any dependence of the anomaly shape on the source body shape can be removed (Fig. 3). Thus, after upward continuation, the shape of an anomaly will reflect only the direction of magnetization of the causative body.

All the abnormally shaped magnetic anomalies discussed here in the Italian area show no change in the orientation of their peak-to-trough axes after upward continuation (Fig. 4; Fedi & Rapolla 1990; Fedi *et al.* 1991; Florio *et al.* 1993), and so it can be concluded that their shapes depend only on the direction of magnetization of the sources. Judging by the shape of these anomalies, these directions seem to be very different from that of the present day magnetic field in the region.

Anisotropy of magnetic susceptibility (AMS)

The hypothesis that a strong AMS could significantly deflect the direction of magnetization, and thus the orientation of the peak-to-trough axis of an anomaly, requires a coherent arrangement of magnetic minerals in the whole volume of the anomaly source. This is not very probable when crustal magnetic anomalies are considered, or where groups of several anomalies of abnormal shape with the same orientation of peak-to-trough axes are present. Even so, an AMS study on samples from exposed sections of continental deep crust was carried out, in order to check the possible influence of magnetic anisotropy on crustal anomaly shape (Fig. 5; Florio *et al.* 1993).

The highest ($P = 3.2$) and the average ($P = 1.4$) measured values of AMS were used in the computation of synthetic magnetic anomalies for a number of different orientations of the maximum susceptibility. The results (Fig. 6) indicate that α angles can reach values as high as 23° using the maximum value of AMS when the azimuth of the maximum susceptibility (measured anticlockwise from north) is 30°, 150°, 210° and 330°. Models using average AMS values show no significant deflection of the peak-to-trough axis, regardless of the orientation of the maximum susceptibility. These results demonstrate that the shape of the magnetic anomalies present in the Italian area

Fig. 3. Upward continuation of a synthetic magnetic anomaly generated by a dyke-like body having a high length-to-width ratio and only induced magnetization (after Fedi & Rapolla 1990). The body shape effect is suppressed by the continuation. Altitudes: (**a**) 3 km; (**b**) 10 km; (**c**) 18 km.

M. FEDI *ET AL.*

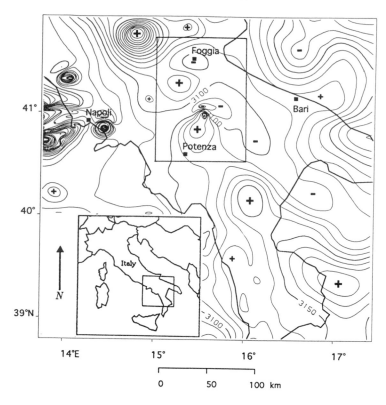

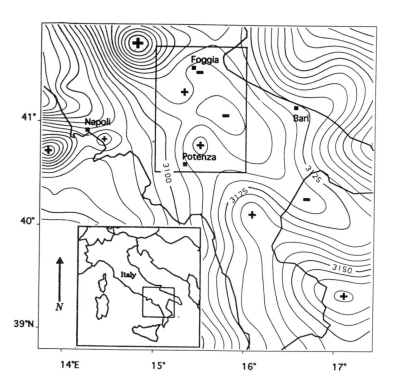

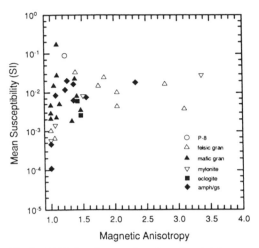

Fig. 5. AMS data for samples of continental deep crust (after Florio *et al.* 1993).

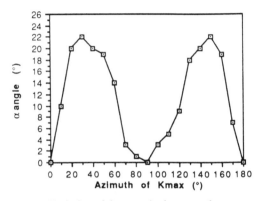

Fig. 6. Variation of the α angle, between the peak-to-trough axis and magnetic north, as a function of the azimuth of maximum susceptibility (after Florio *et al.* 1993). Anisotropy degree $P = 3.2$.

(where α angles are much greater than 23°) cannot be wholly ascribed to a strong and coherent AMS. However, a strong and coherent AMS may in some cases explain the shape of high frequency and isolated magnetic anomalies with α angles smaller than 20–25°.

Presence of a component of remanent magnetization

The presence within an anomaly source body of a remanent magnetization of much higher intensity than the induced magnetization and with a different direction to that of the inducing field can significantly distort the shape of the resulting magnetic anomaly. This appears to be the only phenomenon able to explain the abnormally shaped anomalies of the southern Italian area.

This hypothesis is confirmed by palaeomagnetic study of those parts of the magnetic anomaly source rocks which outcrop at the surface. These studies always result in a good agreement between the measured direction of NRM and the direction of magnetization necessary to produce a good fit with the observed abnormally shaped anomalies (Books 1962; Zietz & Andreasen 1967; Galdeano & Ciminale 1987; Alma-Valdivia *et al.* 1991; Hearst & Morris 1991). This conclusion is particularly interesting because it suggests the presence of strong remanent magnetizations at intermediate to lower crustal depths, a suggestion which is still under debate (Fedi *et al.* 1991 and references therein).

The geodynamic implications of this conclusion are significant, and a number of crustal block rotations can be inferred on the basis of the analysis of magnetic anomalies. The southern Italian region can be divided into several zones on the basis of the orientation of the peak-to-trough axes of groups of aeromagnetic anomalies (Fig. 7; Fedi & Rapolla 1990). The Tyrrhenian area, Sardinia and Sicily appear to be characterized by an anticlockwise rotation, whereas the Ionian and peninsular zones seem to have rotated clockwise. A limitation of this method of identifying rotations is that, in most cases, the age of the source bodies and hence the age of rotation are unknown.

To quantitatively define these rotations, a new method of estimating the direction of magnetization of the sources of magnetic anomalies was proposed (Fedi *et al.* 1994). This method is based on the anomaly distortion produced by the reduction to the pole transformation. The application of this method to anomalies in the Italian peninsular region (those shown in Fig. 4) indicates a direction of magnetization very different to that of the inducing field in that area (present day field: Dec = 2°; Inc = 56°). The inferred inclination of magnetization (65° ± 5°) is not far from the present day field inclination, but the declination (55° ± 5°) appears to be

Fig. 4. Aeromagnetic maps of southern Italy (after Florio *et al.* 1993). The box encloses two crustal anomalies with abnormal shapes. (**a**) upward continued to an altitude of 1.4 km; (**b**) upward continued to an altitude of 14.4 km. The upward continuation does not change the orientation of the peak-to-trough axes of the anomalies, and so it can be concluded that their shapes depend only on the direction of magnetization of the source bodies.

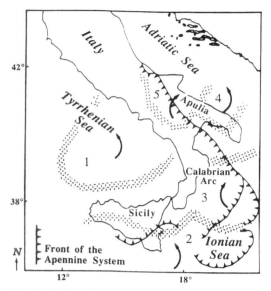

Fig. 7. Regional rotation of crustal blocks in the southern Italian area deduced from analysis of aeromagnetic anomalies (after Fedi & Rapolla 1990). Compare this pattern of rotation with that shown in fig. 3 of Mattei *et al.* (this volume).

significantly different from that of the inducing field. This analysis supports the inferred clockwise rotation of the anomaly sources in the peninsular region.

Conclusions

Accurate study of aeromagnetic anomalies can give important information about the magnetization at depth and can give independent data about the geodynamic history of a region. In particular, the following conclusions can be drawn: (1) the analysis of several aeromagnetic anomalies characterized by an 'abnormal' shape in the southern Italian area supports the hypothesis that remanent magnetization can be dominant at crustal depths; (2) on the basis of the study of anomaly shapes and of source magnetizations, the southern Italian area can be divided into zones that experienced different rotations. The age of the rotations cannot be constrained using this method. The inferred pattern of rotations is in good agreement with that documented by palaeomagnetic analyses in Italy (e.g. see Channell this volume & Mattei *et al.* this

volume). These rotations must be considered when reconstructing the geodynamic history of the region.

References

ALMA-VALDIVIA, L. M., URRUTIA-FUCUGAUCHI, J., BOHNEL, H. & MORAN-ZENTERO, D. J. 1991. Aeromagnetic anomalies and paleomagnetism in Jalisco and Michoacan, southern Mexico continental margin. *Tectonophysics*, **192**, 169–190.

BOOKS, K. G. 1962. Remanent magnetism as a contributor to some aeromagnetic anomalies. *Geophysics*, **27**, 359–375.

CHANNELL, J. E. T. 1996. Palaeomagnetism and palaeogeography of Adria. *This volume.*

FEDI, M. & RAPOLLA, A. 1988. Rotation movements of the Italian peninsula from aeromagnetic evidences. *Physics of the Earth and Planetary Interiors*, **52**, 301–307.

—— & —— 1990. Aeromagnetic anomaly shape analysis in the Southern Italian region for the evaluation of crustal block rotations. *Journal of Geodynamics*, **12**, 149–161.

——, FLORIO, G. & RAPOLLA, A. 1991 The role of remanent magnetization in the southern Italian crust from aeromagnetic anomalies. *Terra Nova*, **2**, 629–637.

——, —— & —— 1994. A method to estimate the total magnetization direction from a distortion analysis of magnetic anomalies. *Geophysical Prospecting*, **42**, 261–274.

FLORIO, G. 1992. *La magnetizzazione della crosta profonda dall'analisi delle anomalie magnetiche: il caso dell'Italia Centro-Meridionale.* PhD Thesis, Italian National Library of Roma and Firenze.

——, FEDI, M., RAPOLLA, A., FOUNTAIN, D. M. & SHIVE, P. N. 1993. Anisotropic magnetic susceptibility in the continental lower crust and its implications for the shape of magnetic anomalies. *Geophysical Research Letters*, **20**, 2623–2626.

GALDEANO, A. & CIMINALE, M. 1987. Aeromagnetic evidence for the rotation of Sardinia (Mediterranean Sea): comparison with paleomagnetic measurements. *Earth and Planetary Science Letters*, **82**, 193–205.

HEARST, R. B. & MORRIS, W. A. 1991. Rock magnetic properties: why bother? *Proceedings of the Society of Exploration Geophysicists. 1991 Meeting*, Houston, 608–610.

MATTEI, M., KISSEL, C., SAGNOTTI, L., FUNICELLO, R. & FACCENNA, C. 1996. Lack of Late Miocene to Present rotation in the Northern Tyrrhenian margin (Italy): a constraint on geodynamic evolution. *This volume.*

ZIETZ, I. & ANDREASEN, G. E. 1967. Remanent magnetization and aeromagnetic interpretation. Mining Geophysics, II, Society of Exploration Geophysicists, Tulsa, Oklahoma, 569–590.

Large scale rotations in North Hungary during the Neogene as indicated by palaeomagnetic data

EMŐ MÁRTON[1] & PÉTER MÁRTON[2]

[1] Eötvös Loránd Geophysical Institute of Hungary, Columbus u. 17–23, H-1145
Budapest, Hungary

[2] Geophysics Department, Eötvös Loránd University, Ludovika tér 2, H-1083 Budapest,
Hungary

Abstract: Tertiary igneous and sedimentary rocks were collected for palaeomagnetic study at about 90 localities from the North Hungarian Central Range. Rock ages range from the late Eocene to the Mid-Miocene but the majority of the samples are Early Miocene in age.
After laboratory treatment 66 localities yielded palaeomagnetic results, which fall into two groups. The first group is characterized by a westerly declination rotation of 70–90° accompanied by shallowed inclinations, and the second group is characterized by a declination rotation of only 30° in the same sense. The first group is older and comprises rocks of late Eocene to early Miocene age, whereas the second group comes from rocks of late Ottnangian to Karpatian age. These results are indicative of two successive tectonic rotations, one in the late Ottnangian and another just preceding the Badenian. The first rotation was accompanied by a northward shift of the area.

In a series of earlier papers (e.g. Márton & Mártonné-Szalay 1968, 1971, 1972; Márton & Szalay 1969*a*, *b*) it was shown that the products of the Miocene andesite volcanism along the North Hungarian Central Range (namely the Börzsöny, Cserhát, Mátra and Tokaji Mountains) possess palaeomagnetic directions nearly parallel or antiparallel to the present direction of the geomagnetic field, implying that the area of the North Central Range did not move since the mid-Miocene.

Later, in another part of the North Central Range, a molluscan clay profile in Eger at the southwestern limb of the Bükk Mountains gave a palaeomagnetic direction for the Egerian (Chattian) stage which showed a westerly declination rotation of 75° and an inclination 15° shallower than the present value. This result was immediately interpreted as an indication of movement of the sampling area prior to the beginning of the andesite volcanism, i.e. in the post-Egerian to pre-Badenian (post-Chattian to pre-Serravalian) period (Márton 1983). In contrast palaeomagnetic declinations in early Oligocene rocks from the Transdanubian Central range were found to be rotated only 35° to the west, with inclinations 10–15° shallower than at present (Márton & Márton 1983). There is therefore, a clear difference between the Cenozoic tectonic evolution of the southwestern and northeastern parts of the Hungarian Central Range (provided that the single Egerian result is confirmed).

Following recognition of the importance of this region, further palaeomagnetic studies were carried out both on sedimentary and igneous rocks from North Hungary for the late Eocene to mid-Miocene period, with the aim of determining the areal extent and timing of the suggested tectonic movements. We present here a concise summary of the results and their implications.

Geology and sampling

The sampling area constitutes the northeastern part of a larger (300 × 150 km) NE–SW-trending intramontane basin which evolved by rifting and subsidence during the Palaeogene (Báldi 1983; Tari *et al.* 1993). The onset of basin formation is marked by igneous activity in the late Eocene (Priabonian). The central part of the basin was formed along the Budapest–Eger axis during the Priabonian and early Kiscellian (Rupelian), and the basin attained its largest extension in the Egerian. By the end of the Eggenburgian (Aquitanian), the Palaeogene basin was completely filled, the final deposit being a terrestrial (lagoonal) clay. At the same time episodic eruptions of acidic composition took place, which culminated in the deposition of the so called 'lower rhyolite tuff' in the Ottnangian (Báldi 1983). This unit is considered to represent the initial phase of the middle Miocene volcanic activity which produced the bulk of the so called Inner Carpathian Volcanic Chain (eg. Börzsöny, Cserhát, Mátra Mountains). The

From Morris, A. & Tarling, D. H. (eds), 1996, *Palaeomagnetism and Tectonics of the Mediterranean Region*, Geological Society Special Publication No. 105, pp. 153–173.

153

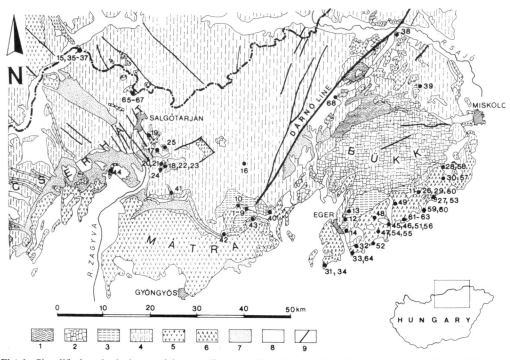

Fig. 1. Simplified geological map of the sampling area. Sampling localities (dots) are numbered as in Table 1. Other symbols are as follows: 1, undifferentiated Palaeozoic rocks; 2, undifferentiated Mesozoic rocks; 3, Eocene–Oligocene sedimentary rocks; 4, Early Miocene sedimentary rocks; 5, Late Eocene andesites in northern Mátra (sampling localities 1–10) and Miocene andesites and tuffs elsewhere; 6, Miocene rhyolites, dacites, tuffs and ignimbrites; 7, Karpatian sediments; 8, Late Miocene and younger sediments; 9, fault lines.

lower rhyolite tuff is sometimes welded (eg. the lower ignimbrites in the Bükk Mountains). This key horizon was extensively sampled throughout the study area, including the southern Bükk and northern Mátra Mountains and the northern foreground of the Mátra and Cserhát Mountains. Immediately overlying the lower rhyolite tuff in the southern Bükk Mountains is the upper dacite tuff which consists of a variety of effusive rocks from lavas to loose, porous tuffs. In contrast to the lower rhyolite tuff, the upper dacite tuff is poorly dated and has been mapped to be of any age between the Ottnangian (late Burdigalian) and Badenian (early Serravalian; Balogh & Rónai 1965). The upper dacite tuff also was sampled in detail in order to see whether the age difference between the upper dacitic and lower rhyolitic horizons is discernible from the palaeomagnetic data. The andesite laccolith of Mt Karancs (Badenian) was sampled because of its central position in the northern foreground of the Mátra and Cserhát Mountains.

The oldest rocks sampled are limestone beds of Triassic age from a deep ore mine at Recsk

which had been contact metamorphosed as a result of magmatic activity during the late Eocene. The products of this magmatism, mainly andesitic intrusions, were sampled at several points both at the surface and at depth in the Recsk ore mining area in the northern Mátra Mountains. Late Eocene limestones at the southern margin of the Bükk Mountains were also sampled. The detrital sedimentary rocks of the Palaeogene basin can be termed as molasse consisting mostly of clastic rocks, from which only the argillaceous (pelitic) deposits are generally considered to be potentially suitable for palaeomagnetic studies. Unfortunately, fresh outcrops of the oldest of these deposits are difficult to locate and the large majority of the different siltstones of Eggenburgian–Ottnangian age that we sampled extensively from fresh outcrops in various brickyards and coal mining pits were found to possess no palaeomagnetic signal. Despite of these difficulties we were able to sample the basin fill from bottom to top from the Priabonian to the Karpatian (upper Burdigalian).

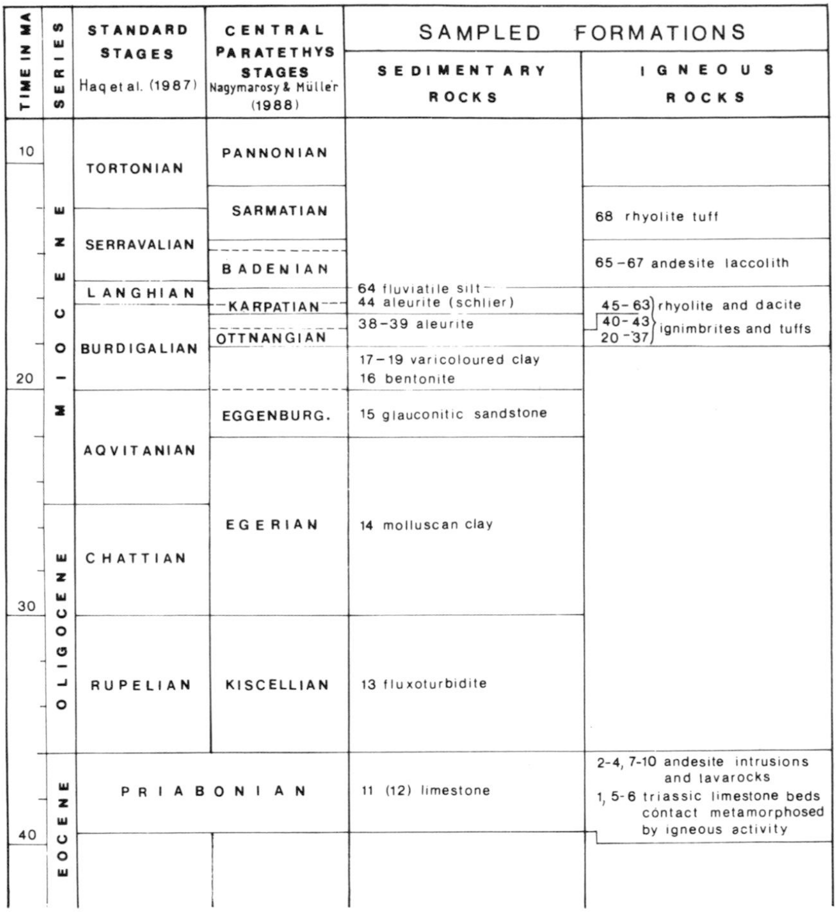

TIME IN MA	SERIES	STANDARD STAGES Haq et al. (1987)	CENTRAL PARATETHYS STAGES Nagymarosy & Müller (1988)	SAMPLED FORMATIONS	
				SEDIMENTARY ROCKS	IGNEOUS ROCKS
10		TORTONIAN	PANNONIAN		
	M	SERRAVALIAN	SARMATIAN		68 rhyolite tuff
			BADENIAN		65–67 andesite laccolith
		LANGHIAN	KARPATIAN	64 fluviatile silt – 44 aleurite (schlier)	45–63 rhyolite and dacite
20	M	BURDIGALIAN	OTTNANGIAN	38–39 aleurite	40–43 / 20–37 ignimbrites and tuffs
				17–19 varicoloured clay	
				16 bentonite	
		AQVITANIAN	EGGENBURG.	15 glauconitic sandstone	
		CHATTIAN	EGERIAN	14 molluscan clay	
30	OLIGOCENE	RUPELIAN	KISCELLIAN	13 fluxoturbidite	
40	EOCENE	PRIABONIAN		11 (12) limestone	2-4, 7-10 andesite intrusions and lavarocks 1, 5-6 triassic limestone beds contact metamorphosed by igneous activity

Fig. 2. Chronostratigraphic positions of the sampled formations. Dashed stage boundaries are based on local radiometric data (Hámor *et al.* 1980). Sampling localities are numbered as in Table 1.

The sampling sites are shown on a simplified geological map in Fig. 1, and the chronostratigraphic positions of the sampled formations are shown in Fig. 2.

The sampling area is dissected by a major, NNE–SSW-trending tectonic line (Darnó-line), which actually is a zone of compressive tectonics (underthrusting) which was active at the end of the Oligocene (Báldi 1983). Younger, minor strike-slip faults have also been identified (Fig. 1).

Laboratory treatments and results

The magnetic measurements were carried out using JR-4 (Jelinek) magnetometers with a noise level of $4 \mu A m^{-1}$ and a two-axis cryogenic magnetometer. The characteristic remanence was determined by analysing orthogonal demagnetization (Zijderveld) diagrams obtained for each specimen after stepwise demagnetization by the thermal or AF method.

Schonstedt AF demagnetizers (100 mT maximum peak field), a high field AF demagnetizer (240 mT maximum peak field) and Schonstedt ovens were used for demagnetization. Possible mineralogical changes on heating were monitored by susceptibility measurements with KLY-1 and KLY-2 susceptibility bridges. Identification of the magnetic minerals was attempted by comparing several magnetic characteristics, such as the decay of NRM (and the behaviour of the susceptibility) on heating, the resistance of NRM to AF demagnetization, IRM acquisition curves and Curie temperature measurements. Examples are shown where both

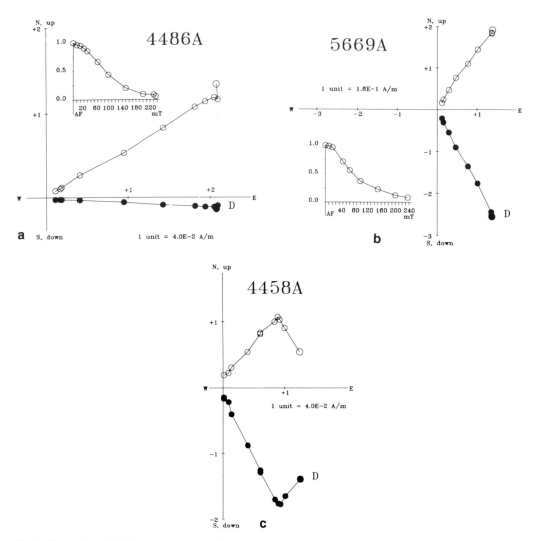

Fig. 3. Examples of AF demagnetization behaviour of effusive rocks from the Bükk Mountains. (**a**) Lower ignimbrite sample from (26); (**b**) upper ignimbrite sample from (59); (**c**) upper dacite tuff sample from (51) (cf. Table 1).

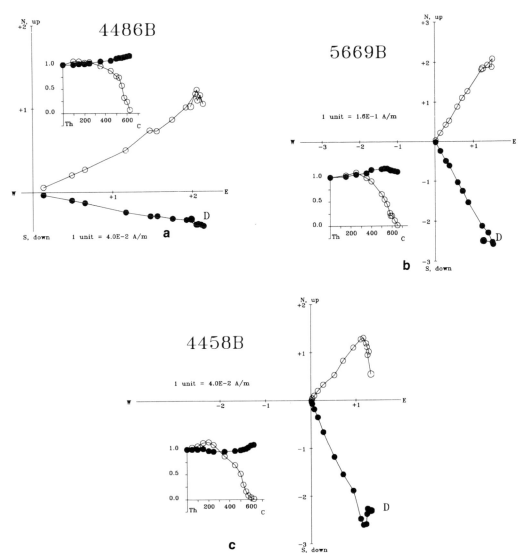

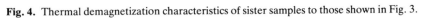

Fig. 4. Thermal demagnetization characteristics of sister samples to those shown in Fig. 3.

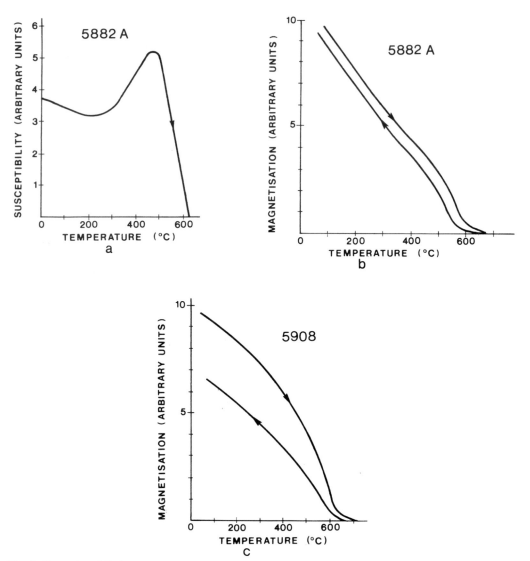

Fig. 5. Examples of Curie temperature measurements of effusive rocks from the Bükk Mountains. (**a**) susceptibility vs. temperature curve for a lower ignimbrite sample from (26); (**b**) magnetization v. temperature curve for the same sample as in (a); (**c**) as in (b) for an upper ignimbrite sample from (60) (cf. Table 1.)

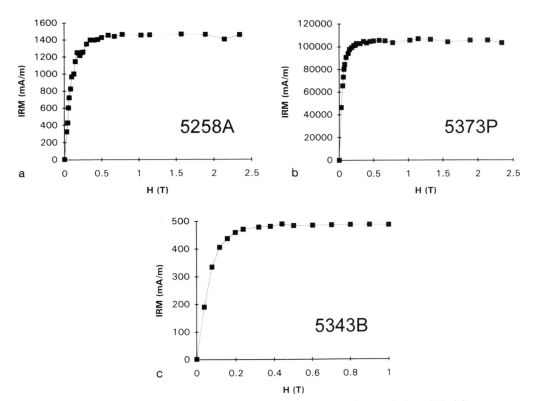

Fig. 6. Some IRM acquisition curves of effusive rocks. (**a**) Lower ignimbrite sample from (29); (**b**) upper ignimbrite sample from (55); (**c**) lower rhyolite (ignimbrite) sample from (25) (cf. Table 1.)

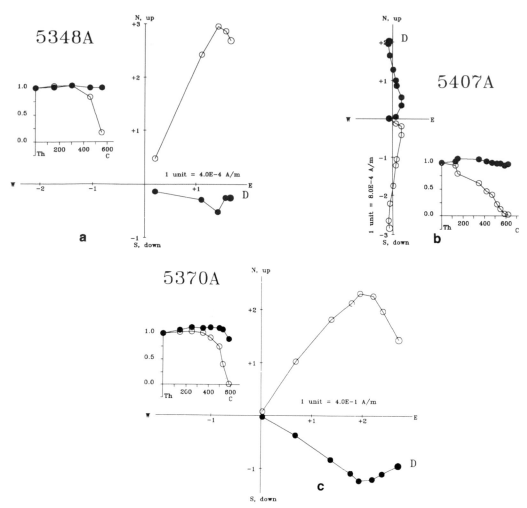

Fig. 7. Other examples of thermal demagnetization of igneous rock samples. (**a**) Lower rhyolite (ignimbrite) sample from (25); (**b**) Lower rhyolite tuff sample from (43); (**c**) late Eocene andesite sample from (7) (cf. Table 1).

AF and thermal demagnetizations were successful, but in order to aid identification of the magnetic minerals, thermal demagnetization was applied more often than AF demagnetization.

Nearly all igneous sites proved to have characteristic remanences, which could very often be isolated with both demagnetization techniques. All igneous samples studied from the southern margin of the Bükk Mountains exhibited remarkably uniform behaviour both on thermal and AF demagnetizations irrespective of their rhyolitic or dacitic composition or their welded or porous nature. On AF treatment they showed extreme hardness (median destructive field around 70 mT) and fields exceeding 200 mT were needed to achieve complete demagnetization (Fig. 3). On thermal demagnetiz-

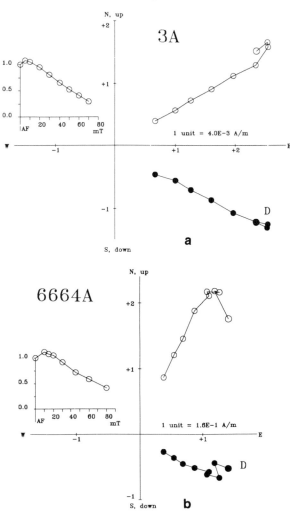

Fig. 8. Examples of AF demagnetization of sedimentary rocks. (**a**) Molluscan clay sample from (14); (**b**) bentonite (smectite) sample from (16) (cf. Table 1).

ation, the NRM signal was lost above 600°C, but below the Curie-temperature of hematite or maghemite (Fig. 4.). Curie-temperature measurements suggest a single-phase mineral with a Curie-temperature at 620–630°C (Fig, 5.). We think, therefore, that the NRM in these rocks is carried by extremely stable oxidized magnetite or Ti-substituted maghemite. This is confirmed also by the IRM acquisition curves rapidly saturating with increasing applied fields

but at higher fields than typical for magnetite (Fig. 6). The NRM of the ignimbrites and tuffs from elsewhere and also of the andesites of late Eocene age from the Recsk area unblock at temperatures at or below the Curie-temperature of magnetite, as shown by Fig. 7. Thus, their carrier mineral must be magnetite with more or less titanium.

Many sampling sites of sedimentary rocks failed on laboratory treatment, partly because of

Table 1. *Palaeomagnetic results from the North Hungarian Central Mountains*

Locality samples	Rock type stage	N/N_o	D/D_c	I/I_c	k	α_{95}	Cleaning	Polarity	Reference
Late Eocene									
1. Recsk −700 m 6605–6609	Anisian limestone contact metamorphosed	5/5	98	−54	56	10.3	250–400°C	R	P. Márton (1991)
2. Recsk −700 m 6616–6621	Diorite–porphyrite intrusion	6/6	119	−59	64	8.4	30–100 mT	R	P. Márton (1991)
3. Recsk −700 m ST101–106	Andesite dyke	6/6	94	−49	78	7.6	≥500°C	R	P. Márton (1991)
4. Recsk −700 m ST107–112	Quartz diorite porphyrite dyke	6/6	113	−64	46	10.0	≥500°C	R	P. Márton (1991)
5. Recsk −700 m 6742–6746	Baked Anisian limestone	5/5	97	−65	213	5.3	250–500°C	R	Present paper
6. Recsk −700 m 6727–6745	Baked Anisian limestone	12/19	111	−55	190	3.2	250–500°C	R	Present paper
Recsk, −700 m average		6/6	105	−58	92	7.0		R	Present paper
7. Recsk Lahóca hegy, North 5367–5376	Augite-andesite dyke	9/10	122	−52	104	5.1	≥500°C	R	P. Márton (1991)
8. Recsk Lahóca hegy 585–590	Andesite	5/6	110	−21	70	9.0	AF > 40 mT	R	Márton & Márton (1969)
9. Recsk Nagykő 5393–5402	Andesite	10/10	105 / 104	0 / −39	37 / 37	8.1 / 8.1	530–585°C		P. Márton (1991)
10. Mátrade–recske Valériahegy ST68–80	Andesite dyke	12/13	109	−26	20	10.0	450–585°C	R	P. Márton (1991)
11. Kács Farkaskő 4956–4987 3 sites	Limestone	18/20	110 / 92	−44 / −46	20 / 14	8.2 / 9.2	150–200°C	R+N	E. Márton (1989)
12. Kiseged 4516–4522 4927–4937	Limestone	12/20	148 / (175)	−46 / −80	79 / 79	4.9 / (4.9)	300–325°C	R	E. Márton (1989) Secondary magnetization!
Oligocene									
13. Noszvaj 4448–4455	Fluxoturbidite Kiscellian	4/8	309 / 262	48 / 69	22 / 22	20.2 / 20.2	150–500°C	R+N	E. Márton (1989)
14. Eger, Wind-brickyard W1–101	Molluscan clay Egerian (24–28 Ma)	75/101	110 / 105	−24 / −48	20 / 20	3.7 / 3.7	≥40 mT	R	P. Márton (1983)

Table 1. *Continued*

Locality samples	Rock type stage	N/N_0	D/D_C	I/I_C	k	α_{95}	Cleaning	Polarity	Reference
Early Miocene									
15. Ipolytarnóc 6669–6680	Glauconitic sandstone Eggenburgian	13/23	277 / 255	47 / 51	27 / 46	8.0 / 10.0	150–400°C	N	Present paper
Nógrád Basin									
16. Pétervására 6644–6669	Bentonite Eggenburgian	18/26	124 / 123	−61 / −54	95 / 95	3.6 / 3.6	20–80 mt after 110–200°C	R	Present paper
17. Kazár 6714–6726	Varicoloured clay Eggenburgian	11/13	121 / 117	−26 / −35	31 / 31	8.4 / 8.4	400–560°C	R	Present paper
18. Nemti 6687–6697	Varicoloured clay Eggenburgian	6/11	137 / 151	−49 / −46	28 / 28	12.9 / 12.9	≥600°C, resp. ≥80 mT	R	Present paper
19. Zagyvapálfalva, brickyard 6705–6713	Varicoloured clay Eggenburgian	7/9	303 / 284	55 / 43	43 / 43	9.3 / 9.3	150–560°C, resp. ≥10 mT	N	Present paper
Nógrád basin sediments, average		4	126 / 123	−48 / −46	26 / 25	18.3 / 18.6		N+R	
20. Rákóczitelep 5202–5209	Rhyolite tuff Ottnangian	8/8	91	−61	28	10.6	≥50 mT	N+R	P. Márton (1990a)
21. Gyulakeszi 3 subsites KT8–10 5180–5188 5200 5254–5255	Rhyolite Ottnangian	12/15	97	−56	20	10.0	25–50 mT	N+R	Present paper
22. Nemti 5189–5196 6670–6673	Ignimbrite Ottnangian	11/12	83	−53	57	6.1	10–50 mT	R	Present paper
23. Nemti 6674–6681	Rhyolite tuff Ottnangian	7/8	60	−35	35	10.3	300–520°C	R	Present paper
24. Kisterenye-Nemti 5238–5249	Rhyolite tuff Ottnangian	11/12	73	−48	25	9.3	≥50 mT	N+R	Present paper
25. Mátraszele Kazári-quarry 5341–5354	Ignimbrite Ottnangian	14/14	103	−64	117	3.7	450–550°C	R	P. Márton (1990b)
Nógrád basin ignimbrites + tuffs, average		6	82	−54	32	11.0		N+R	

Table 1. *Continued*

Locality samples	Rock type stage	N/N_o	D/D_C	I/I_C	k	α_{95}	Cleaning	Polarity	Reference
Southern Bükk Mts									
26. Kács I. 4486–4500 5882–5889	Ignimbrite Ottnangian	13/15	111	−36	29	7.8	150 mT	R	E. Márton (1990)
27. Sály I. 2 subsites 4500–4508	Ignimbrite Ottnangian	8/8	101	−45	81	6.2	150 mT	R	E. Márton (1990)
28. Kisgyőr I. 4 subsites 4523–4530 4938–4955	Ignimbrite Ottnangian	25/26	88	−50	139	2.5	NRM-70 mT	R	E. Márton (1990)
29. Kács II. (Isten-mezeje) 2 subsites 5258–5277	Ignimbrite Ottnangian	16/20	106	−51	128	3.3	70 mT	R	E. Márton (1990)
30. Pusztamocsolyás II. 3 subsites 5626–5642	Rhyolite tuff Ottnangian	15/17	91	−55	391	1.9	NRM-70 mT	R	E. Márton (1990)
31. Demjén Pünkösdhegy I. 2 subsites 5795–5801 5719–5723	Ignimbrite Ottnangian	10/10	116	−49	109	4.6	350°C	R	E. Márton (1990)
32. Ostoros 6019–6024	Rhyolite tuff Ottnangian	6/6	119	−45	193	4.8	NRM	R	Present paper
33. Eger-Andornaktálya 6308–6312	Rhyolite tuff Ottnangian?	5/5	115	−55	86	8.3	10 mT	R	Present paper
S. Bükk ignimbrites and tuffs, average		8	106	−49	68	6.7		R	
34. Demjén Nagyeresztvény 5278–5299	Rhyolite tuff Ottnangian?	21/22	188	−50	92	3.3	400°C	R	E. Márton (1990)
North of Cserhát Mts									
35. Ipolytarnóc Borókás árok 5300–5305	Ignimbrite Ottnangian	6/6	325	62	135	5.8	300–500°C	N	P. Márton (1990b)

Table 1. *Continued*

Locality samples	Rock type stage	N/N_o	D/D_C	I/I_C	k	α_{95}	Cleaning	Polarity	Reference
36. Ipolytarnóc Puhakő-quarry 5322–5329	Rhyolite tuff Ottnangian	8/8	339	62	396	2.8	300–550°C	N	P. Márton (1990*b*)
37. Ipolytarnóc Botos-árok 5330–5340	Ignimbrite (rhyolite tuff) Ottnangian	9/11	334	59	77	5.9	300–550°C	N	P. Márton (1990*b*)
Ipolytarnóc tuffs + ignimbrites average		3	333	61	475	5.7		N	
North of Bükk Mts									
38. Vadna sand pit V1–24	Aleurite Ottnangian	23/24	337 337	56 50	46 46	4.5 4.5	150°C		P. Márton (1987)
39. Varbó 6278–6285	Aleurite Ottnangian	5/8	308	60	14	21.0		N	Present paper
Northeast Mátra Mts									
40. Recsk-Siroki-út 5306–5321	Rhyolite tuff Ottnangian	7/16	335	57	468	2.8	110–250°C	N	Present paper
41. Szuha 5262–5264 5266–5271	Rhyolite tuff Ottnangian	9/9	319	60	33	9.1	25–50 mT	N	P. Márton (1990*a*)
42. Ördöggátak 5377–5386	Rhyolite tuff Ottnangian	9/12	314	48	39	8.4	250–500°C, resp. 40–80 mT	N	Present paper
43. Recsk Miklósvölgy 5403–5416	Rhyolite tuff Ottnangian	14/14	328	62	64	5.0	130–520°C	N	Present paper
Northeast Mátra rhyolite tuffs average		4	323 (297	57 63	100 100	9.2 9.2)		N	
Cserhát Mts									
44. Sámsonháza-Márkháza 6946–6954	Aleurite Karpatian	9/9	324 321	57 60	36 36	9.0 9.0	NRM-10 mT 10 mT–150°C	N	Present paper

Table 1. *Continued*

Southern Bükk Mts

Locality samples	Rock type stage	N/N_0	D/D_C	I/I_C	k	α_{95}	Cleaning	Polarity	Reference
45. Bogács-N 2 subsites 5890–5901	Ignimbrite	10/12	156	−26	676	1.9	400°C	R	Present paper
46. Bogács-NW 3 subsites 5970–5983	Ignimbrite	14/14	151	−26	187	2.9	400°C	R	Present paper
47. Szomolya III. 5984–989	Ignimbrite	6/6	154	−24	168	5.2	400°C	R	Present paper
48. Noszvaj-Szomolya 2 subsites 5990–6001	Ignimbrite	12/12	159	−30	222	3.2	400°C	R	Present paper
49. Cserépfalu 6002–6007	Ignimbrite	6/6	164	−26	396	3.4	400°C	R	Present paper
50. Kács III 6014–6018	Ignimbrite	4/5	155	−38	168	7.1	400°C	R	Present paper
51. Bogács I 4456–4485	Ignimbrite	30/30	156	−34	115	2.5	350°C	R	E. Márton (1990)
52. Novaj 3 subsites 5300–5314	Rhyolite tuff	14/15	155	−26	168	3.1	400°C	R	E. Márton (1990)
53. Sály II 2 subsites 5325–5337	Ignimbrite	13/13	162	−46	114	3.9	70 mT–400°C	R	E. Márton (1990)
54. Szomolya I. 2 subsites 5355–5366	Ignimbrite	12/12	158	−30	222	2.9	400°C	R	E. Márton (1990)
55. Szomolya II. 2 subsites 5367–5378	Lava?	12/12	157	−35	672	1.8	400°C	R	E. Márton (1990)
56. Bogács-W 3 subsites 5506–5511 5705–5718	Ignimbrite	19/20	155	−36	294	2.0	350°C	R	E. Márton (1990)
57. Pusztamocsolyás I. 2 subsites 5613–5625	Rhyolite tuff	13/13	153	−39	110	4.0	70 mT	R	E. Márton (1990)
58. Kisgyőr, Veresbánya 4 subsites 5643–5666	Ignimbrite	24/24	157	−40	133	2.6	NRM–350°C	R	E. Márton (1990)

Table 1. *Continued*

Locality samples	Rock type stage	N/N_o	D/D_C	I/I_C	k	α_{95}	Cleaning	Polarity	Reference
59. Tibold-daróc I. 2 subsites 5667–5678	Ignimbrite and tuff	10/12	160	−39	203	3.4	500°C	R	E. Márton (1990)
60. Tibold-daróc II. 4 subsites 5680–5704 5908–5911	Ignimbrite	22/25	148	−31	64	3.9	500°C	R	E. Márton (1990)
61. Tard I. 3 subsites 5802–5820	Ignimbrite	18/18	155	−28	130	3.0	550°C	R	E. Márton (1990)
62. Tard II. 2 subsites 5821–831	Rhyolite tuff	11/11	148	−27	171	3.5	550°C	R	E. Márton (1990)
63. Tard III. 2 subsites 5832–841	Ignimbrite	9/10	149	−22	168	4.0	350°C	R	E. Márton (1990)
S. Bükk, ignimbrites and tuffs (Karpatian?) average		19	155	−32	115	3.1		R	
64. Eger-Andornaktálya 6301–6307	Fluviatile silt Karpatian–Badenian boundary	6/7	326 339	48 55	148 148	5.5 5.5	30 mT	N	Present paper
Northern Cserhát Mts									
65. Karancs I. 5276–5283	Andesite	7/7	151	−75	57	8.0	≥40 mT	R	P. Márton (1990*b*)
66. Karancs II. 5284–5291	Andesite	8/8	201	−77	94	6.0	≥40 mT	R	P. Márton (1990*b*)
67. Karancs III. 5292–5299	Andesite	8/8	138	−70	250	4.0	≥40 mT	R	P. Márton (1990*b*)
Karancs average	Laccolith (Badenian)	3	158	−76	76	14.2		R	P. Márton (1990*b*)
Northern Bükk Mts									
68. Lénárd-daróc 6292–6300	Rhyolite tuff Sarmatian	8/9	354	23	33	10.0	150°C	N	Present paper

Explanation of symbols: Locality/samples, location of sampling site with sample codes (numbers); Rock type/stage, description of sampled rock and geological stage; N/N_o, number of samples used/collected; D/D_C, mean declination/mean declination corrected for tectonic tilt; I/I_C, mean inclination/mean inclination corrected for tectonic tilt; k, Fisher precision parameter (Fisher 1953); α_{95}, angle of confidence at 95% level (Fisher 1953); Cleaning, interval or value of AF/thermal demagnetization used for the definition of the characteristic magnetization; Polarity, N – normal, R – reversed.

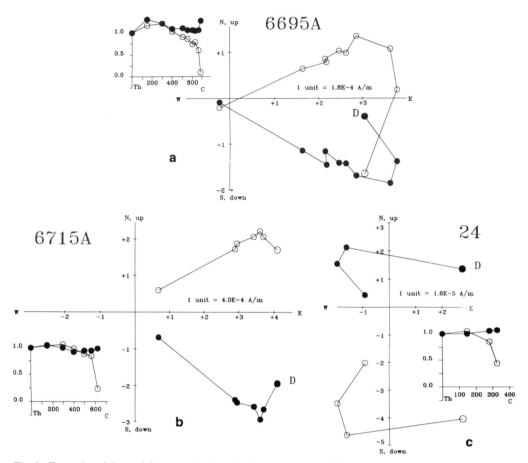

Fig. 9. Examples of thermal demagnetization of sedimentary rocks. (a) Varicoloured clay sample from (18); (b) varicoloured clay sample from (17); (c) bentonite sample from (16) (cf. Table 1).

extremely weak and quickly demagnetizing NRMs (mostly in limestones), and partly because of unstable behaviour on demagnetization (in many clastic sediments). However, a reasonable number of sedimentary localities of different lithologies yielded characteristic magnetizations after treatment.

Figures 8–11 show examples of demagnetization behaviour during AF and thermal treatment, IRM acquisition characteristics and Curie-temperature measurements. These suggest that in most sampled sedimentary rocks the NRM resides in the magnetite phase (Figs, 9a, b and 10c) which is sometimes oxidized and/or may contain some titanium (except in the terrestial clays where the remanence is carried predominantly by haematite). The glauconitic

sandstone from Ipolytarnóc contains pyrite (Fig. 11c.), the presence of which explains why the isolation of the characteristic remanence was successful in less samples than collected.

Table 1 summarizes the palaeomagnetic results in chronological order and other information pertaining to them. The remanence directions were corrected for tectonic tilt where available. For all sedimentary rocks the correction was based on the dip of the sampled strata. As the tectonic attitude of the igneous intrusions and lavas of the Recsk area was not discernible in the field, we studied the anisotropy of susceptibility of these rocks for magnetic foliation. This might be expected to parallel the horizontal plane during crystallization provided that the rock was not stressed significantly at

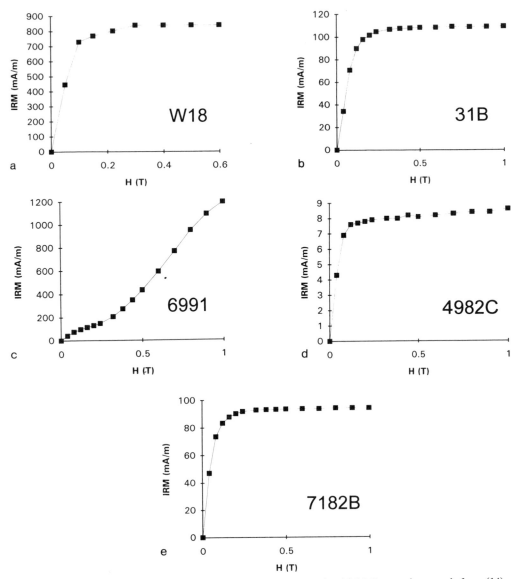

Fig. 10. Examples of IRM acquisition behaviour of sedimentary rocks. (**a**) Molluscan clay sample from (14); (**b**) bentonite sample from (16); (**c**) varicoloured clay sample from (18); (**d**) limestone sample from (11); (**e**) glauconitic sandstone sample from (15) (cf. Table 1).

later times. However, predominant foliation was found only for three superficial sites, and even from these only the lavas of Nagykő showed a foliation plane significantly different from the present horizontal plane. Untilting by the measured attitude of foliation resulted in a mean remanence direction for Nagykő ($D = 104°$, $I = -39°$) which is more consistent with those of the other igneous rocks and baked limestones in the Recsk area (see Table 1).

In the eruptive ignimbrites and tuffs of the southern Bükk area both rock texture and magnetic fabric are foliated and both kinds of foliation are more or less parallel to each other, making an angle of less than 20° with the horizontal plane. Correction for the foliation,

Table 2. *Newly obtained potassium-argon ages for the lower rhyolite and upper dacite tuff horizons (Pécskay, private comm.).*

Locality (cf. Table 1)	K–Ar age (Ma)
26	16.8 ± 0.8
28	17.5 ± 0.5
23	17.6 ± 0.8
45	17.2 ± 2.3
59	17.0 ± 0.7

however, increases the scatter of the site mean remanence directions. This suggests that the tilting of the foliation planes is of depositional origin. For this reason tilt correction for these eruptive volcanics was not applied.

Evidence for tectonic tilt of the volcanic complex of the Mátra Mountains is available both from geological observations (Varga 1975) and palaeomagnetic data (Márton & Márton 1969; Márton 1977). The palaeomagnetically inferred dip of the middle Miocene volcanic complex is 16° to 200°, but it is not clear whether the lower rhyolite tuffs of the eastern Mátra (which we sampled) are tilted by the same amount or by less, because their outcrops run practically along the northern boundary of the tilted block of the complex.

Owing to the uncertainty of the age of the upper dacite horizon in the Bükk Mountains, a number of age determinations were made on selected samples both for the lower rhyolite and upper dacite tuff horizons. Suitable samples for K/Ar dating were obtained by examining a large number of thin sections for fresh biotite under the microscope. However, the results are rather uniform (Table 2) and do not allow distinction of the two horizons in question on the basis of the K/Ar ages.

Discussion

It is convenient to discuss the palaeomagnetic directions in chronological order (cf. Fig. 2).

The late Eocene (Priabonian) is represented by results from two areas that are separated by the Darnó tectonic line. Thus, the corresponding palaeomagnetic data pertain to two different

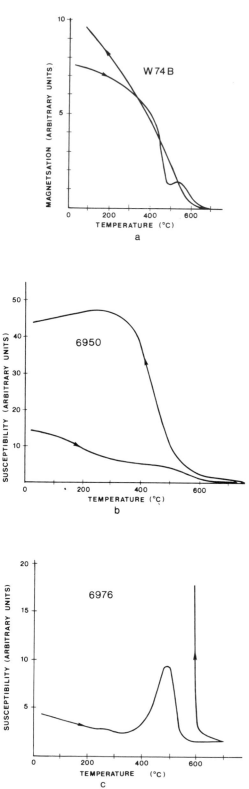

Fig. 11. Some examples of Curie-temperature measurements of sedimentary rocks. (**a**) Magnetization v. temperature for a molluscan clay sample from (14); (**b**) susceptibility v. temperature for a Karpatian siltstone sample from (44); (**c**) same for a glauconitic sandstone sample from (15) (cf. Table 1).

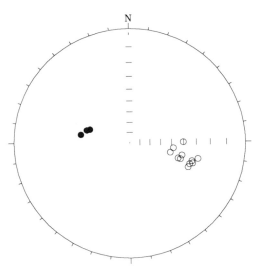

Fig. 12. Sterographic plot of the directions of characteristic magnetizations of individual samples from (11) (cf Table 1). Normal and reversed magnetizations are exactly antiparallel.

tectonic domains. They are: (1) from igneous intrusions, limestones baked by the intrusions and lavas from Recsk in the north central Mátra Mountains; and (2) from limestones in the southern Bükk Mountains. Despite the different provenances, the palaeomagnetic results are rather uniform. All show reversed polarity (except three samples from Farkaskő which, however, pass the reversal test as shown in Fig. 12), a 70–90° declination rotation and an inclination some 10–15° shallower than expected locally in a geocentric axial dipole field.

Two results are available for the Oligocene from the southern Bükk area, of which the Kiscellian (Rupelian) result clearly needs better definition. The Egerian result is of excellent quality and is the first of the present series of data from North Hungary showing westerly declination rotation.

The combination of shallowed inclinations and large westerly rotations of the declinations appears to continue into the early Miocene, as shown by the palaeomagnetic results for the Eggenburgian (upper Aquitanian) sediments and the overlying Ottnangian (Burdigalian) rhyolite tuffs/ignimbrites of the Nógrád basin, north of the Mátra Mountains, as well as for the Ottnangian lower ignimbrites in the southern Bükk area. The lower rhyolite tuffs of the Mátra Mountains also fit in this palaeomagnetic pattern if the application of the full tilt correction is correct. Note also that all tuffs/ignimbrites of the Nógrád basin and all ignimbrites of the southern

Bükk are of reversed polarity, whereas the lower rhyolite tuffs of the Mátra Mountains are of normal polarity. This indicates that the so called lower rhyolite tuff is not strictly contemporaneous all over North Hungary.

It is also possible, however, that the lower rhyolite tuffs of the Mátra Mountains are not tilted at all. In that case their palaeomagnetic direction and normal polarity are in close agreement with those for the same effusive horizon at Ipolytarnóc, north of the Cserhát Mountains. Here the declination rotation to the west is only 30°. This smaller rotation is corroborated by results from two sedimentary sites of Ottnangian (Burdigalian) age (but definitely younger than the tuffs) in the northern foreground of the Bükk Mountains. It is noteworthy that the late Ottnangian to Karpatian upper dacite tuffs of the southern Bükk Mountains and a Karpatian siltstone site in the Cserhát Mountains show the same declination rotation (*c.* 30° to west). Finally, the Karancs laccolith, which is Badenian (late Langhian–early Serravalian) in age, shows no significant palaeomagnetic rotation and this result is in close agreement with our former palaeomagnetic data for the Badenian from the Börzsöny, etc. Mountains.

To summarize, two palaeomagnetic groups are recognized in the study area, each consisting of geographically distributed localities both of sedimentary and igneous rocks. The first group is characterized by a declination rotation of 70–90° to the west and the second is characterized by a lower rotation of about 30° in the same sense. The first group is older, and comprises rocks from late Eocene through Ottnangian (Burdigalian) in age, and the second represents a period of time between the late Ottnangian and the Karpatian (late Burdigalian–Langhian). Both groups are characterized by inclinations significantly shallower than the present value, although the second group also includes the igneous sites from the southern margin of the Bükk Mountains which show a large inclination error. Although the inclination extrapolated for zero foliation is 45° (Fig. 13), this is still shallower than that for the older, lower ignimbrites and definitely shallower than that of the Karpatian sediment from the Cserhát and the tuff from the Ipolytarnóc site.

Conclusions

Based on the palaeomagnetic data of the present study we suggest that the time from the late Eocene to the Ottnangian was uneventful in the tectonic history of the North Hungarian Central

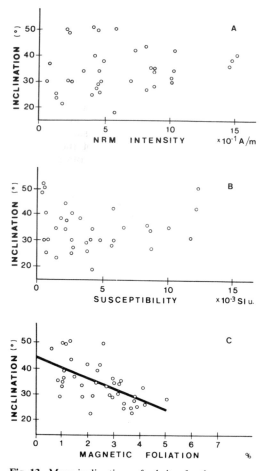

Fig. 13. Mean inclinations of subsites for the upper dacite horizon in the Bükk Mountains (45–63) (cf. Table 1) as a function of: (a) NRM intensity; (b) low field susceptibility; (c) magnetic foliation. The inclination only shows correlation with the foliation.

those between e.g. the Ottnangian sediments and ignimbrites north of the Mátra Mountains.

This work was partially supported by an OTKA research grant 2128/342 1113 'Study of Tertiary movement history of Inner Carpathian tectonic units by palaeomagnetism and other methods'. Improvement of the manuscript by the editors is gratefully acknowledged.

References

BÁLDI, T. 1983. *Magyarországi oligocén és alsó miocén formációk.* Akadémiai Kiadó. Budapest.

BALOGH, K. & RÓNAI, A. 1965. *Magyarázó Magyarország 200000-es földtani térképsorozatához L-34-III. Eger.* Magyar Állami Földtani Intézet, Budapest. 1965.

FISHER, R. A. 1953. Dispersion on a sphere. *Proceedings of the Royal Society, London,* **217**, 295–305.

HÁMOR, G., RAVASZ-BARANYAI, L., BALOGH, K. & ÁRVA-SOÓS, E. 1980. Magyarországi miocén riolittufa szintek radiometrikus kora. *MÁFI Évi jelentése az 1978. Évről.* 65–74. [Radiometric Ages of the Miocene rhyolite tuffs in Hungary]. *Annual Report Hungarian Geological Survey for the year 1978.*

HAQ, B. U., HARDENBOL, J. & VAIL, P. R. 1987. Chronology of fluctuating sea levels since the Triassic. *Science,* **235**, 1156–1167.

MÁRTON, E. 1989. A new constraint on the tectonic models for the Pannonian Basin. *Terra Abstracts,* **1**, 86.

—— 1990. Palaeomagnetic studies on the Miocene volcanic horizons at the southern margin of the Bükk Mountains. *Annual report of the Eötvös Loránd Geophysical Institute of Hungary for 1988–1989,* 307–308 (English) and 211–217 (Hungarian).

—— & MÁRTON, P. 1983. A refined apparent polar wander curve for the Transdanubian Central Mountains and its bearing on the Mediterrranean tectonic history. *Tectonophysics,* **98**, 43–57.

MÁRTON, P. 1977. A paleomágneses szerkezetkutatás alapjairól. *Magyar Geofizika,* **18**, 161–165.

—— 1983. *Északmagyarországi oligocén formációk paleomágneses vizsgálata.* Internal Report MÁFI 3416/81.

—— 1987. *Északmagyarországi eocén-középső miocén formációk paleomágneses vizsgálata.* Internal Report MÁFI 4585/87.

—— 1990a. *Paleomágneses vizsgálatok Észak-Magyarországon.* Internal Report KFH 778/89.

—— 1990b. *Paleomágneses vizsgálatok Észak-Magyarországon II.* Internal Report KFH 976/1990.

—— 1991. *Paleomágneses vizsgálatok Észak-Magyarországon III.* Internal Report KFH 29/1991.

—— & MÁRTONNÉ-SZALAY, E. 1968. Cserhát-hegységi andezitek áttekintö paleomágneses vizsgálata. *Magyar Geofizika,* **9**, 224–230.

—— & —— 1971. Paleomágneses vizsgálatok a

Range. During this period, the whole area was situated much further to the south than today. It was then abruptly involved in a large-scale displacement, which manifested itself in a large (about 40–60°) counterclockwise rotation and probable northward shift in the Ottnangian (Burdigalian). This was followed by a second phase of deformation which produced an additional 30° counterclockwise rotation that finally emplaced the area before the calc-alcaline volcanism began, since the Badenian sites are not rotated.

Admittedly, differences exist in coeval site mean directions when the southern Bükk results are compared with those from other areas in the North Hungarian Range. However, these differences are not systematic and are smaller than

Börzsöny- hegységben. *Magyar Geofizika,* **12**, 77–83.

—— & —— 1972. Paleomágneses vizsgálatok a Tokaj-hegységben. *Magyar Geofizika,* **13**, 219–226.

—— & SZALAY, E. 1969a. Áttekintő paleomágneses vizsgálatok Mátra-hegységi andeziteken. *Földtani Közlöny,* **99**, 166–180.

—— & —— 1969b. Geologische Verwendungen der Paläomagnetischen Forschungen in Ungarn. *Geofizikai Közlemények,* **18**, 79–84.

NAGYMAROSY, A. & MÜLLER, P. 1988. Some aspects of Neogene biostratigraphy in the Pannonian Basin. *In*: ROYDEN, L. H. & HORVÁTH, F. (eds) *The Pannonian Basin.* American Association of Petroleum Geologists, Memoirs, **45**, 66–77.

TARI, G., BÁLDI, T. & BEKE-BÁLDI, M. 1993. Paleogene retroarc flexural basin beneath the Neogene Pannonian Basin: a geodynamic model. *Tectonophysics,* **226**, 433–455.

VARGA, GY., CSILLAG-TEPLÁNSZKY, E., FÉLEGYHÁZI, Zs. 1975. A Mátra hegység földtana [Geology of the Mátra Mountains], *MÁFI Évkönyve*, **57–1**. Müszaki Könyvkiadó, Budapest.

Palaeomagnetism and palaeogeography of the Western Carpathians from the Permian to the Neogene

MIROSLAV KRS, MARTA KRSOVÁ & PETR PRUNER

Geological Institute, Academy of Sciences of the Czech Republic, Rozvojová 135, 165 00 Prague 6, Czech Republic

Abstract: Geodynamic models for the Western Carpathians require evaluation of palaeomagnetic data from the Outer Carpathian flysch belt, from limestones of the Klippen Belt and from volcanic and sedimentary rocks of the Inner Carpathians. Palaeomagnetic data for the Permian to the Neogene are evaluated. The data indicate marked, mostly anticlockwise tectonic rotations of larger rock complexes and nappes. The distribution of palaeomagnetic poles is characteristic of a collision zone, and crosses the apparent polar wandering path for the African plate. Tectonic rotations are observed in both the Inner Carpathians and the Klippen Belt as well as in the Outer Carpathian flysch belt. Most drift occurred during the Permian to Triassic interval, with northward movement from initial southern equatorial latitudes. Drift decelerated from the Jurassic to the Neogene.

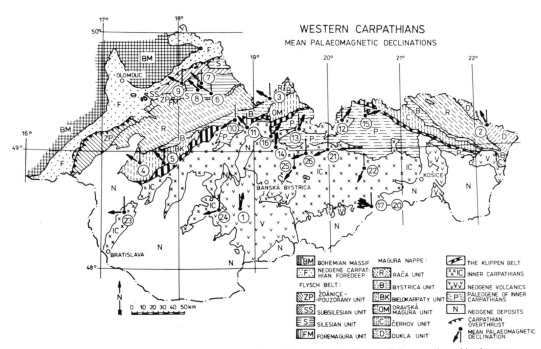

Fig. 1. Simplified geological map of the Western Carpathians, showing the localities mentioned in the text.

The Carpathian Mountains represent a complex Alpine fold and thrust belt which is over 1500 km long and has a pronounced arcuate shape. Figure 1 shows that part of the Western Carpathians within the territories of Slovakia and Moravia. The system can be subdivided into three main tectonostratigraphic zones, namely the Outer Zone or Outer Carpathian flysch belt, the Klippen Belt and the Inner Zone or Inner Carpathians. This paper summarizes palaeomagnetic data obtained in the Western Carpathians from units of Permian to Neogene age by various authors. These data come from flysch deposits of the Outer Carpathians, limestones of the Klippen Belt, and volcanic and sedimentary rocks of the Inner Carpathians, and mainly come

From Morris, A. & Tarling, D. H. (eds), 1996, *Palaeomagnetism and Tectonics of the Mediterranean Region*, Geological Society Special Publication No. 105, pp. 175–184.

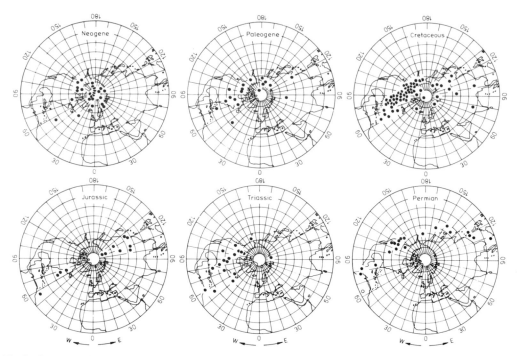

Fig. 2. Stereographic projection of pole positions for the Alpine–Carpathian–Pannonian zone from the Permian to the Neogene.

from Slovakia, east Moravia and north Hungary.

The first palaeomagnetic paper on this region suggested the presence of tectonic rotations of varying magnitudes and in different senses (Kotásek & Krs 1965). Subsequent syntheses explained the observed declinations in terms of rotational deformation of nappes or blocks during Alpine folding (Krs et al. 1982). A statistical evaluation of Phanerozoic palaeomagnetic data for the Eurasian and African continents outlined striking rotations about vertical axes for the whole Alpine belt. It was noted that Cretaceous rocks to the west of the Western Carpathians had experienced rotations in a predominant clockwise sense (Krs 1982). Subsequent syntheses of palaeomagnetic data from the region (Márton 1987; Márton et al. 1987; Márton & Mauritsch 1990) resulted in documentation of tectonic rotations in different parts of the Alpine-Carpathian-Pannonian zone.

The present paper summarises data presented in several publications (Mauritsch & Becke 1987; Irving et al. 1976; Márton et al. 1988; Pätrascu et al. 1990), and in others which are not easily obtainable outside the Czech Republic (e.g. Krs et al. 1982; Krs 1982; and internal reports of the Geological Institute of the Academy of Sciences and the Geofyzika Brno Company). In this review, only mean pole positions with values of $\alpha_{95} < 15°$ and $k > 5$ are considered to be statistically well defined. On this basis, reliable poles have been defined for nappes in the Northern Apennines, the Western, Southern, Eastern and Northeastern Alps, Istria, the Transdanubian Mountains, the Outer, Western and Eastern Carpathians, and the Inner Carpathians (including the Little Carpathian Mountains). These data verify the Gondwanan affinity of the Carpathian area during the Mesozoic. Palaeogeographic affinity to the African plate can also be proven for the Northern Apennines, Southern Alps, Istria and the Transdanubian Mountains, on the basis of sense of rotation (prevailingly anticlockwise) and on absolute values of palaeolatitudes. Some other areas exhibit similar affinity, such as the Eastern Alps and the Inner Carpathians, although the statistical reliability is lower. The Northern Apennines, Southern Alps and Istria show similar declinations from the Permian until the Cretaceous. An anomalous clockwise rotation occurs in the Northeastern Alps (shown by data from Jurassic and Cretaceous units) and the Outer Eastern Carpathians (shown by data from Jurassic units, and tentatively by data from

Cretaceous rocks). Tectonic rotations give rise to scatter of palaeomagnetic pole positions. Figure 2 shows the pole positions for the Alpine–Carpathian–Pannonian zone from the Permian to the Neogene, as inferred by different authors.

The results obtained from nappe systems in the Western Carpathians are interpreted within this framework of deformation of the wider Alpine–Carpathian–Pannonian system, where tectonic rotations are known to be a characteristic feature. These rotations can be simulated by a model for the entire Alpine–Carpathian–Pannonian zone and for the Western Carpathians.

Palaeomagnetic properties of rocks from the Western Carpathians

All samples analysed at our laboratory were subjected to detailed magnetic cleaning, generally by thermal demagnetization using a MA-VACS system (which guarantees a high magnetic vacuum during heating). Pilot samples were studied petromagnetically to define magnetic carriers. Account was taken of the effect of local tectonics on the deviation of palaeomagnetic directions, and samples were not collected from highly tectonized nappes. Remanences were studied by multi-component analysis (Kirschvink 1980; Kent et al. 1983; McFadden & Schmidt 1986).

Palaeomagnetic studies in the Outer Carpathian flysch belt have focused on pelitic sediments, red and grey claystones and fine-grained sandstones. These rocks display important secondary components of magnetization. In most of these rocks, the magnetic carrier is fine-grained magnetite, with haematite being frequent in grey sediments. In most cases, the sampled units of the Outer Western Carpathians exhibit anticlockwise rotations. Rocks of the teschenite association and their contact margins record thermo-remanent magnetizations which also indicate anticlockwise rotation (References 2 to 9 in Table 1).

The subhorizontal nappe of the White Carpathians has exposures of relatively undeformed sandstones, siltstones and mudstones. Magnetization components from these rocks display both normal and reverse polarities, which are defined with a high degree of confidence. Figure 3 shows typical results of thermal demagnetization, in which hematite is the magnetic carrier. Figure 4 gives mean palaeomagnetic directions derived by progressive thermal demagnetization and multi-component analysis. The data are given in Table 1 (Reference 5, White Carpathians, western Slovakia), after inverting

reversed polarities through the origin, and again indicate an anticlockwise tectonic rotation.

The reliability of the interpretation of tectonic rotations can be enhanced by systematic magnetostratigraphic investigations. The magnetostratigraphy of the Tithonian-Berriasian boundary strata has been investigated at Brodno (near Žilina, W. Slovakia), in the Klippen Belt (see Houša et al. this volume). Palaeomagnetic directions from this unit indicate large anticlockwise tectonic rotation (Reference 10 in Table 1).

Samples from several rock complexes in the Western Carpathians indicate syn-tectonic remagnetization during Alpine folding, produced by long-lived metamorphism. The remanence of the Silica nappe (northern Hungary) has also been attributed to this mechanism (Márton et al. 1991). Folding causes both chemical remanent magnetization and thermo-viscous magnetization. A typical example is the Meliata series, which consists of Triassic and Jurassic radiolarian rocks, Ladinian and Carnian corneo-limestones, radiolarian rocks, grey slates and red limestones of Dogger age, and limestones (olistolites?) of Triassic age. Magnetic carriers with relatively low unblocking temperatures are observed, except in the red radiolarites and corneo-limestones, in which hematite is the magnetic carrier. Primary palaeomagnetic components are not preserved. For example, Fig. 5 gives the results of the mean directions of separated components of remanence in the temperature interval of 150–400°C for six sites. Similar results were obtained for the other 11 sites. Mean virtual poles for the Meliata series show an increase in dispersion after tilt correction (in situ: $\alpha_{95} = 8.5°$, $k = 18.6$; tilt corrected: $\alpha_{95} = 13.6°$, $k = 7.9$; $N = 17$). Thus, the separated remanence components in this temperature interval are post-folding in age and are of secondary origin. The pole position computed from the in situ mean directions (palaeolatitude = 46.5°N; palaeolongitude = 299.0°; $\alpha_{95} = 10.8°$; $k = 23.8$) is close to the pole positions derived for Cretaceous rocks in the Western Carpathians (see Table 1). All directions in the Meliata series are of normal polarity, suggesting that remagnetization occurred in the Cretaceous long normal polarity epoch, which persisted from the Aptian to the Santonian. The magnetization is probably of syntectonic origin and again indicates an anticlockwise tectonic rotation, in this case in the period after remagnetization.

Kruczyk et al. (1992) prove pre-tectonic magnetization components in Jurassic carbonates from the Krížná nappe, using magnetomineralogical and other analyses. In contrast to the majority of data from the Western

Table 1. *Palaeomagnetic data from the Western Carpathians.*

	Region	Age and lithology	Location		Mean direction						Pole position		Confidence		Reference
			Lat°N	Long°E	Dec	Inc	α_{95}	k	N	n	Lat (φ_P)	Long (λ_P)	δ_m	δ_P	
1	Central Slovakia	L Mio volcanics	48.5	18.8	9.2	64.7	7.7	53	70	280	83.5°N	122.3°E	12.4	10.0	Nairn in Irving et al. (1976)
2	Dukla, E Slovakia	E-M Eoc red clsts.	49.16	22.19	158.7	-40.1	3.6	10.4	5	165	58.8°S	62.4°E	4.3	2.6	Koráb et al. (1981)
3	Oravská Magura, N Slovakia	L Pal-M Eoc. clsts & ssts	49.42	19.21	122.9	-58.8	9.8	5.3	4	49	49.1°S	117.7°E	14.6	10.7	Krs et al. (1991)
4	SE Moravia (Louka, Nivnice)	L Pal ssts & clsts	48.94	17.55	121.0	-41.6	11.4	46.2	5	31	38.0°S	101.1°E	13.9	8.5	Krs et al. (1993)
5	White Carpathians W Slovakia	L Sen ssts & clsts	48.9	17.8	320.6	43.7	5.6	86.0	9	75	51.6°N	95.1°W	7.0	4.4	Krs et al. (1993)
6	NE Moravia, Silesia	Con grey ssts	49.48	18.43	300.3	46.0	10.0	3.1		94	39.8°N	74.7°W	12.8	8.2	Krs et al. (1978)
7	NE Moravia, Silesia Ondřejník Mt	Cen-E Tur red clsts	49.57	18.30	317.7	72.6	4.5	10.9		101	64.1°N	36.7°W	7.9	7.1	Krs et al. (1977)
8	NE Moravia, Silesia	Cen-E Tur red clsts	49.52	18.27	312.7	52.7	3.3	8.8		228	51.8°N	78.7°W	4.6	3.2	Krs et al. (1978)
9	NE Moravia, Silesia	Haut-Barr teschenites	49.57	18.07	295.1	55.5	5.6	6.4		116	42.2°N	63.3°W	8.1	5.8	Krs & Šmíd (1979)
10	W. Slovakia, Brodno near Žilina	Tith-Berr carbs	49.26	18.75	236.3	45.4	5.6	9.8		104	1.1°N	29.2°W	7.1	4.5	
11	Malá Fatra, N Slovakia	M-L JUR carbs	49.25	20.2	321	44	3	96	21	173	58°N	253°E	11.9	8.9	Kruczyk et al. (1992)
12	Belanski Tatra, N Slovakia	M L JUR carbs			40	59	8	118	4	17	61°N	113°E			Kruczyk et al. (1992)
13	W Tatra, N Slovakia	M L JUR carbs	49.03	19.28	22	59	4.3	198	7	29	71.7°N	132.2°E	20.1	14.5	Kruczyk et al. (1992)
14	Low Tatra, N Slovakia	M JUR carbs			2	56	14	73	3	12	78°N	192°E	15.3	9.8	Kruczyk et al. (1992)
15	Spišská Magura, N Slovakia	M JUR carbs	49.28	20.53	75	46	12	114	3		30°N	102°E			Kruczyk et al. (1992)
16	Choč Hills, N Slovakia	E JUR carbs	49.12	19.23	39	63	12	104	3	15	63°N	105°E	18.9	14.8	Kruczyk et al. (1992)
17	Aggtelek Mts, Silica nappe, N Hungary	L TRI carbs	48.5	20.6	289	59	11	38	6	45	40.2°N	51.4°W	16.4	12.3	Márton et al. (1988)
18	Aggtelek Mts, Silica nappe, N Hungary	M TRI carbs	48.5	20.6	272	40	16	34	4	22	18.2°N	55.3°W	19.3	11.6	Márton et al. (1988)
19	Aggtelek Mts, Silica nappe, N Hungary	E TRI carbs	48.5	20.6	294	24	23	31	3	20	25.2°N	79.2°W	24.6	13.1	Márton et al. (1988)
20	Rudabánya Mts, Bódva nappe, N Hungary	M TRI carbs	48.5	20.6	298	43	17	20	5	40	36.8°N	72.2°W	21.1	13.1	Márton et al. (1988)
21	Choč nappe, Central Slovakia	L PER red shales	49.0	20.0	71.8	19.8	6.4	4.5	13	141	19.6°N	117.0°E	6.6	3.5	Kotásek & Krs (1965)
22	S of Sp. N. Ves, NW of Košice, E Slovakia	L PER red shales	48.83	20.50	29.2	16.9	5.3	4.6	9	195	42.9°N	159.3°E	5.5	2.8	Kotásek & Krs (1965) Krs (1966)
23	Little Carpathian Mts, W Slovakia	PER melaphyres	48.47	17.3	269.4	-2.3	18.9	13.5	6	46	1.2°S	73.0°W	18.9	9.5	Krs et al. (1982)
24	Tríbeč Mts, W Slovakia	PER melaphyres	48.49	18.53	254.8	-18.0	9.9	8.1	3	29	16.9°S	66.2°W	10.3	5.3	Krs et al. (1982)
25	South Low Tatra Mts, N Slovakia	PER melaphyres	48.85	19.55	222.8	-13.2	18.2	5.1	16	121	34.6°S	35.4°W	18.6	9.5	Krs et al. (1982)
26	North Low Tatra Mts, N Slovakia	PER melaphyres	48.97	19.70	249.7	-16.2	11.4	6.1	32	300	19.5°S	60.2°W	11.8	6.1	Krs et al. (1982)

Mio, Miocene; Pal, Palaeocene; Eoc, Eocene; Sen, Senonian; Con, Coniacian; Cen, Cenomanian; Tur, Turonian; Haut, Hauterivian; Barr, Barremian; Berr, Berriasian; JUR, Jurassic; TRI, Triassic; PER, Permian; L, Late; M, Middle; E, Early; carbs, carbonates; clsts, claystones; ssts, sandstones.

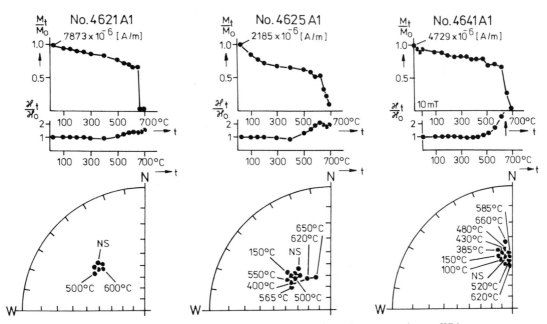

Fig. 3. Thermal demagnetization of representative samples of Late Senonian grey sandstone, White Carpathians. western Slovakia.

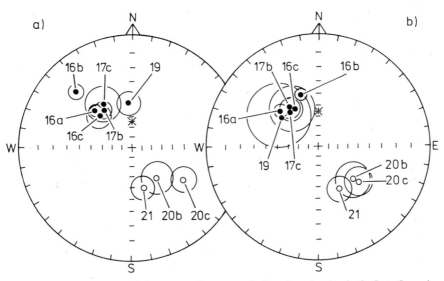

Fig. 4. Stereographic projection of the mean palaeomagnetic directions for sites in the Late Senonian sandstones and claystones of the Javorina Formation, White Carpathians, western Slovakia. (**a**) Mean directions with the optimum Fisherian grouping during progressive thermal demagnetization; (**b**) mean directions derived by multi-component analysis.

Carpathians, both anticlockwise and clockwise tectonic rotations occur in different parts of this nappe.

In the Aggtelek–Rudabánya Mountains, in the southernmost part of the Inner Carpathians on the territory of North Hungary, anticlockwise tectonic rotation occurred in the period following the Triassic, probably in the Neogene

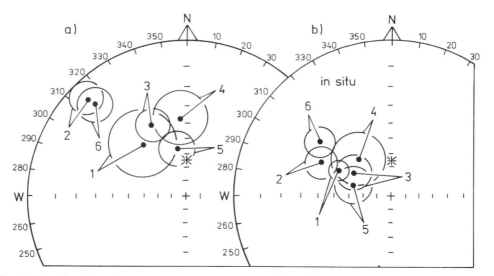

Fig. 5. Mean directions of remanence components separated by multi-component analysis for 6 sites in the Meliata series, southern Slovakia. (**a**) tilt corrected results; (**b**) *in situ* results.

(Márton *et al.* 1988). The remanence components used in this interpretation were derived in higher temperature intervals, show both normal and reverse polarities and pass a fold test (References 17–20 in Table 1).

Summary of palaeolatitudes and rotations

Figure 6 presents values of palaeomagnetic declination, inclination and palaeolatitude plotted against age. Permian to the Early–Mid-Eocene rocks from the Western Carpathians experienced predominantly anticlockwise rotations. Lower inclination values were generally found, indicating that the rocks were deposited at lower palaeolatitudes. Because most studied rocks of the Western Carpathians are from the flysch formation, and so were deposited on inclined sedimentation surfaces, the magnetic inclination of these rocks is weighted with an error. The tilt of the sedimentation surfaces, however, attained only a few degrees (M. Rakús and M. Potfaj, pers. comm.). Large changes in palaeolatitude occurred during the Permian and Triassic. Similar changes in palaeolatitude also occurred for the European plate in the Permian and Triassic, and were caused by the drift of the Laurasian plate (cf. Dercourt *et al.* 1993).

Anticlockwise rotations prevail throughout the Western Carpathians in Slovakia, East Moravia and North Hungary, except in the Jurassic rocks of the Krížná nappe. Up to 110° rotation occurred in the Permian rocks and the flysch formation shows about 60° anticlockwise rotation.

Separation of components of tectonic rotation

Figure 2 shows that palaeomagnetic pole positions for rocks of the same or similar age from the Alpine-Carpathian-Pannonian zone display specific distributions, indicating rotation of the whole zone as suggested by Márton (1987), Márton & Mauritsch (1990) and Mauritsch & Becke (1987). Similar results were obtained in the Western Carpathians (Kotásek & Krs 1965; Krs *et al.* 1982; Márton *et al.* 1991).

The scatter of poles can be explained using a model in which movements are partitioned into two components: the first relating to rotation of the major plate to which the unit is attached (rotation about a distant rotation pole); the second relating to rotation during Alpine collision of the smaller-scale tectonic block containing the unit (rotation about a proximal pole of rotation). We computed parameters of small circles centred on the Western Carpathians (50°N, 13°E). These represent the loci of poles of equal distance from the study area. Poles lying on a certain circle may differ in declination but not in inclination. Localised rotations without translation will result in a dispersion of poles along a small circle, whereas large-scale movements will also produce movement of poles from one small circle to another. Six pole trajectories

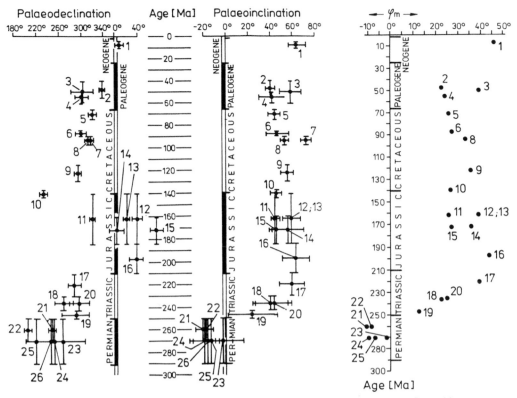

Fig. 6. Variation of declination, inclination and palaeolatitude with age for the Western Carpathians.

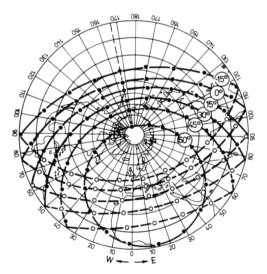

Fig. 7. Model distribution of pole positions calculated for rocks with different values of inclination = −15°, 0°, 15°, 30°, 45° and 60°, roughly corresponding to the time span from the Permian to the Neogene. Solid (dashed) lines and solid (open) symbols indicate projection onto the upper (lower) hemisphere.

were calculated (Fig. 7), with inclinations corresponding to −15°, 0°, 15°, 30°, 45° and 60°. This range of inclinations corresponds to that observed from the Permian to the youngest rocks studied.

Figure 8 shows pole positions for Permian rocks of the Inner Carpathians relative to pole positions from Permian rocks in other parts of the Alpine–Carpathian–Pannonian zone. This indicates a translation of Permian rocks from equatorial and subequatorial zones, and also a rotation of nappes resulting from Alpine collision. Figure 9 shows pole positions derived from Jurassic carbonates from the Krížná nappe (References 11 to 16, Table 1) and a pole position derived from the Tithonian–Berriasian boundary at Brodno, near Žilina (Reference 10, Table 1). Although of widely different position, poles from both areas approach the theoretical trajectory for inclination = 45°, indicating the influence of vertical axis rotation. Kruczyk et al. (1992) suggested this difference results from oroclinal bending of the Inner Western Carpathians. Though seemingly anomalous, the pole position derived in the Tithonian-Berriasian carbonates adheres to the path for Jurassic

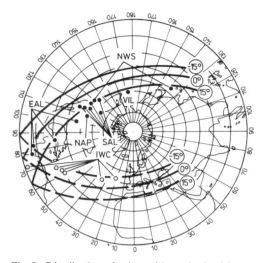

Fig. 8. Distribution of pole positions obtained from Permian rocks in the Inner Western Carpathians (IWC), the region of Villany (VIL), the Southern Alps (SAL), NW Slavonia (NWS), the Northern Apennines (NAP) and the Eastern Alps (EAL). Solid (dashed) lines and solid (open) symbols indicate projection onto the upper (lower) hemisphere.

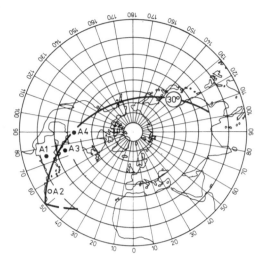

Fig. 10. Distribution of pole positions obtained from Early Triassic carbonates from the Aggtelek–Rudabánya Mountains. Solid (dashed) line indicates theoretical distribution of pole positions for rocks with inclination = 30°.

Tithonian–Berriasian carbonates at Štramberk, North Moravia (Outer Western Carpathians).

This analysis suggests tectonic rotation on the scale of entire nappe systems. To judge the possibility of tectonic rotations within a smaller area, virtual pole positions were calculated for two areas, and their distributions compared with the modelled small circles. The data of Márton *et al.* (1988) were used to compute virtual pole positions for Early Triassic carbonates of the Aggtelek Mountains. These poles lie close to the small circle corresponding to inclination = 30°, and are distributed along the circle as a result of small-scale tectonic rotations (Fig. 10). Similarly, virtual pole positions for Late Senonian rocks of the White Carpathians (Reference 5, Table 1) show a distribution which also suggests significant small-scale tectonic rotations (Fig. 11), even within the single sampled nappe which contains subhorizontal beds.

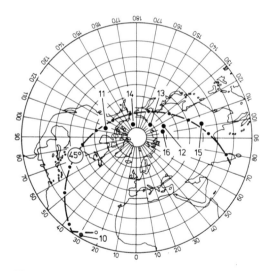

Fig. 9. Distribution of pole positions obtained from Jurassic carbonates from the Križná nappe system (Reference 11–16, Table 1) and in Tithonian-Berriasian limestones at the locality of Brodno, near Žilina (Reference 10, Table 1). Solid (dashed) line indicates theoretical distribution of pole positions for rocks with inclination = 45°.

rocks, and indicates a distinct anticlockwise tectonic rotation with respect to the Križná nappe. Similar rotation has been deduced from another magnetostratigraphic profile across

Conclusions

The palaeomagnetic data from the Western Carpathians indicate a marked tectonic rotation of larger rock units, predominantly in an anticlockwise sense. These rotations are observed in the rocks of the Inner Carpathians, the Klippen Belt, and the Outer Carpathian flysch belt. Tectonic rotations lead to the formation of a specific pole distribution which runs across the apparent polar wander path (in this case for the African plate). A similar distribution also occurs

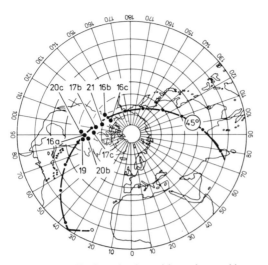

Fig. 11. Distribution of pole positions observed in Late Senonian sediments from the White Carpathians, western Slovakia, within subhorizontal beds in the same nappe. The mean pole position for this nappe is given under Reference 5 in Table 1. Solid (dashed) line indicates theoretical distribution of pole positions for rocks with inclination = 45°.

in the West European and Central European Hercynian belt (Edel 1987; Krs *et al.* 1995). In the Western Carpathians, tectonic rotations occur on a range of scales, from those affecting the whole region to those affecting single nappes/parts of nappes.

The most pronounced palaeolatitudinal drift occurred in the Permian and Triassic, and the drift began decelerating from the Jurassic until the Neogene (Fig. 6). Similar variations have been established for other regions of the Tethyan realm, e.g. the Iberian Meseta and adjacent mobile belts, Corsica and Sardinia, Italy including Sicily and the adjacent parts of the Alps, Greece and Southern Bulgaria, the Transdanubian Mountains in Hungary, and Turkey including the eastern Aegean territory and Cyprus (Van der Voo 1993).

Rotations of individual fault blocks induce large scatters of palaeomagnetic pole positions even though they may affect only small units. In contrast, changes in pole positions due to drift of larger units are generally smaller, though the translations involved are appreciable. For instance, for the Permian rocks of the Western Carpathians these translations reach values of up to 6000 km since the Permian to the present time. These differences in the magnitude of pole movements reflect two different types of rotation, i.e. those around rotation poles close to and far removed from the study area, respectively.

The authors wish to thank A. Morris for reviewing the paper, for help in preparation and editing the final version of the paper, and for improvement of the English. They are also grateful to E. Márton and D. Peacock for reviews and helpful suggestions.

References

DERCOURT, J., RICOU, L. E. & VRIELYNCK, B. (eds) 1993. *Atlas, Tethys, Palaeoenvironmental maps, explanatory notes.* CCGM, CGMW, Paris.

EDEL, J. B. 1987. Paleopositions of the western Europe Hercynides during the Late Carboniferous deduced from paleomagnetic data: consequences for "stable" Europe. *Tectonophysics,* **139**, 31–41.

HOUŠA, V., KRS, M., KRSOVÁ, M. & PRUNER, P. 1996. Magnetostratigraphy of Jurassic–Cretaceous Limestones in the Western Carpathians. *This volume.*

IRVING, E., TANCZYK, E. & HASTIE, J. 1976. Catalogue of paleomagnetic directions and poles. Energy, Mines and Resources, Ottawa. *Geomagnetic Series Number 6,* **10**, 1–70.

KENT, J. T., BRIDEN, J. C. & MARDIA, K. V. 1983. Linear and planar structure in ordered multivariate data as applied to progressive demagnetization of palaeomagnetic remanence. *Geophysical Journal of the Royal Astronomical Society,* **75**, 593–621.

KIRSCHVINK, J. L. 1980. The least-squares line and plane and the analysis of palaeomagnetic data. *Geophysical Journal of the Royal Astronomical Society,* **62**, 699–718.

KORÁB, T., KRS, M., KRSOVÁ, M. & PAGÁČ, P. 1981. Palaeomagnetic investigations of Albian (?)–Paleocene to Lower Oligocene sediments from the Dukla unit, East Slovakian Flysch, Czechoslovakia. *Západné Karpaty, séria geológia,* Geological Institute of Dionýz Štúr, Bratislava, **7**, 127–149.

KOTÁSEK, J. & KRS, M. 1965. Palaeomagnetic study of tectonic rotation in the Carpathian Mountains of Czechoslovakia. *Palaeogeography, Palaeoclimatology, Palaeoecology,* **1**, 39–49.

KRS, M. 1966. Palaeomagnetic pole position for the Lower Triassic of East Slovakia (Czechoslovakia). *Vestník, Geological Survey, Prague,* XLI, **4**, 287–290.

—— 1982. Implication of statistical evaluation of Phanerozoic palaeomagnetic data (Eurasia, Africa). *Rozpravy ČSAV, řada matematických a přírodních věd,* Academia, Prague, **92**, 3, 1–86.

—— & ŠMÍD, B. 1979. Palaeomagnetism of Cretaceous rocks of the teschenite association, Outer West Carpathians of Czechoslovakia. *Applied Geophysics, Prague, Journal of Geological Sciences,* **16**, 7–25.

——, KRSOVÁ, M., CHVOJKA, R. & POTFAJ, M. 1991. Paleomagnetic investigations of the flysch belt in the Orava region, Magura unit, Czechoslovak Western Carpathians. *Geologické práce, Geological Institute of Dionýz Štúr, Bratislava,* **92**, 135–151.

——, —— & PRUNER, P. 1995. Palaeomagnetism and palaeography of Variscan formations of the Bohemian Massif: a comparison with other regions in Europe. *Studia geophysica et geodaetica, Academia, Prague,* **39**, 309–319.

——, ——, ——, CHVOJKA, R. & POTFAJ, M. 1993. Palaeomagnetic investigations in the Biele Karpaty Mountains unit, Flysch Belt of the Western Carpathians. *Geologica Carpathica, Bratislava,* **45**, 35–43.

——, —— & ROTH, Z. 1977. A palaeomagnetic study of Cenomanian–Lower Turonian sediments in the Moravskoslezské Beskydy Mountains. *Věstník, Geological Survey, Prague,* **52**, 323–332.

——, MUŠKA, P. & PAGÁČ, P. 1982. Review of palaeomagnetic investigations in the West Carpathians of Czechoslovakia. *Geologické práce, Správy, Geological Institute of Dionýz Štúr, Bratislava,* **78**, 39–58.

——, PRUNER, P. & ROTH, Z. 1978. Palaeotectonics and palaeomagnetism of Cretaceous rocks of the Outer West Carpathians of Czechoslovakia. *Geofyzikální sborník ČSAV Prague,* XXVI, **512**, 269–291.

KRUCZYK, J., KADZIALKO-HOFMOKL, M., LEFELD, J., PAGÁČ, P. & TUNYI, I. 1992. Paleomagnetism of Jurassic sediments as evidence of oroclinal bending of the Inner West Carpathians. *Tectonophysics,* **206**, 315–324.

MÁRTON, E. 1987. Paleomagnetism and tectonics in the Mediterranean region. *Journal of Geodynamics,* **7**, 33–57.

——, MÁRTON, P. & LESS, G. 1988. Palaeomagnetic evidence of tectonic rotations in the southern margin of the Inner West Carpathians. *Physics of the Earth and Planetary Interiors,* **52**, 256–266.

—— & MAURITSCH, H. J. 1990. Structural applications and discussion of a paleomagnetic post-Paleozoic data-base for the Central Mediterranean. *Physics of the Earth and Planetary Interiors,* **62**, 48–59.

——, —— & TARLING, D. H. 1987. Pre-Alpine paleomagnetic results of the Alpine–Mediterranean belt. *In:* FLÜGEL, SASSI & GRECULA (eds) *Pre-Variscan and Variscan events in the Alpine–Mediterranean mountain belts.* Mineralia Slovaca Monograph, Bratislava, 351–360.

MÁRTON, P., ROZLOŽNÍK, L. & SASVÁRI, T. 1991. Implications of a palaeomagnetic study of the Silica nappe, Slovakia. *Geophysical Journal International,* **107**, 67–75.

MAURITSCH, H. J. & BECKE, M. 1987. Paleomagnetic investigations in the Eastern Alps and the Southern border zone. *In:* FLÜGEL, H. W. & FAUPL, P. (eds) *Geodynamics of the Eastern Alps.* Vienna, 282–308.

MCFADDEN, P. L. & SCHMIDT, P. W. 1986. The accumulation of palaeomagnetic results from multi-component analysis. *Geophysical Journal of the Royal Astronomical Society,* **86**, 965–979.

PÄTRASCU, S., BLEAHU, M. & PANAIOTU, C. 1990. Tectonic implications of paleomagnetic research into Upper Cretaceous magmatic rocks in the Apuseni Mountains, Romania. *Tectonophysics,* **180**, 309–322.

VAN DER VOO, R. 1993. *Paleomagnetism of the Atlantic, Tethys and Iapetus Oceans.* Cambridge University Press.

Magnetostratigraphy of Jurassic–Cretaceous limestones in the Western Carpathians

VÁCLAV HOUŠA, MIROSLAV KRS, MARTA KRSOVÁ & PETR PRUNER

Geological Institute, Academy of Sciences of the Czech Republic, Rozvojová 135, 165 00
Prague 6 – Suchdol, Czech Republic

Abstract: Magnetostratigraphic studies were carried out on limestones spanning the Tithonian–Berriasian boundary at Brodno (near the town of Žilina, west Slovakia) and Štramberk (north Moravia) in the Western Carpathians. The magnetic carrier in both sections was found to be magnetite. The observed pattern of normal and reverse polarity magnetozones corresponds well to magnetostratigraphic profiles in other areas of the Tethyan realm. Calpionellids are abundant in both profiles, while ammonites are almost completely missing. The base of the *Calpionella* zone, provisionally considered to represent the Jurassic–Cretaceous boundary in the Tethyan realm, was established in magnetozone M19n. The palaeolatitude of 27°N derived from the Brodno profile corresponds well to those derived from Jurassic rocks of the Križná nappe, on the northern rim of the Inner Carpathians. A pronounced anticlockwise tectonic rotation of 124° is established.

The Jurassic–Cretaceous (Tithonian–Berriasian) boundary is relatively well-defined on the basis of biostratigraphy (Remane *et al.* 1986). In the Tethyan and Boreal realms it is based on ammonite zonation, but taxa common to both realms are absent and the boundary is therefore defined using different ammonite taxa in each realm. As a consequence, the boundary evidently lies at different time horizons in the two realms (Zeiss 1986; Hoedemaeker 1987), and is therefore provisional. In the Tethyan realm, calpionellids are also used to define the biostratigraphy of Jurassic–Cretaceous boundary strata, because in many areas ammonites are relatively rare or completely absent. The boundary in the Tethyan realm is situated at the base of the Jacobi–Grandis ammonite zone, and this level is treated as practically identical to the base of the Calpionella Standard Zone (Remane *et al.* 1986). In the Boreal realm, calpionellids are absent and the Jurassic–Cretaceous boundary is situated at the base of the Sibiricus–Maynci ammonite zone. This position is younger than that adopted in the Tethyan realm. Exact biostratigraphic correlation of these boundaries between realms is impossible due to a lack of common index species and a poorly developed biostratigraphic record in transitional areas.

Magnetostratigraphic dating offers a reliable alternative method of identifying chronologically identical sections in distant regions and can potentially be used to correlate globally biostratigraphic zonations near the Jurassic–Cretaceous boundary. In the Tethyan realm some earlier studies placed the base of the Calpionella Standard Zone in different positions, e.g. in the central part of magnetozone M19n (Ogg 1983 *in* Galbrun *et al.* 1990), in the older part of M19n (Ogg *et al.* 1984), in the basal part of M17 (Lowrie & Channell 1984), and in the younger part of M19n (Mazaud *et al.* 1986).

In order to address this problem of the exact position within the geomagnetic polarity timescale of the biostratigraphically defined Jurassic–Cretaceous boundary in the Tethyan realm, this paper presents detailed magnetostratigraphic results from the boundary strata of Tithonian–Berriasian limestones from two localities in the Western Carpathians. Preliminary studies were carried out at five localities (Fig. 1): (1) the 'Kotouč' quarry at Štramberk, northern Moravia; (2) the 'Horní Skalka' quarry at Štramberk, northern Moravia; (3) an abandoned quarry at Brodno, near Žilina, western Slovakia; (4) at Strážovce, between the villages of Čičmany and Zliechov, western Slovakia; and (5) Hlboč, near Smolenice, western Slovakia. The Kotouč and Brodno localities were selected from these for more detailed study (Houša 1990; Michalík *et al.* 1990), on the basis of the preliminary palaeomagnetic results and other geological and palaeontological criteria. In addition to the magnetostratigraphic results, we briefly discuss the tectonic implications of the magnetization directions obtained in the sections.

Magnetic mineralogy

Remanent magnetization and volume magnetic susceptibility were measured using JR-4 and

From Morris, A. & Tarling, D. H. (eds), 1996, *Palaeomagnetism and Tectonics of the Mediterranean Region*, Geological Society Special Publication No. 105, pp. 185–194.

Fig. 1. Sketch map showing the five localities of Jurassic–Cretaceous boundary strata investigated in this study.

Table 1. *Natural remanent magnetization* (J_n) *and volume magnetic susceptibility* (χ_n) *of pilot samples from the five localities of Jurassic–Cretaceous boundary strata sampled intially in the Western Carpathians*

Locality	J_n ($\times\ 10^{-6}\,A\,m^{-1}$)	χ_n ($\times\ 10^{-6}\,SI$)	No. of samples
'Kotouč' quarry, Štramberk, northern Moravia	98 ± 75	$|8| \pm |3|$	24
'Horní skalka' quarry, Štramberk, northern Moravia	373 ± 462	$|7| \pm |7|$	7
Brodno, near Žilina, western Slovakia	758 ± 315	14 ± 8	10
Strážovce, between Čičmany & Zliechov, western Slovakia	542 ± 289	60 ± 16	10
Hĺboč, near Smolenice, western Slovakia	1800 ± 627	351 ± 335	5

JR-5 spinner magnetometers and a KLY-2 kappa-bridge (Jelínek l966, 1973). Alternating field (AF) demagnetization was carried out using a Schonstedt GSD-1 apparatus and thermal demagnetization using a MAVACS system (which produces a high magnetic vacuum during heating). The magnetic measurements were combined with X-ray diffraction studies to identify magnetic carriers in the extremely weakly magnetized rocks.

Table 1 gives the mean values of moduli of remanent magnetization (J_n) and of magnetic susceptibility of pilot samples (χ_n) of the Tithonian–Berriasian limestones in their natural state from the five sampled localities. The samples from the Kotouč quarry at Štramberk have extremely low magnetizations, and mainly negative magnetic susceptibilities (resulting from the dominance of diamagnetic calcite). The limestones at Štramberk were rapidly deposited in a dynamic peri-reefal environment. Similar low magnetizations of Jurassic–Cretaceous limestones have also been reported in other Tethyan localities (e.g. Ogg et al. 1984, 1988, 1991; Lowrie & Channell 1984; Galbrun 1985; Moreau et al. 1992; Márton 1986). Coercivity spectra show that the magnetic carriers are of medium to high magnetic hardness. Figure 2 shows an example of a wide coercivity spectrum obtained from a limestone sample from the Štramberk locality. The sample reaches saturation at fields higher than 300 mT. This, combined with unblocking temperatures of about 540°C, suggests the presence of fine-grained magnetite.

Samples from the Štramberk and Brodno localities were given an IRM in a 500 mT field and progressively thermally demagnetized (Figs 3 & 4). The unblocking temperatures of between 540° and 560°C for samples from the two localities suggest the presence of magnetite. The samples from the Brodno locality also

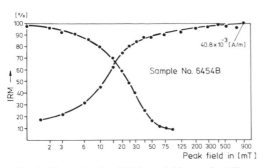

Fig. 2. Example of an IRM acquisition and a AF demagnetization curve. Sample of Late Tithonian limestone, Kotouč quarry locality, Štramberk, north Moravia.

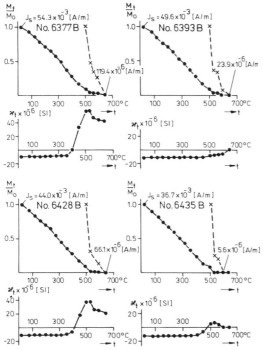

Fig. 3. Results of thermal demagnetization of IRM for limestone samples from the Štramberk locality, north Moravia. J_s is saturation remanent magnetization; M_t is the remanent magnetic moment of a sample demagnetized at temperature t; M_o is the sample moment in the saturated state; M_t/M_o and χ_t/χ_o are normalized values of the remanent magnetic moment and of volume magnetic susceptibility respectively.

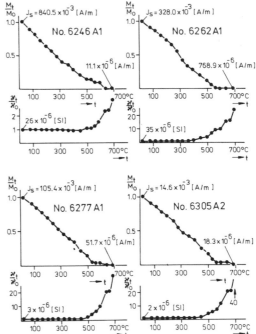

Fig. 4. Results of thermal demagnetization of IRM for limestone samples from the Brodno locality, near Žilina, west Slovakia. J_s is saturation remanent magnetization; M_t is the remanent magnetic moment of a sample demagnetized at temperature t; M_o is the sample moment in the saturated state; M_t/M_o and χ_t/χ_o are normalized values of the remanent magnetic moment and of volume magnetic susceptibility respectively.

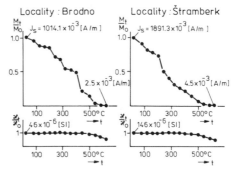

Fig. 5. Results of thermal demagnetization of isothermally magnetized samples of magnetic concentrates prepared from bulk samples for the localities of Brodno and Štramberk. These concentrates were X-ray analysed. M_t is the remanent magnetic moment of a sample demagnetized at temperature t; M_o is the sample moment in the saturated state; M_t/M_o and χ_t/χ_o are normalized values of the remanent magnetic moment and of volume magnetic susceptibility respectively.

exhibit a small fraction of a mineral with an unblocking temperature below 680°C (see samples Nos 6277 A1 and 6305 A2 in Fig. 4), probably due to a small admixture of hematite in the samples collected from surface outcrops. Phase changes of magnetic minerals at higher temperatures during progressive thermal demagnetization are shown by increases in magnetic susceptibility.

The concentration and grain-size of the magnetic minerals in these limestones are so low that they could not be studied in polished sections. X-ray diffraction studies were, therefore, performed to confirm the presence of fine-grained magnetite. The concentration of ferrimagnetic minerals in the limestones under study is low, which is shown by anomalously low values of magnetic susceptibility and remanent magnetization (Table l). It was therefore necessary to prepare concentrates of ferrimagnetic particles for X-ray analysis. To prevent undesirable phase changes or contamination of individual fractions during concentrate preparation, we used a procedure based on magneto-mineralogical identification of ferrimagnetics in the original limestone, in successively prepared sludge fractions, and eventually in the concentrate itself, which was then X-ray analysed. To separate the magnetic minerals, large limestone sample charges were crushed and dissolved in acetic acid of 10% concentration for 2–3 days. Acetic acid was used to prevent possible disturbance of the magnetic minerals. As a check, the samples were also dissolved in HCl acid of 10% concentration for 2–3 hours.

The Štramberk sample shows a higher magnetic content after being dissolved in acetic acid than after dissolving in HCl. The magnetite content after treatment in HCl was *c.* $0.3\,g\,t^{-1}$. Irregular, seldom isometric, and occasionally spherolithoidal magnetite particles have dimensions between 10 and 20 μm. Grains of 3–10 μm size (mainly spherolites) were also present . In the Brodno sample, the magnetic mineral content after treatment in the acetic acid ($8.9\,g\,t^{-1}$) was again higher than after treatment in HCl ($2.2\,g\,t^{-1}$). The size of magnetite grains ranges between 3 and 10 μm, with grains greater than 10 μm being rare. The magnetite is in the form of isometric grains to imperfect crystals, predominantly of octahedral habit, with rare fine spherolites. Apart from magnetite, hematite is also present in the Brodno sample, along with rare minute crystals of pyrite (which are partly limonitized in places).

Figure 5 shows results of thermal demagnetization of the magnetic concentrates. Unblocking temperatures in the range of 540–560°C correspond to the values established in the original limestone samples. This demonstrates neither contamination nor phase changes of ferrimagnetics occurred during preparation of the magnetic

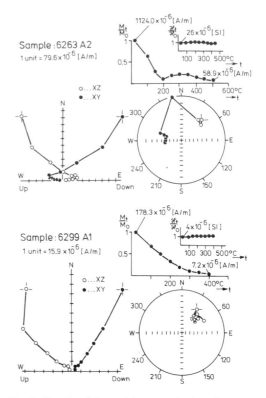

Fig. 6. Results of thermal demagnetization of limestone samples with normal and reverse polarities from the Brodno locality.

concentrate from these exceptionally weakly magnetic samples. The presence of magnetite was proven by both indirect magnetic mineralogy experiments and X-ray analyses.

Palaeomagnetic results

Alternating field demagnetization proved to be non-effective on these samples, but good results were obtained by thermal demagnetization. All samples were, therefore, subjected to thermal demagnetization and changes in magnetic susceptibility with temperature were monitored. Examples of thermal demagnetization of normally and reversely magnetized samples are given in Figs 6 and 7. Good results were obtained for all samples at the Brodno locality and for many samples at Štramberk. The magnetization of samples from some sections in Štramberk is very low, with NRM intensities in the natural state (J_n) of the order of $10^{-5}\,A\,m^{-1}$, decreasing to $5 \times 10^{-6}\,A\,m^{-1}$ after demagnetization (i.e. close to the noise level of the spinner magnetometers used). Results were obtained from a

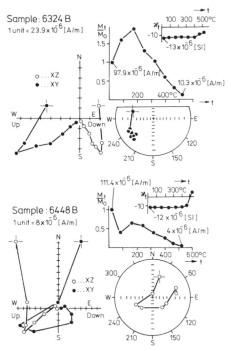

Sample: 6324 B
1 unit = 23.9×10⁶ [A/m]

o ... XZ
● ... XY

Sample: 6448 B
1 unit = 8×10⁶ [A/m]

o ... XZ
● ... XY

Fig. 7. Results of thermal demagnetization of limestone samples with normal and reverse polarities from the Štramberk locality.

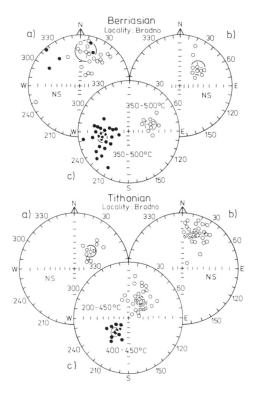

Fig. 8. Stereographic projection of remanence directions of Tithonian–Berriasian boundary limestone samples from the Brodno locality, near Žilina (west Slovakia). Upper stereonets are Berriasian samples; Lower stereonets are Tithonian samples. (**a**) NRM directions of samples which yielded normal polarities of magnetization after thermal demagnetization; (**b**) NRM directions of samples which yielded reverse polarities of magnetization after thermal demagnetization; (**c**) directions obtained by multicomponent analyses of thermal demagnetization data.

total of 524 samples, which were all subjected to multi-component analysis (Kirschvink 1980).

All samples showed a significant low un-blocking temperature component of magnetiz-ation, attributed to viscous magnetization and/or chemical remanent magnetization ac-quired during weathering. In the majority of samples it was possible to isolate a more stable component with unblocking temperatures be-tween 520° and 580°C. This component is therefore probably carried by magnetite, in agreement with the petromagnetic studies de-scribed above, and with the results obtained by other authors in other Tethyan limestone locali-ties.

Figure 8 shows stereographic projections of NRM directions of samples in their natural state, and of palaeomagnetic directions identified by thermal demagnetization and multi-component analysis for samples collected from a part of the section close to the Tithonian–Berriasian boundary at Brodno. The mean direction for the Brodno locality after demagnetization is:

Dec = 236.3°; Inc = 45.4°; α_{95} = 5.6°; k = 9.8; n = 104

with a corresponding palaeopole position at

1.13°N, 29.16°W (dp = 4.52°, dm = 7.11°). The anomalous palaeomagnetic declination suggests an anticlockwise tectonic rotation of about 124°. The pole position appears highly anomalous, but after taking the effects of tectonic rotation into account it is in agreement with those obtained from rocks of similar age from parts of the Križná nappe (Kruczyk et al. 1992; see Krs et al. this volume, fig. 9). The identification of the pronounced anticlockwise tectonic rotation of the Brodno locality is supported by the presence of normal and reversely polarized magnetozones along the magnetostratigraphic profile. Similar results were also obtained at Štramberk where, however, the bedding is not so clear and the bedding orientation has to be extrapolated in places.

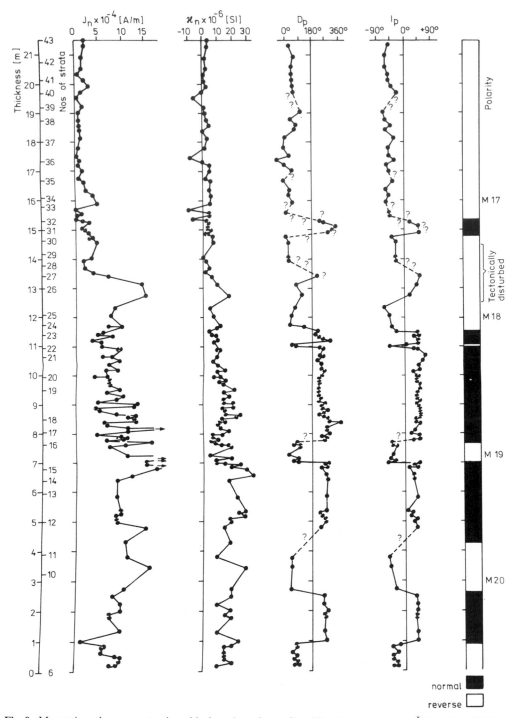

Fig. 9. Magnetic and magnetostratigraphic data along the profile of Brodno quarry, near Žilina. J_n = NRM intensity; χ_n = magnetic susceptibility; D_p = declination; I_p = inclination. Black shows normal polarity magnetozones; White shows reverse polarity magnetozones.

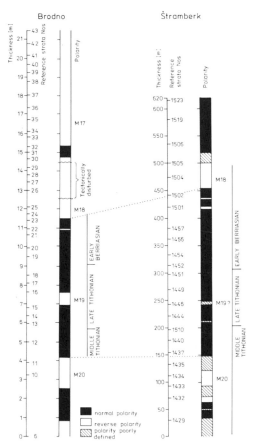

Fig. 10. Resultant magnetostratigraphic profiles across the Tithonian–Berriasian boundary strata at the Brodno and Štramberk localities.

Magnetostratigraphy

Magnetostratigraphic profiles of the Tithonian–Berriasian boundary strata were compiled for both the localities. Sedimentation at Štramberk was rapid, and took place in the peri-reef zone. In contrast, sedimentation at the Brodno locality took place in a quiet basin, making the locality ideal for magnetostratigraphic study.

The final magnetostratigraphic profile obtained for beds 6 to 43 at Brodno is shown in Fig. 9, while Fig. 10 shows the correlation between the Brodno and Štramberk profiles. In both profiles it was found that magnetic susceptibility shows no correlation with the boundaries of normal and reversed magnetozones, which reflects the local to regional origin of magnetic susceptibility and the global origin of the geomagnetic field. In order to correlate the Brodno data (slow, quiet sedimentation) with

the Štramberk data (fast, dynamic sedimentation), the sedimentation rate must be considered approximately 42 times higher at Štramberk than at Brodno.

The difficulties in defining the Tithonian–Berriasian boundary arise from the lack of correlation between the biostratigraphic scales in different realms. However, the geomagnetic polarity timescale may be used to reliably define the boundary. The Brodno results presented in Fig. 9 give a clear picture of magnetozones spanning the Tithonian–Berriasian boundary. The combined magnetostratigraphic and biostratigraphic results (see below) indicate that the boundary occurs in magnetozone M19n. Other useful magnetostratigraphic profiles with clearly defined magnetozones have been previously published for southern Spain (Ogg *et al.* 1984) and for Umbria, central Italy (Lowrie & Channell 1984). In Umbria, white, micritic Tithonian to Hauterivian limestones of the Maiolica Formation have also been studied in the Fonte del Giordano river section (Cirilli *et al.* 1984). The results, which pass a fold test, are comparable to those obtained in pelagic limestones at Foza in the Southern Alps. However, the critical section around the M19 magnetozone could not be sampled (Cirilli *et al.* 1984). In contrast to the results obtained here, some earlier works place the Tithonian–Berriasian boundary above the M19 magnetozone, whereas Lowrie & Channell (1984) place it close to the base of M17. Márton (1986) also suggested a location in zone M17. The pattern of magnetozones and the corresponding marine magnetic anomalies correlate with data from the Boreal realm, namely from limestones of the Purbeck formation, Dorset, UK (Ogg *et al.* 1991). The existing results support the interpretation by Kirschvink (see Lowrie & Channell 1984), that marine biological changes were not synchronous with palaeomagnetic field transitions.

Biostratigraphy

The possibility of correlating biostratigraphic and palaeomagnetic data is of interest. The potential exists for correlating possibly diachronous biostratigraphic events across the Tethyan and Boreal realms using the global pattern of geomagnetic reversals. Biostratigraphic data can be used to initially identify the broad period of the magnetic polarity timescale covered by a section, and the magnetostratigraphy then used to refine the dating of the biostratigraphic events. To this end, ammonites have been of little use in the present study, since they are absent from the Brodno profile and are rare at Štramberk. In contrast, calpionellids are abundant in both the profiles, particularly at Brodno where the section containing the Jurassic–Cretaceous boundary is virtually continual. Dense sampling was carried out in the key sections of this profile to allow the boundaries of single biozones and magnetozones to be located with great accuracy, and to observe their possible mutual dependence as closely as possible.

In Tethyan profiles, calpionellids do not begin appearing until the Tithonian. *Chitinoidellids* are known from the younger part of the Early Tithonian (i.e. from the Mid-Tithonian, if distinguished). Their appearance is very gradual, and the earliest small forms belong to species *Ch. dobeni, Ch. colomi, Ch. slovenica* and *Ch. tithonica* (from the so-called Dobeni sub-zone). Following their spread, however, the whole association became diversified, and the small forms were replaced by larger ones of the species *Ch. boneti, Ch. bermudezi, Ch. cubensis, Ch. pinarensis, Ch. insueta* etc. (of the so-called Boneti sub-zone). The end of this sub-zone is distinguished by the practically sole occurrence of the species *Ch. boneti*, which was very abundant (the so-called acme *Ch. boneti*). This sub-zone is followed by occurrences of the first calpionellids with hyaline tests. The transitional form, *Praetintinnopsella andrusovi*, only appears at this boundary. This pronounced and sharp change, namely the substitution of chitinoidellids by calpionellids with hyaline tests (genera *Tintinnopsella* and *Crassicollaria*) is presently regarded as identical with the boundary between the Early and Late Tithonian in the Tethyan realm.

All these sub-zones were identified at the Brodno locality. In particular, the acme of *Ch. boneti* is very well developed, with the respective species being represented in abundance. The first representatives of the genus *Chitinoidella* appear in the younger part of reverse magnetozone M20, whereas acme of *Ch. boneti*, characteristic of the end of the *Chitinoidella* biozone (i.e. the youngest Early (or youngest Mid-) Tithonian) lies approximately in the middle of magnetozone M20n. At this level, *Praetintinnopsella andrusovi* was found together with *Ch. boneti*. This indicates that this level corresponds exactly to the boundary of the Early (or Mid-, if distinguished) and the Late Tithonian. All the other younger samples contain calpionellids with hyaline tests only.

In Štramberk, chitinoidellids are less abundant, but their first appearance again occurs in the youngest part of magnetozone M 20. The acme of *Ch. boneti* is easily discernible, and

again appears in magnetozone M20n in this profile. Within this magnetozone, however, its position cannot be determined with certainty, for in the Štramberk profile it was not possible to unambiguously identify magnetozone M19, and consequently the boundary of normal magnetozones M20n and M19n is uncertain.

The evolution of calpionellid associations during the *Crassicollaria* zone (Late Tithonian) is very well documented at Brodno, where we can observe gradual changes in the calpionellid associations (especially toward the close of this biozone) as far as the base of *Calpionella* zone, which is the most distinct and well-known boundary in the calpionellid zonation. This boundary marks the extinction of three lineages of genus *Crassicollaria*, which were previously undergoing characteristic morphological changes. A distinct appearance of small spheroidal calpionellids of species *Calpionella alpina*, which replace the large Tithonian representatives of species *C. grandalpina*, make this boundary evident throughout the Tethyan realm. This event is treated as the base of the Calpionella Zone and is presently used as the provisional Jurassic–Cretaceous boundary in the Tethyan realm. In the Brodno profile, this important biohorizon lies in the younger part of the older half of magnetozone M19n. This biohorizon can also be unambiguously identified in the Štramberk profile, again in magnetozone M19n (but its exact position within M19n cannot be determined for the reasons mentioned above).

The remaining, younger part of the two profiles belongs to the zone with *Calpionella* (Early Berriasian). At Brodno, the sedimentation changes in the older sub-zone of this biozone (in sub-zone *Alpina*) and characteristic limestones of Maiolica-type appear, which indicates an acceleration of the sedimentation. This is also responsible for the relatively great thickness of the reverse part of magnetozone M17, which occupies the rest of the profile under study. In the younger part of the studied profile at Brodno, however, *Calpionella elliptica* was identified, i.e. a species characteristic of the following sub-zone *Elliptica*. The studied profile terminates at this point. Sedimentation also accelerated at Štramberk in the *Calpionella* zone, even though no lithological change occurred. In the youngest parts of this profile only *C. alpina* occurs, and the sub-zone with *C. elliptica* is not reached.

Conclusions

The pattern of normal and reverse polarity magnetozones at the Brodno and Štramberk localities has been derived from limestones spanning the Tithonian–Berriasian boundary. The magnetostratigraphy corresponds well to other reliable profiles from other areas of the Tethyan realm (chiefly Foza and Bosso, Italy), and correlates with oceanic magnetic anomalies.

Biostratigraphic data are shown to be useful for identifying magnetozones with respect to the magnetic polarity timescale. Ammonites are largely absent from the two studied profiles, but calpionellids are present and are extremely abundant at the Brodno locality. In both profiles the period from the Late Tithonian to the Late Berriasian was documented in great detail. In this time interval, there are two very distinct biohorizons based on calpionellid faunas. The first is the end of the *Chitinoidella* zone, which lies approximately beyond the boundary of the Early (Mid-) and Late Tithonian. It is distinguished by mass occurrence (acme) of *Ch. boneti* and the end of the zone is defined as the end of this acme. At Brodno, the first representatives of genus *Chitinoidella* appear as early as in the youngest parts of the reverse section of magnetozone M20. The mass occurrence of *Ch. boneti*, characteristic for the close of the *Chitinoidella* biozone (latest Mid-Tithonian) is situated approximately halfway through magnetozone M20n. The younger half of this magnetozone thus belongs to the Late Tithonian. In Štramberk, where calpionellids are less abundant, the acme of the *Ch. boneti* zone is also well documented and is also located in magnetozone M20n.

The other distinct boundary, the base of the *Calpionella* zone, i. e. the presently accepted provisional Jurassic–Cretaceous boundary in the Tethyan realm, was established at Brodno in the younger part of the older half of magnetozone M19n. This boundary is also unambiguously identified within M19n at Štramberk, but its position inside the normal magnetozone cannot be determined with certainty because the position of the reverse part of magnetozone M19 could not be unambiguously identified. Therefore, the normal parts of magnetozones M20 and M19 cannot be reliably separated.

The mineralogical identification of ferrimagnetics in the limestone samples from the Štramberk and Brodno localities was carried out by F. Novák and J. Jansa of the Institute for Raw Materials in Kutná Hora, Czech Republic. We wish to thank E. Márton and A. Morris for helpful reviews. We also thank A. Morris for help in preparation and editing the final version of the paper, and for improvement of the English.

References

CIRILLI, S., MÁRTON, P. & VIGLI, L. 1984. Implications of a combined biostratigraphic and palaeomagnetic study of the Umbrian Maiolica Formation. *Earth and Planetary Science Letters*, **69**, 203–214.

GALBRUN, B. 1985. Magnetostratigraphy of the Berriasian stratotype section (Berrias, France). *Earth and Planetary Science Letters*, **74**, 130–136.

——, BERTHOU, P.-Y., MOUSSIN, C. & AZÉMA, J. 1990. Magnétostratigraphie de la limite Jurassique-Crétacé en facies de la plate-forme carbonatée: la coupe de Bias do Norte (Algarve, Portugal). *Bulletin Societé Géologique de France*, **6**, 133–143.

HOEDEMAEKER, PH. J. 1987. Correlation possibilities around the Jurassic/Cretaceous boundary. *Scripta Geologica*, **84**, 1–55.

HOUŠA, V. 1990. Stratigraphy and calpionellid zonation of the Štramberg Limestone and associated Lower Cretaceous beds. *Atti II. Convegno Internazionale 'Fossili, Evoluzione, Ambiente'*, Pergola, 365–370.

JELÍNEK, V. 1966. A high sensitivity spinner magnetometer. *Studia geophysica et geodaetica, Academia, Prague*, **10**, 58–78.

—— 1973. Precision A.C. bridge set for measuring magnetic susceptibility and its anisotropy. *Studia geophysica et geodaetica, Academia, Prague*, **17**, 36–48.

KIRSCHVINK, J. L. 1980. The least-squares line and plane and the analysis of palaeomagnetic data. *Geophysical Journal of the Royal Astronomical Society*, **62**, 699–718.

KRS, M., KRSOVA, M. & PRUNER, P. 1996 Palaeomagnetism and palaeogeography of the Western Carpathians from the Permian to the Neogene. *This volume*.

KRUCZYK, J., KADZIALKO-HOFMOKL, M., LEFELD, J., PAGÁČ, P. & TUNYI, I. 1992. Paleomagnetism of Jurassic sediments as evidence of oroclinal bending of the Inner West Carpathians. *Tectonophysics*, **206**, 315–324.

LOWRIE, W. & CHANNELL, J. E. T. 1984. Magnetostratigraphy of the Jurassic-Cretaceous boundary in the Maiolica limestone (Umbria, Italy). *Geology*, **12**, 44–47.

MÁRTON, E. 1986. The problems of correlation between magnetozones and calpionellid zones in Late Jurassic-Early Cretaceous sections. *Acta Geologica Hungarica*, **29**, 125–131.

MAZAUD, A., GALBRUN, B., AZÉMA, J., ENAY, R., FOURCADE, E. & RASPLUS, L. 1986. Données magnétostratigraphiques sur le Jurassique supérieur et le Berriasien du NE des Cordilleres bétiques. *Comptes Rendus de l'Acadamie des Sciences, Paris*, **302**, 1165–1170.

MICHALÍK, J., REHÁKOVÁ, D. & PETERČÁKOVÁ, M. 1990. To the stratigraphy of Jurassic-Cretaceous boundary beds in the Kysuca sequence of the West Carpathian Klippen belt, Brodno section near Žilina. *Knihovna Zemního plynu a nafty, Hodonín*, **9 b**, 57–71 (In Slovakian with English abstract).

194

V. HOUŠA *ET AL.*

Moreau, M. G., Canérot, J. & Malod, J. A. 1992. Paleomagnetic study of Mesozoic sediments from the Iberian Chain (Spain). Suggestions for Barremian remagnetization and implications for the rotation of Iberia. *Bulletin de la Societé Géologique de France,* **163**, 4, 393–402.

Ogg, J. G., Hasenyager, R. W., Wimbledon, W. A., Channell, J. E. T. & Bralower, T. J. 1991. Magnetostratigraphy of the Jurassic-Cretaceous boundary interval – Tethyan and English faunal realms. *Cretaceous Research,* **12**, 455–482.

——, Steiner, M. B., Company, M. & Tavera, J. M. 1988. Magnetostratigraphy across the Berriasian–Valanginian stage boundary (Early Cretaceous), at Cehegin (Murcia Province, southern Spain). *Earth and Planetary Science Letters,* **87**, 205–215.

——, ——, Oloriz, F. & Tavera, J. M. 1984. Jurassic magnetostratigraphy, 1. Kimmeridgian-Tithonian of Sierra Gorda and Carcabuey, southern Spain. *Earth and Planetary Science Letters,* **71**, 147–162.

Remane, J., Bakalova-Ivanova, D., Borza, K., Kmauer, J., Nagy, I. & Pop, G. 1986. Agreement on the subdivision of the Standard Calpionellid Zones defined at the IInd Planktonic Conference, Roma, 1970. *Acta Geologica Hungarica,* **29**, 5–14.

Zeiss, A. 1986. Comments on a tentative correlation chart for the most important marine provinces at the Jurassic/Cretaceous boundary. *Acta Geologica Hungarica,* **29**, 27–30.

Ultra-fine magnetostratigraphy of Cretaceous shallow water carbonates, Monte Raggeto, southern Italy

M. IORIO[1,2], D. H. TARLING[2], B. D'ARGENIO[1,3] & G. NARDI[3]

[1]Geomare Sud, Institute of Marine Science, CNR, Via Vespucci 10, Napoli, Italy

[2]Department of Geological Sciences, University of Plymouth, PL48AA, UK

[3]Dipartimento di Scienze della Terra, Università di Napoli, Largo San Marcellino 10, Napoli, Italy

Abstract: The sedimentological and palaeomagnetic parameters of bore cores of shallow-water Hauterivian and Barremian carbonates have been defined at 1 cm and 2 cm intervals, respectively. The stability of remanence to alternating fields has been established and a composite 88 m sequence compiled. This shows fine structure polarity changes, with the Hauterivian section being almost entirely of Normal polarity and the Barremian including thick zones of both predominantly Normal and Reversed polarity. Narrow zones of opposite polarity occur throughout. The sedimentological and palaeomagnetic properties are found to exhibit clear cyclicities, which, when normalized, show an extremely high correlation to each other and to the predicted orbital variations for this period. This enables a study of polarity transitions and secular variations using a chronology in which 1 cm is equivalent, on average, to 285 ± 8 years.

Palaeomagnetic studies have been undertaken previously in the Southern Apennines, but such studies have been largely concerned with structural problems and particularly block rotations within this region (e.g. Catalano *et al.* 1976; Incoronato *et al.* 1985; Iorio & Nardi 1992; also Iorio *et al.* and Channell this volume). Recently periodicities of the low-field susceptibility and intensities of magnetic remanence have been noticed in pelagic limestones, e.g. Robinson (1986), Bloemendal *et al.* (1988*b*) and Tarduno *et al.* (1991), while Napoleone *et al.* (1989) observed cyclicities in directions of remanence in an 8 m section (*c.* 400 000 years). However, most such observations were from sequences known to have been affected by varying amounts of chemical changes of unclear age, although generally attributed to the reduction of magnetite during prolonged diagenesis. No previous detailed studies have been made of Cretaceous shallow-water (mostly peritidal) carbonates, as it had been assumed that the absence of terrigenous detritus would imply that they would have no measurable magnetic remanence. The recognition of the importance of magnetite producing organisms in such environments (e.g. Vali *et al.* 1987; McNeil 1990) raised the possibility of the very early authigenic formation of magnetic crystals which would likely to be preserved during the very rapid cementation associated with peritidal environments. This suggested that such lithologies may well contain a magnetic record that could be related to the time of either deposition or very early diagenesis and that these were, in any case, virtually synchronous. On this basis, initial studies were made to establish if a measurable remanence was present. These confirmed the presence of magnetic minerals which are stable to both thermal and alternating field demagnetisation. A detailed regional study was, therefore, initiated to define the optimum location for a series of bore cores that could provide a continuous record for lithological, sedimentological, palaeontological and palaeomagnetic study.

The results of the palaeomagnetic study of hand samples are reported here, together with the initial results of the magnetostratigraphic study of two of three bore cores specifically drilled to establish a continuous record for a part of the Cretaceous period. It is emphasised that the results from this investigation are still being analysed, but preliminary analyses indicate the presence of an ultra-fine sequence of magnetic polarity sequences that have been preserved with a precision that will enable a study of the directional changes in the geomagnetic field during a reversal and during polarity zones at a scale comparable to that of secular variation, i.e. some 200–300 years. In addition, clear periodicity has been established between the sedimentological and palaeomagnetic properties which, in turn, show a very high correlation with

From Morris, A. & Tarling, D. H. (eds), 1996, *Palaeomagnetism and Tectonics of the Mediterranean Region*, Geological Society Special Publication No. 105, pp. 195–203.

the predicted Milankovitch cyclicities for the Hauterivian–Barremian. Each of these will be considered separately following an outline of the geology involved and the techniques employed in the palaeomagnetic study.

Geology of the region and location of the bore cores

The Apennines are a complex fold and thrust belt which includes thick carbonate rock bodies, mostly of Mesozoic age. In the central and southern Apennines, these carbonates outcrop as well-stratified units that were deposited on platforms within formerly extensive shelf areas that also included widespread platform and basin systems of Bahamian type (D'Argenio 1970, 1976; Bernoulli & Jenkyns 1974; Laubscher & Benoulli 1977, D'Argenio & Alvarez 1980). In the late Tertiary, the previous extensional regime changed to a compressional form, resulting in formation of nappe systems, foredeeps and piggy-back basins gradually deforming towards the North and East. At the same time, the Tyrrhenian basin opened and further deformation, extending into the Pleistocene, created the present high relief (Oldow *et al.* 1993; Marsella *et al.* 1995). The Cretaceous carbonate sequences discussed in this paper were mostly deposited in peritidal waters at depths of less than 10 m, and regional studies suggest that the original deposition was then part of a mature, passive continental margin that was characterised by steady subsidence rates (D'Argenio & Alvarez 1980).

The sequence studied belongs to the Matese–Monte Maggiore tectonic unit which includes a carbonate platform sequence extending from the Upper Triassic to the Miocene, with only a few major stratigraphic gaps. Within this extensive unit a detailed study was made to determine the optimum location where a continuous section of peritidal carbonates could best be sampled. On this basis, Monte Raggeto, some 60 km north of Naples, was chosen (Fig. 1) as it contains a sequence little affected by tectonism and where a 300 m thick succession outcrops, extending from the Lower Barremian to the Lower Cenomanian, of which the interval from the Hauterivian–Barremian boundary to the early Aptian was particularly well exposed in several large quarries. At these locations, the beds were uniformly dipping (southerly at 16°) and virtually unaffected by faults. This area was then studied in detail, including a palaeomagnetic study of surface outcrops (see below). Only two minor faults were observed in the quarries with displacements of only 2 m and 20 cm. On

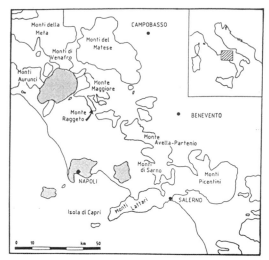

Fig. 1. The Monte Raggeto area near Naples. The solid contour encloses mountains predominantly formed by Mesozoic carbonate platform strata. The shading indicates volcanic areas.

the basis of the lithological, sedimentological and palaeomagnetic studies, as well as practical problems, two vertical bore cores were initially planned in order to ensure stratigraphic continuity with an adequate overlap to enable thorough checks of the stratigraphic and lateral continuity of features within each core (subsequently a third core was drilled to give increased stratigraphical coverage, but this has not yet been studied.) The two bore-cores (S1 and S2) were drilled 300 m laterally apart at the base of a rock cliff in quarries (Fig. 2) that were not in production at that time (Winter 1990). The cores were 10 cm in diameter and were recovered in up to 2 m lengths, each of which was oriented by means of its primary sedimentary structures, including the very uniform bedding planes. A total of 115 m of core was drilled, representing 88 m stratigraphic coverage and 14 m of overlap. The recovery was excellent, averaging at 98%, with the notable exception of 50% recovery for the bottom 10 m of well S2 (where the drill encountered phreatic waters). Each of the cores were then cut into quadrants, enabling separate analysis of sedimentology, biostratigraphy and magnetic stratigraphy. Preliminary biostratigraphic study of the outcrops had shown that the cores would be of late Hauterivian to middle Barremian age. This has been confirmed from biostratigraphical analyses of the cores (Raika Radoicic, pers. comm.), which have also shown that the Hauterivian to

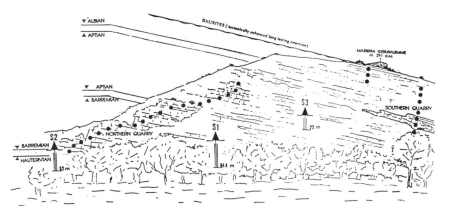

Fig. 2. Sketch section of the quarries and bore-core locations. The Barremian–Aptian and Aptian–Albian boundaries are defined biostratigraphically (modified from D'Argenio *et al.* 1993).

Barremian boundary occurs between about 3800 and 4600 cm on the combined core (Fig. 3).

Palaeomagnetic parameters

A total of 20 hand samples were taken from 10 positions within the quarry outcrops where the boreholes were later drilled. The samples were drilled in the laboratory to provide 40 standard specimens that were then analysed using the JR4 spinner magnetometers at the Eotvös Lorand Geophysical Institute (Budapest) and at the University of Plymouth. The noise-level of the JR4 is just below $10 \, \mu Am^{-1}$ for standard-sized specimens. One specimen from each site was subjected to detailed thermal demagnetization (usually 16 heating steps) and all other specimens were subjected to detailed alternating field (AF) demagnetization (usually 10 steps). The initial remanences of all specimens were extremely low and, for two sites, close the noise level of measurement. The remanences in the other 8 sites were measurable at most demagnetization levels and showed remarkably consistent directions and only one component after the first or second demagnetization step (Fig. 4). In all sites where the magnetization was measurable up to at least 400°C or greater than 30 mT there was good agreement between all specimens taken from the same locality ($\alpha_{95} < 12°$ except one site, 19°). While no magnetic mineralogy study has been undertaken, the remanence observed was consistent up to temperatures above 500°C, but not above 600°C. The coercivity was also high, although the magnetic noise associated with AF demagnetization commonly began to affect the definition of the high

coercivity components. The behaviour is therefore consistent with a magnetic mineralogy comprising mostly single domain magnetite, with no evidence for a detectable remanence attributable to haematite. Nonetheless, such a conclusion must not be regarded as established until further analyses have been undertaken.

The core quadrant for palaeomagnetic study was cut into lengths of up to 1.5 m for measurement in the '2-G Enterprises' long-core cryogenic magnetometer at the Department of Oceanography, Southampton University. The susceptibility of each quadrant was also measured using a Bartington loop sensor of 10 cm diameter, but all cores were diamagnetic (-10 to -40×10^{-6} SI) with variations in the concentration of any ferromagnetic fraction presumed to be dominated by changes in the diamagnetic (and paramagnetic) mineralogies. There was no systematic variation in the observed signal, suggesting that the magnetic susceptibility of the core was close to or below the sensitivity of the susceptibility meter. The initial remanence of each quadrant core was then measured at 2 cm intervals and then subjected to AF partial demagnetization in 5 mT steps up to 40 mT; a few cores were taken as high as 55 mT peak field. The remanence was again measured at 2 cm intervals after each step. The arithmetic average initial remanence, before demagnetization, was $47.2 \, \mu Am^{-1}$, which is two orders of magnitude greater than the noise measured when no core was present (0.9–$1.0 \, \mu Am^{-1}$). The initial intensity for most 1.5 m core lengths decreased gradually by 40 to 80% during demagnetization to 40 mT and in the vast majority of cores, it was decided that the vectors

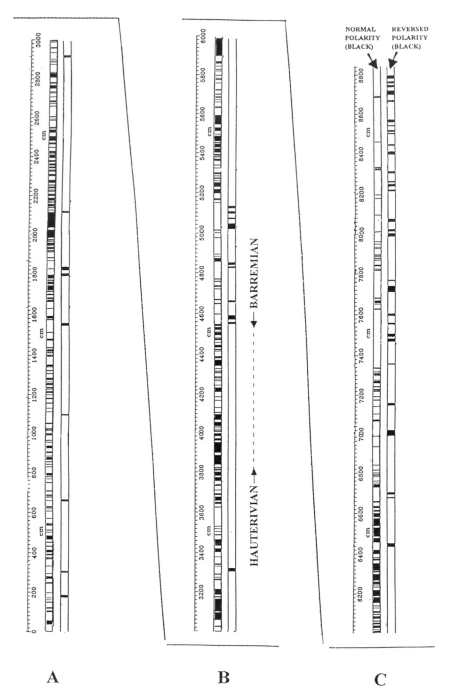

Fig. 3. The Polarity Sequence for Bore Core S1 and the non-overlapping part of Bore Core S2. Note that the left hand column marks the Normal intervals as black, while the right hand column marks the Reversed intervals as black. In both cases, white areas correspond to the opposite polarity and transitional polarities. The left scale corresponds to the vertical height in cms.

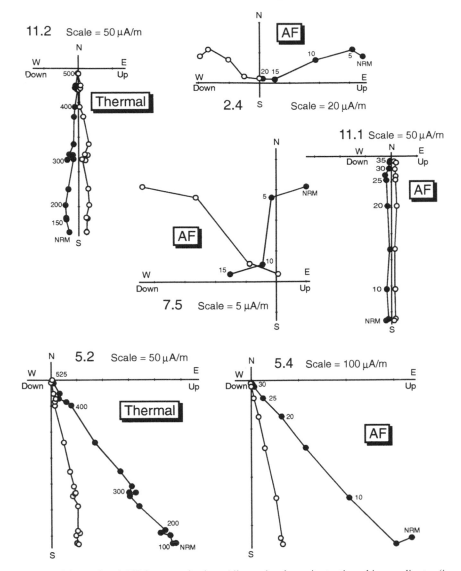

Fig. 4. Examples of thermal and AF demagnetisation. All samples shown in stratigraphic coordinates (i.e. after tilt correction).

isolated at 40 mT represented the characteristic magnetization of the bore-cores. However, in a few core lengths, the intensity of remanence had dropped close to the noise-level of measurement at lower demagnetization levels and the characteristic magnetization was consequently defined at lower demagnetization levels (2 at 20 mT, 1 at 25 mT, 1 at 30 mT, and 10 at 35 mT peak applied alternating field). The characteristic remanences were then used to construct a composite sequence (Fig. 3) in which the two cores were matched lithologically and palaeomagnetically

and zone of overlap was taken entirely from the core from well S1.

The magnetostratigraphic record

The definition of Normal and Reversed polarity is not always simple in a zone of tectonic deformation in which rotations about both vertical and horizontal axes may have occurred. However, it is reasonable to assume that the entire sequence studied has behaved as a single unit and that, as elsewhere in the Southern

Apennines, the large tectonic rotations occurred mostly in Late Tertiary, or even more recently, and affected all of the sequence by the same amount. Thus the mean declination of remanence, after correction for bedding, should be a reasonable representation of original north-south direction from which the net tectonic rotation could be inferred. The declination values for the combined data after correction for bedding tilt showed a clear Gaussian-like distribution about 220°N with a very weak peak around 0°N, but no distinct peak around 40°N, the expected antipodal direction to the main distribution. The inclination values showed highest occurrences around +40° and −40°. On this basis, the median declination and inclination values were, for all directions within 40° of 220°N and of 40°N were 222.6°N +42.1° and 30.9°N, −42.1° respectively. As the negative inclination grouping was probably affected by the small cluster of directions around 0°N, it was decided that the mean directions could be taken as Normal, $D = 222.6°$, $I = 42.1°$ with the corresponding Reversed direction being $D = 42.6°$, $I = -42.1°$.

The next problem was to define which directions should be considered to be of Intermediate polarity. The conventional definition is to accept those directions as Normal or Reversed whose calculated virtual pole positions lie within 45° of the mean Normal or Reversed pole position (Wilson 1962), while Fuller *et al.* (1979) used a limit 60° from the means. Both definitions assume that the virtual poles have a Fisherian distribution about the mean pole so that directions must have a non-Fisherian distribution. In this study, it was more convenient to assume that the directions had a Fisherian distribution and so that, while the error limits on inclination remain constant, the error in declination is a function of the palaeolatitude of the site. As the mean inclination, 42.1°, corresponds to a palaeolatitude of 20°, the expected circular standard deviation, based on the present geomagnetic field (Creer 1962), would be ±15.0° (solid angle). This would correspond to 63% of the declinations being within 20.2° of the mean declination and 63% of the inclinations within in 15.0° of the mean inclination. Using this boundary to define Normal polarity as having a declination of $223 ± 20°$ and inclination of $+42 ± 15°$, and Reversed as having a declination of $43 ± 20°$ and inclination of $−42 ± 15°$ gives a ratio of Normal to Reversed polarities of approximately 10:1 (Table 1), but this is based on only 20% of the total data available. If the definition is 'loosened' to correspond to that used for distinguishing between specific polarities and intermediate

directions in late Tertiary times, i.e. ± 40° from the mean direction then the N : R ratio becomes 5.9:1. However, while further assessment is still being undertaken, the Gaussian distribution observed for the Normal directions would suggest that a definition of ±25° would provide a sensible compromise between the other two definitions, as this would also exclude results within the minor peak of declinations at 0°N which probably represent the few unstable observations.

Using the polarity definition of ±25° (solid angle), it is immediately clear (Fig. 3) that most of the polarities defined this way are Normal. It is also evident that continuous polarity sequences (≥ 20 cm = 5 or more consecutive readings) are almost all of Normal polarity with relatively few Reversed zones (Fig. 3). It is also clear that Reversed polarities are most common above 7400 cm, while there are extensive zones containing only Normal and Intermediate polarities below this level, e.g. 2140 to 2880 cm and 3620 to 4540 cm. Clearly, at this stage, it would be desirable to reconsider the definition of those directions considered to be of Intermediate polarity as statistical considerations means that about one third of the directions attributable to the Normal or Reversed groups are included within the Intermediate group. It is also evident, from visual inspection, that there is systematic directional behaviour within the Intermediate data, but as such work is currently in progress, it is not yet possible to derive objective criteria for distinguishing Normal, Reversed and Transitional directions from those that may merely represent greater amplitude secular variations. However, these preliminary studies do suggest that the axial dipole component of the field was not so dominant as it appears to have been during much of the last million years or so.

As discussed below, there are clear cyclicities in the observations. Visual inspection also suggests that there are systematic changes in directions associated with several of the polarity transitions, many of which seem to show regular behaviour of their high inclinations (falling into the Transitional category as used here). It is thus necessary to complete an evaluation of each of these transitions, and also of systematic directional changes within the observed polarity zones to determine a more satisfactory definition of the polarity zones for the time represented by these cores. Provisional assessment would suggest that the Reversed polarities immediately above the Barremian boundary, i.e. *c.* 4560–5160 cm, probably correspond to anomalies M5 and M4 and the Reversed polarities around 8200–8800 may correspond to either Anomaly

Table 1. *Polarity frequencies as a function of their definition*

	Scatter		Polarity			
dI	*dD*	*N*	*R*	*I*	*R/N%*	
5.0	6.7	147	4	4249	2.7	
10.0	13.5	424	35	3941	8.3	
15.0	20.2	782	70	3548	9.0	
20.0	26.9	1184	110	3106	9.3	
25.0	33.7	1546	168	2686	10.9	
30.0	40.0	1887	259	2254	13.7	
35.0	47.2	2159	314	1927	14.5	
40.0	53.9	2411	349	1640	14.5	
45.0	60.6	2623	377	1400	14.4	

dI and *dD* are the boundaries for an oval about the mean directions; *dI* is equivalent to the radius of a circular boundary around the mean pole position. *N* and *R* are the number of vectors defined as normal (*N*) or reversed (*R*) on this definition of intermediate (*I*) or clear polarity.

Table 2. *Periodicities in the lithological and palaeomagnetic parameters*

Text. (cm)	Dolom. (cm)	Intensity (cm)	Inclin. (cm)	Declin. (cm)	Astron. (years)
53	56	52	54	–	18 350
68	68	65	69	61	22 230
100	106	103	118	–	38 200
145	146	144	142	–	48 750
247	249	258	260	268	95 800
921	921	–	1073	982	403 800
Normalized					
1.0	1.0	1.0	1.0	–	1.0
1.3	1.2	1.2	1.3	–	1.2
1.9	1.9	2.0	2.2	–	2.0
2.7	2.6	2.7	2.6	–	2.6
4.6	4.4	4.9	4.8	–	5.2
17.3	16.4	–	19.8	–	22.0

Text., texture; Dolom., dolomitization (%); Inclin., inclination; Declin., declination; Astron., predicted astronomical periodocities for the Hauterivian and Barremian (Berger *et al.* 1989, 1992)

M3 or M1, depending mainly on the magnitude of the stratigraphic breaks in this part of the sequence. These breaks are evidenced in the cyclical emergences shown in the lithostratigraphy (Longo *et al.* 1994).

Periodicity in the magnetic record

Analysis of the lithological and sedimentological properties, undertaken at 1 cm intervals, have shown very clear periodicities in their stratigraphic occurrence (Longo *et al.* 1994). These periodicities can be correlated with the estimated time sequences for changes in the Earth's orbital parameters for the Cretaceous (Berger *et al.* 1989; Berger *et al.* 1992), i.e. the Milankovitch cycles (Table 2). Such a correlation is not, in itself, surprising as correlations between

such parameters have been established for different sediments of various pre-Quaternary periods (see articles in *Terra Nova*, 1989, **1**, 402–479; also Fisher *et al.* 1990; Fisher & Bottjer 1991; Larson *et al.* 1992; De Boer & Smith 1994). However, this is the first time that such high frequency cycles, representing the full spectrum, have been established in such very shallow-water deposits. The same cyclicities can be observed in both the intensity and inclination of remanence (Table 2) and some indication for similar cycles are suggested in the declination values (Iorio *et al.* 1995). These cyclicities are present at all demagnetization levels, and the details are currently been examined further. Nonetheless, the agreement between the time series and the stratigraphic thickness series is remarkable strong, the *lowest* correlation coefficient, r, between the normalised parameters being 98% (Table 2). This suggests that the Milankovitch time series can be used to derive the relative durations of the stratigraphic sequences from their lithological, sedimentological and palaeomagnetic parameters. One consequence of this is that the average rate of accumulation (present-day thickness) can be quantified as $3.5\,cm\,ka^{-1}$, with a standard deviation of $\pm\,0.1\,cm\,ka^{-1}$. Consequently, as the stratigraphic correlations in these sequences can be established on at least a 1 cm level, such correlations have a relative precision equivalent to an average of $1\,cm = 285 \pm 8$ years (as emergence during disposition was also periodic (Longo *et al.* 1994), it will also be possible to further refine this precision.) As such precision applies equally to the palaeomagnetic record, it has clear implications for the rate of acquisition of the observed remanence and consequently the ultra-fine time-scales at which geomagnetic parameters can be studied. For example, the durations of polarity transitions are commonly estimated to be some 10 000 years duration for directions and 30 000 years for the geomagnetic field strength and the average rock thickness equivalent would be 30 and 90 cm respectively.

It is assumed, at this stage, that the remanence is carried by single domain magnetite and hence is likely to be organogenic (McNeill 1990). This assumption still requires verification but is distinctly suggested by the thermal and coercivity properties currently studied. This also implies that the remanence is unlikely to be detrital but is more likely to be associated with the locking in of the biogenetic particles during diagenesis, which would generally occur very rapidly after deposition in such peritidal carbonates. Indeed, the precision of the cyclicity suggests that this happens very shortly after the original deposition. Such early cementation would also mean that the sediments have not suffered significant compaction since, i.e. dewatering occurred simultaneously with diagenesis and that the spherical, minute magnetic particles would not be affected by compaction, i.e. the cyclicity in inclination in not likely to be due to differential compaction, but is more likely to be of geomagnetic origin. The intensity variations may, as in deep-sea sediments, be associated with either variations in the concentration of magnetic particles or their degree of alignment, the latter implying a correlation between higher intensities and stronger geomagnetic fields. On the basis of provisional analyses, there appears to be no correlation between the lithologies and the intensity of remanence, which suggests that

geomagnetic influences may be more important, but this requires further study. However, the very high correlation coefficients for the periodicities of both palaeomagnetic and sedimentological parameters suggest that the locking-in mechanism for the remanence occurs very rapidly during early diagenesis, probably within less than 200–300 years.

Conclusions

While it must be emphasized that the results presented here are preliminary, it is evident that shallow-water carbonates, formed in water depths less than a few tens of metres, carry a detailed record of the behaviour of the past geomagnetic field. Although the record is not strongly magnetic, this is compensated, in these rocks at least, by very high stability. This has begun to establish a detailed polarity chronology for the Hauterivian and Barremian that has a potential for correlation at a centimetre-scale, corresponding on average to 285 ± 8 calendar years. The record is carried in a variety of carbonate deposits, but these do not appear to correlate with the different textures, i.e. it seems that the palaeomagnetic record is predominantly controlled by geomagnetic factors. On-going study is expected to clarify the detailed behaviour of the geomagnetic field, some 120 Ma ago, in terms of both polarity transitions and secular variation. The observation that the cyclicities expressed by most of the parameters, of whatever origin, correlate with those expected from orbital variations also suggests that these data can be calibrated in terms of a time-scale based on calendar years for this part of the Cretaceous.

This work is part of a long-term multidisciplinary study on the shallow-water carbonate rocks of Mesozoic Southern Italy, carried out by the Research Institute Geomare sud, CNR, Naples. We thank E. A. Hailwood and E. Márton for their help in providing facilities and discussion, and many other colleagues, particularly V. Ferreri (Dipt. Scienze della Terra, Univ. Naples), for her sedimentological assistance, and R. Radoicic (Belgrade) for biostratigraphic information, G. Longo (Astronomical Observatory, Capodimonte, Naples). and N. Pelosi, for their assistance in FFT analysis. We also acknowledge the Geomare sud, CNR (National Research Council of Italy) for funding this research.

References

BERGER, A., LOUTRE, M. F. & DEHANT, V. 1989. Astronomical frequencies for pre-Quaternary palaeoclimate studies. *Terra Nova*, **1**, 474–479.
——, —— & LASKAR, J. 1992. Stability of the astronomical Frequencies over the Earth's history for paleoclimate studies. *Science*, **255**, 560–566.

BERNOULLI, D. & JENKYNS, H. C. 1974. Alpine Mediterranean and Central Atlantic Mesozoic facies in relation to the evolution of the Tethys. *In*: DOTT, R. H. JNR. & SHAVER, R. H. (eds) *Modern and Ancient Geosynclinal Sedimentation*. Society of Economic Paleontologist and Mineralogist, Special Publications, **19**, 129–160.

BLOEMENDAL, J., TAUXE, L., VALET, J.-P. & Shipboard Scientific Party. 1988. High-resolution, whole-core magnetic susceptibility logs from leg 108. *In*: RUDDIMAN, W., SARNTHEIN, M., BALDAUF, J. *et al.*, *Proceedings of the ODP, Initial Reports*, **108**, 1005–1013.

CATALANO, R., CHANNELL, J. E. T., D'ARGENIO, B. & NAPLEONE, G. 1976. Mesozoic paleogeography of the Southern Apennines and Sicily: Problems of paleotectonics and paleomagnetism. *Memorie della Società Geologica Italiana*, **15**, 85–118.

CHANNELL, J. E. T. 1996. Palaeomagnetism and palaeogeography of Adria. *This Volume*.

CREER, K. M. 1962. The dispersion of the geomagnetic field due to secular variation and its determination for remote times from palaeomagnetic data. *Journal of Geophysical Research*, **67**, 3461–3476.

D'ARGENIO, B. 1970. Evoluzione geotectonica comparata tra alcune piattaforme carbonatiche dei Mediterranei europeo ed americano. *Attidella Accademia Pontaniana.*, **20**, 3–34.

—— 1976. Le piattaforme carbonatiche periadriatiche. *Memorie della Società Geologica Italiana*, **13**, 137–160.

—— & ALVAREZ, W. 1980. Stratigraphic evidence for crustal thickness changes on the Southern Tethyan margin during the Alpine cycle. *Geological Society of America, Bulletin*, **91**, 2558–2587.

——, FERRANTI, L., MARSELLA, E., PAPPONE, G. & SACCHI, M. 1993. From the lost Lagonegro Basin to the present Tyrrhenian: The Southern Apennines between compression and extension. *International Lithosphere Program, 4th Workshop of the Task Force: 'Origin of Sedimentary Basins'*, Benevento Italy, 143.

——, FERRERI, B., ARDILLO, F. & BUONOCUNTO, F. P. 1993. Microstratigrafia e stratigrafia sequenziale. Studi sui depositi de piattaforma carbonatica nel Cretacico del Monte Maggiore (Appenino Meridionale). *Bollettino della Società Geologica Italiana*, **112**, 739–749.

DE BOER, P. L. & SMITH, D. G. (eds) 1994. *Orbital Forcing of Cyclical Sequences*. International Association of Sedimentologists, Special Publications, **19**, Blackwell, Oxford.

FISHER, A. G. & BOTTJER, D. 1991. Orbital forcing and sedimentary sequences. *Journal of Sedimentary Petrology*, **61**, 7.

——, DE BOER, P. L. & PREMOLI SILVA, I. 1990. Cyclostratigraphy. *In*: GINSBURG, R. N. & BEADOIN, B. (eds) *Cretaceous Resources, Events and Rhythms: Background and Plans for Research*. NATO ASI Series, Dordrecht, Klewer, 139–172.

FULLER, M. D., WILLIAMS, I. & HOFFMAN, K. A. 1979. Paleomagnetic records of geomagnetic field reversals and the morphology of the transitional fields. *Reviews of Geophysics and Space Physics*, **17**, 179–203.

INCORONATO, A., TARLING, D. H. & NARDI, G. 1985. Palaeomagnetic study of an allochthonous terrain: the Scisti Siicei Formation, Lagonegro Basin, Southern Italy. *Geophysical Journal of the Royal Astronomical Society*, **83**, 721–729.

IORIO, M. & NARDI, G. 1992. Studi paleomagnetici sul Mesozoico del Matesse Occidentale. *Memorie della Società Geologica Italiana*, **41**, 1253–1261.

——, NARDI, G., PIERRATINI, D. & TARLING, D. H. 1996. Block rotation in the Matese Mountains, Southern Apennines, Southern Italy. *This volume*.

——, TARLING, D. H., D'ARGENIO, B., NARDI, G. & HAILWOOD, A. E. 1995. Milankovitch cyclicity of magnetic directions in Cretaceous shallow water carbonate rocks. Southern Italy. *Bollettino Geofisica Teorica ed Applicata*, **37**, 109–118.

LARSON, R. L., FISHER, A. G., ERBA, E. & PREMOLI SILVA, I. (eds) 1992. *Apticore–Albicore: workshop report on Global Events and Rhythms of the mid-Cretaceous*. 4–9 October Perugia, Italy.

LAUBSCHER, H. & BERNOULLI, D. 1977. Mediterranean and Tethys. *In*: NAIRN, A. E. M. *et al.* (eds) *The Ocean Basins and Margins*, 1–28.

LONGO, G., D'ARGENIO, B., FERRERI, V. & IORIO, M. 1994. Fourier evidence for astronomical cycles recorded in lower Cretaceous platform strata. *In*: DE BOER, P. L. & SMITH, D. G. *q.v.*, 77–85.

MARSELLA, E., BALLY, A. W., CIPPITELLI, G., D'ARGENIO, B. & PAPPONE, G. 1995. The tectonic history of the Lagonegro domain and Southern Apennines thrust belt evolution. *Tectonophysics*, in press.

MCNEILL, D. F. 1990. Biogenic magnetite from surface Holocene carbonate sediments, Great Bahama bank. *Journal of Geophysical Research*, **95**, 4363–4371.

NAPOLEONE, G. & RIPEPE, M. 1989. Cyclic geomagnetic changes in Mid-Cretaceous rhythmites, Italy. *Terra Nova*, **1**, 437–442.

OLDOW, J. S., D'ARGENIO, B., FERRANTI, L., MARSELLA, E. & SACCHI, M. 1993. Large-scale longitudinal extension in the Southern Apennines contractional belt, Italy. *Geology*, **21**, 1123–1126.

ROBINSON, S. G. 1986. The late Pleistocene palaeoclimatic record of North Atlantic deep-sea sediments revealed by mineral-magnetic measurements. *Physics of the Earth and Planetary Interiors*, **42**, 22–47.

TARDUNO, J. A., MAYER, L. A., MUSGRAVE, R. & Shipboard Scientific Party. 1991. High-resolution, whole-core magnetic susceptibility data from leg 130, Ontong Java plateau. *In*: KROENKE, L. W., BERGER, W. H., JANECEK, T. R. *et al. Proceedings of ODP, Initial Reports*, **130**, 541–548.

VALI, H., FORSTER, O., AMARANTIDIS, G. & PETERSEN, N. 1987. Magnetotactic bacteria and their magnetofossils in sediments. *Earth and Planetary Science Letters*, **86**, 389–400.

WILSON, R. L. 1962. The palaeomagnetism of baked contact rocks and reversals of the Earth's magnetic field. *Geophysical Journal of the Royal Astronomical Society*, **7**, 194–202.

A magnetostratigraphic study of the onset of the Mediterranean Messinian salility crisis; Caltanissetta Basin, Sicily

E. McCLELLAND[1], B. FINEGAN[1] & R. W. H. BUTLER[2]

[1]*Department of Earth Sciences, Oxford University, Oxford OX1 3PR, UK*
[2]*Department of Earth Sciences, Leeds University, Leeds LS2 9JT, UK*

Abstract: The Messinian 'salinity crisis' marked a dramatic climatic change in the Mediterranean region. The timing and extent of this event remain controversial, with conflicting models ranging from predictions of rapid and catastrophic dessication of all of the Mediterranean basin at once, to progressive draw-down of sea-level causing gradual onset of evaporite formation starting at the margins of the Mediterranean Sea. In this paper the preliminary magnetostratigraphy is described of two sections in the Neogene Caltanissetta basin, Sicily, containing the early Messinian Tripoli Formation, which pre-dates evaporite formation, and the overlying Calcare di Base shallow water carbonates that mark the first evaporite phase. The magnetisation of the sediments has been extensively overprinted by recent weathering, but primary or early diagenetic remanences which pass reversal and fold tests have been isolated. This has been possible because the sections have been tilted by Pliocene tectonic movements, thus rotating the pre-tilt remanence direction away from the later overprint direction, allowing resolution of the components by great circle analysis. The sections have been correlated with Krijsman *et al.*'s revision of the geomagnetic polarity time scale (GPTS) of Cande & Kent. This was achieved by comparing variations in the sediment rate derived from the alternative matches with the GPTS with the observed proportions of diatomite (relatively fast deposition) and clay (slow deposition). The proposed correlation puts the base of both sections at or close to the start of the Messinian, but predicts diachroneity in the onset of Calcare di Base deposition between the two sections, as the base of the Calcare di Base falls in opposite polarity chrons in the two sections. The sections lie in two separate sub-basins controlled by thrust anticlines of the Maghrebian thrust belt. Structural and sedimentological evidence suggests that the onset of evaporite formation in these perched basins was strongly controlled by the extent and development of the emerging anticlines. The magnetostratigraphy provides a first positive test of this prediction of tectonically controlled diachroneity of facies. Further palaeomagnetic work is in progress to assess the relative timing of evaporite formation on Sicily in more detail.

During the Messinian, the Mediterranean was affected by a dramatic 'salinity crisis' leading to large-scale deposition of evaporites (Hsü *et al.* 1978). Details of the extent and timing of this event remain controversial (e.g. Sonnenfeld 1985; Kastens 1992) and its relationship to global sealevel change is still a matter of debate (e.g. Aharon *et al.* 1993). Many authors have considered that the Messinian evaporites on Sicily were formed at a deep level in the Mediterranean basin as they were both preceded and followed by deep water facies, and that they have been subsequently uplifted in the Plio-Quaternary. These models assume that the evaporites were essentially contemporaneous throughout the Mediterranean and were related to a catastrophic climatic change due to the isolation of the Mediterranean from the Atlantic. New integrated structural and stratigraphic work contradicts this catastrophic view (Grasso

& Pedley 1988; Butler & Grasso 1993; Butler *et al.* 1994). These new models suggest that late Miocene and early Pliocene sediments of the Caltanissetta basin, Sicily, lie in satellite perched basins formed by deformation in the frontal part of the Maghrebian thrust belt. The onset of evaporite formation on Sicily should therefore considerably predate that in the deep Mediterranean (Butler *et al.* 1994), and a magnetostratigraphic study is underway, designed to date the first evaporite formation in a number of the perched basins. In this paper, the techniques required to determine magnetostratigraphy in this setting are described and preliminary results presented from two such sections.

Structural and sedimentological setting

The Caltanissetta basin covers most of central Sicily, and contains a thick pile of Neogene

From Morris, A. & Tarling, D. H. (eds), 1996, *Palaeomagnetism and Tectonics of the Mediterranean Region*, Geological Society Special Publication No. 105, pp. 205–217.

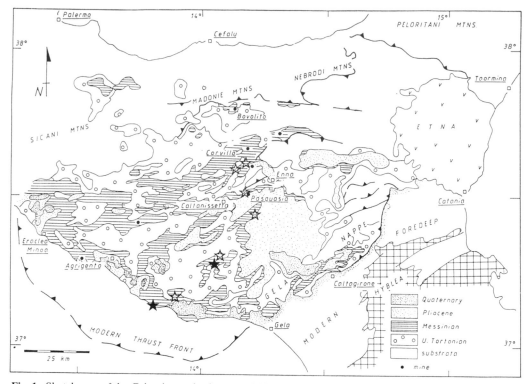

Fig. 1. Sketch map of the Caltanissetta basin, central Sicily. The location of the lower Messinian sections discussed in this paper are shown as solid stars and sections where work is in progress are marked as open stars.

sediments. Late Tortonian sediments record a marine transgression across a delta, with progressively deeper water facies being deposited in basins perched above developing thrusts (Grasso & Pedley 1988). The early Messinian Tripoli Formation diatomites were deposited when perched basins became isolated from the rest of the Mediterranean by developing thrust anticlines (Pedley & Grasso 1993). The Tripoli Formation is characterized by multiple laminated beds of diatomites with virtually no bioturbation present. There are abundant, well preserved fish fossils in these sediments. The Tripoli Formation is found extensively over Italy and North Africa and has widely varying thickness. On Sicily, thicknesses in the Caltanisseta Basin vary from 80 m to zero, depending on location within the basin structure. These sediments were deposited as the basins deepened during tectonic subsidence as the lithosphere flexed under the load of the developing orogenic wedge to the north (Grasso *et al.* 1990). Pedley & Grasso (1993) have documented a number of cycles within the Tripoli Formation which began with diverse benthonic and planktonic fauna, but this diversity rapidly

decreased during each cycle leading to the eventual loss of all fauna except diatoms. They relate this cyclicity to the interaction of eustatic changes and tectonically driven cyclic isolation of the basins; the cyclicity cannot be correlated between basins due to local variations in tectonic control. Periods of marine connection flushed the basins with nutrient rich water, followed by increasing isolation of the basins from the Mediterranean with consequent algal blooms resulting in a general loss of foraminifera due to anoxia. No evidence is found from faunal diversity of increasing salinity during the deposition of the Tripoli Formation.

The Tripoli diatomites pass up transitionally into the 'first cycle' Messinian evaporites of the Gessoso Solfifera Formation. These evaporites are characterised by thick accumulations of gypsum, halite and potash salts in basin depocentres and thin shallow water carbonates (the Calcare di Base member) on the structural highs. The Calcare di Base consists of lime mudstones that have generally experienced *in situ* collapse (probably by dissolution of evaporite crystals) to form autobreccia units. Subsequent dolomitization occurred after the autobreccia formation. These autobreccias are interbedded with clays and shales containing abundant gypsum. Marine fauna are found in the lowest levels of the Calcare di Base member indicating periodic flushing of the basin with marine water, but are absent in most of the units. It is likely that early deposition of the Calcare di Base on basin flanks may be contemporaneous with deposition of the Tripoli Formation in other deeper and less saline basins. Following the deposition of the Gessoso Solfifera Formation, a second cycle of evaporite development laid down thick sequences of gypsum. The 'salinity crisis' ended at the end of the Messinian with a return to fully marine conditions and the deposition of the Trubi chalky marls.

Palaeomagnetic sampling has been undertaken at eight sections through the Tripoli Formation up into the Calcare di Base, within the Caltanissetta basin (Fig. 1). Deformation continued within this area until the Plio-Pleistocene and so the sections are all tilted to varying degrees. In this paper, the magnetic properties and polarity sequence will be described that were obtained from Tripoli diatomites and Calcare di Base lime mudstones and shales from the Trabia–Tallarita section (SNRI) of Pedley & Grasso (1993) near Riesi, and from a coastal section near Licata (SNL2).

The Trabia–Tallarita section

The basinal sediments of the Caltanissetta basin have been uplifted by later Pliocene tectonics. Our first section (SNR1) is exposed on the southern limb of a syncline near Riesi (37° 17' 15"N, 14° 2'E). Late Tortonian clays outcrop at the base of this section, and are conformably overlain by 12m of Tripoli Formation proper containing alternating diatom and clay beds (Fig. 2). Pedley and Grasso take the first occurrence of a lime mudstone autobreccia to mark the top of the Tripoli, although diatomite beds occur above this level and microfaunal populations only decline 10 m above this level, just below the occurrence of massive autobreccia units in the Calcare di Base. The transition between Tripoli and Calcare di Base is evidently highly transitional here. The material is not ideal for palaeomagnetic sampling, being highly laminated and fissile. In situ drilling of cores was only successful in the diatomites and lime muds; no satisfactory samples were collected from the interbedded clay layers. A total of 60 cores were collected in the bedded units below the lowest massive auto breccia unit, and a total of 41 cores from the overlying autobreccia and lime muds.

Deformation in this part of Sicily continued

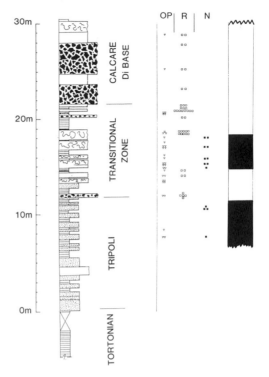

Fig. 2. Sketch stratigraphic log for the Trabia–Tallarita section (SNR1). Identified remanence components are shown at correct stratigraphic height; OP and open triangles indicate only overprint remanence has been identified; R and open squares indicate reversed polarity; N and solid squares indicate normal polarity. The number of symbols at a given level indicates the number of sub-samples. The final column is the inferred magnetostratigraphy; Black indicates normal polarity intervals, white indicates reversed polarity intervals.

until the Plio-Pleistocene, and the beds of this section are steeply tilted (bedding of 220°/ 70°NW). Such tilt is an advantage for magnetostratigraphic study as there is often a real problem in separating recent magnetic overprinting from 'primary' magnetization in rocks that are of Miocene age or younger as the direction of the ancient ambient field is the same as the recent overprinting field. Thus it can often be difficult to decide when to accept a magnetic direction as primary. When the strata have been tilted, as is generally the case in these sections, any primary remanence remaining in the rocks will be rotated away from the present magnetic field direction, and recent overprinting and primary (or at least pre-tilt) magnetization can be separated. In the Trabia–Tallarita section, the expected primary magnetization directions are therefore 077°/+56° for normal polarity

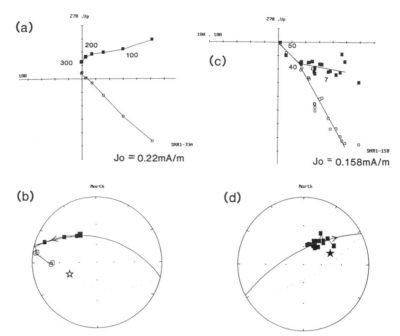

Fig. 3. Demagnetization data from the Trabia–Tallarita section (SNR1). Reversed pre-tilt remanence is identified by a swing of the vector to the west on demagnetization. (**a**) Vector plot with temperature steps (°C). (**b**) Equal-angle projection of great circle demagnetization trajectory of a reversed sample. Normal remanence is identified by a swing to the east. (**c**) Vector plot with AF field treatment in mT. (**d**) Equal-angle projection of great circle demagnetization trajectory of normal sample. The predicted direction of a tilted primary magnetization is shown as an open star (reversed) or a closed star (normal).

Miocene rocks, and 257°/−56° for reversed polarity.

Most cores yielded two or three 11 cm³ sub-specimens. Where possible, one sub-specimen from each core was treated by AF demagnetization and another by step-wise thermal demagnetization. The total NRM intensity was generally weak for the Tripoli laminites (average 0.19 mA m⁻¹), and even weaker for the Calcare di Base material (average 0.013 mA m⁻¹). All magnetizations were measured on a Cryogenic Consultants GM400 cryogenic magnetometer which gives satisfactory results for 11 cm³ samples with magnetization down to 0.002 mA m⁻¹; repeat measurements were made when the intensity fell below 0.1 mA m⁻¹. Magnetic susceptibility was monitored using a KLY2 Kappa Bridge and a Minisep susceptibility meter at each thermal demagnetization step to look for chemical alteration of the samples during the experiments.

Both AF and thermal demagnetization revealed significant magnetic overprinting directed along the recent Earth's magnetic field. In 35 sub-specimens, the overprinting was total,

and no estimate of a pre-tilt magnetization could be made. In eight sub-specimens, apparently all of the overprint could be demagnetized to reveal a stable end point. Three of these samples revealed a normal magnetization (034°/+43°, $\alpha_{95} = 19°$) and five revealed a reversed magnetization (275°/−29°, $\alpha_{95} = 22°$). These components are not anti-parallel, indicating that the overprint direction was not completely removed. Most of the Tripoli and Calcare di Base sediments contained overprint and pre-tilt magnetization with overlapping coercivity and blocking temperature spectra. In these samples, the trajectory of the remanence vector traces out a great circle path from the overprint direction towards the pre-tilt component, but does not reach a stable end point (Fig. 3), so the direction of the characteristic, pre-tilt magnetization could not be determined accurately. However, in the majority of these, the polarity of the pre-tilt magnetization could easily be identified by the direction of the great circle path traced out during demagnetization. At this site, reversed polarity pre-tilt magnetizations were recognized by demagnetization paths that swung westwards and progressively shallowed in inclination as the

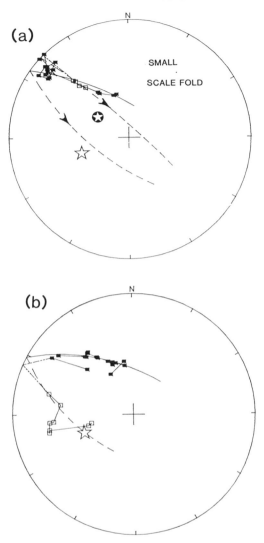

(a)

SMALL SCALE FOLD

(b)

Fig. 4. Fold test in the Trabia-Tallarita section.
(**a**) Equal-angle projection of great circle demagnetization trajectory for sample in the steep limb of small fold structure. The open star indicates the predicted direction of a tilted primary magnetization, and open star in solid circle indicates the predicted direction of a tilted and folded primary magnetization. Above 150°C the trajectory swings towards the tilted and folded reference direction.
(**b**) Demagnetization trajectories from outside the small fold structure.

257/−56 component became more dominant (Fig. 3b). Similarly, normal polarity pre-tilt magnetizations were manifested by demagnetization paths that swung eastwards towards 077/+56 with little change in inclination (Fig. 3d). Unfortunately, it is more difficult to

differentiate between normal pre-tilt remanence and normal overprinting than between reversed pre-tilt and normal overprinting. The total length of the great circle path is much shorter in the former case than the latter, so a reversed pre-tilt can be identified in a sample that has a larger overlapping overprint component than can a normal pre-tilt remanence.

Both thermal and AF treatments were generally successful at revealing a great circle trajectory for all lithologies studied. Magnetic instability usually started once the samples had been subjected to temperatures of between 300°C and 350°C, due to alteration of the magnetic mineralogy. The initial magnetic mineralogy before heating, as indicated by progressive acquisition of IRM, is predominantly magnetite. To test the age of magnetization, samples were studied from an approximately 10 cm thick lime-mud bed that had been subjected to syn-sedimentary deformation. The deformation is believed to be due to active local thrusting (W. H. Lickorish, pers. comm.) rather than any slumping mechanism. The fold is only approximately 10 cm across, and six cores were collected around the structure, two with maximum deflection of bedding. Figure 4b shows the demagnetization trajectories for two samples with typical bedding from the underlying undeformed layer which swing towards the reference tilted reversed direction. The demagnetization trajectory is also shown (Fig. 4a) for one sample with maximum bedding deflection of 33° from the typical. The initial part of the demagnetization trajectory follows those with typical bedding up to 150°C but above this temperature the trajectory changes direction and swings towards the deflected (folded and tilted) reference reversed direction. This is interpreted as support for the high Tb/Hc component in the laminated units being primary, with a lower Tb/Hc post-depositional PDRM having been overprinted during diagenesis but before tilting. Unfortunately, not enough samples could be taken to provide a statistical test of this intepretation.

The polarity sequence determined using the methods described above can then be matched to the lithological log of the section (Fig. 2) from Pedley & Grasso (1993).

The 'Licata' Section

A section has been sampled on the coast near Licata [GR 3915 41123] on the southern limb of a wide synclinorium trending roughly E–W. The SNL2 section contains Tripoli Formation laminites and Calcare di Base autobreccias. There

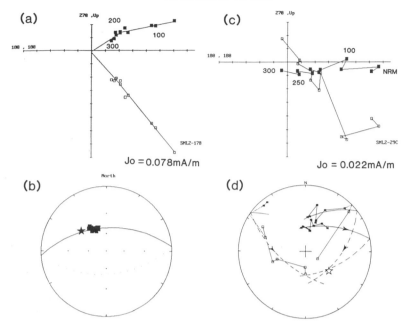

Fig. 5. Demagnetization data from the Licata section (SNL2). Normal pre-tilt remanence is identified from a swing of the vector to the west on demagnetization. (**a**) Vector plot with temperature steps in °C. (**b**) Equal angle projection of great circle demagnetization trajectory of normal sample. Reversed remanence is identified by a much larger swing away from the recent field direction. (**c**) vector plot with temperature steps in °C. (**d**) Equal angle projection of great circle demagnetization trajectory of three reversed samples showing the wide distribution of the great circle trajectories. The open star is the intersection point of all great circles indicating reversed remanence. The predicted direction of a tilted primary magnetization is shown as a closed star for the normal component, which is anti-parallel to the great circle intersection point.

are 45 m of Tripoli Formation in the section, but the material was extremely difficult to sample. 30 samples were collected in the underlying Tripoli Formation. The transition between the two units is much more sharp here, and the base of the Calcare di Base lies at 45 m above the base of the section. The section has an average bedding tilt of 170°/45°W. The expected primary magnetization directions at this locality are therefore 310°/+40° for normal polarity rocks and 130°/−40° for reversed polarity. Only data from the Tripoli sediments are presented here.

The magnetic behaviour of the Tripoli sediments is similar to the behaviour seen in the Trabia–Tallarita section. The total NRM intensity was generally very weak for the Tripoli laminites (average $0.06\,\mathrm{mA\,m^{-1}}$) and is dominated by recent overprint magnetizations, but the overprint could be completely removed in a higher proportion of samples here than in SNR1. A Normal stable end point is reached in seven samples (323°/+40°, $\alpha_{95} = 9°$). In most samples, the pre-tilt direction can only be determined from the great circle trajectory during demagnetization. Figure 5 shows demagnetization

trajectories and vector plots of thermal and AF demagnetization of N and R magnetizations from site SNL2. Unlike demagnetization trajectories for site SNR1, where the trajectories are sub-parallel for each polarity, the reversed trajectories for site SNL2 are widely scattered (Fig. 5d) and the tilted reversed reference direction (130/40) is more antiparallel to the recent overprint direction than for SNR1. The constraint for the demagnetization trajectories heading for an almost antiparallel vector to lie along sub-parallel great circles is much less, given the intrinsic errors involved in measuring magnetization in such weakly magnetized specimens. The intersection of the great circle paths (142°/−36°, Fig. 5d) lies close to the tilted reference direction and is antiparallel to the calculated mean Normal direction. Since the tilted reference directions differ considerably between the two sites due to their different structural position, the data from SRN1 and SNL2 constitute a qualitative fold test. This gives strong support for the high coercivity/ blocking temperature directions being primary with a valid record of the ancient field reversal pattern.

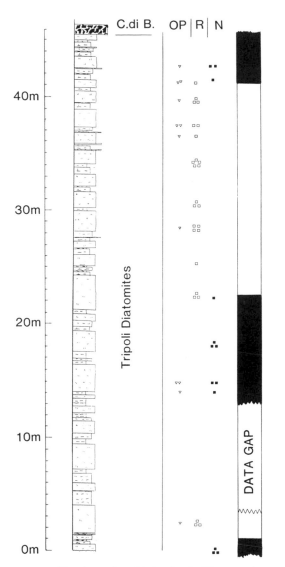

C.di B. OP | R | N

Tripoli Diatomites

DATA GAP

Fig. 6. Sketch stratigraphic log for the Licata section (SNL2). Identified remanence components are shown at correct stratigraphic height; OP and open triangles indicate only overprint remanence has been identified; R and open squares indicate reversed polarity; N and solid squares indicate normal polarity. The number of symbols at a given level indicates the number of sub-samples. The final column is the inferred magnetostratigraphy; Black indicates normal polarity intervals, white indicates reversed polarity intervals.

Figure 6 shows the polarity sequence determined using the methods described above, matched to the lithological log of the section from W. H. Lickorish (pers. comm.). There is a data gap in the interval from 4 m above the base of section to 14 m above base, because the material was too friable to sample, but there is actually no missing section in this interval.

Polarity correlation and sedimentation rate calculations

A polarity reversal pattern was obtained in both sections. The most striking feature of our data is that the top of the Tripoli Formation is of opposite polarity in the two localities (SNR1 and SNL2). Below the top normal interval in SNL2, a long (18 m) reversed interval is preceded by shorter N/R/N intervals going down section. This same pattern is seen in SNR1, where the top reversed interval is long (at least 18 m of laminates and Calcare di Base), and is preceded by shorter N/R/N intervals. The geomagnetic polarity time scale (GPTS) of Krijsman *et al.* (1994) which revises that of Cande & Kent (1992) can be compared to this observed polarity sequence. The sections are known to lie within the Messinian interval, and so there are four possible time matches with the GPTS for each section (Fig. 7); correlations SNR-A to SNR-D and SNL-A to SNL-D. In order to discriminate between these options, the sedimentation rate for each polarity interval was calculated for each match (Table 1). Since there is a 10 m data gap in the SNL section due to sampling problems, the information about the lengths of the polarity intervals that span this gap is limited. Consequently the length of the combined normal/reversed interval was used to calculate sedimentation rates for the maximum and minimum possible length of each polarity interval.

The Licata section contains a sharp transition between the Tripoli laminites and the overlying Calcare di Base, so its sedimentary history is simpler than that of the Trabia–Tallarita section and so the correlation of the Licata section will be considered first. The youngest two possible correlations with the GPTS (Fig 7) put the top of the Tripoli into the Pliocene (chron 3n.3n for SNL-A and chron 3n.4n for SNL-B). These are clearly impossible, as they overlap with the highly constrained magnetostratigraphy of Hilgen (1991) and Langereis & Hilgen (1991) on the younger Trubi marls. Their very detailed work places the base of the Trubi marls, that overlie the top of the Messinian evaporitic sequence, just at the top of chron 3r. Rejecting SNL-A and SNL-B, the remaining two possible correlations give opposing variations of sedimentation rate down section. Correlation SNL-C gives an increase in sedimentation rate with time from 30–34 m/Ma for combined chrons 3Ar and

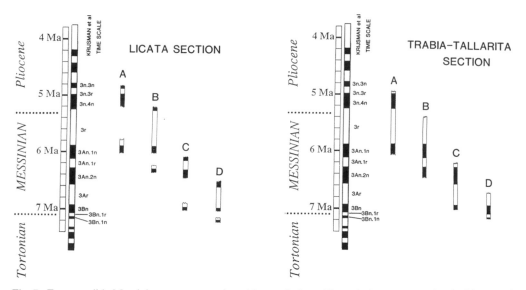

Fig. 7. Four possible Messinian magnetostratigraphic correlations. The polarity sequences for the Licata and Trabia–Tallarita sections are compared with the Geomagnetic Polarity Time Scale of Krijsman *et al.* (1994). The preferred correlation in both cases is D (see text).

Table 1. *Sedimentation rates for four possible correlations with the Krijsman* et al. *(1994) revision of the Geomagnetic Polarity Time Scale of Cande & Kent (1992), and the durations of the polarity intervals, for sections SNR1 and SNL2*

		SNR1				SNL2	
		Sedimentation rate (m Ma⁻¹)				Sedimentation rate (m Ma⁻¹)	
Duration		Max	Min	Duration		Max	Min
Youngest correlation (A)							
R1	0.09 Ma	>36.6	>32.2	N1	0.09 Ma	>13.3	
N1	0.25 Ma	25.8	24.6	R1	0.09 Ma	208.8	
R2	0.62 Ma	6.5	4.3	N2 + R2	0.87 Ma	25.7	23.3
N2	0.24 Ma	>17.9	>12.5	N3	0.24 Ma	>9.6	
Correlation (B)							
R1	0.62 Ma	>5.3	>4.6	N1	0.18 Ma	>6.6	
N1	0.24 Ma	15.4	13.3	R1	0.62 Ma	30.3	
R2	0.16 Ma	25.0	16.8	N2 + R2	0.40 Ma	56.0	50.7
N2	0.29 Ma	>14.8	>10.3	N3	0.29 Ma	>7.9	
Correlation (C)							
R1	0.16 Ma	>20.6	>18.1	N1	0.24 Ma	>5	
N1	0.29 Ma	12.7	11.0	R1	0.16 Ma	117.5	
R2	0.37 Ma	10.8	7.3	N2 + R2	0.66 Ma	33.9	30.7
N2	0.14 Ma	>30.7	>21.4	N3	0.14 Ma	>16.5	
Oldest correlation (D)							
R1	0.37 Ma	>8.9	>7.8	N1	0.29 Ma	>4.1	
N1	0.14 Ma	26.4	22.8	R1	0.37 Ma	50.8	
R2	0.06 Ma	66.6	45.0	N2 + R2	0.20 Ma	112	101.0
N2	0.04 Ma	>107.5	>75.0	N3	0.04 Ma	>57.5	

3An.2n to a sedimentation rate of 117 m Ma⁻¹ for chron 3An.1r. Only minimum rates can be calculated for the top (>5 m Ma⁻¹) and the bottom (>16 m Ma⁻¹) normal intervals. Correlation SNL-D gives a decrease in sedimentation rate with time from 100–112 m/Ma for

chrons 3Br.1r and 3Bn combined to a rate of 51 m Ma⁻¹ for chron 3Ar. The minimum rates for the top and the bottom normal intervals are >4 m Ma⁻¹ and >57 m Ma⁻¹ respectively.

The available sedimentological evidence has to be assessed to choose which of these correlations is most plausible. Ogniben (1957) suggested that the laminations in the Tripoli diatomites are varves, and estimated that sedimentation rates were between 100 to 500 m Ma^{-1}. McKenzie et al. (1979/1980) point out that the intervening clay-rich layers would have been deposited at a much lower sedimentation rate. The stratigraphic log (Lickorish, pers. comm.) indicates that the proportion of clay-rich layers increases up section, and that the thickness of individual diatomite beds increases down section. This suggests that the sedimentation rate is likely to have decreased with time, favouring the oldest correlation, SNL-D, and placing this section at the base of the Messinian. Biostratigraphic investigations of these sections is underway (Pedley & Grasso, pers. comm.), but is not at a stage to help with correlation.

It is more difficult to constrain the age of the Trabia–Tallarita section by matching the polarity pattern and using sedimentation rates, as the lithological variation in the section is more pronounced, and the middle part of the section is a transitional facies between the diatomites and the overlying evaporites. However, SNR-D is the only correlation that gives high sedimentation rates in any part of the section. This interpretation gives >75–107 m Ma^{-1} for the basal part of the sequence that is dominated by diatomitic units, 45–66 m Ma^{-1} for the alternating diatomitic and lime-mud units and 22–26 m Ma^{-1} for the upper part dominated by lime-mud beds. None of the other correlations (SNR-A to C) give variations in sedimentation rate that can be linked to the sedimentary evidence, so these are rejected on this basis. Furthermore, in this section the Tripoli sediments conformably overlie Tortonian sediments and correlation SNR-D puts the base of this section at the Tortonian–Messinian boundary. This oldest correlation also puts the onset of Calcare di Base formation at a similar time in both sections, which is most plausible. However, a clear diachroneity is evident in the onset of the first evaporite phase, as the base of the Calcare di Base is in normal chron 3An.2n at SNL2 (6.28–6.57 Ma) and in the older reversed chron 3Ar at SNR1 (6.57–6.94 Ma).

Comparison with other Mediterranean Messinian magnetostratigraphies

Gautier et al. (1994) recently published a brief report on magnetostratigraphy from Messinian sections from the Caltanissetta basin. From their two sections that contain the top of the Tripoli Formation, their work suggests that the top of the Tripoli at Contrada Gaspa and Capodarco

lies in a reversed chron. They use a more complete section from southern Spain to constrain the age of this reversed chron to be 3r (5.26–5.88 Ma), i.e. in the late Messinian. Such a correlation for these sections must be treated cautiously for two reasons. Firstly, Gautier et al. rely heavily on biostratigraphic correlation to constrain the age of their sections but there is an ongoing debate about the chronology of the evolution of Mediterranean faunas (Bizon & Glaçon 1978; Pedley, pers. comm.). This makes it imperative that magnetostratigraphic correlations should be constructed independently of biostratigraphy. Secondly, the differing tectonic settings of the Spanish section and those on Sicily mean that diachroneity in the onset of evaporite formation might be expected. The older age proposed here for the start of the 'salinity crisis' on Sicily is preferred as this is independent of biostratigraphy and correlated on plausible variations of sedimentation rate. Also, the preferred stratigraphy starts at the base of the Messinian in both sections which ties in with the Messinian–Tortonian boundary exposed in the SNR1 section.

It is valuable to compare our magnetostratigraphic interpretation with magnetostratigraphy from ODP drill sites 652 and 654 in the Tyrrhenian Sea (Channell et al. 1990). Structural models of Butler et al. (1994) predict that the onset of evaporite deposition in the Sicilian perched basins should considerably predate deposition in the open Mediterranean. Channell et al. interpret their data from the ODP holes to suggest that the evaporite formation in these locations occurred during reversed chron 3r (5.26–5.88 Ma), although the sections are not continuous, and other interpretations that place the evaporite sequence in an older reversed interval cannot be precluded. Channell et al.'s interpretation puts the evaporite deposition in the Tyrrhenian Sea some 0.4–1.1 Ma later than in the Caltanissetta basin, according to the interpretation preferred here. It is suggested in this paper that evaporite formation in the two holes is not synchronous, but correlates with Sicilian evaporites at 6.94–6.57 Ma in hole 654. Details of the lithologies and palaeomagnetic data from the Tyrrhenian Sea are given below, to document the uncertainties in magnetic correlation in these cores.

Both ODP holes contain an extensive Messinian succession (Tripoli Formation or equivalent and evaporites), although the expression is different in each hole. At hole 654, deep marine late Tortonian sediments are overlain by 35 m of dark shales, deposited in a shallow-marine environment (Borsetti et al. 1990), which are correlated with the Tripoli formation. The lower evaporites marking the start of the 'salinity crisis' are missing in this section, and the upper evaporites start with a thick gypsum layer. The evaporite sequence is made up of approximately 70 m of thick evaporite layers, dolomitic mudstones and gypsiferous sands, with occasional occurrences of marine forams indicating that intermittent marine reflooding events occurred (Robertson *et al.* 1990). Pliocene nannofossil oozes overlying the evaporites mark the return to fully marine conditions.

Sediments recovered from hole 652 are probably Tortonian at the base and it is likely they are unconformably overlain by the Messinian sequence. A 250 m thick, shaly sequence overlies Tortonian (?) sandstones and is correlated with the Tripoli Formation, with a strong turbitite component indicating a deep, subsiding environment. The transition into evaporite facies is gradual with the appearance of diffuse anhydrite laminae or nodules. Only a few approximately 50 mm thick anhydrite layers are seen within the 480 m thickness of Messinian clays, mudstones, siltstones and sandstones (Borsetti *et al.* 1990). A thick conglomerate unit marks a boundary between the underlying anhydrite and dark shaley horizons and the overlying light coloured mudstone/siltstones (130 m thick) containing no anhydrite. Recovery is very poor at this boundary. The depositional environment is interpreted to be a closed, inward draining lake of variable salinity (Robertson *et al.* 1990) on the northern margins of 'Italy'. Only brackish water algae and forams are found, implying no marine connection.

Channell *et al.* (1990) discussed the magnetostratigraphy of these two sections. The return to fully marine conditions at the end of the Messinian in hole 652 was correlated by Channell *et al.* to lie just within the top of chron 3r of Cande & Kent (1992). Data from hole 654 do not contradict this interpretation, although data are not available from the evaporite–marl boundary. This is consistent with the work of Hilgen (1991) and others of the Utrecht group on the Trubi marls of southern Sicily, which also places the Miocene–Pliocene boundary (the end of the Messinian) just at the top of chron 3r. At hole 652, a continuous section from the Pliocene oozes down to Messinian mudstones and siltstones above the evaporite sequence has apparently been recovered. The reversal pattern of the lower Matuyama, Gauss and Gilbert chrons have been indentified in the Pliocene oozes. The whole of the rest of the underlying sequence is mostly of reversed polarity. Channell *et al.* discounted the few normally magnetized samples because they correlate with high organic carbon content and have placed the entire lower section in chron 3r. If this interpretation is correct then the lowermost sand units that were tentatively identified as Tortonian are actually late Messinian in age, and the whole of the section sampled in hole 652 is younger than 5.88 Ma. The base of the evaporitic sequence can be calculated to be 5.53 Ma or younger, using the minimum sedimentation rate of 871 m/Ma for this sequence calculated by Channell *et al.* (1990).

At hole 654, the polarity signature of the Gilbert and Gauss chrons has been identified in the Pliocene sediments by Channell *et al.* (1990). The boundary between the overlying Pliocene oozes and the evaporites correlated with the Sicilian Gessoso Solfifera Formation was not recovered. All of the evaporitic sequence gave reversed polarity magnetizations and have been correlated with chron 3r. Few reliable palaeomagnetic inclinations were recovered from the Tripoli Formation shales, but the data from the top of the Tripoli indicate that the magnetization changes polarity from reversed to normal downwards in the top few metres of the unit. The magnetization is reversed towards the base of the formation. The underlying Tortonian (?) sediments show a R/N/R/N/R/N sequence going down section, which Channell *et al.* have correlated with chrons 3b and 4.

Borsetti *et al.* (1990) point out that the evaporites in hole 652 could either represent the same evaporite sequence as in hole 654, or a later episode of remobilisation, in which case the entire Messinian section in hole 652 could post-date all of the Messinian in hole 654. Unfortunately, the palaeomagnetic data do not discriminate between these options. The setting of hole 654 during the Messinian is similar to that of the Sicilian perched basins. This locality is believed to have been on a tilted fault block on the Sardinian margin. Deposition of Tripoli shales indicates anoxic conditions due to limited access to fully marine water, and periodic influxes of seawater continued during the evaporite deposition. It is therefore possible that the timing of onset of evaporite formation may be similar to that in the Sicilian basins. This would put the reversely magnetized evaporites of hole 654 into chron 3Ar, and the underlying Tripoli into chrons 3Ar and 3Bn, more-or-less synchronous with the Trabia–Tallarita section. The structural setting of hole 652 in the Messinian is unusual; a rapidly subsiding basin that was not connected to marine waters is postulated. It is possible that subsidence in this area had not created an environment in which sediment could accumulate at the time that evaporites were forming off the Sardinian margin (hole 654) and on Sicily. In this case, Channell et al.'s assignment of an upper Messinian age for the hole 652 sequence can be accepted, which is then considerably later than the lower Messinian sediments at the localities considered in this paper.

Unfortunately, no satisfactory Messinian magnetostratigraphy was obtained from the cores collected from the deep Mediterranean on DSDP leg 42 (Hamilton et al. 1978), so that the chronology of the onset of the 'salinity crisis' in the marginal basins cannot be compared to the chronology in the deep Mediterranean at present.

Summary

It has been possible to construct comparable magnetostratigraphies from two sections from the Caltanissetta basin, Sicily. These are apparently almost complete sections through the lower Messinian Tripoli Formation and the base of the Calcare di Base, which marks the start of the Mediterranean salinity crisis. These magnetostratigraphies have been constructed despite extensive remagnetization of the sediments. This has been possible because the sections have been tilted by Pliocene tectonic movements, thus rotating the pre-tilt remanence direction away from the later overprint direction, allowing resolution of the components by great circle analysis. The sections have been correlated with the revision by Krijsman et al. (1994) of the geomagnetic polarity time scale (GPTS) of Cande & Kent (1992) by comparing the sedimentation rate variation implied by the alternative matches with the GPTS with the observed proportions of diatomite (relatively fast deposition) and clay (slow deposition). Our proposed correlation puts the base of both sections at or close to the start of the Messinian. This is independently corroborated by the presence of Tortonian clays conformably underlying Tripoli diatomites at the Trabia–Tallarita section.

The origin and timing of the Mediterranean Messinian 'salinity crisis' is still a matter of debate. The magnetostratigraphy presented here suggests that the base of the Calcare di Base which represents the start of the 'salinity crisis' in Sicily lies in chron 3Ar (between 6.57 Ma and 6.94 Ma) in the Trabia–Tallarita section (SNR1) and in chron 3An.2n (between 6.28 Ma and 6.57 Ma) in the Licata section (SNL2). These dates are considerably older than (i) the 5.26–5.88 Ma proposed by Gautier et al. (1994), (ii) the c. 6 Ma (recalculated to the Cande & Kent

(1992) time scale) suggested by Colalongo et al. (1979) and (ii) the c. 6.15 Ma (recalculated) proposed by Müller & Mueller (1991) using Sr isotope data for the lower evaporite boundary on Sicily. It is suggested above that there may be problems with the correlation by Gautier et al., partly because of uncertainties in dating biostratigraphic faunal zones on Sicily. Furthermore, Gautier et al.'s sections on Sicily are not dated directly by their reversal stratigraphy, but by inferring synchroneity between the Sicilian basal evaporites and those from a magnetostratigraphically dated section in Southern Spain. There is an inconsistency in the argument used in the dating proposed by Müller & Mueller. They find Sr isotope ratios of 0.708892–0.708930 for the lower gypsum and main salt from the Cattolica Eraclea mine on Sicily, and they rightly point out that these ratios plot on the ocean seawater curve of Hodell et al. (1989) in an interval where the Sr ratios were essentially constant, from 6 Ma to 8 Ma (recalculated dates). Hodell et al. showed that there was a rapid increase in Sr ratio from 0.70892 to 0.70902 in the younger interval from 6 Ma to 4.9 Ma (recalculated) which provides a good independent dating tool. Müller & Mueller (1991) find Sr ratios of 0.708955 for the main salt in ODP hole 374 from the Messina Abyssal Plain, which correlates to a date of c. 5.5 Ma (recalculated). However, they then go on to use the regression relationship of Hodell et al., which had been developed for use only in the region of rapid Sr isotopic change back to 6 Ma, to calculate an age for the Sicilian salt of c. 6.15 Ma. These low Sr ratios cannot be used to accurately constrain the age, but instead bracket the onset of evaporite formation only to within 6 Ma to 8 Ma. These data therefore do not conflict with the dates proposed here for the start of the 'salinity crisis' on Sicily.

Several lines of evidence discussed in this paper suggest diachroneity in the onset of evaporite formation across the Mediterranean region. The magnetostratigraphy obtained clearly shows that evaporites formed later in the more southerly basin, in which the Licata section is situated, than in the more central basin containing the Trabia–Tallarita section. An on-going structural and sedimentological study of the Caltanissetta basin indicates that deformation propagated towards the south, so tectonically controlled facies changes should have occurred later in the more southerly sub-basins. This model of facies development is supported by these new, but preliminary, magnetostratigraphic data, but a robust assessment of this model will have to await completion of work now in progress on six other sections. New models of the early Messinian marine regression suggest that the draw-down of sea-level should have affected the marginal basins earlier than the deep Mediterranean (Butler *et al.* 1994). The evidence of Müller & Mueller (1991) suggests that the main salt in the deep Mediterranean basins formed significantly later (perhaps by 0.5 Ma to 1 Ma) than the salt in the marginal Sicilian basins. It is believed that these magneto-stratigraphic studies on early Messinian sediments in the Caltanissetta basin will help to produce a high resolution record of the environmental change that took place in the Mediterranean, and to gain a deeper understanding of how such thick sequences of evaporites developed.

Research funds for this project have been provided by a research grant from the Natural Environment Research Council (GR3/8172) and by a research fellowship to E. M. from the Royal Society of London. We wish to thank Mario Grasso, Martyn Pedley, Henry Lickorish and Ed Jones for valuable discussion and field assistance.

References

AHARON, P., GOLDSTEIN, S. L., WHEELER, C. W. & JACOBSON, G. 1993. Sea-level events in the South Pacific linked with the Messinian salinity crisis. *Geology*, 21, 771–775.

BIZON, G. & GLAÇON, G. 1978. Morphological investigations on the genus *Globorotalia* from site 372. *In*: Hsü *et al.* (eds) *Initial Reports of the Deep Sea Drilling Project*, 42. NSF, Washington, 687–707.

BORSETTI, A. M., CURZI, P. V., LANDUZZI, V., MUTTI, M., RICCI LUCCHI, F., SARTORI, R., TOMADIN, L. & ZUFFA, G. G. 1990. Messinian and pre-Messinian sediments from ODP leg 107 sites 652 and 654 in the Tyrrhenian Sea: sedimentological and petrographic study and possible comparisons with Italian sequences. *In*: KASTENS, K. A., MASCLE, J., *et al.* (eds) *Proceedings of the Ocean Drilling Program, Scientific Results*. 107, College Station, TX(Ocean Drilling Program) 169–186.

BUTLER, R. W. H. & GRASSO, M. 1993. Tectonic controls on base-level variations and depositional sequences within thrust-top and foredeep basins: examples from the Neogene thrust belt of central Sicily. *Basin Research*, 5, 137–151.

BUTLER, R. W. H., LICKORISH, W. H., GRASSO, M., PEDLEY, H. M. & RAMBERTI, L. 1995. Tectonics and sequence stratigraphy in Messinian basins, Sicily: constraints on the initiation and termination of the Mediterranean salinity crisis. *Geological Society of America Bulletin*, 107, 425–439.

CANDE, S. C. & KENT, D. V. 1992. A new Geomagnetic Polarity Time Scale for the Late Cretaceous and Cenozoic. *Journal of Geophysical Research*, 97, 13 917–13 951.

CHANNELL, J. E. T., TORII, M. & HAWTHORNE, T. 1990. Magnetostratigraphy of sediments recovered at sites 650, 651, 652 and 654 (Leg 107, Tyrrhenian Sea). *In*: KASTENS, K. A., MASCLE, J., *et al.* (eds) *Proceedings of the Ocean Drilling Program, Scientific Results*, 107. College Station, TX(Ocean Drilling Program), 335–346.

COLALONGO, M. L., DI GRANDE, A., D'ONOFRIO, S., GIANNELLI, L., ICCARINO, S., MAZZEI, R., ROMEO, M. & SALVATORINI, G. 1979. Stratigraphy of the late Miocene Italian sections straddling the Tortonian/Messinian boundary. *Bollettino della Società Paleontologica Italiana*, 18, 258–302.

GAUTIER, F., CLAUZON, G., SUC, J-P., CRAVATTE, J. & VIOLANTI, D. 1994. Age et dureé de la crise de salinité messinienne. *Stratigraphie*, 318, série II, 1103–1109.

GRASSO, M. & PEDLEY, H. M. 1988. The sedimentology and development of Terravecchia Formation carbonates (Upper Miocene) of north-central Sicily: possible eustatic influence on facies development. *Sedimentary Geology*, 57, 131–149.

——, —— & ROMEO, M. 1990. The Messinian Tripoli formation of north-central Sicily: palaeoenvironmental interpretation based on sedimentological, micropalaeontological and regional tectonic studies. *Paléobiologie Continentale*, 17, 189–204.

HAMILTON, N., HAILWOOD, E. A. & KIDD, R. B. 1978. Preliminary palaeomagnetic chronology of Cenozoic sediments from DSDP sites 372, 374 and 376 of the Mediterranean Sea. *In*: Hsü *et al.* (eds) *Initial Reports of the Deep Sea Drilling Project*, 42. NSF, Washington, 873–880.

HILGEN, F. J. 1991. Extension of the astronomically calibrated (polarity) time scale to the Miocene/Pliocene boundary. *Earth and Planetary Science Letters*, 107, 349–368.

HODELL, D. A., MUELLER, P. A., McKENZIE, J. A. & MEAD, G. A. 1989. Strontium isotope stratigraphy and geochemistry of the late Neogene ocean. *Earth and Planetary Science Letters*, 92, 165–178.

HSÜ, K. J., MONTADERT, L., BERNOULLI, D., CITA, M. B., ERICKSON, A., GARRISON, R. E., KIDD, R. B., MÉLIÈRES, F., MUELLER, C. & WRIGHT, R. 1978. History of the Mediterranean Salinity crisis. *In*: Hsü *et al.* (eds) *Initial Reports of the Deep Sea Drilling Project*, 42. NSF, Washington, 1053–1078.

KASTENS, K. A. 1992. Did glacio-eustatic sea level drop trigger the Messinian salinity crisis? New evidence from Ocean Drilling Program site 654 in the Tyrrhenian Sea. *Paleoceanography*, 7, 333–356.

KRIJSMAN, W., HILGEN, F. J., LANGEREIS, C. G. & ZACHARIASSE, W. J. 1994. The age of the Tortonian/Messinian boundary. *Earth and Planetary Science Letters*, 121, 533–547.

LANGEREIS, C. G. & HILGEN, F. J. 1991. The Rossello composite: a Mediterranean and global reference section for the Early to early Late Pliocene. *Earth and Planetary Science Letters*, 104, 211–225.

McKENZIE, J. A., JENKYNS, H. C. & BENNET, G. G. 1979/1980. Stable isotope study of the cyclic diatomite-claystones from the Tripoli Formation, Sicily: a prelude to the Messinian salinity crisis.

Palaeogeography, Palaeoclimatology, Palaeo-ecology, **29**, 125–141.

MÜLLER, D. W. & MUELLER, P. A. 1991. Origin and age of the Mediterranean Messinian evaporites: implications from Sr isotopes. *Earth and Planetary Science Letters*, **107**, 1–12.

OGNIBEN, L. 1957. Petrografia della Serie Solfifera Siciliana e considerazioni geologiche relative. *Memorie Descrittive Carta Geologica d'Italia*, **33**.

PEDLEY, H. M. & GRASSO, M. 1993. Controls on faunal and sediment cyclicity within the Tripoli and Calcare di Base basins (Late Miocene) of central Sicily. *Palaeogeography, Palaeoclimatology, Palaeoecology*, **105**, 337–360.

ROBERTSON, A., HEIKE, W., MASCLE, G., McCOY, F.,

McKENZIE, J., REHAULT, J-P. & SARTORI, R. 1990. Summary and Synthesis of Late Miocene to Recent sedimentary and paleoceanographic evolution of the Tyrrhenian Sea, Western Mediterranean: Leg 107 of the Ocean Drilling Program. *In*: KASTENS, K. A., MASCLE, J. *et al.* (eds) *Proceedings of the Ocean Drilling Program, Scientific Results*, **107**. College Station, TX (Ocean Drilling Program) 639–663.

SONNENFELD, P. 1985. Models of Upper Miocene evaporite genesis in the Mediterranean region. *In*: STANLEY, D. J. & WEZEL, F.-C. (eds) *Geological Evolution of the Mediterranean Basin*. Springer-Verlag, New York, 323–346.

Revised magnetostratigraphy and rock magnetism of Pliocene sediments from Valle Ricca (Rome, Italy)

F. FLORINDO & L. SAGNOTTI

Istituto Nazionale di Geofisica, Via di Vigna Murata 605, 00143, Roma, Italy

Abstract: This is an extended abstract of the paper 'Palaeomagnetism and rock magnetism in the upper Pliocene Valle Ricca (Rome, Italy) section' by Florindo & Sagnotti (1995). It is shown that a normal polarity event reported from a sedimentary section in the Tini quarry is associated with a different mineralogy than the rest of the sequence. Combined thermal and AF demagnetization reveals that there is a pervasive reversed polarity component throughout the sequence, but this is obscured by a strong normal component in sediments close to an ash bed. A variety of techniques show that the reversed component is carried by magnetite and likely to be primary. The normal component is associated with greigite and is likely to be of late diagenetic origin or possibly associated with self-reversal behaviour. The previous correlation between the normal polarity zone and the Reunion Event is therefore invalid and the entire sequence is considered to correlate with the C2r.1r chron, between the Olduvai and Reunion Events.

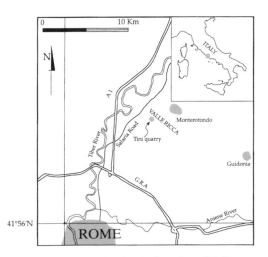

Fig. 1. Sketch map showing the location of Valle Ricca and the Tini quarry section.

The sediment section exposed at the Tini quarry in Valle Ricca, Italy (Fig. 1) has been the subject of several stratigraphic studies which have led to controversial interpretations. The magnetostratigraphy of the Valle Ricca sections, including both the Tini and Silpa quarries, was firstly studied by Arias *et al.* (1980, 1990). These studies used the behaviour of a few pilot specimens during stepwise demagnetization in alternating field up to 50 mT to undertake blanket demagnetisation of specimens throughout the sequence at 20 mT peak field. In the Tini section, this led to the identification of a normal polarity interval around a volcanic ash layer within a sequence of marly-clays of reversed polarity. This normal polarity interval was correlated with the Reunion sub-chron. This interpretation has recently been questioned by Carboni *et al.* (1993) and Mary *et al.* (1993) on the basis of new biostratigraphic and palaeomagnetic data. In particular, the identification of the normal polarity was ascribed to inadequate demagnetization. In order to resolve this controversy a dense sampling has been made which extends over 10 m of the sequence and includes clays and ashes ascribed to the Reunion sub-chron. These samples have been subjected to a wide range of magnetic measurements and mineralogical analyses in order to identify any differences in magnetic mineral composition and concentration.

Palaeomagnetism and magnetostratigraphy

Oriented cores were drilled *in situ* and then sliced into standard cylindrical specimens (25 mm diameter × 22 mm height). All magnetic measurements were performed inside a magnetically shielded room at the palaeomagnetic laboratory of the Istituto Nazionale di Geofisica. Low-field magnetic susceptibility (k) was measured on a KLY-2 kappabridge and the remanent magnetizations were measured on a JR-4 spinner magnetometer. Alternating field demagnetization was accomplished using a Molspin tumbling demagnetizer. An electrical oven was used for thermal demagnetization.

From Morris, A. & Tarling, D. H. (eds), 1996, *Palaeomagnetism and Tectonics of the Mediterranean Region*, Geological Society Special Publication No. 105, pp. 219–223.

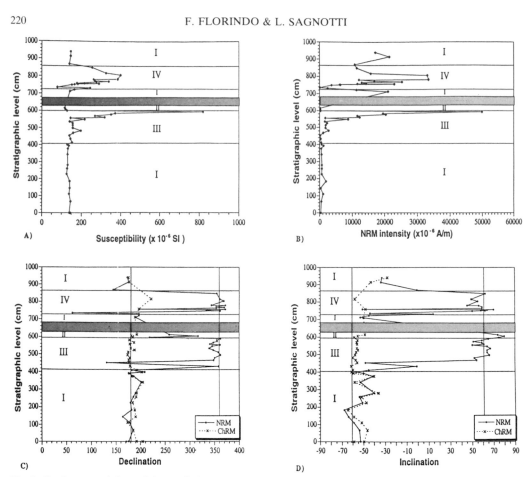

Fig. 2. Basic magnetic logs of the studied interval. The stippled level represents the volcanic ash layer. Specimens were all taken from marly clays. Roman numbers refer to the zonation described in the text.

The magnetic properties clearly show strong changes around the ash layer (Fig. 2). The inclinations of natural remanent magnetization (NRM) suggest that two changes of polarity are possible within the section, but the NRM intensity and susceptibility values also show that strong changes in the magnetic mineralogy occur over the same parts of the section. In particular, the intervals characterized by positive NRM inclinations correspond closely to the peak values in the NRM intensities and low-field magnetic susceptibilities. Pilot specimens were stepwise demagnetized by AF or by heating; in the latter case variations of bulk susceptibility were monitored after each temperature increment. These demagnetization data were then evaluated by principal component analysis (Kirschvink 1980). The thermal treatment resulted in a sharp change in the remanence of sediment samples from close to the volcanic ash layer after weak to moderate heating (120–360°C), accompanied by the complete removal of the

normal polarity component. However, the AF treatment was ineffective in completely removing this normal component in the clays. Samples far from the ash layer showed a single, stable reversed component of magnetization during both thermal and AF treatment. In all samples, a strong increase in susceptibility (2.5 to 10 times the original) occurred around 380–400°C. On the basis of these properties, it was decided to undertake a combined treatment of all remaining specimens, i.e. thermal demagnetization until the normal component (if present) was removed, and then AF demagnetization up to the limit of reproducible results.

Four main zones could be identified after combined stepwise demagnetization, each characterised by distinct palaeomagnetic behaviour. Specimens from Zone I are characterized by a single reversed component throughout treatment. In Zone II, a normal component was completely removed after the

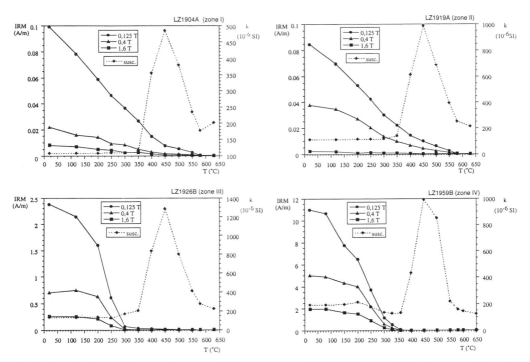

Fig. 3. Stepwise thermal demagnetization of a three component IRM. Bulk susceptibility changes on thermal demagnetization are also shown.

first heating step (120°C), with a single reverse component remaining thereafter. Zone III shows a strong normal component that is completely removed after heating to 250°C, when there is a sudden jump into the upper hemisphere and a single reversed component remains thereafter. In Zone IV, the intensities of remanence are generally one to two orders of magnitude higher than in Zones I and II. The removal of the strong normal component is more gradual than in Zone III, but is mostly completed by 330°C. After this the directions of remanence move towards the upper hemisphere, but generally do not reach stable end-points. The interpretation of the demagnetisation data in terms of declination and inclination of the characteristic remanent magnetization is shown in Fig. 2 (c, d).

Magnetic mineralogy

The nature of the magnetic carriers and the origin of the strong normal component from Zones III and IV were investigated in representative specimens using: (i) isothermal remanent magnetisation (IRM) acquisition curves up to 1.6 T; (ii) remanence coercive force (Hcr), determined by back-field application to the saturation IRM; and (iii) thermal demagnetisation of IRMs imparted along the three sample axes (Lowrie 1990). Two additional analyses were undertaken on a few specimens. These were construction of hysteresis loops in fields up to 1 T, using a Molspin vibrating sample magnetometer at the Department of Geology and Geophysics of Edinburgh University, and thermomagnetic curves constructed using an horizontal Weiss-Curie force balance at the ETH laboratory, Zurich. Magnetic concentrates were also made from seven representative specimens using a high-gradient magnetic extraction system on a 'Box-mag rapid' electromagnet in a steady magnetic field of 2.4 T at the Centro per il trattamento dei minerali (Roma, Italy). Mineralogical analyses of these concentrates were then made using an optical microscope, a scanning electron microscope (SEM), electron probe microanalysis (EPMA) and X-ray diffraction.

IRM acquisition and back-field curves showed little variation in all samples; saturation is reached in relatively small fields (around 300 mT) and the coercivity of remanence has similar values (40–70 mT). In contrast, strong differences appear during thermal demagnetization of the three artificial IRMs (Fig. 3).

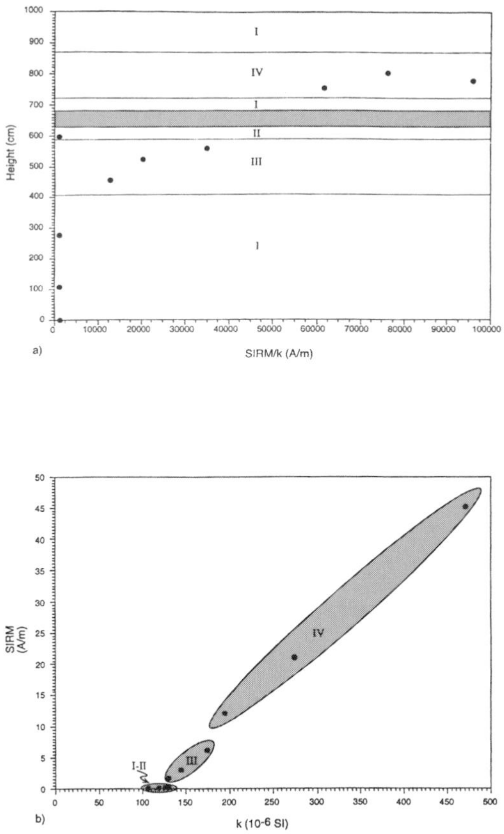

Fig. 4. (a) Variation of SIRM/*k* with stratigraphic height for ten representative specimens. The stippled level represents the volcanic ash layer. (b) SIRM v. *k* plot. While values for specimens from Zones I and II plot close to the abscissa axis, there is a linear relationship between the two values for specimens from Zones III and IV. Larger values of both parameters correspond to higher concentrations of magnetic iron sulphides.

around 360°C. The difference in demagnetization behaviour between the four zones is particularly evident when the SIRM/susceptibility ratio is plotted against the stratigraphic height (Fig. 4a). Zones III and IV have ratios that are about two orders of magnitude greater than those of Zones I and II. The SIRM and susceptibility values are also linearly related for specimens from Zones III and IV, but lie close to the abscissa for Zones I and II (Fig. 3b). The S-ratio (Snowball 1991) is similar throughout the section, with values around -0.5 to -0.6, with only a minimum of -0.9 in Zone III.

The optical microscope, SEM and EPMA analyses showed that framboidal aggregates of iron sulphides dominated the magnetic separates from Zones III and IV, and are often concentrated within the shells of foraminifera. The individual crystals were very small, commonly in the range 0.5-2-3 μm. Iron sulphides were much more scarce in separates from Zones I and II. Stochiometric determinations (obtained by EDS on SEM and by WDS on the microprobe) were hindered by the small size of the iron sulphide grains, but indicated the presence of pyrite. A large number of Fe-rich chromites were also identified, mostly in separates from Zone IV. Iron sulphides and chromites are absent in the volcanic ash layer, in which several magnetite grains were observed.

X-ray diffraction of the magnetic separates only showed peaks of diamagnetic and paramagnetic minerals (quartz, calcite, and clay minerals). The absence of peaks corresponding to ferromagnetic minerals reflects the difficulty in obtaining pure magnetic separates, while the lack of iron sulphides peaks indicates that they are very small, below or close to the detection limit.

Conclusions

The new palaeomagnetic data indicate that the entire section carries a component of reversed polarity, but that this component is obscured at certain levels close to the ash layer by a strong secondary component associated with a different magnetic mineralogy. Four zones with differing magnetic mineralogy have been identified. In Zones I and II the main magnetic carrier is magnetite, while in Zones III and IV the main magnetic carrier is the iron sulphide greigite (Fe_3S_4), which has previously been reported in clays from Valle Ricca (Bracci *et al.* 1985). An additional magnetic phase could be present in Zone IV, with a maximum blocking temperature of 360°C, but this has not been fully analysed. The magnetic carrier is magnetite where only the

Specimens from Zone I have a remanence that is largely held by the low coercivity fraction and their intensities decay in a quasi-linear fashion from room temperature to 580°C. Zone II specimens behave similarly, but have a more pronounced intermediate coercivity component and a weak inflexion around 300°C. The behaviour of specimens from Zone III is very different. While the low coercivity fraction still dominates the remanence, all three fractions show a sharp drop around 300°C. In Zone IV, the high coercivity fraction is more important than in the other zones and two major drops can be distinguished in all three fractions; a marked drop around 300°C, and a less pronounced drop reversed component is present, but near the ash layer the dominant magnetisation is a normal polarity component carried by greigite. This greigite component remains dominant during AF demagnetization but is destroyed during thermal demagnetization. The origin and age of this normal polarity component are not clear. It may have originated by late diagenesis during a normal polarity period, or it could have arisen by a self-reversal process during early diagenesis. The former origin seems unlikely, as greigite is usually formed during the very early stages of diagenesis during the reduction of anoxic marine sediments at depths greater than 1m below the sediment–water interface (Berner 1969; Leslie *et al.* 1990; Roberts & Turner 1993). The self-reversal mechanism is similarly unlikely as it would imply a growth-CRM setting antiparallel to the Earth's magnetic field, although Snowball & Sandgren (1995) have recently reported such behaviour in greigite-bearing late Weichselian clays. Irrespective of the origin, it is clear that the entire section was deposited during a period of reversed polarity and that the reported normal polarity event is not real, as suggested by Mary *et al.* (1993). Consequently any correlation with the Reunion polarity event is invalid. The available stratigraphic constraints, recently reviewed by Carboni *et al.* (1993), suggest that this reversed interval correlates with the C2r.1r chron, between the Olduvai and Reunion Event of the Geomagnetic Polarity Time Scale.

References

ARIAS, C., AZZAROLI, A., BIGAZZI, G. & BONADONNA, F. P. 1980. Magnetostratigraphy and Pliocene–Pleistocene boundary in Italy. *Quaternary Research*, **13**, 65–74.

—— , BIGAZZI, G., BONADONNA, F. P., IACCARINO, S., URBAN, B., DAL MOLIN, M., DAL MONTE, L. & MARTOLINI, M. 1990. Valle Ricca Late Neogene stratigraphy (Lazio region, central Italy). *Palèobiologie Continentale*, **17**, 61–68.

BERNER, R. A. 1969. Migration of iron and sulfur within anaerobic sediments during early diagenesis. *American Journal of Science*, **267**, 19–42.

BRACCI, G., DALENA, D. & ORLANDI, P. 1985. La Greigite di Mentana, Lazio. *Rendiconti della Società Italiana di Mineralogia e Petrologia*, **40**, 295–298.

CARBONI, M. G., DI BELLA, L. & GIROTTI, O. 1993. Nuovi dati sul Pleistocene di Valle Ricca (Monterotondo, Roma). *Il Quaternario*, **6**, 39–48.

FLORINDO, F. & SAGNOTTI, L. 1995. Palaeomagnetism and rock magnetism in the upper Pliocene Valle Ricca (Rome, Italy) section. *Geophysical Journal International*, in press.

KIRSCHVINK, J. L. 1980. The least-squares line and plane and the analysis of paleomagnetic data. *Geophysical Journal of the Royal Astronomical Society*, **62**, 699–718.

LESLIE, B. W., HAMMOND, D. E., BERELSON, W. M. & LUND, S. P. 1990. Diagenesis in anoxic sediments from the California Continental Borderland and its influence on iron, sulfur, and magnetite behaviour. *Journal of Geophysical Research*, **95**, 4453–4470.

LOWRIE, W. 1990. Identification of ferromagnetic minerals in a rock by coercivity and unblocking temperature properties. *Geophysical Research Letters*, **17**, 159–162.

MARY, C., IACCARINO, S., COURTILLOT, V., BESSE, J. & AISSAOUI, D. M. 1993. Magnetostratigraphy of Pliocene sediments from the Stirone River (Po Valley). *Geophysical Journal International*, **112**, 359–380.

ROBERTS, A. P. & TURNER, G. M. 1993. Diagenetic formation of ferrimagnetic iron sulphide minerals in rapidly deposited marine sediments, South Island, New Zealand. *Earth and Planetary Science Letters*, **115**, 257–273.

SNOWBALL, I. F. 1991. Magnetic hysteresis properties of greigite (Fe_3S_4) and a new occurrence in Holocene sediments from Swedish Lappland. *Physics of the Earth and Planetary Interiors*, **68**, 32–40.

—— & SANDGREN, P. 1995. Reversed chemical remanent magnetization in late Weichselian clays: a possible cause of anomalous excursions in sediments. *Annales Geophysicae*, Supplement I to Volume 13, C72.

Palaeomagnetic database: the effect of quality filtering for geodynamic studies

GIANCARLO SCALERA[1], PAOLO FAVALI[1,2] & FABIO FLORINDO[1]

[1] *Istituto Nazionale di Geofisica, Via di Vigna Murata, 605-00143 Roma, Italy*

[2] *Centro di Studi Avanzati per la Geodinamica, Università degli Studi della Basilicata-Potenza, Italy*

Abstract: The Global Palaeomagnetic Database (GPMDB), now updated to 1992, contains about 7000 palaeomagnetic data, which are fundamental tools to define regional and global geodynamic models. A software developed at the Istituto Nazionale di Geofisica allows the selection of data on the basis of space, time, and quality. Six quality classes have been proposed. The African and European Apparent Polar Wander Paths (APWPs) have been computed and the role of the statistical uncertainties is discussed. Some examples from the Tethys Belt have been chosen to demonstrate the effect of the quality filtering in geodynamic studies.

In the last few years, the International Association of Geomagnetism and Aeronomy (IAGA) has officially encouraged the development of several palaeomagnetic and rock-magnetic databases, which fulfil the need of storing and easily handling the increasing amount of data coming from several palaeomagnetic disciplines: among others rock-magnetism, archaeomagnetism and magnetostratigraphy. In particular, five regional databases of directions and palaeopoles were compiled by Khramov & Pisarevsky (Russian Federation), Pesonen (Fennoscandia), Enkin (Canada), Luyendyk & Butler (USA) and Westphal (Europe). Besides the regional databases, IAGA was the sponsor of the world-wide database, the Global Palaeomagnetic Database (GPMDB), co-ordinated and published by McElhinny & Lock (McElhinny & Lock 1990a, b, 1993; Lock & McElhinny 1991). IAGA also encouraged cross checking of the regional databases with the global database, so that errors and omissions were avoided. The GPMDB synthesises all the palaeopole parameters and their quality, and its updated version contains about 7000 palaeopoles. The collection for the period 1989–1992 has been performed by Van der Voo. The GPMDB has a complex file structure, and includes data produced for completely different aims, such as magnetostratigraphy, determination of virtual geomagnetic poles (VGP), averaged palaeomagnetic poles etc. Consequently the data must be selected and weighted according to the scientific field in which the palaeopoles are to be used. The problem of quality filtering is the subject of this paper.

The tectonic framework

Data originating from the GPMDB are referred to specific areas (Fig. 1) which include Africa (excluding Madagascar), Europe to the Urals and part of Asia (Middle East, Caucasus and Arabia). Traces of various orogenic episodes can be found in Africa, Europe and Asia: the two Palaeozoic events (Caledonian, 570–370 Ma; Variscan–Hercynian, 370–220 Ma) and the Alpine Mesozoic–Cenozoic orogeny (Early Mesozoic Alpine, 220–65 Ma; Mid-Cenozoic Alpine, 65–20 Ma; Late Cenozoic Alpine, 20 Ma to present) (UNESCO 1976; Bally *et al.* 1985). The Caledonian orogeny was active in the Mauritanides (Africa) and in Scandinavia, Scotland, Wales and the Ardennes (Europe). Cratonic basins, located mainly in continental pre-Mesozoic lithosphere and shields deformed by Pre-Cambrian orogenic episodes, can be found alongside the Caledonian orogenic belts. Most of Africa has a cratonic structure, such as in western Africa, Congo, Tanzania and Kalahari, with other old deformed areas like Katanga. In Europe we can notice both old deformed zones, like the Karelian system and Fennosarmatian shield, and wide craton basins (e.g., the North Sea Basin and the Russian–Ukrainian Basin). The Variscan-Hercynian orogeny is strongly marked in Morocco, southern African Cape, in Europe in the Central Massif, the Variscides and the Urals. In Africa the Mesozoic–Cenozoic Alpine orogeny affected the Maghrebides and the Cape Range. The rift-valleys of East Africa were formed during the Neogene and are still volcanically active.

From Morris, A. & Tarling, D. H. (eds), 1996, *Palaeomagnetism and Tectonics of the Mediterranean Region*, Geological Society Special Publication No. 105, pp. 225–237.

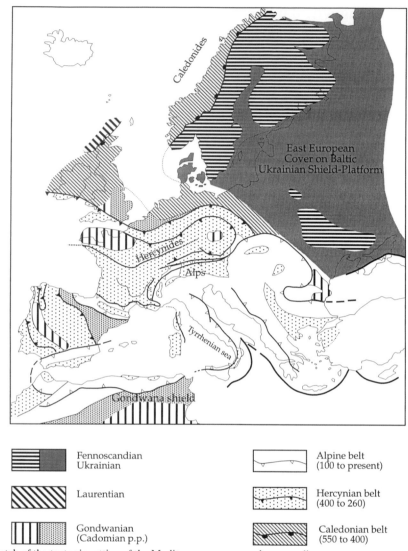

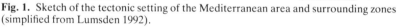

Fig. 1. Sketch of the tectonic setting of the Mediterranean area and surrounding zones (simplified from Lumsden 1992).

Further volcanic activity took place in the Ahaggar and the Tibesti during the Tertiary and Quaternary. In Europe, the Alpine orogeny deformed the mountain belts of the Betics, Pyrenees, Alps, Apennines, Dinarides, Hellenides, Carpathians, Crimean peninsula, Pontides, Taurides, Zagros, Greater and Lesser Caucasus, Elzburg and the Kopet–Dag (East of Caspian Sea). Close to these belts, some recent oceanised or relict ocean basins are recognised, like the Alboran Sea, Tyrrhenian Sea, Aegean Sea, Pannonian basin and Black Sea. In the Alpine orogenic zone (the so-called Tethys Belt), the complex interaction between Africa and Europe caused a fragmentation into several microplates that have undergone relative movements (e.g., Kissel & Laj 1989; Şengör 1989; Stöcklin 1989; Favali *et al.* 1993*a*, *b*). It is thus necessary to separate the areas of Africa and Europe affected by older orogenies (defined stable) from those of the mobile Tethys Belt.

Data filtering

After separating the data on a regional basis, it is then necessary to select them on the basis of

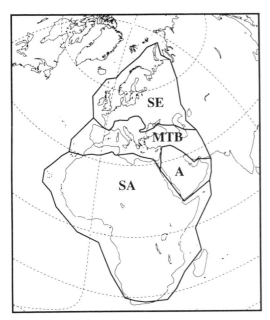

Fig. 2. Map of the polygonals used in this study for extracting palaeopoles, separating data in stable areas from data in mobile ones (SE, Stable Europe; MTB, Mobile Tethys Belt; SA, Stable Africa; A, Arabia). These polygonals have been drawn on the basis of the following maps and papers: Merla *et al.* (1973); Stöcklin & Nabavi (1973); UNESCO (1976); Martin *et al.* (1978); Bartov (1979); BRGM (1980); AA.VV. (1981); Nairn *et al.* (1981); Boccaletti & Dainelli (1982); Boccaletti *et al.* (1985); Salmon *et al.* (1988); Condie (1989); Kampunzu & Lubala (1991); Kruczyk *et al.* (1992); Lumsden (1992).

Table 1. *Time windows in Ma used to subdivide the data, together with approximate corresponding positions in the epochs of the Geological Time Scale*

Time windows (Ma)	Epochs	Uncertainty (Ma)
0–5	Pliocene–Quaternary	5
5–15	Mid–Late Miocene	10
15–25	Early Miocene	10
25–35	Oligocene	10
35–45	Mid–Late Eocene	10
45–55	Early–Mid-Eocene	10
55–65	Palaeocene	10
65–100	Late Cretaceous	30
100–145	Early Cretaceous	30
145–180	Mid–Late Jurassic	30
180–210	Early Jurassic	30
210–245	Triassic	30
245–290	Permian	35
290–325	Late Carboniferous	35
325–360	Early Carboniferous	35
360–410	Devonian	35

The uncertainties in Ma are those required for the better quality data falling in each time window. See text for discussion.

several different parameters linked to their palaeomagnetic quality. For both of these purposes, software has been developed by the Istituto Nazionale di Geofisica (Florindo *et al.* 1994), based on the ACCESS version of the GPMDB. This software uses 20 criteria in space, time and quality, enabling immediate evaluation of the pole distributions by means of a mapping program.

Geography – chronology

Within each geological–tectonic framework, the palaeopoles are grouped on the basis of the geographical co-ordinates of the sites. Three main polygonals have been drawn to include Stable Africa (SA), Stable Europe (SE) and the Mobile Tethys Belt (MTB) (Fig. 2), based on several tectonic maps and publications (references in Fig. 2 legend). A fourth polygonal, the Arabian plate (A in Fig. 2), was also considered as the geodynamics of this region are of great

interest, but the shortage of available data prevented further consideration. The palaeopoles were then selected on the basis of their rock magnetization ages using non-superimposed time windows (first column of Table 1). The width of the time windows has been calibrated using the epochs and periods in the Geological Time Table (Harland *et al.* 1990), as used by the compilers of the GPMDB. The first time window (5 Ma wide) includes the Pliocene and Quaternary, the six following windows (each 10 Ma wide) cover from Miocene to Palaeocene time, the next eight windows (30–45 Ma wide) extend from the Cretaceous to the Early Carboniferous. Finally the last window (50 Ma wide) represents the Devonian. Each datum was assigned to a time window on the basis of the mean of its maximum and minimum magnetization ages. The assignation of a datum was unequivocal; if the mean value fell at a window boundary, then it was assigned to the younger window.

Quality filtering

Clearly, laboratory and analytical procedures (Bazhenov & Shipunov 1991) are vital, as are the number of sites and number of samples per site, as these affect the statistical results (Tarling 1983). Multiple sites within a given rock unit are needed to provide an adequate time sampling of

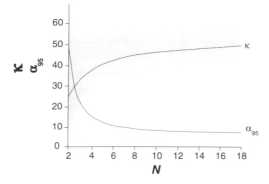

Fig. 3. Precision parameter (K) and α_{95} versus the number of observations N (redrawn from Tarling 1983). These parameters are important to define the quality of the data. It can be seen from the figure that the values of K and α_{95} tend to stabilize and become acceptable for number N of observations above 6–8. These limits have been assumed in fixing the criteria for the two higher quality classes (see text).

the geomagnetic field, and at least six to eight observations are required per site for an acceptable estimate of precision (K) and α_{95} in Fisherian statistics (Fisher 1953) (Fig. 3). Many authors have attempted to define quality criteria, using for example: the number of sites and samples per site; the lack of remagnetization; the age determination method; and the structural control.

Van der Voo & French (1974) considered reliable only those data characterised by successful removal of all possible secondary components of magnetization and based on a considerable number of samples. Moreover, they consider adequate structural control and a well-determined age to be essential, assuming that the magnetization has about the same age as the rock. Briden & Duff (1981) suggested that palaeomagnetic data could be divided into four quality classes of increasing severity. Their main criterion was that A class key-poles had known magnetization ages. May & Butler (1986) suggested that reliable poles should have at least 10 sites per pole, with Virtual Geomagnetic Pole (VGP) precision parameters (K) between 20 and 150 and α_{95} radii less than 15°. Their maximum accepted age uncertainty was 10 Ma and an appropriate understanding of the structural corrections was also a prerequisite. Van der Voo (1990) proposed a set of seven reliability criteria: (1) a well-determined age of the rocks and the presumption that the magnetization has the same age; (2) sufficient samples (N $\geqslant$ 24), $K \geqslant$

10 and $\alpha_{95} \lesssim 16°$; (3) adequate demagnetization, including demonstrable vector subtraction; (4) field tests limiting the age of magnetization; (5) structural control and tectonic coherence; (6) the presence of reversals; (7) no resemblance of the palaeopole position to that of a pole of younger age by more than a Period. The quality index (Q from 0 to 7) is assigned according to the number of criteria that have been satisfied. Besse & Courtillot (1991) assumed the following minimum reliability criteria: at least 6 sites per pole and 6 samples per site, $\alpha_{95} < 15°$, evidence for successful alternating field (AF) and/or thermal demagnetization and maximum date uncertainty of 15 Ma. They have tried to identify signs of remagnetization, to check quality of the age determination and to assess the tectonic structure, including fold tests. Li & Powell (1993), following and developing the approach of Briden & Duff (1981), assigned data to five quality classes (A to E). They considered 'key-poles' as those of A and B classes, which had a well-determined magnetization age (less than 30 and 60 Ma uncertainty) and with palaeohorizontal control. The C class poles must have palaeohorizontal control but lack tight age control. The D and E class poles were considered unreliable.

In view of the need for a careful selection of palaeopoles, six different quality classes (A, B, C, D, M and U; see Table 2) are suggested here. The uncertainty of the magnetization age plays a special role. For geodynamic purposes, it is possible to describe the evolution through time of a pre-determined area by using only data with a degree of limited uncertainty, otherwise the approach would be unreliable. On the other hand, the real uncertainty of the limits among the epochs and of the age determined by isotopic methods both increase further back through time, as shown in Fig. 4, based on data given by Harland *et al.* (1990). For higher quality data, the uncertainty values should never be greater than the width of the time windows (first column of Table 1); on the other hand too strict a time uncertainty criterion can lead to an unjustified rejection. This choice is subjective and, in this study, the acceptable age uncertainties are equal or slightly smaller than the width of the time windows (5 Ma for Pliocene-Quaternary, 10 Ma for Tertiary, 30 Ma for Mesozoic, and 35 Ma for Palaeozoic; column 3, Table 1).

We define six quality classes using the following criteria.

A quality is characterized by: an uncertainty of magnetization age not greater than the width of time windows; at least eight sites per pole and eight samples per site; an α_{95} confidence interval

Table 2. *The six quality classes (A, B, C, D, M and U) proposed in this study*

	Sites	Samples	α_{95}	Laboratory and analytical procedures*	Conclusive field test
A	≥ 8	≥ 8	≤ 15	≥ 3	Yes
B	≥ 6	≥ 6	≤ 15	≥ 2	Yes
C	≥ 4	≥ 4	≤ 20	≥ 2	No
D	< 4	< 4	> 20	< 2	No
M	Magnetostratigraphy (?); samples/sites > 30				
U	Undefined number of sites and/or samples				

Although the uncertainty of magnetization ages does not appear in this table, its role for the definition of the quality of the data is important, as discussed in the text.
* Codes of the laboratory and analytical procedures (Lock & McElhinny 1991); 0, No demagnetization; 1, demagnetization only on some pilot samples; 2, all samples treated, with blanket treatment only; 3, all samples demagnetized step by step, stereonets with J/J_o or vector plots provided (Zijderveld 1967); 4, principal component analysis (PCA) (Kirschvink 1980) and stereonets with J/J_o or vector plots; 5, PCA and vector plots and multiple demagnetization treatments which are successfully isolating different vectors (e.g., AF and thermal, or thermal and chemical)

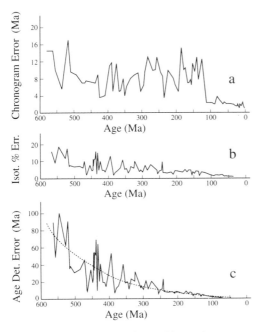

Fig. 4. The uncertainties of quantities against geological time. (**a**) the error (Ma) in defining the age limit between two stages in the Geological Time Scale; (**b**) the percentage error in measuring specimen age by isotopic methods; and (**c**) the uncertainty in Ma in measuring specimen age by isotopic methods.

less than or equal to 15°; a laboratory and analytical procedure code (Lock & McElhinny 1991) greater than or equal to 3; and at least a conclusive field test. Such A class palaeopoles are considered as 'key-poles'.

B quality is characterized by: an uncertainty of

magnetization age not greater than the width of the time windows; at least six sites per pole and six samples per site; the same α_{95} as for the A class; laboratory and analytical procedure codes greater or equal to 2; and at least a conclusive field test.

C quality data have: an uncertainty of magnetization age 50% greater than B class; at least four sites per pole and four samples per site; α_{95} less than or equal to 20°; laboratory and analytical procedure codes greater or equal to 2; but a field test is not required. Poles in this intermediate class are not suitable for the computation of APWPs. When no higher quality data are available, C palaeopoles can only be used to indicate possible trends.

D quality includes data with: less than four sites per pole and four samples per site; α_{95} greater than 20°; and laboratory and analytical procedure codes less than 2. These data are of limited use in geodynamic studies. In this class, the uncertainty in magnetization age is irrelevant.

M quality includes all those studies with samples/sites ratios greater than 30. This class could be suitable for magnetostratigraphic studies.

U quality includes all studies with an undefined number of samples and sites.

Where the GPMDB does not have the α_{95}, this was derived from the polar semi-axis errors, i.e. $(\delta_p . \delta_m)^{1/2}$ (Khramov 1987).

It should be stressed that a different choice of time window widths and of the associated uncertainties in magnetization age strongly influences the quality class assigned to each datum by the extraction program. For example, an uncertainty of 30 Ma is required to resolve the

Table 3. *The lithology of the European and African sites which have been selected to construct APWPs*

Time windows (Ma)	Stable Europe APWP lithology		Stable Africa APWP lithology	
	Quality A	Quality B	Quality A	Quality B
0–5	2e	1s, 2e	1e	2e
5–15		2e		3e
15–25				
25–35				
35–45			1s	
45–55				
55–65		3i, 1e		
65–100	1s		2s, 1i	1i, 1e
100–145				1e
145–180		4s		
180–210		2s		2i
210–245		4s		
245–290	2i, 1e	4s		
290–325	1s, 2i, 1e	1s		1s
325–360	1e	1e		
360–410	1s, 1e	3s, 2e		1e

The subdivision follows the time windows of Table 1, and is further divided into the two higher quality classes, A and B, which have been used in the APWPs of Fig. 5. *s*, sedimentary; *i*, intrusive; *e*, extrusive.

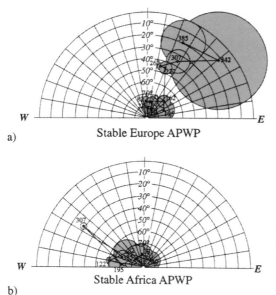

Fig. 5. Apparent Polar Wander Paths: (**a**) Stable Europe; (**b**) Stable Africa. The circles filled in grey represent the uncertainties in degrees. The numbers indicate the mid-point of the time windows.

Early Jurassic (180–210 Ma) palaeolatitudes whereas an uncertainty of about 10 Ma is required for the Sinemurian (194.5–203.5 Ma) data. It is recommended that the width of a time window (W) is never less than the chronogram

error (E_c; Harland *et al.* 1990) of the geological period or stage under consideration.

African and European APWPs

The APWPs of Africa and Europe have been computed using the non-superimposing time windows defined in Table 1, and using only A and B quality data (Table 4). The lithologies involved are shown in Table 3. A 'key-pole' was selected in only one time window (325–360 Ma, Stable Europe) in which the A and B class poles had very different azimuths (*c.* 50°). In many cases the low number of data did not allow the computation of a statistically well-defined mean pole. In other cases the density of averaged poles was particularly high, for instance from 0 to 200 Ma for Stable Europe (Fig. 5a) and Stable Africa (Fig. 5b), but the circles of uncertainty (based on very few reliable palaeomagnetic poles) were very large, and did not permit definitive conclusions. The European APWP (Fig. 5a) shows a regular progression between 2.5 and 82 Ma, with two cusps at 82 and 195 Ma. There was a fast, clear displacement from high to the middle latitudes from the Jurassic to Triassic, linked to some important tectonic events, which can be interpreted in terms of plate tectonics, expanding Earth theory (Scalera 1988, 1990) or other frameworks (Chatterjee & Hotton 1992). The African APWP (Fig. 5b) shows a recent small loop between 2.5 and

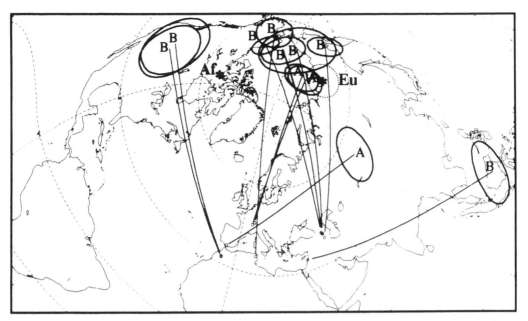

Fig. 6. Mobile Tethys Belt: sampling sites and their poles with confidence ellipses for the Mid–Late Jurassic (145–180 Ma). The stars represent the coeval African (Af) and European (Eu) reference poles and their confidence ellipse (dotted). In this time window it was possible to compute only the European confidence ellipse (see text for the explanation).

40 Ma, and a southwards trend which ends in a cusp near 122 Ma. Subsequently the path comes back towards the present higher latitudes. The next mean pole (307 Ma) is very far away from the other mean poles, and is based only on one B quality pole. Finally the APWP comes back again towards the present latitudes, giving rise to a sharp cusp.

Examples from the Tethys Belt

The poles constituting the APWPs of the major plates are used as 'reference poles' to detect motions (vertical-axis rotations and latitudinal movements) of crustal blocks. The latitudinal transport (Beck 1976, 1980; Demarest 1983; Beck *et al*. 1986; Butler 1992) is given by $L = l_o - l_r$ where l_o represents the great-circle distance in degrees between the sampling site and the observed pole and l_r is the great-circle distance in degrees between the site and the reference pole. L is positive if the block has moved towards the reference pole. The vertical-axis rotation in degrees (Beck 1976, 1980; Demarest 1983; Beck *et al*. 1986; Butler 1992) is more complex:

$$R = \cos^{-1}\left(\frac{\cos s - \cos l_o \cos l_r}{\sin l_o \sin l_r}\right)$$

where s is the great circle distance in degrees between the observed and reference pole.

A total of 1047 palaeopoles come from the Mobile Tethys Belt (MTB in Fig. 2), but only 77 (7.6 %) fall in A and B quality classes (Table 4). Two time windows (Mid–Late Jurassic and Late Cretaceous) were chosen as examples for the entire Mobile Tethys Belt as they are rich in high quality data (Figs 6 and 7). For the Mid–Late Jurassic window, the African pole is derived from the mean between the poles of the two adjacent windows (Early Jurassic and Early Cretaceous; 195 and 120 Ma in Fig. 5b) because they are close to each other. The European subset of Mid–Late Jurassic palaeopoles (two A and four B quality data, excluding the data from Lebanon, Tunisia, Morocco and Southern Spain) has a low dispersion distribution compared to the European reference pole (Fig. 6). The palaeopoles from the Caucasus show an elongated, arcuate distribution indicative of rotations among small-scale blocks of different widths (MacDonald 1980). The Tunisian pole (Nairn *et al*. 1981) has a clockwise rotation relative to the African pole and is on the edge of the European palaeopoles. This fact could be a clue to understanding its evolution as being linked to the European plate. Another anomalous result in this time window is

Table 4. *Number of data subdivided for quality and extracted on the basis of the time windows and the uncertainties of Table 2*

Quality classes	Age (Ma)															
	0–5	5–15	15–25	25–35	35–45	45–55	55–65	65–100	100–145	145–180	180–210	210–245	245–290	290–325	325–360	360–410
Stable Europe																
A	2	0	0	0	0	0	0	1	0	0	0	0	3	4	1	2
B	3	2	0	0	0	0	4	0	0	4	2	4	4	1	1	5
C	7	3	1	1	0	3	4	2	0	1	2	2	16	13	11	16
D	56	7	4	7	2	7	5	6	5	7	4	42	166	53	48	105
M	7	3	0	1	0	1	1	0	0	0	0	5	19	8	9	9
U	8	4	4	2	0	3	3	2	4	3	5	15	79	25	11	38
Stable Africa																
A	1	0	0	0	1	0	0	3	0	0	0	0	0	0	0	0
B	2	3	0	0	0	0	0	2	1	0	2	0	0	1	0	1
C	7	2	0	1	1	0	0	8	2	3	5	2	6	1	0	3
D	21	10	7	3	1	3	1	3	6	9	5	4	8	2	2	2
M	0	0	0	1	0	0	0	0	0	0	0	0	0	0	0	0
U	2	7	1	1	1	1	0	4	1	1	0	1	3	1	1	0
Mobile Tethys Belt																
A	4	3	3	0	1	0	0	1	0	3	3	0	3	0	0	1
B	11	7	0	1	4	0	0	10	4	8	3	1	5	0	0	1
C	18	16	3	5	4	4	2	41	18	15	5	13	27	2	5	2
D	112	38	12	15	13	18	15	73	48	71	24	33	43	5	12	10
M	15	6	3	0	4	2	1	2	23	10	1	1	1	0	0	6
U	11	9	2	4	2	6	4	38	12	28	11	11	21	2	1	11

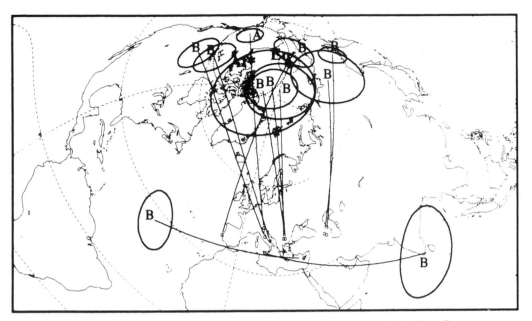

Fig. 7. Mobile Tethys Belt: sampling sites and their poles with confidence ellipses for the Late Cretaceous (65–100 Ma). The stars represent the coeval African (Af) and European (Eu) reference poles and their confidence ellipses (dotted).

from the Late Jurassic volcanic rocks from Lebanon, near the Jordan–Dead Sea (Gregor *et al.* 1974), which shows important rotations with respect to both the European ($R_{\text{Europe}}^{\text{Lebanon}} \pm \Delta R$ =72.3° ± 10°) and the African reference poles ($R_{\text{Africa}}^{\text{Lebanon}}$ = 112°). This can be interpreted as a result of block rotations associated with the evolution of the Jordan–Dead Sea left-lateral transform fault. Its evolution is recent, starting about 30 Ma ago (Westphal *et al.* 1986), and it is possible to estimate from geophysical and geological field studies a total left-lateral movement of about 100 km after the Cretaceous (Jaradat 1990). Only one pole of A quality (Platzman & Lowrie 1992) belongs to the Iberian mobile southern belt (Betic Cordillera) and this shows a very strong clockwise rotation relative to both the European and African reference poles (Fig. 6). This rotation, which is opposite to the well known anticlockwise rotation of the Iberian microplate, could be explained by right-lateral transcurrent movement along the boundary between Africa and Iberia (e.g., Udias 1982; Grimison & Chen 1986).

The Late Cretaceous MTB palaeopoles show large rotations of $R_{\text{Europe}}^{\text{Troodos}} \pm \Delta R = -85.1° \pm 8.3°$ and $R_{\text{Africa}}^{\text{Troodos}} \pm \Delta R = -71.3° \pm 14.0°$ of the Troodos microplate (Cyprus; Fig. 7; see Morris, this volume). The Sicilian pole shows a large

clockwise rotation with respect to both European and African coeval reference poles, $R_{\text{Europe}}^{\text{Sicily}} \pm \Delta R = 96.0° \pm 12.0°$ and $R_{\text{Africa}}^{\text{Sicily}} \pm \Delta R = 110.9° \pm 16.5°$. These data come from structural stratigraphic units derived from the Sicani basin (Western Central Sicily). This large tectonic rotation has been interpreted as the probable result of the emplacement of the Calabrian–Peloritani structure onto the Sicilian continental margin.

Italy

Several palaeomagnetic studies have been carried out in the last 15 years on the Italian Peninsula (e.g. Vandenberg *et al.* 1978; Channell *et al.* 1978, 1980; Horner & Lowrie 1981; Tauxe *et al.* 1983; Besse *et al.* 1984; Mattei *et al.* 1992; Sagnotti 1992; Sagnotti *et al.* 1994; other papers in this volume). Many authors have also published critical reviews of the existing data (e.g., Vandenberg & Zijderveld 1982; Lowrie 1986; Van der Voo 1993; Channell, this volume). In these papers anticlockwise rotations have been determined for Italy since the Early Cretaceous, but the data are concentrated in space and time and some windows of Tertiary and Quaternary age contain no high class data (Table 5). The northern Apennines, central Italy up to Campania, Puglia and the Messina

Table 5. *Number of data subdivided for quality and extracted on the basis of the time windows and the uncertainties of Table 2*

Quality classes								Age (Ma)								
	0–5	5–15	15–25	25–35	35–45	45–55	55–65	65–100	100–145	145–180	180–210	210–245	245–290	290–325	325–360	360–410
Italy																
A	3	0	0	0	0	0	0	1	0	0	1	0	1	0	0	0
B	3	1	0	1	4	0	0	3	1	0	0	0	0	0	0	0
C	4	1	1	2	1	1	0	18	5	1	0	3	10	1	0	0
D	7	4	3	1	2	2	3	21	7	6	1	3	8	1	2	0
M	0	0	0	0	0	0	0	1	3	2	0	0	0	0	0	0
U	2	0	1	4	1	2	2	4	3	4	2	1	3	0	0	0
Sardinia–Corsica block																
A	0	0	0	0	0	0	0	0	0	0	0	0	0	0	0	0
B	0	0	0	0	0	0	0	0	0	0	0	0	0	0	0	0
C	1	1	2	1	0	0	0	0	0	1	0	0	7	0	0	0
D	2	2	2	1	0	1	0	0	0	0	1	1	3	0	1	0
M	0	0	0	0	0	0	0	0	0	0	0	0	0	0	0	0
U	0	0	1	1	0	0	0	0	0	0	0	0	2	0	0	0

Strait are particularly lacking in sampling sites. More data are present in the central-eastern Alps, Umbrian–Marche Apennines, Sardinia and Sicily, but few high quality data are present (Table 5). This uneven distribution of data prevents the definition of the main deformation phases of these Italian units.

One of the more evident effects of quality filtering can be seen in the Late Cretaceous window (65–100 Ma), in which all the A, B and C data (except three palaeopoles from Sicily) show anticlockwise rotations of up to 40° with respect to the European reference pole (Figs 7 and 8, and Table 5). The dispersion of the A and B (and more markedly C) data can be a consequence of two main concomitant causes: intrinsic uncertainties (Bazhenov & Shipunov 1991) and differential amounts of tectonic rotation (MacDonald 1980). This last cause could be explained by the location of the sites in completely different structural-geologic domains along the entire peninsula. On the other hand, the wide time span of the window (35 Ma) has to be considered. The palaeopole positions are so distant in time at the edges of the time window that they are representative of different geological ages. Consequently, no 'absolute' rotation of the peninsula should be computed, only the magnitude of the anticlockwise rotations of each A and B pole (Channell & Tarling 1975; Channell 1977; Channell *et al.* 1992) with respect to the European reference poles:

$$R_{Gargano} \pm \Delta R = -29.1° \pm 7.4°,$$
$$R_{Umbria} \pm \Delta R = -34.3° \pm 7.1°,$$
$$R_{Venetian\ Alps} \pm \Delta R = -16.3° \pm 5.3°.$$

With respect to the African reference pole, this group of poles show no significant rotation (Fig. 7) suggesting a connection between the kinematics of the African plate and the Italian crustal fragments (see Channell this volume). Such small rotations prevent palaeogeographic reconstructions of the closure of the Tyrrhenian and Alboran Seas which require nearly 90° of rotation in some models. Notwithstanding these limitations many authors have assumed an anticlockwise rotation of the entire peninsula from Cretaceous to present (e.g. Vandenberg *et al.* 1978).

Conclusions

The GPMDB has allowed a methodological, objective approach to using palaeomagnetic data in geodynamic studies, although it does not exclude the necessity for reading the original papers. The quality of a datum is dependent on the length of the period of geological time under

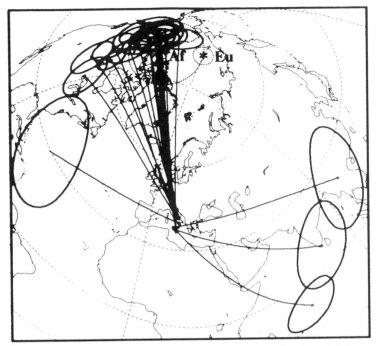

Fig. 8. Late Cretaceous (65–100 Ma) Italian palaeopoles (A, B and C quality classes). The stars represent the coeval African (Af) and European (Eu) reference poles and their confidence ellipses (dotted).

study. A lower limit for time window amplitudes has been proposed. The trend of the African and European APWPs presented here show similar patterns to previously published results, although the imprecision impedes any definitive interpretations, particularly in the narrowest loops of the paths. The firmest conclusion is the existence of a large jump in the European APWP during the Jurassic and Triassic. In the Mid–Late Jurassic and Late Cretaceous time windows there are several examples of very marked block rotations (e.g., Lebanon and Betics for Jurassic; Troodos, Cyprus for Cretaceous).

In recent years, higher quality palaeomagnetic results have been obtained following improvements in data acquisition and analysis methods. Most studies are now aimed at understanding the geodynamics of terranes in mobile zones. The contents of the database suggest that there is also a need to obtain additional data from stable areas. APWPs are in need of improvement, by provision of data for empty time windows and increasing the number of observations in others. We suggest that some tectonic models now need to be reconsidered to take new, well-constrained palaeomagnetic data into account.

The authors wish to thank M. Marani and A. Meloni for helpful suggestions and discussions. We also thank Z. X. Li, M. W. McElhinny, A. Morris and D. H. Tarling for revisions to the manuscript. Appreciation is expressed to Daniela Riposati for drawing some of the figures.

References

AA.VV. 1981. *Tectonic map of Italy (1:1 500 000)*. CNR-PFG, Publication no. 269, GEO, Roma.

BALLY, A. W., CATALANO, R. & OLDOW, J. 1985. *Elementi di Tettonica Regionale*. Pitagora Editrice, Bologna.

BARTOV, Y. 1979. *Geological map of Israel (1:500 000)*. Geological Survey of Israel, Tel Aviv.

BAZHENOV, M. L. & SHIPUNOV, S. V. 1991. Dispersion of paleomagnetic data. *Izvestya, Earth Physics*, **27**, 392–401.

BECK, M. E. 1976. Discordant paleomagnetic pole position as evidence of regional shear in the Western Cordillera of North America. *American Journal of Science*, **276**, 694–712.

—— 1980. Paleomagnetic record of plate-margin tectonic processes along the western edge of North America. *Journal of Geophysical Research*, **85**, 7115–7131.

——, BURMESTER, R. F., CRAIG, D. E., GROMMÉ, C. S. & WELLS, R. E. 1986. Paleomagnetism of Middle Tertiary volcanic rocks from the Western Cascade series, Northern California. Timing and scale of rotation in the Southern Cascades and Klamath Mountains. *Journal of Geophysical Research,* **91,** 8219–8230.

BESSE, J. & COURTILLOT, V. 1991. Revised and Synthetic Apparent Polar Wander Paths of the African, Eurasian, North America and Indian Plates, and True Polar Wander Since 200 Ma. *Journal of Geophysical Research,* **96,** 4029–4050.

——, POZZI, J. P., MASCLE, G. & FEINBERG, H. 1984. Paleomagnetic study of Sicily: consequences for the deformation of Italian and African margins over the last 100 Ma. *Earth and Planetary Science Letters,* **67,** 377–390.

BOCCALETTI, M. & DAINELLI, P. 1982. Il Sistema Regmatico Neogenico-Quaternario nell'area Mediterranea: esempio di deformazione plastico/rigida postcollisionale. *Memorie delle Società Geologica Italiana,* **24,** 465–482.

——, CONEDERA, C., DAINELLI, P. & GOCEV, P. 1985. *Tectonic map of the western Mediterranean area (1 : 2,500,000).* CNR, Publication no. **158,** Roma.

BRGM 1980. *Carte geologique de la France et de la marge continentale (1 : 1,500,000).* Bureau de Recherches Géologiques et Minières, Orleans.

BRIDEN, J. C. & DUFF, B. A. 1981. Pre-Carboniferous paleomagnetism of Europe north of the Alpine orogenic belt. *In:* McELHINNY, M. W. & VALENCIO, D. A. (eds) *Paleoreconstruction of the continents.* American Geophysical Union – Geological Society of America, Geodynamics Series, **2,** 137–149.

BUTLER, R. F. 1992. *Paleomagnetism: Magnetic Domains to Geological Terranes.* Blackwell, Oxford.

CHANNELL, J. E. T. 1977. Paleomagnetism of limestones from the Gargano Peninsula (Italy), and the implications of these data. *Geophysical Journal of the Royal Astronomical Society,* **51,** 605–616.

—— & TARLING, D. H. 1975. Paleomagnetism and the rotation of Italy. *Earth and Planetary Science Letters,* **25,** 177–188.

—— 1996. Palaeomagnetism and palaeogeography of Adria. *This volume.*

——, CATALANO, R. & D'ARGENIO, B. 1980. Paleomagnetism of the Mesozoic continental margin in Sicily. *Tectonophysics,* **61,** 391–407.

——, DOGLIONI, C. & STONER, J. S. 1992. Jurassic and Cretaceous paleomagnetic data from the Southern Alps (Italy). *Tectonics,* **11,** 811–822.

——, LOWRIE, W., MEDIZZA, F. & ALVAREZ, W. 1978. Paleomagnetism and tectonics in Umbria, Italy. *Earth and Planetary Science Letters,* **39,** 199–210.

CHATTERJEE, S. & HOTTON, N. (eds) 1992. *New concepts in global tectonics.* Texas Technical University Press, Lubbock.

CONDIE, K. C. 1989. *Tectonic map of the Earth (1 : 36,000,000).* Williams & Heintz Map Corp., Capitol Heights.

DEMAREST, H. H. 1983. Error analysis of the determination of tectonic rotation from paleomagnetic data. *Journal of Geophysical Research,* **88,** 4321–4328.

FAVALI, P., FUNICIELLO, R., MELE, G., MATTIETTI, G. & SALVINI, F. 1993a. An active margin across the Adriatic Sea (central Mediterranean Sea). *Tectonophysics,* **219,** 109–117.

——, —— & SALVINI, F. 1993b. Geological and Seismological evidence of strike-slip displacement along the E–W Adriatic–Central Apennines belt. *In:* BOSCHI, E., MANTOVANI, E. & MORELLI, A. (eds) *Recent evolution and Seismicity of the Mediterranean region.* Kluwer Academic Publicshers NATO ASI Series, **402,** 333–346.

FISHER, R. A. 1953. Dispersion on a sphere. *Proceedings of the Royal Astronomical Society,* **A217,** 295–305.

FLORINDO, F., SAGNOTTI, L. & SCALERA, G. 1994. Using the ASCII version of the Global Paleomagnetic Database. *EOS,* **75/21,** 236–237.

GREGOR, C. B., MERTZMAN, S., NAIRN, A. E. M. & NEGENDANK, J. 1974. The paleomagnetism of some Mesozoic and Cenozoic volcanic rocks from the Lebanon. *Tectonophysics,* **21,** 375–395.

GRIMISON, N. L. & CHEN, W-P. 1986. The Azores-Gibraltar plate boundary: focal mechanisms, depths of earthquakes, and their tectonic implications. *Journal of Geophysical Research,* **91,** 2229–2047.

HARLAND, W. B., AMSTRONG, R. L., COX, A. V., CRAIG, L. G., SMITH, A. G. & SMITH, D. G. 1990. *A geologic time scale 1989.* Cambridge University Press.

HORNER, F. & LOWRIE, W. 1981. Paleomagnetic evidence from Mesozoic carbonate rocks for the rotation of Sardinia. *Journal of Geophysics,* **49,** 11–19.

JARADAT, M. 1990. The Jordan Seismological Observatory and Seismicity in Jordan. *In:* BOSCHI, E., GIARDINI, D. & MORELLI, A. (eds) *MedNet. The broad-band seismic network for the Mediterranean.* Ed. Il Cigno-G. Galilei, Roma, 426–443.

KAMPUNZU, A. B. & LUBALA, R. T. 1991. *Magmatism in extensional structural setting.* Springer-Verlag, Berlin.

KIRSCHVINK, J. 1980. The least-square line and plane and the analysis of paleomagnetic data. *Geophysical Journal of Royal Astronomical Society,* **62,** 699–718.

KISSEL, C. & LAJ, C. (eds) 1989. *Paleomagnetic rotations and continental deformation.* Kluwer Academic Publishers, NATO ASI Series, **254,** 516.

KHRAMOV, A. N. 1987. *Paleomagnetology.* Springer-Verlag, Berlin.

KRUCZYK, J., KADZIALKO-HOFMOKL, M., LEFELD, J., PAGAC, P. & TUNYI, I. 1992. Paleomagnetism of Jurassic sediments as evidence for oroclinal bending of the Inner West Carpathians. *Tectonophysics,* **206,** 315–324.

LI, Z. X. & POWELL, C. McA. 1993. Late Proterozoic to Early Paleozoic paleomagnetism and the formation of Gondwanaland. *In:* FINDLAY, R. H., UNRUG, R., BANKS, M. R. & VEEVERS, J. J. (eds) *Gondwana Eight. Assembly, evolution and dispersal.* Balkema, Rotterdam, Brookfield, 9–22.

LOCK, J. & McELHINNY, M. W. 1991. The Global Paleomagnetic Data Base: Design, installation and use with ORACLE. *Surveys in Geophysics,* **12,** 317–491.

LOWRIE, W. 1986. Paleomagnetism and the Adriatic promontory: a reappraisal. *Tectonics,* **5,** 797–807.

LUMSDEN, G. I. (ed.) 1992. *Geology and the Environment in Western Europe*. Clarendon Press, Oxford.

MACDONALD, W. D. 1980. Net tectonic rotation, apparent tectonic rotation, and the structural tilt correction in paleomagnetic studies. *Journal of Geophysical Research*, **85**, 3659–3669.

MARTIN, D. L., NAIRN, A. E. M., NOLTIMIER, H. C., PETTY, M. H. & SCHMITT, T. J. 1978. Paleozoic and Mesozoic paleomagnetic results from Morocco. *Tectonophysics*, **44**, 91–114.

MATTEI, M., FUNICIELLO, R., KISSEL, C. & LAJ, C. 1992. Rotazione di blocchi crostali neogenici nell'Appennino centrale: analisi paleomagnetiche e di anisotropia della suscettività magnetica (AMS). *Studi Geologica Camerti, Special Publications. 1991/2*, 221–229.

MAY, S. R. & BUTLER, R. F. 1986. North-American Jurassic apparent polar wander: implications for plate motions, paleogeography and Cordilleran tectonics. *Journal of Geophysical Research*, **91**, 11519–11544.

MCELHINNY, M. W. & LOCK, J. 1990a. Global Paleomagnetic Data Base Project. *Physics of the Earth and Planetary Interiors*, **63**, 1–6.

—— & —— 1990b. IAGA Global Paleomagnetic Data Base. *Geophysical Journal International*, **101**, 763–766.

—— & —— 1993. Global Paleomagnetic Database Supplement number one: Update to 1992. *Surveys in Geophysics*, **14**, 303–329.

MERLA, G., ABBATE, E., CANUTI, P., SAGRI, M. & TACCONI, P. 1973. *Geologic map of Ethiopia and Somalia (1 : 2,000,000)*. CNR, Roma.

MORRIS, A. 1996. A review of palaeomagnetic research in the Troodos ophiolite, Cyprus. *This volume*.

NAIRN, A. E. M., SCHMITT, T. J. & SMITHWICK, M. E. 1981. A paleomagnetic study of the Upper Mesozoic succession in the northern Tunisia. *Geophysical Journal of the Royal Astronomical Society*, **65**, 1–18.

PLATZMAN, E. & LOWRIE, W. 1992. Paleomagnetic evidence for rotation of the Iberian Peninsula and the external Betic Cordillera, Southern Spain. *Earth and Planetary Science Letters*, **108**, 45–60.

SAGNOTTI, L. 1992. Paleomagnetic evidence for a pleistocene counterclockwise rotation of the Sant'Arcangelo basin, Southern Italy. *Geophysical Research Letters*, **19**, 135–138.

——, MATTEI, M., FACCENNA, C. & FUNICIELLO, R. 1994. Paleomagnetic evidence for no tectonic rotation of the Central Italy Tyrrhenian margin since Upper Pliocene. *Geophysical Research Letters*, **21**, 481–484.

SALMON, E., EDEL, J. B., PIQUE, A. & WESTPHAL, M. 1988. Possible origins of Permian remagnetizations in Devonian and Carboniferous limestones from the Moroccan Anti-Atlas (Tafilalet) and Meseta. *Physics of the Earth and Planetary Interiors*, **52**, 339–351.

SCALERA, G. 1988. Nonconventional Pangea reconstructions: new evidence for an expanding Earth. *Tectonophysics*, **146**, 365–383.

—— 1990. Paleopoles on an expanding Earth: a comparison between synthetic and real data sets. *Physics of the Earth and Planetary Interiors*, **62**, 126–140.

ŞENGÖR, A. M. C. (ed.) 1989. *Tectonic evolution of the Tethyan region*. KLUWER Academic Publishers, NATO ASI Series, **259**.

STÖCKLIN, J. 1989. Tethys evolution in the Afghanistan–Pamir–Pakistan region. *In*: ŞENGÖR, A. M. C. (ed.) *Tectonic evolution of the Tethyan region*. Kluwer Academic Publishers, NATO ASI Series, **259**, 241–264.

—— & NABAVI, M. H. 1973. *Tectonic map of Iran (1 : 2,500,000)*. Geological Survey of Iran, Tehran.

TARLING, D. H. 1983. *Palaeomagnetism*. Chapman & Hall, London.

TAUXE, L., OPDYKE, N. D., PASINI, G. & ELMI, C. 1983. Age of the Plio-Pleistocene boundary in the Vrica section, Southern Italy. *Nature*, **304**, 125–129.

UDIAS, A. 1982. Seismicity and Seismotectonic stress field in the Alpine-Mediterranean region. *In*: BERCKHEMER, H. & HSÜ, K. (eds) *Alpine–Mediterranean Geodynamics*. American Geophysical Union – Geological Society of America, Geodynamics Series, **7**, 75–82.

UNESCO 1976. *Geological World Atlas*. Commission for the Geological map of the World (CGMW), Paris.

VANDENBERG, J. & ZIJDERVELD, J. D. A. 1982. Paleomagnetism in the Mediterranean area. *In*: BERCKHEMER, H. & HSÜ, K. (eds) *Alpine–Mediterranean Geodynamics*. American Geophysical Union – Geological Society of America, Geodynamics Series, **7**, 83–112.

——, KLOOTWIJK, C. T. & WONDERS, A. A. H. 1978. Late Mesozoic and Cenozoic movements of the Italian peninsula: further paleomagnetic data from the Umbrian sequence. *Geological Society of America Bulletin*, **89**, 133–150.

VAN DER VOO, R. 1990. Phanerozoic paleomagnetic poles from Europe and North America and comparison with continental reconstructions. *Reviews in Geophysics*, **28**, 167–206.

—— 1993. *Paleomagnetism of the Atlantic, Tethys and Iapetus Oceans*. Cambridge University Press.

—— & FRENCH, R .B. 1974. Apparent polar wandering for the Atlantic-bordering continents: Late Carboniferous to Eocene. *Earth Science Reviews*, **10**, 99–119.

WESTPHAL, M., BAZHENOV, M. L., LAUER, J. P., PECHERSKY, D. M. & SIBUET, J. 1986. Paleomagnetic implications of the evolution of the Tethys Belt from the Atlantic Ocean to the Pamirs since the Triassic. *Tectonophysics*, **123**, 37–82.

ZIJDERVED, J. D. A. 1967. A. C. demagnetization of rocks: analysis of results. *In*: RUNCORN, S.K., CREER, K. M. & COLLINSON, D. W. (eds) *Methods in Palaeomagnetism*. Elsevier, Amsterdam, 254–286.

Alternative tectonic models for the Late Palaeozoic–Early Tertiary development of Tethys in the Eastern Mediterranean region

A. H. F. ROBERTSON[1], J. E. DIXON[1], S. BROWN[1], A. COLLINS[1], A. MORRIS[3], E. PICKETT[1], I. SHARP[1] & T. USTAÖMER[2]

[1]*Department of Geology and Geophysics, University of Edinburgh, West Mains Road, Edinburgh EH9 3JW, UK*
[2]*Department of Geology, Istanbul University, Avcilar, 34850 Istanbul, Turkey*
[3]*Department of Geological Sciences, University of Plymouth, Drake Circus, Plymouth, Devon PL4 8AA, UK*

Abstract: A summary and discussion is given of alternative models of the tectonic evolution of the Tethyan orogenic belt in the Eastern Mediterranean region, based on recent information.

Model 1 (Robertson & Dixon 1984). A single Tethyan ocean continuously existed in the Eastern Mediterranean region, at least from Late Palaeozoic onwards. The dominant influences were episodic northward subduction of Tethyan oceanic crust beneath Eurasia, and the northward drift of continental fragments, from Gondwana towards Eurasia. During the Mesozoic, the south Tethyan area was interspersed with Gondwana-derived microcontinents and small ocean basins. Ophiolites formed mainly by spreading above subduction zones in both northerly (internal) and southerly (external) oceanic basins during times of regional plate convergence, and were mainly emplaced as a result of trench-passive margin collisions. In a related model, Stampfli *et al.* (1991) argued for spreading along the North African margin in the Late Permian.

Model 2A (Dercourt *et al.* 1986). Only one evolving Tethys existed. Triassic–Jurassic oceanic crust (Neotethys) formed in a single Tethyan ocean basin located north of Gondwana-related units. Spreading later formed a small ocean basin in the present Eastern Mediterranean Sea area during the Cretaceous. Jurassic and Cretaceous ophiolites formed at spreading ridges and record times of regional plate divergence. In an updated version, *Model 2B* (Dercourt *et al.* 1993), spreading extended along the northern margin of Gondwana, with an arm extending through the south Aegean, splitting off a large microcontinent. Further spreading in the Cretaceous then opened the Eastern Mediterranean basin and fragmented pre-existing carbonate platforms. The Mesozoic ophiolites were seen as being mainly far-travelled from northerly (i.e. internal) orogenic areas.

Model 3 (Şengör *et al.* 1984). Subduction in the Late Palaeozoic was dominantly southwards, beneath the northern margin of Gondwana in the Eastern Mediterranean. This subduction led to opening of Triassic backarc basins; and a rifted Gondwana fragment (Cimmeria) drifted across a pre-existing Tethys (Palaeo-Tethys) to collide with a passive Eurasian margin. In their model, a backarc basin (Karakaya Basin) rifted and then closed prior to collision of a Cimmerian microcontinent in the Mid Jurassic, and this was followed by renewed rifting of a small ocean basin in the Early Jurassic. Mesozoic ophiolites mainly formed above subduction zones; they were variously seen as far-travelled (in the 'Greek area'), or more locally rooted (in the 'Turkish area').

Recent evidence shows that difficulties exist in detail with all three models. However, four key elements are met in Model 1: dominantly northward subduction in the north; multiple ocean basins from Triassic onwards in the south; supra-subduction spreading of the major ophiolites; and emplacement from both northerly and southerly Mesozoic oceanic basins. Palaeomagnetism has played an important role, in setting the large-scale Africa–Eurasia relative motion framework and in providing tests for the tectonic affinities of smaller units, but such smaller-scale studies have often been compromised by the geological complexity and by the remagnetisation of tectonically thickened units.

This paper presents a summary and discussion of the validity of alternative reconstructions, of the tectonic evolution of the Tethys ocean in the Eastern Mediterranean area in the light of important information gained over the last decade or so (Fig. 1). The reader is referred to the main syntheses cited below for a review and references to earlier literature (i.e. pre-med-

From Morris, A. & Tarling, D. H. (eds), 1996, *Palaeomagnetism and Tectonics of the Mediterranean Region*, Geological Society Special Publication No. 105, pp. 239–263.

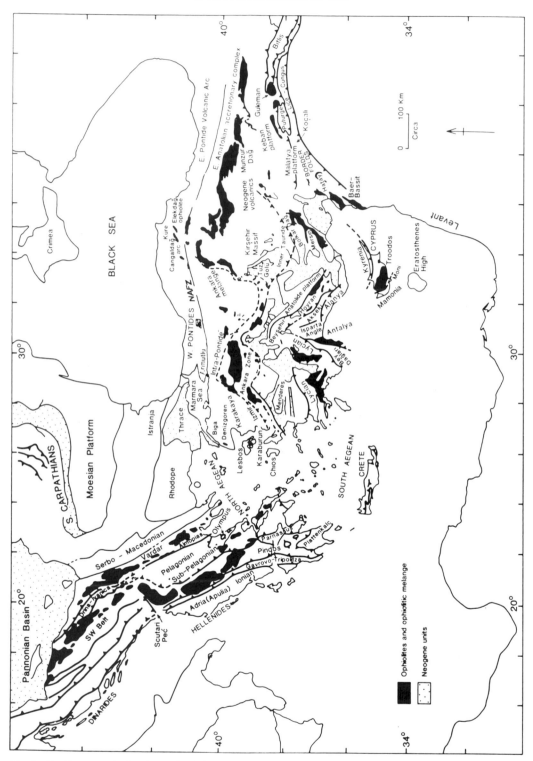

Fig. 1. Outline tectonic sketch map including the main tectonic units and localities mentioned in the text. Note: neotectonic features are omitted.

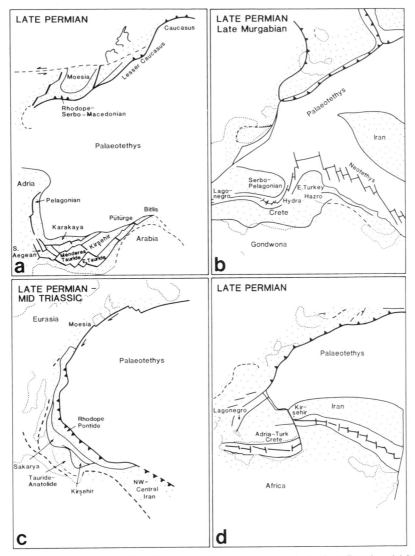

Fig. 2. Alternative reconstructions of the Eastern Mediterranean Tethys in the Late Permian. (**a**) Model 1 – Robertson & Dixon (1984); (**b**) Model 2B – Dercourt *et al.* (1993); (**c**) Model 3 – Şengör *et al.* (1984); (**d**) Stampfli *et al.* (1991). See text for discussion.

1980s). We will highlight the components of earlier syntheses that have been supported by more recent data and indicate those that can now be discounted. The discussion is intended to clear the way for future, improved tectonic reconstructions, in which palaeomagnetic data will continue to play an important role.

The terms Tethys, Palaeotethys (or Palaeo-Tethys) and Neotethys (or Neo-Tethys) have been used in many different ways, that are often model dependent. Below, the terms used by individual authors are retained when their individual models are discussed. However, in

our own discussion we simply refer to Tethys as the ocean basin system that was present in the Eastern Mediterranean region at any given time (e.g. Late Palaeozoic Tethys, Early Cretaceous Tethys etc.), with no palaeogeographic implications. To simplify discussion we refer to the orogenic assemblage in the west, including Greece, Albania and former Yugoslavia as the 'Greek area' and that in the east, including Turkey, Cyprus and northern Syria as the 'Turkish area'.

Recent reconstructions assume a wedge-shaped Tethyan embayment in the Eastern

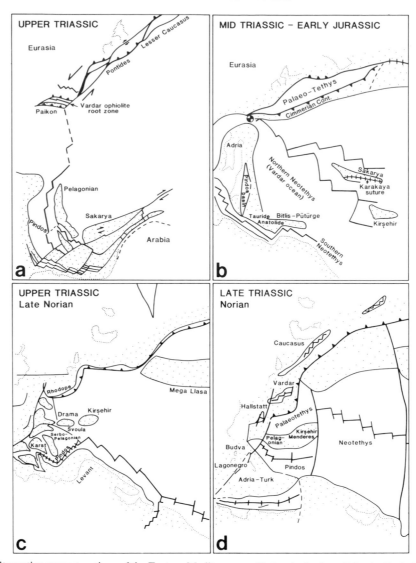

Fig. 3. Alternative reconstructions of the Eastern Mediterranean Tethys in the Late Triassic–Early Jurassic. (**a**) Model 1 – Robertson & Dixon (1984); (**b**) Model 3 – Şengör *et al.* (1984); (**c**) Model 2B – Dercourt *et al.* (1993); (**d**) Alternative model of Stampfli *et al.* (1991). See text for discussion.

Mediterranean, at least from Late Permian to Early Tertiary time. The main uncertainty concerns the role of possible mega-shear between Gondwana and Laurasia during Permo-Triassic time (Livermore *et al.* 1986). The Eurasia–Africa fit is relatively well constrained from Early Jurassic onwards, based on palaeomagnetic data and ocean floor magnetic anomaly correlations (Livermore & Smith 1984; Dercourt *et al.* 1986).

Alternative tectonic models

Over the last ten years, three main alternative models for the Eastern Mediterranean region as a whole have been proposed.

Model 1: single evolving Tethys model

Robertson & Dixon (1984) argued that Tethys in the Eastern Mediterranean region existed as a wide ocean that developed continuously from late Palaeozoic to Recent time (Figs 2a, 3a, 4c, 5a & 6a). The south margin of Eurasia was seen as an active continental margin undergoing mainly northward subduction. The southern, Gondwana margin (i.e. Africa) was seen as passive, at least from the Late Palaeozoic

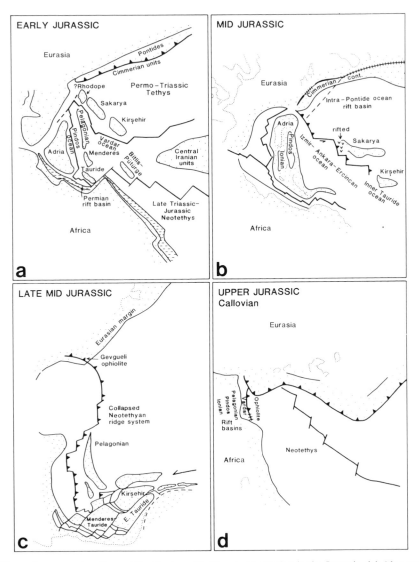

Fig. 4. Alternative reconstructions of the Eastern Mediterranean Tethys in the Jurassic. (a) Alternative model of Robertson *et al.* (1991); (b) Model 3 – Şengör *et al.* (1984) (c) Model 1 – Robertson & Dixon (1984); (d) Model 2A – Dercourt *et al.* (1986). See text for discussion.

onwards. During the Permo-Triassic, microcontinents were rifted from Gondwana, then drifted variable distances into the Mesozoic Tethys ocean, followed in due course by amalgamation with an Eurasian active margin to the north. The large continental area of Adria (i.e. Apulia) was shown as a promontory of Gondwana. Large ophiolites, of Jurassic age in the westerly 'Greek area' and of Late Cretaceous age in the easterly 'Turkish area' were seen as a consequence of spreading above oceanic subduction zones, initiated by regional plate convergence (Pearce *et al.* 1984).

Robertson *et al.* (1991) further interpreted the south-Tethyan area of the Eastern Mediterranean as Mesozoic oceanic crust interspersed with continental fragments, rifted from Gondwana (Fig. 4a). These fragments included the large Adrian microcontinent (Apulia), the Pelagonian zone in Greece and the Tauride carbonate platforms of Turkey. It was also suggested that rifting was initiated in the Late Permian to form an open seaway adjacent to the present North African continental margin in the Eastern Mediterranean.

Stampfli *et al.* (1991) proposed a somewhat similar model involving northward subduction of Palaeotethys and early rifting of a southerly

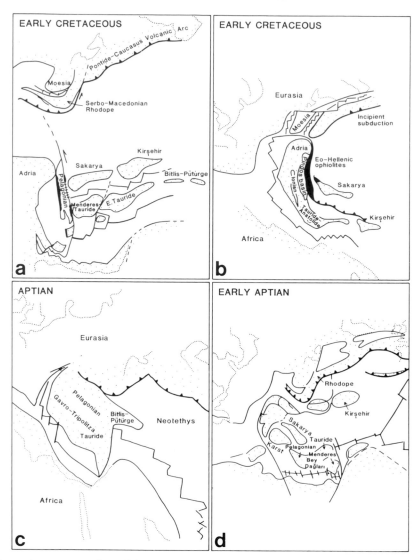

Fig. 5. Alternative reconstructions of the Eastern Mediterranean Tethys in the Early Cretaceous. (**a**) Model 1 – Robertson & Dixon (1984); (**b**) Model 3 – Şengör *et al.* (1984); (**c**) Model 2A – Dercourt *et al.* (1986); (**d**) Model 2B – Dercourt *et al.* (1993). See text for discussion.

Neotethys (Figs 2d &3d). It was argued that spreading took place in Late Permian time to form a south-Tethyan small ocean basin, extending along the northern margin of Gondwana from North Africa (Libya–Tunisia) through the present Eastern Mediterranean Sea to the Oman region. Permian spreading separated an Adrian–Turk microcontinent from an Iranian microcontinent (extending eastwards to include the Kirşehir massif). Most of the spreading in the southerly, Eastern Mediterranean basin had ended by the Triassic, in contrast to Model 1. For Late Triassic (Norian), time a complex model was envisaged in which Palaeotethyan

(i.e. pre-Permian) and Neotethyan (i.e. Permo-Triassic) oceanic crust were separated by long, north-south transforms (e.g. in central Turkey). Palaeotethys in the west was subducted northwards, opening the Caucasus, Vardar and Hallstatt small ocean basins.

Model 2: single Mesozoic Tethys model

One basically similar model evolved from the mid-1980s to early 1990s (Models 2A and 2B). Model 2A (Dercourt *et al.* 1986) envisaged a simple palaeogeography of Tethyan oceanic crust in the Eastern Mediterranean, whereas in

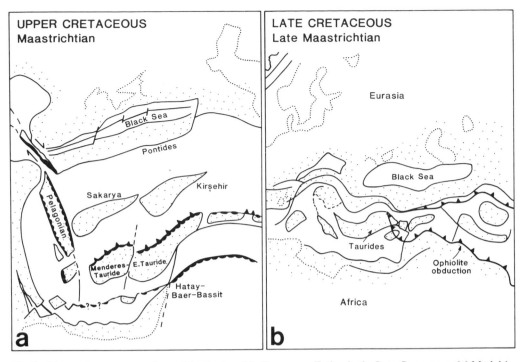

Fig. 6. Alternative reconstructions of the Eastern Mediterranean Tethys in the Late Cretaceous. (**a**) Model 1 – Robertson & Dixon (1984); (**b**) Model 2B – Dercourt *et al.* (1993). See text for discussion.

Model 2B (Dercourt *et al.* 1993) this has become more varied, particularly for Cretaceous–Early Tertiary time.

In Model 2A, Dercourt *et al.* (1986) argued that only one Tethys ocean basin existed during Early Mesozoic time; a second smaller ocean basin then opened in the south of the area in the Cretaceous (Figs 4d & 5a). As in Model 1, above, the Gondwana margin was passive, while the Eurasian margin was of active, subduction-related type. The Mesozoic ophiolites were seen as forming at mid-ocean ridges during times of regional plate separation. These ophiolites all were assumed to have been rooted in a single Mesozoic Tethyan oceanic basin (Neotethys). The Jurassic 'Greek' ophiolites (e.g. Pindos) were, thus, rooted to the north-east, in the Vardar (Axios) zone or beyond, while the Cretaceous 'Turkish' ophiolites (e.g. Lycian and Antalya) were derived from the north, from a single Neotethyan ocean basin. During Permian to Jurassic time, the north Gondwana margin was faulted to form intracontinental rift basins (e.g. Pindos basin). Adria (Apulia) was considered as the largest of these rifted units; it was only detached, leaving a small ocean basin behind it during the Cretaceous. On the other hand, large units of Eastern Turkey (e.g. Bitlis–Pütürge) were seen as part of a promontory of Gondwana.

Model 2B of Dercourt *et al.* (1993) adopted the framework of the earlier Model A (Dercourt *et al.* 1986), but with several new elements (Figs 2b & 5c). The concept of 'transit plates' was introduced. These were units of 'combined oceanic and continental crust' and were deemed to have moved continuously northward from near Gondwana to collide, in turn, with an Eurasian active margin to the north. In several cases, these 'transit plates' partly correspond to known geological units (i.e. Kirşehir Massif, Turkey; Mega–Lhasa further east). However, the Drama plate, in the 'Greek area' is not clearly defined as a geological entity. Transit plates were apparently introduced mainly to satisfy kinematic constraints of the model. The south Eurasian margin is seen as mainly active with northward subduction, as in Model 1. Neotethyan crust formed at a spreading ridge system in the south from the Late Permian onwards. All of southern Turkey (e.g. the Taurides) and the south Aegean (e.g. Gavrovo–Tripolitza zone) were seen as part of Gondwana.

In contrast to Model 2A, in which only one Neotethyan basin was present, an arm of the Permian Neotethys is inferred to have extended through the south Aegean (e.g. through Hydra island) to connect with the Lagonegro zone further west, thus splitting a Serbo-Pelagonian microcontinent (Fig. 3c). By the Late Triassic (late Norian), the Drama transit plate had moved northeastwards from the Adrian microplate, opening a Vardar ocean between a Serbo-Pelagonian microcontinent and the Drama plate. The Jurassic Svoula flysch, exposed in the Vardar zone of northeastern Greece, accumulated in this basin and was later thrust southwestwards to its present position. In other words, the Svoula flysch is considered to be grossly allochthonous in this model, since it was deposited to the north of the Serbo-Macedonian zone, but now crops out to the south of it. The Jurassic ophiolites of the 'Greek area' are, likewise, seen as being very substantially allochthonous. They were rooted within the Vardar ocean (north of the Serbo-Macedonian microcontinent) and were later thrust over the Serbo-Macedonian zone, along with the Svoula flysch, ending up partly within the Vardar zone in northeastern Greece (i.e. Vardar ophiolites) and partly within the Sub-Pelagonian and Pindos zones further west (e.g. Dağları Vourinos and Pindos ophiolites). During the Early Cretaceous, in Model 2A (Fig. 5c), a southerly Neotethyan basin then opened, rifting off a huge microcontinental block for the first time, encompassing most of the Gondwana-derived units in the Eastern Mediterranean (including Adria, Taurides etc.). By contrast, in Model 2B this same continental area was envisaged as fragmenting into a number of smaller microplates, notably Sakarya, a combined Pelagonian–Menderes–Tauride microplate and a small Bey Dağları microplate in the Cretaceous (Fig. 5d). In the latest Cretaceous (Campanian–Maastrichtian) the ophiolites of the 'Turkish area' were thrust from a northerly Neotethyan ocean basin, leaving a collage of continental blocks to collide with Eurasia in the Early Tertiary. In summary, the key differences between Models 1 and 2B are the presence, versus absence, of a Permian–Triassic south Tethyan oceanic basin system, and the contrasting inferred origins of the 'Greek' and 'Turkish' ophiolites (i.e. Model 1, partly southerly; or Model 2, entirely northerly derived).

Model 3: southward-subduction model

Şengör *et al.* (1984) envisaged the south margin of Eurasia as being passive in the Late Palaeozoic–Early Mesozoic, whilst the northern margin of Gondwana was active (Fig. 2c). Southward subduction beneath the Gondwana margin lead to rifting of a single elongate Gondwana-derived continental fragment, Cimmeria, that then drifted northward opening a new (i.e. Neo-Tethyan) back-arc basin as the older (i.e. Palaeo-Tethyan) ocean was subducted (Figs 3b & 4b). Eventually, the rifted Cimmerian fragment collided with the Eurasian passive margin, followed by 'Tibetan-type' magmatism, (Şengör *et al.* 1994). Prior to this collision, a backarc basin opened within the Cimmerian fragment (Karakaya basin) and then closed during collision. Renewed rifting then took place within the Palaeo-Tethyan suture to open a new Neotethyan ocean basin in the Early Jurassic (i.e. Intra-Pontide ocean). Rifted microcontinental fragments in the south Tethyan region (e.g. Menderes, SW Turkey) were separated by Triassic Neotethyan oceanic crust, of backarc origin. Most of the ophiolites were of supra-subduction zone type, rather than mid ocean ridge type (e.g. Troodos). A northerly (internal) origin was accepted for the Jurassic ophiolites of the 'Greek area', while the Cretaceous ophiolites of the 'Turkish area' were seen as having been emplaced from several different ocean basins (e.g. Ankara–İzmir–Erzincan, Inner Tauride and Intra-Pontide; Şengör & Yılmaz 1981). Cretaceous–Early Tertiary arc magmatism documented the active margin history of the south margin of Eurasia (Şengör *et al.* 1991). On a much wider scale Şengör (1992) has gone on to identify Palaeo-Tethyan units and sutures throughout Asia.

Each of these models involves movement of tectonic units (whether ophiolitic or microcontinental) relative to the major bounding plates of Gondwana and Eurasia. Palaeomagnetism has played a key role in documenting the motion histories of the major plates. Under ideal circumstances, palaeomagnetic apparent polar wander paths for microcontinental units could be compared to those of the bounding continents, in order to clarify the history of separation, motion and docking. Unfortunately, it has proved impossible to define apparent polar wander paths for most such units, because of a lack of suitable lithologies of a range of ages for palaeomagnetic study. However, in several cases inclination data have been used successfully to determine the proximity of individual tectonic units to the palaeo-position of the bounding plates (i.e. Gondwanan versus Eurasian affinities). For example, Sarıbudak *et al.* (1989) obtained palaeolatitudes within Lower to Middle Triassic units of the western Pontides which were consistent with those expected along the southern edge of Eurasia (in line with the location of the Pontide units in Models 1 and 2).

It should also be mentioned that it has been suggested that some, or all, of the tectonically transported units within the Eastern Mediterranean region might be allochthonous terranes that originated outside the area. For example, Lauer (1984) used palaeomagnetic data to suggest that Turkey consists of three blocks (Pontides, Western Taurides and Eastern Taurides) which were originally located off the southeastern Arabian margin during the Triassic and which moved independently northwestwards until Neogene collision. Subsequent studies within Lauer's Western Tauride block (in SW Turkey) have documented either Neogene remagnetization (also affecting some of the units sampled by Lauer (1984); Morris & Robertson (1993)), or apparently primary Late Triassic magnetostratigraphies which indicate differing palaeolatitudes of 14°N (from the Bölücetktaşı section (Gallet et al. 1992), which is consistent with a location in the 'Gondwanan embayment') and 17°S (from the Kavur Tepe section (Gallet et al. 1993), which suggests a location close to the northern margin of India during the late Triassic). There is no obvious geological evidence for such contrasting origins of these units. Clearly, further detailed palaeomagnetic studies of the Mesozoic Tauride successions are necessary to resolve this controversy.

Current controversies and new evidence

A number of important disagreements still exist between the main alternative tectonic models outlined above. Key points to be resolved include the following. (i) Was the south Eurasian margin active or passive in the Late Palaeozoic–Early Mesozoic? (ii) Were the south Tethyan Triassic basins intra-continental rifts, Red Sea-type small ocean basins, or backarc basins? (iii) Did the large Mesozoic ophiolites form at mid-ocean ridges, or by spreading above subduction zones? (iv) What was the size, distribution and timing of rifting of Gondwana-derived continental fragments?

Over the last decade important new information on these, and other, questions has become available, especially for northern and western Turkey, northeastern Greece, Albania and the former Yugoslavia. Some implications of these results for alternative tectonic interpretations are now outlined. Some of this evidence has recently been summarized in the context of definition and recognition of tectonic facies exposed in the Eastern Mediterranean area (Robertson 1994).

South Eurasian margin: active or passive?

In Models 1 and 2, the southern margin of Eurasia was episodically active, with northward subduction along all, or part, of its length in the Eastern Mediterranean region from Late Permian to Early Tertiary time (Fig. 2a & b). Specifically, in Model 2B northward subduction was inferred for the entire Late Permian to Early Tertiary period, except possibly for the Late Jurassic. In Model 3, the Eurasian margin was passive in the Late Palaeozoic–Early Mesozoic, but was active in the Cretaceous–Early Tertiary, with the south margin of Eurasia as the site of rift volcanism in the Early Jurassic (Şengör & Yılmaz 1981). In line with Models 1 and 2, Stampfli et al. (1991) stressed the role of ophiolitic melange (e.g. Karakaya; Tekeli 1981) and olistostromes (Weidmann et al. 1992) as evidence of active margin processes associated with an active Eurasian margin (Figs 2d & 3d). All of the region north of their Pindos ocean was seen as being composed of accretionary terranes of Gondwanian origin (e.g. Serbo-Macedonian zone). It was specifically envisaged that melange in the island of Chios (NE Aegean), including Palaeozoic exotic blocks in a partly Permian clastic matrix, represented evidence of a Palaeotethyan subduction-accretion complex (Papanikolaou & Sideris 1983; Baud et al. 1991).

Discussion. Much of the detailed evidence for the Late Palaeozoic–Early Mesozoic tectonic history of the Eurasian margin is to be found in the central Pontide area of northern Turkey (Fig. 1). The structure of this area was initially mapped and interpreted in terms of southward-subduction, as in Model 3 (Tüysüz 1990). A re-investigation then led to a re-interpretation (Ustaömer & Robertson 1993, 1994, in press). What was interpreted in the southward subduction model (Model 3) as the main Palaeo-Tethyan suture, was reinterpreted as a latest Palaeozoic–Early Mesozoic backarc basin (Küre basin), remnants of which can also be found in other areas, including the Crimea (Ustaömer & Robertson 1993). A unit previously seen as a Palaeo-Tethyan ophiolite was identified as an oceanic volcanic arc unit, with an oceanic basement (Çangaldağ unit). Remnants of the main Palaeotethys (Permo-Triassic?) were represented by a major north-dipping tectonic slice complex, including blueschists, and were interpreted as a subduction-accretion complex. In addition, a Permian carbonate platform unit in the south (Kargı unit) was seen as an accreted Gondwana-derived fragment. In summary, the evidence from the central Pontides and adjacent areas, including the Crimea, Caucasus and Dobrogea is compatible with Models 1 and 2, involving mainly northward subduction in the Late Permian–Triassic. A more complicated picture, however, emerges from study of associated melange terranes in the northwestern Pontides.

A large tract of central northern Anatolia, stretching from the Aegean coast to eastern Turkey, is composed of mainly low-grade meta-sedimentary and meta-volcanic rocks, forming a melange terrain known as the Karakaya Complex (Karakaya Formation of Bingol *et al.* 1973). Meta-ophiolites are present on the island of Lesbos (Lesbos ophiolite) and the adjacent mainland (Denizgören ophiolite). In the southerly dipping subduction model (Model 3), the Karakaya Complex was interpreted as a Triassic marginal basin that opened within the Cimmerian crustal fragment, as it migrated across Palaeotethys, and then closed by the Early Jurassic. The key area to test this hypothesis is the type area of the Karakaya Complex in northwestern Turkey. Okay *et al.* (1991) mapped much of this area, including the Biga Peninsula, and established a tectono-stratigraphy of low-angle thrusted units, which they interpreted generally in line with Model 3 (i.e. as a backarc basin). They also observed that thrust vergence was generally northwards.

Much additional work was recently completed on this area (Pickett 1994; Pickett *et al.* in press). The Karakaya Complex shows the characteristics of an accretionary prism related to subduction. The accreted units include: basalts and volcaniclastics of within-plate type; basalts of mid-ocean ridge type overlain by radiolarites and turbiditic sandstones; and Permian shallow-water limestones associated with terrigenous sediment. The accretionary prism is unconformably overlain by little-deformed siliciclastic and shelf carbonates of Late Triassic to Jurassic age. The large thrust sheets of ultramafic rocks (represented by the Lesbos and Denizgören ophiolites) exhibit local metamorphic soles and underlying tectonic melange (on Lesbos) and were emplaced onto Permian carbonate platforms. Structural evidence, including outcrop-scale shear fabrics and folds, and evidence of large-scale thrusting of units (including the ophiolites) indicate an original direction of thrusting towards the north. In summary, the recent evidence from the Karakaya Complex from northwestern Turkey supports its interpretation as a subduction-accretion complex, as in Models 1 and 2, rather than a backarc basin origin (Model 3). However, it appears that this subduction was southwards, rather than northwards (as in Models 1 and 2). This is not in itself evidence in support of Model 3, however, as in this case the Palaeo-Tethyan subduction zone was seen as being located much further north, in the Pontides–Caucasus region (Fig. 1). Instead,

the new evidence implies that, while subduction of Palaeotethys can be seen as mainly northwards in the Late Palaeozoic–Early Mesozoic (based on evidence from the central Pontides), some amount of southward subduction also took place, at least in the western Pontides. The nature of this subduction remains unclear. It is possible, however, that southward subduction was initiated relatively locally to accommodate rapid northward drift of Gondwana-derived continental fragments (i.e. Taurides–Anatolides), as oceanic basins opened further south in the Triassic (e.g. Antalya, Mamonia).

The Jurassic history of the Eurasian margin in the Pontides is particularly controversial. In Model 2B the Eurasian margin was again active by the Early Jurassic, with supra-subduction zone rifting from the Dobrogea eastwards. This includes volcanism in a possible backarc basin west of Moesia (Sandulescu 1989). By contrast, in Model 3 the Eurasian margin rifted in the Early Jurassic to form an Intra-Pontide ocean to the south, followed by passive margin subsidence. There is a general consensus that northward subduction then took place in the Cretaceous–Early Tertiary, and continued until stopped by collisions of Gondwana-derived continental units with the Eurasian active margin.

South Tethyan evolution

It is generally agreed that southerly areas of the Tethyan orogenic belt in the Eastern Mediterranean are dominated by crust by Gondwanian origin. However, disagreements remain particularly as to the timing of rifting, size and distribution of microcontinents; as to whether intervening basins were floored by stretched continental crust or oceanic crust (or both); and also as to the settings of ophiolite genesis and emplacement.

Permian rifting and spreading

Each of the Models 1–3 identified the Permo-Triassic as a time of important rifting of microcontinents from Gondwana. In Model 1 (Fig. 3a), rifting of external units took place in the Permo-Triassic in the northerly (i.e. more internal) units (e.g. Vardar), but not until the Triassic in southerly (i.e. external) units (e.g. Pindos zone and Antalya). In Model 2A, rifting to create Neotethys took place in the Triassic, wholly within Palaeotethys, close to its Gondwana margin (Fig. 2b). However, in Model 2B (Fig. 3c), Permian deep-water sediments in the central Aegean area (i.e. on Hydra, E Peloponnese) were taken as evidence of an early oceanic connection (Pindos ocean of Model 2B) through the central Aegean from a Permian Neotethys to the east. The introduction of this connection represented a considerable departure from the previous concept (Model 2A) of a continuous and intact Gondwana margin in an area comprising the whole of Greece and the former Yugoslavia region (i.e. a large part of Adria). Permian deep-water sediments of Crete and S Peloponnese (i.e. Phyllite–Quartzite unit) were seen as derived from this Pindos ocean. To reach their present southerly position substantial southward thrusting was envisaged. In strong contrast, in Model 3 rifting of the entire north Gondwana margin in the Eastern Mediterranean took place in the Triassic to form a backarc basin system (Fig. 3b).

Discussion. A number of lines of evidence indicate that important rifting *did* take place along the north margin of Gondwana in the Late Permian, earlier than previously envisaged (Stampfli *et al.* 1991; Robertson *et al.* 1991). Doming of Permian age in the Levant (Helez) and eastern Turkey (Hazro) may record rift-related uplift of thermal origin (Gvirtzman & Weissbrod 1984). Seismic interpretation and evidence from wells along the north margin of Africa (e.g. in Tunisia, Libya) and the Levant (e.g. Palmyra rift) document Late Permian rifting (Garfunkel & Derin 1984; M'Rabat *et al.* 1989; Burolet *et al.* 1978; May 1991). Deep-water sediments of Permian age, in Sicily, Crete and the SE Peloponnese also imply the existence of a deep basin adjacent to the present north African margin (Kozur 1991; Krahl 1992). The presence of a cosmopolitan radiolarian fauna indicates that this basin had an open-marine connection from east to west. Two alternatives are that: (i) rifting in the Late Permian gave rise to a broad rift floored by stretched continental crust, with actual spreading delayed until the Triassic (Robertson *et al.* 1991); or (ii) that spreading began in the Late Permian, forming a small ocean basin along the northern margin of Gondwana, and was followed by thermal subsidence in the Triassic (Stampfli *et al.* 1991). By contrast, Kozur (1991) envisaged a wide southerly Pindos oceanic basin, accommodating most of the gap between Africa and Eurasia in the Permo-Triassic. This seems unlikely, however, especially as ophiolites (or other evidence of oceanic crust) are not known to be associated with the Permian outcrops (e.g. the Phyllite–Quartzite unit). In summary, it is concluded that a large continental fragment rifted from Gondwana in the Late Permian, and was separated from Gondwana by a broad rift, or narrow ocean basin, running along the North African margin to the Levant.

Triassic rifting and spreading

There is widespread evidence of rifting in the Early–Mid-Triassic, mainly from allochthonous units, including the Budva (former Yugoslavia), the Cukali (Albania), the Pindos and the Sub-Pelagonian (Greece), SE Aegean (Harbury & Hall 1988) the Antalya and Lycian (SW Turkey), Tauride (and Turkey) and Koçali (E Turkey) units. In Model 1, rifting was followed by genesis of small oceanic basins of Red Sea type in the Late Triassic (Carnian–Norian), with genesis of mid ocean ridge-type basalt (e.g. in Pindos; Jones & Robertson in press). Spreading was active (e.g. Pindos, Antalya), both in westerly and southerly (i.e. external basins), and in more northerly (i.e. internal) oceanic areas (e.g. Vardar zone and N of Tauride platforms). In the southerly (external) basins, spreading was more extensive in the westerly 'Greek area' (e.g. Pindos ocean) at this stage relative to the 'Turkish area' (e.g. Antalya; Fig. 3a). This is suggested particularly by the existence of substantial subduction-accretion complexes recording closure of a wide Mesozoic–Early Tertiary Pindos ocean in the 'Greek area', as in Model 1 (Jones & Robertson 1991, in press; Degnan & Robertson 1990, in press). Alternatively, similar-sized oceanic basins might have formed in both areas in the Late Triassic (Robertson *et al.* 1991; Fig. 4a). Support for the existence of a westerly (external) ocean basin in the 'Greek area' has come from numerous authors, including Karamata (1988) for Serbia, Mercier & Vergely (1986), Smith (1993) and Doutsos *et al.* (1993) for Greece, Poisson (1984) for SW Turkey and Fourcade *et al.* (1991) for SE Turkey. By contrast, in Model 2A these units were all rooted in a single northerly Neotethyan ocean basin. In the derivative Model 2B, a Kırşehir 'transit plate' was rifted from Gondwana in the Early Triassic. Also in Model 2B, the Jurassic Svoula flysch in NW Greece was seen as having accumulated within a Vardar (Axios) ocean near the western margin of a Drama transit plate (Fig. 3c). In Models 2A & 2B tectonic movements in the southern areas (e.g. North African margin) were restricted to block faulting and formation of rift basins, including the Ionian trough in Greece and Albania and the Kızılıca Çorak Göl basin in SW Turkey (Poisson 1984). All such rifting in the Permian–Jurassic was merely a precursor to spreading in the Cretaceous. In Model 3 (Fig. 3b), southerly back-arc basins (e.g. Antalya, Mamonia) opened in the Triassic to accommodate the full separation of Africa and Eurasia, as Palaeo-Tethys was eliminated to the north.

Discussion. Recent field and geochemical evidence strongly supports interpretation of the Pindos zone (i.e. represented by the Pindos-Olonos nappes; Fleury 1980) as being oceanic (Fig. 3a), rather than merely a rift basin (as in Models 2B and 3). Tholeiitic (rather than alkaline-type) volcanics are present beneath Triassic deep-sea sediments in the Peloponnese (Degnan 1992). Triassic to Early Tertiary sedimentary thrust sheets (Pindos–Olonos nappes) are much more consistently interpreted as the deep-water (i.e. thousands of metres) passive margin of Adria (Apulia), rather than as relatively shallow-water rift sediments (Degnan & Robertson 1990). Structural unstacking of the Pindos–Olonos nappes indicates that at least 300 km of basement have disappeared; this is most easily explained by subduction of oceanic crust within a Pindos oceanic basin in the Early Tertiary (Degnan & Robertson in press). Similar arguments apply to the Antalya region, where basalts of Late Triassic age range from transitional to MORB type (Robertson & Waldron 1990), although the scale of the accretionary units is relatively small compared to the Pindos zone in Greece. The evidence to distinguish a Red Sea from a backarc basin origin remains uncertain, in that in the 'Turkish area' (e.g. Antalya) Triassic basalts contain no identifiable subduction component (Robertson & Waldron 1990), whereas in some parts of the 'Greek area' (e.g. Peloponnese, Pe-Piper & Piper 1990) a subduction signal is present. However, it is possible that this does not indicate contemporaneous arc volcanism but instead an inherited lithospheric mantle signature, especially as associated sediments are mainly terrigenous and are therefore unlikely to be arc derived. In addition, allochthonous units that in Models 2A & 2B had an internal (northerly) origin, in fact, show structural evidence of initial thrust displacement generally northwards from southerly (external) ocean basins. This includes the Sub-Pelagonian zone, Greece (see Smith 1993) and the NE area of the Antalya Complex (Waldron 1984). Restoration of thrust sheets in SE Anatolia also supports the existence of a southerly Early Mesozoic ocean basin in this area adjacent to the Arabian margin (Aktaş & Robertson 1990; Yılmaz 1991; Fourcade *et al.* 1991).

Related to the question of the origin of southerly oceanic basins, Dilek & Rowland (1993) supported a southerly (external) origin of the allochthonous Antalya units from within the Isparta angle area. However, they suggested that rifting to form a small oceanic basin did not take place until the Cenomanian, when the carbonate platform subsided and broke up.

This, however, contradicts the clear evidence that the deep-water passive margin successions (e.g. *Halobia* limestones and radiolarian cherts) bordering the adjacent carbonate platforms (e.g. Bey Dağları; Karacahisar) span at least Upper Triassic to locally Upper Cretaceous with no major break. Rifting and initial spreading to form these passive margins took place much earlier, mainly in the Late Triassic. The Cenomanian platform subsidence may instead relate to the inferred onset of supra-subduction zone spreading (i.e. rear subduction zone extension), possibly coupled with a near global demise of carbonate platforms around this time.

Promontories or microcontinents?

A long-standing question is whether Adria (Apulia) and the Bitlis-Pütürge massifs persisted as promontories of Gondwana (Channell *et al.* 1979), as in Model 1 (Fig. 2a), or became completely detached in the Mesozoic, as in Models 2A, 2B. Palaeomagnetic studies have so far been inconclusive (see Channell this volume).

Discussion. Recent field and seismic evidence now support the 'fully detached' interpretation (Model 2A) of Dercourt *et al.* (1986), although this separation probably dates from the Permo-Triassic rather than the Cretaceous (see Channell *et al.* 1979). In Models 1 and 3, the large Bitlis and Pütürge metamorphic massifs in SE Turkey were seen as separate microcontinents within Neotethys, while in Models 2A & 2B they were still attached to Gondwana as a large irregular promontory. Geochemical evidence of Triassic low-grade meta-lavas in the Bitlis and Pütürge massifs favours rifting from Gondwana (Fig. 4a). Also, recent work in SE Turkey allows interpretation of units located structurally between the Bitlis and Pütürge massifs and the Arabian foreland beneath as remnants of a sutured Mesozoic ocean basin; this separated the Bitlis and Pütürge microcontinents from Gondwana during the Mesozoic (Aktaş & Robertson 1990; Yılmaz 1991; Fourcade *et al.* 1991).

Cretaceous spreading of southerly basins

In Model 1, further, spreading is envisaged to have taken place in the southern part of the Eastern Mediterranean region in the Upper Jurassic-Lower Cretaceous (Fig. 5a). This widened pre-existing, relatively narrow, southerly Triassic ocean basins (e.g. Antalya; Mamonia). Space was created for this spreading by partial closure of the Pindos and Vardar oceans.

In Model 2A (Fig. 5c), by contrast, separation of a Gondwanian continental fragment is envisaged as occurring entirely within the Cretaceous to form the present Eastern Mediterranean oceanic basin. In Model 2B (Fig. 5d) the Serbo-Pelagonian block rifted into several smaller fragments, possibly including a small Parnassus platform in central Greece. Space was created by subduction of the Vardar (Axios) ocean (north of the Serbo-Pelagonian block).

Discussion. There is, indeed, some evidence of volcanism in the Late Jurassic–Early Cretaceous in the southerly 'Turkish area'. This includes: (i) tholeiitic volcanism of Late Jurassic–Early Cretaceous age in the Antalya region of SW Turkey (Robertson & Waldron 1990); (ii) reactivation of passive margin units, with rift related alkaline volcanism in the Baër-Bassit region of northern Syria (Delaune-Mayere 1984); (iii) increased rates of subsidence of the deep-water passive margin of the Levant and related volcanism, as evidenced from seismic, well and outcrop studies (e.g. Mart 1987). Such volcanism marks a reactivation of pre-existing passive margins, rather than initial rifting. Related uplift may have been the cause of an unconformity near the Jurassic–Cretaceous boundary in the subsurface of the coastal plain.

In an extreme alternative view no spreading took place, even in the Cretaceous, and the Eastern Mediterranean is entirely floored by continental crust (Hirsch 1984; Hirsch *et al.* 1995). All the allochthonous units were thrust a long distance from the northern margin of Gondwana, which would be located near the Black Sea. However, this is highly unrealistic, especially as it would require present northward subduction in the south Aegean to be entirely fuelled by subduction of continental crust. This interpretation of a continentally-floored Eastern Mediterranean Sea can safely be discounted and is not considered further here.

Origin of southerly ophiolites

In Models 1 & 3, the large Mesozoic ophiolites, including those of Jurassic age in the 'Greek area' and Upper Cretaceous age in the 'Turkish area', were seen as being of supra-subduction zone origin, whereas in Models 2A and 2B these ophiolites formed at 'normal' spreading ocean ridges (Dercourt *et al.* 1986, 1993). This issue clearly has a profound influence on the location and nature of inferred plate boundaries.

Discussion. The field evidence is most compelling in the case of the Jurassic 'Eastern ophiolite' of Albania (Shallo *et al.* 1990), where a complete ophiolite succession is overlain by calc-alkaline extrusives and is cut by associated intrusives (e.g. trondhemites). Elsewhere, this upper, arc-type unit has not been reported and evidence supporting an above-subduction zone genesis is mainly geochemical (e.g. Pindos ophiolite; Jones *et al.* 1991). In the case of the Upper Cretaceous Troodos ophiolite (recent review: Robertson 1993), the extrusives include unaltered volcanic glass, on which whole-rock geochemical and microprobe analysis indicate classic orogenic andesitic compositions (Robinson & Malpas 1990). In addition, highly 'depleted' extrusives are locally present that have been compared to the high-magnesium andesites (or boninites) from SW Pacific forearcs. Interpretation of other ophiolites in the Eastern Mediterranean is mainly based on 'immobile element' geochemistry. This utilises elements that have not changed during weathering and low-grade metamorphism (Pearce *et al.* 1984), and also microprobe analysis of plutonic ophiolitic rocks (e.g. chromites).

Applying these criteria to the Eastern Mediterranean ophiolites as a whole, subduction-related origins are identified for the majority of the large ophiolites, including the northerly Elekdağ ophiolite of the Central Pontides; (Ustaömer 1993; Ustaömer & Robertson in press) and the Denizgören and Lesbos ophiolites in the northeastern Aegean (Pickett 1994). This origin also applies to many of the Jurassic ophiolites, including Pindos, Vourinos, the 'Eastern-type' ophiolite of Albania and those in the Southwestern Zone of Serbia (Robertson & Karamata 1994). Supra-subduction zone settings are also identified for most of the Upper Cretaceous ophiolites of the 'Turkish area', including the Lycian, Mersin, Troodos, Hatay, Baër-Bassit, Koçali and Guleman ophiolites (Robertson *et al.* 1991). These ophiolites are of 'pre-arc type', as a related arc is not present. The reason for this is that subduction was never sufficiently extensive for the early stages of ophiolite genesis above a subduction zone to be followed by genesis of a thick arc unit above, as is seen in the SW Pacific region. Comparisons of the Tethyan ophiolites with modern settings were recently discussed elsewhere (Robertson 1994).

By contrast, several of the (generally smaller) Neotethyan ophiolites are geochemically of mid-ocean ridge type (Robertson *et al.* 1991; Jones & Robertson in press). Such ophiolites formed during spreading, especially in the Late Triassic. Most of this crust was later destroyed. However, remnants remain: notably the intact

'Western-type' ophiolites of Albania (Shallo *et al.* 1990); the more fragmentary ophiolites in the Othris (Sipetorrema Lava) and Pindos (Avdella Complex) mountains; and extrusives of inferred ophiolitic origin in SW Cyprus (Dhiarizos Group).

A third type of ophiolitic setting, that of the backarc marginal basin lying behind a fully developed arc, is represented by the Late Triassic–Early Jurassic Küre ophiolite of the Central Pontides, which developed behind the inferred Çangaldağ arc (Ustaömer & Robertson 1993). Another example is the Mid–Late Jurassic Guevgeli Chalkidiki and Oreokastro ophiolites (Haenel-Rémy & Bebien 1985) marginal basin that probably opened behind an inferred Jurassic Paikon arc (Bebien *et al.* 1980; Brown & Robertson in press *b*), possibly as a series of discontinuous pull-aparts. An alternative, less likely, suggestion is that the Chalkidiki ophiolite in this area opened above a westward–dipping subduction zone, behind a 'Chortiatis' volcanic arc of Mid–Late Jurassic age (Mussallam 1991). However, it now appears that the Guevgeli, Chalkidiki and Oreokastro ophiolites, together with the meta-igneous Volvi Complex form remnants of a related intracontinental marginal basin.

An additional type of small oceanic basin is represented by the Latest Jurassic–Cretaceous Meglenitsa basin of the Vardar (Axios) zone in NW Greece; this appears to have developed as a pull-apart basin following suturing of the Early Mesozoic Vardar (Axios) oceanic basin in this area (Sharp 1994; Sharp & Robertson in press *b*).

Inner orogenic units

Additional important terrains, dominated by ophiolites or ophiolite-related rocks of Mesozoic age, are located between the southern margin of Eurasia and Gondwana-derived microcontinental units located further south. These units include the Vardar (Axios) Zone in Greece, and a number of potential suture zones in Turkey, including the Intra-Pontide suture, Ankara–İzmir–Erzincan suture and the Inner Tauride suture (see Şengör & Yılmaz 1981).

In Turkey, the northern margins of the leading Gondwana-derived continental fragments (e.g. Menderes) were seen in Model 1 as Palaeotethyan passive margins, facing into Palaeotethys and later into Neotethys (Fig. 3a). The Late Palaeozoic reconstruction implied that the margins were as old as Palaeotethys itself. In Model 1, Palaeotethys remained open and continued to evolve into the Mesozoic as a Neotethyan ocean strand, only closing finally in

the Late Mesozoic–Early Tertiary (Fig. 4b). During the Late Jurassic–Early Cretaceous, the inferred successor to Palaeotethys, the Vardar (Axios) zone in Greece, was seen as undergoing incipient collision and left-lateral mega-shear, but remained essentially open as an oceanic basin until the latest Cretaceous–Early Tertiary.

In Model 2 (i.e. single Mesozoic Tethys), renewed spreading took place during the Late Permian-Triassic at mid ocean ridges located within Palaeotethys near the northern margin of Gondwana to form a Neotethyan ocean (Figs 2b & 3c). There is thus the same problem as in Model 1, that older Palaeotethyan margins of Gondwana are not preserved, for example in allochthonous units thrust to the south (e.g. Lycian and Beyşehir–Hoyran nappes). In Model 2B, the Cretaceous history of the Vardar zone is admitted to be unclear. It is suggested that a marginal sea might have formed and then closed between the Pelagonian zone and the Serbo-Macedonian zone, although this is inconsistent with the concept of a single Serbo-Pelagonian block for Late Permian–Jurassic time. On the other hand, Dercourt *et al.* (1993) acknowledge that the Vardar ocean might have remained open as the main separation between Eurasia and Adria (Apulia), as in Model 1.

In Model 3, several different sutures were envisaged as being entirely of Mesozoic age (i.e. Neo-Tethyan). Palaeo-Tethys was a separate, earlier, ocean basin located further north. Of these Neo-Tethyan ocean basins, those in the south were seen as part of the Triassic back-arc basin system that opened above a southward-dipping subduction zone (İzmir–Ankara–Erzincan ocean; Fig. 3b). Eastwards, this ocean basin split into two strands separated by the Kirsehir Massif, with the Inner Tauride ocean to the south and an eastward extension of the Ankara–İzmir–Erzincan ocean in the north. Further north, the Palaeo-Tethyan suture was almost immediately split by rifting in the Early Jurassic, following collision of the Cimmerian fragment with Eurasia, to form a new small ocean basin (Intra-Pontide ocean; Şengör & Yılmaz 1981).

A further alternative model is that a subduction zone dipped southwards from a remnant Palaeotethys, opening up a Cretaceous backarc basin in the Eastern Mediterranean Sea area, including the Troodos ophiolite (Dilek *et al.* 1990). This scenario is similar to Model 3, except that backarc spreading took place in the Late Cretaceous, rather than the Triassic. This is, however, not supported by evidence from the southerly terrains (e.g. Mamonia, Cyprus) that genesis of mid-ocean ridge type extrusives in these areas took place in the Triassic. In addition, where Upper Jurassic–Early Cretaceous oceanic volcanics have been identified, as in the northeastern area of the Antalya Complex they lack any identifiable subduction component (Robertson & Waldron 1990). In addition, as yet there are no structural data from Late Cretaceous accretionary terrains of central Anatolia (i.e. Ankara Melange) to support southward subduction along the north margin of the Tauride–Anatolide carbonate platforms in the Late Cretaceous.

Discussion. Recent studies have not confirmed the existence of very old (i.e. pre-Permian) passive margins along the northern margin of the Tauride–Anatolide carbonate platforms and related allochthonous units in Turkey, as implied by Models 1 & 2. Instead, the available evidence documents rifting in the Late Permian (e.g. Teke Dere unit of the Lycian nappes) and the Triassic (e.g. north margin of the Bolkar Dağ). On the other hand, fragments of Late Permian carbonate platforms of Gondwanian affinities have been identified in the central Pontides. The implication is that the presently preserved inner orogenic units of central and northern Turkey (i.e. Anatolides) date only from the Late Permian–Triassic, rather than from an earlier Palaeotethys. This observation is consistent with Model 3, in which backarc rifting accompanied the demise of Palaeo-Tethys further south. However, a Permian–Triassic rifting history could be explained in several other ways. First, the Pangean assembly might have evolved by large-scale strike-slip to yield essentially new margins in the Permo-Triassic; i.e. that Palaeotethys was itself newly created in the Permian. This is, however, difficult to reconcile with the long-lived active margin history of Eurasia, as documented in Models 1 and 2B (e.g. preserved in the Pontides). Secondly, the lack of a Late Palaeozoic passive margin along the northern margin of the Tauride–Anatolide carbonate platform might indicate an earlier history of rifting and fragmentation of Gondwana, with microcontinents drifting across Palaeotethys to be amalgamated with Eurasia, and with corresponding renewal of the southern Tethyan margin. Multiple Palaeozoic–Mesozoic rift events have, indeed, been reported from Gondwanian-derived units in the Taurides (Demirtaşlı 1984). Also, the Palaeozoic of Istanbul in the NW Pontides could represent a Gondwanian fragment that was amalgamated with Eurasia in pre-Late Carboniferous time. The Serbo-Macedonian zone in northern Greece could have a similar history.

In summary, we infer that the northward margin of Gondwana episodically fragmented during the Palaeozoic, with microcontinents crossing Palaeotethys to become amalgamated with Eurasia. Mesozoic (i.e. Neotethyan) oceanic crust formed within the remnant Tethys by spreading, both at mid ocean ridges and above subduction zones. The history of oceanic spreading in the Late Palaeozoic is poorly understood, but may include genesis of the Lesbos and Denizgören ophiolites in the NE Aegean (mentioned earlier). Oceanic crust was formed above subduction zones in the inner orogenic zones in the 'Turkish area', at least in the Late Cretaceous, and was later preferentially preserved as emplaced ophiolites (e.g. Lycian ophiolites). By contrast, mid ocean-ridge type crust of Palaeozoic and/or Mesozoic age was almost entirely subducted and is preserved only as fragments in subduction/accretion complexes (Fig. 6a).

Lateral correlations

One long-standing question relates to whether Tethys in the 'Turkish area' can be traced through into the 'Greek area' at any given time. Until now correlations across the Aegean region have remained unconvincing (e.g. correlation of the Pelagonian zone and Menderes massif). In part, this is because of limited exposure. However, it should also be noted that some of the major tectonic events in the Greek and Turkish areas, including the timing of creation and genesis of ophiolites, were not synchronous and thus 'continuity' as it appears at present was either never present or is likely to have been obscured.

Discussion. A particular case is whether the Vardar (Axios) zone in northern Greece represents an extension of the Palaeozoic–Early Mesozoic Tethys in the northerly (internal) units of the 'Turkish area'. From evidence in Serbia, Karamata *et al.* (1992) argued for a continuation of the Palaeozoic Tethys westwards into the Vardar Zone in this area. However, in northern Greece the inference that the Vardar (Axios) zone could represent Palaeotethys, as in Model 1 (Fig. 3a), has been criticised in view of the apparent absence of Palaeotethyan oceanic units (e.g. ophiolites). Smith (1993) regarded the Almopias (Vardar) ocean as a relatively minor oceanic basin, in contrast to a much wider inferred Mesozoic Pindos ocean further southwest (in present co-ordinates). Ferrière & Stais (1994), Dimitriadis & Asvesta (1993) and Sharp (1994) interpreted the Vardar (Axios) zone as a Triassic–Jurassic small ocean basin of Red Sea type, although earlier (i.e. Late Permian) rifting cannot be entirely ruled out. Alternatively, the main Late Palaeozoic (Palaeotethyan) suture could lie within a stack of metamorphic nappes in the Rhodope massif, but be largely obscured by deformation (Sandulescu 1989). A potential candidate is a zone of highly strained, eclogitized mafic rocks and serpentinites at the base of the Lower Arda 2 unit in Bulgaria (Burg *et al.* 1990). If so, the Serbo-Macedonian zone could have rifted from Gondwana in the Late Permian, and drifted across the Late Palaeozoic Tethys to open a wide Permo-Triassic Vardar ocean, prior to collision and amalgamation with the Eurasian margin. The Rhodope zone further north is generally correlated with Eurasian continental crust of the Moesian platform (at least from Late Palaeozoic time onwards), for example, as exposed in the Strandja massif (see Şengör *et al.* 1984) and the (e.g. Armutlu Peninsula (Yılmaz 1991). Further support for a relatively wide, Early Mesozoic Vardar ocean comes from the Paikon arc in the centre of the Vardar (Axios) zone in NE Greece. This includes thick meta-volcanics and related volcaniclastics of Late Jurassic age (pre-Kimmeridgian–Tithonian; Mercier 1968) that overlie continentally derived meta-sedimentary lithologies in the Paikon unit. There is general agreement that these volcanics relate to subduction during the Mid-Jurassic time (Baroz *et al.* 1987; Bebien 1994; Brown & Robertson in press *b*). One interpretation is that these volcanics were generated above a subduction zone dipping eastwards from the Pindos ocean (Smith 1993). However, it is more probable that the subduction zone was located actually within the Vardar (Axios) ocean, implying that a significantly wide ocean existed (>500 km?). The Guevgueli, Chalkidiki and Oreokastro ophiolites (Eastern Vardar–Axios zone) are identified as the remains of a single, narrow and effectively *in situ* Jurassic marginal basin that formed above an eastward-dipping subduction zone at the southwestward margin of the Serbo-Macedonian zone.

Closure of the Vardar ocean

The classic view is that the Vardar (Axios) ocean finally closed in the Late Jurassic, with associated westward emplacement of ophiolites onto the Pelagonian zone (Aubouin *et al.* 1970). Similarly, Ziegler (1990), while not being concerned with the Eastern Mediterranean in detail, nevertheless was of the view that Tethyan oceanic crust had disappeared throughout the entire area (including Turkey) by the Early Cretaceous, with the Late Cretaceous ophiolites of the 'Turkish area' forming in local pull-apart basins. Evidence of the existence of Cretaceous extension and volcanism in the central Aegean (e.g. Skyros; Jacobshagen & Wallbrecher 1984) was taken to indicate re-opening of the Vardar (Axios) ocean in the Cretaceous, followed by final closure in the Early Tertiary.

Discussion. Recent work in the westerly Vardar (Axios) zone of NE Greece confirms that westward ophiolite emplacement onto the Pelagonian zone did indeed take place in Late Jurassic (i.e. post Oxfordian to Kimmeridgian) (Sharp *et al.* 1991; Sharp 1994), possibly related to closure of a Vardar oceanic basin in this area. This was followed by re-opening of a Late Jurassic to essentially Cretaceous small oceanic basin, represented by the Meglenitsa ophiolite (Sharp 1994; Sharp & Robertson in press *a*). Remnants of Cretaceous oceanic crust are also preserved as ophiolitic rocks derived from the Vardar (Axios) zone further south in Greece (on Euboea and in Argolis), as summarized by Robertson *et al.* (1991). However, in SE Greece, there is no evidence of closure of the Almopias ocean in the Late Jurassic, as indicated by the lack of corresponding deformation and metamorphism of the Pelagonian zone. In this more southerly Greek area, the Vardar (Axios) ocean remained open into the Cretaceous as a remnant basin, finally closing only in Early Tertiary times (Clift & Robertson 1990; Clift 1992). A remnant Vardar ocean basin probably also persisted in Serbia (Karamata *et al.* 1992), where the Pelagonian zone and equivalents (Drina–Ivanica unit) also escaped regional metamorphism associated with ophiolite emplacement during the Late Jurassic (Robertson & Karamata 1994).

In summary, it is probable that the segments of the leading edge of the Pelagonian microcontinent collided with the south margin of the Eurasia, represented by the Rhodope zone (by then augmented by the accretion of the Serbo-Macedonian block) while a remnant Vardar (Axios) ocean basin remained open in adjacent areas (as in Model 1). Ocean crust was then created in the Cretaceous, both within the local suture zone and the remnant Vardar (Axios) ocean (e.g. Meglenitsa ophiolite). The driving force was probably strike-slip motion between Gondwana and Eurasia during Late Jurassic–Late Cretaceous time (e.g. Livermore & Smith 1984). Remnants of Mesozoic oceanic crust located to the north of the Gondwana-derived continental fragments in the 'Turkish area' are preserved in the Ankara–İzmir–Erzincan and Inner-Tauride zones (Fig. 1; Şengör & Yılmaz 1981; Okay & Siyako 1993; and our unpublished data).

Another key question further south is whether the Mesozoic Lycian nappes, including ophiolites, were rooted to the north or to the south of the Menderes Massif, which is generally agreed to be a Gondwana-derived continental fragment. A northerly origin was assumed in each of Models 1, 2 and 3. However, some workers have argued that the dominantly Mesozoic carbonate thrust slices were either wholly or partly rooted to the south, as represented by the Koycegiz nappe of Poisson (1984), and the Köycegiz and Elmalı thrust sheets of Özkaya (1990). These workers postulated the existence of a rifted basin (i.e. Kızılca–Çorak göl; Poisson 1984) separating two continental fragments, now represented by the Menderes metamorphic complex in the NW and the Bey Dağları carbonate platform to the SE. Kinematic evidence from units beneath the emplaced nappe stack, from within the body of the thrust sheets, and from the sole of the overriding Lycian ophiolites all indicate that the Lycian nappes (i.e. upper units) were thrust from the west to northwest of their present location (Okay 1989; Collins & Robertson 1995). Recent structural evidence, and analysis of the sedimentary processes preserved in the Lycian sedimentary thrust sheets additionally support the existence of a rift basin between the Menderes Massif and the Bey Dağları carbonate platform, as shown in Model 2B (Fig. 5d).

Restoring the Lycian nappes and related units (e.g. Beyşehir–Hadim nappes) to a northerly position leads to their interpretation as passive margins and oceanic units facing northwards into the Mesozoic Tethys ocean. Evidence of Permian rifting is preserved in the sedimentary Lycian nappes (i.e. at Teke Dere; De Graciansky 1972; Collins & Robertson, unpublished data), and there is also extensive evidence of Triassic rifting along the north margin of the Tauride–Anatolide carbonate platforms (e.g. Bolkar Dağ). Such rift events could represent detachment and northward drift of continental fragments (e.g. Kırşehir), as in Model 1. During regional plate convergence in the Late Cretaceous, Tethyan oceanic crust to the north was subducted northwards, while the Mesozoic deep-sea sedimentary cover was accreted. The subduction complex and overriding Lycian ophiolites were then emplaced southeastwards over the northern Menderes Massif in the earliest Tertiary (Palaeocene), thus allowing shallow-water carbonate sedimentation to persist further south, until further southeastward overthrusting emplaced the Lycian nappes in their final position during the Late Miocene.

Mesozoic ophiolites of at least partly Late Cretaceous age are also located further north,

within the Pontides (Yılmaz 1991). In the NW Pontides (e.g. Armutlu Peninsula), meta-ophiolites and related rocks are reported as having been thrust southwards in the latest Cretaceous (Yılmaz 1991). These ophiolites are located to the north of the Palaeotethyan Karakaya accretionary complex, discussed earlier (Fig. 1). Early Jurassic volcanics along the length of the Pontides were interpreted as evidence of rifting to form a northerly, Mesozoic small oceanic basin (intra-Pontide ocean) in Model 3 (Şengör & Yılmaz 1981). An alternative possibility is that the Early Jurassic Pontide volcanics and volcaniclastics are instead related to continuing arc volcanism, and that the intra-Pontide oceanic crust formed in a backarc basin and/or pull-apart basin within the southern margin of Eurasia. In addition, further east, in the Central Pontides, unmetamorphosed ophiolitic rocks were thrust northwards in the latest Cretaceous from a Mesozoic Tethyan ocean to the south (Tüysüz 1990; Ustaömer 1993). The available evidence is limited and contradictory and further work is urgently needed on these complex units.

In summary, oceanic regions between the Gondwana-derived fragments to the south and the Eurasian margin to the north mark the site of a Permo-Triassic Tethys ocean. In this region the Upper Permian Tethys evolved into the Mesozoic Tethys by a combination of subduction, northward drift of Gondwana-derived fragments (e.g. Kırşehir) and spreading of Mesozoic oceanic crust.

Ophiolitic root zones

In Model 1, both the Jurassic and Cretaceous ophiolites were derived from several different Neotethyan ocean basins. The Jurassic ophiolites of the 'Greek area' were mainly derived from a westerly (external) Pindos ocean and a more easterly (internal) Vardar ocean (e.g. Mountrakis 1984; Fig. 5a). Similarly, the Cretaceous ophiolites of the 'Turkish area' were derived from both southerly (e.g. Troodos) and northerly (e.g. Lycian) Neotethyan ocean basins (Okay 1989; Fig. 6a). The Late Triassic–Early Jurassic Pindos ocean basin collapsed, possibly driven by regional compression related to opening of the North Atlantic (Livermore & Smith 1984). Oceanic basins in the 'Turkish area' were not deformed until the Upper Cretaceous, related to opening of the South Atlantic.

In Model 2A (Dercourt et al. 1986), all of the ophiolites were derived from a single Neotethyan ocean basin located to the north of Gondwana (Fig. 4d). All of the Cretaceous ophiolites of the 'Turkish area' (e.g. Troodos) were derived from north of the Tauride carbonate platforms (Ricou *et al.* 1984).

Dercourt *et al.* (1993) took the view (Model 2B) that the Jurassic ophiolites of the 'Greek area' were of 'ultra-internal origin', having been derived from a Vardar ocean basin rooted to the NE of a Serbo-Pelagonian block (e.g. Papanikolaou 1989). In practice, this would mean that ophiolites of the Vardar (Axios) zone (i.e. Almopias and Guevgueli) were thrust over the Serbo-Macedonian zone, while the now more westerly ophiolites (e.g. Pindos and Vourinos ophiolites) were thrust even further, over the Pelagonian Zone as well. To explain the apparent absence of Jurassic emplacement in the 'Turkish area' a major transform offset of the obduction front was inferred in Model 2B.

In an alternative (traditional) model involving large-distance thrusting, the ophiolites were rooted in the Vardar zone, rather than to the NE of the Serbo-Macedonian zone and were then thrust over the Pelagonian zone to the west (i.e. Aubouin *et al.* 1970; Ferrière 1982; Jacobshagen & Wallbrecher 1984).

In Model 3 (Şengör *et al.* 1984; Fig. 5b), the Jurassic ophiolites of the 'Greek area' were all rooted in the Vardar (Axios) oceanic basin to the NE (i.e. from between the Pelagonian and Serbo-Macedonian zones), whereas separate origins of the Cretaceous ophiolites, both north (e.g. Lycian ophiolite) and south (e.g. Antalya ophiolites) of Gondwana-derived microcontinents were envisaged.

Discussion. The palaeomagnetic technique can only effectively discriminate between magnetisation directions from different units where they diverge by more than 5–10°. This places a lower limit of 500–800 km discernible relative palaeolatitudinal movements, equivalent to the inferred width and spacing of many Neotethyan basins. In addition, east–west movements cannot be detected palaeomagnetically. Palaeomagnetic inclination data are, thus, of limited value in distinguishing between alternative ophiolite root zones. On the other hand, structural data on ophiolite emplacement directions must be corrected for *localized* postemplacement rotations determined from palaeomagnetic declination data. *Regional* rotations should be taken into account when producing palinspastic reconstructions, but do not compromise structural emplacement data so long as both the ophiolite root zone and relative autochthons experienced equal rotation. For example, Morris (1995) demonstrated that a discrepancy between sedimentological (Baum-

gartner 1985) and structural (Clift 1990) data on the emplacement direction of the small Migdhalitsa ophiolite unit of southern Argolis, Greece, resulted from a Palaeogene relative rotation across the Migdhalitsa graben. The inferred emplacement direction corrected for this localized rotation is from (present) NE to SW, suggesting transport from the Vardar side of the Pelagonian Zone. This conclusion is not affected by removal of a later Neogene rotation (Kissel & Laj 1988; Kondopoulou *et al.* this volume) of regional extent. In the 'Greek area' most structural studies favour derivation of the Pindos, Othris and Vourinos ophiolites from the Pindos ocean, from within the Pindos zone, with emplacement northeastwards (present coordinates) onto a Pelagonian microcontinent (Robertson *et al.* 1991; Smith 1993; Doutsos *et al.* 1993). The traditional view of the ophiolites as rooted in the Vardar zone (i.e. between the Pelagonian and Serbo-Macedonian zone) is, however, applicable to some of the ophiolites emplaced along the eastern margin of the Pelagonian zone (i.e. western Almopias ophiolites; Mercier, 1968; Sharp, 1994). The 'ultrainternal hypothesis' (Model 2B – i.e. from NE of the Serbo-Macedonian zone) can now be conclusively ruled out, based on the detailed geological history of the supposed Vardar (Axios) root zone in NE Greece. Notably, within the Vardar (Axios) zone, the Jurassic Svoula flysch and the Guevgueli ophiolite cannot have been emplaced by thrusting over the Serbo-Macedonian zone. The Jurassic Svoula flysch is in depositional contact with older, Late Triassic shelf carbonates, that both formed part of the relatively autochthonous SW margin of the Serbo-Macedonian zone (Stais & Ferrière 1991; our unpublished data). The Guevgueli ophiolite and the related Volvi Complex are well established as exhibiting primary magmatic contacts with adjacent Serbo-Macedonian zone metamorphics (Dixon & Dimitriadis 1984; Remy 1984; Sidhiropoulos & Dimitriadis 1989; de Wet 1989) and are thus rooted where they now occur, within the eastern Vardar (Axios) zone.

The situation is also clear-cut in the 'Turkish area', where Mesozoic carbonate platforms exposed between outcrops of supposedly once continuous, single ophiolite nappes (e.g. the Antalya and Lycian ophiolites) include sedimentary successions that extend in age beyond the time of emplacement of those same southerly ophiolites (Şengör & Yılmaz 1981). In other words, the ophiolites could not have been transported across coeval sedimentary succession in their path without leaving any trace.

To enable such emplacement to take place, Ricou *et al.* (1979, 1984) proposed a complex model in which ophiolites were first thrust onto the Mesozoic carbonate platform in the north (i.e. Bey Dağları) in the latest Cretaceous. The northerly part of the platform complete, with its already emplaced units, was then thrust southwards in the Early Tertiary carrying ophiolites to a final southerly position (e.g. Antalya). More recent field studies have not confirmed any such major thrust discontinuity within the Bey Dağları carbonate platform (Robertson 1993). In addition, the Troodos ophiolite in Cyprus (Gass 1990; Robertson & Xenophontos 1993) clearly cannot have been thrust far over a platform to the south in the latest Cretaceous, since the deep-water sedimentary cover continues unbroken from the Campanian (Perapedhi Formation) into the Lower Tertiary (Lefkara Formation) (e.g. Robertson *et al.* 1991).

In summary, it is now clear that the (Vardar–Axios) zone in Greece and former Yugoslavia does, indeed, represent an important suture that was the site of oceanic crust of Early Mesozoic (and possibly earlier) age. Mesozoic oceanic crust existed further east in the Turkish area south of the Eurasian margin and surrounding the most northerly derived Gondwana-derived fragments (e.g. Kırşehir).

Closure of Tethys

Each of the alternative Models 1, 2 and 3 envisage progressive closure of Tethys in the Eastern Mediterranean during Late Cretaceous–Early Tertiary time (Fig. 6a & b). Some oceanic crust must have persisted into the Tertiary in view of the remaining separation between Africa and Eurasia. However, this separation is seen as being less significant in Model 2B than Models 1 & 3. Disregarding the evidence of such a continuing oceanic gap, some have argued that ophiolite emplacement, for example onto the Arabian passive margin in the latest Cretaceous, marks the final closure of Tethys, (e.g. in Eastern Turkey; Yazgan 1984). However, this hypothesis can now be discounted.

The chief difference between the three alternative models is that in Models 1 and 3, the Upper Cretaceous ophiolites of the 'Turkish area' are seen as having formed above northward-dipping subduction zones. By contrast, in Model 2 the Late Cretaceous 'Turkish' ophiolites represent evidence of renewed spreading at mid ocean ridges within a single Mesozoic Tethys. The Late Cretaceous ophiolites were emplaced when subduction zones collided with, both microcontinental (e.g. Bitlis–Pütürge) and the main

Arabian passive margins to the south (Aktaş & Robertson 1990; Yılmaz 1991). Emplacement in the east (e.g. Hatay, Baër-Bassit) was mainly in the Campanian-Maastrichtian, while in the west (e.g. Antalya) final emplacement over the marginal platforms was delayed until the Late Palaeocene–Early Eocene (Poisson 1984; Robertson 1993). Similarly, the Pindos and Vardar (Axios) oceans in the 'Greek area' were finally closed in Palaeocene–Eocene time (Robertson *et al.* 1991). As noted earlier, the Troodos was an exception in that it remained within a remnant Mesozoic Tethyan oceanic basin until uplift during Plio-Quaternary time.

Discussion. In the Greek area, in line with the 'ultra-internal' ophiolite derivation in Model 2B but in conflict with the original work in the area (Mercier 1968; Mercier & Vergely 1994), Godfriaux & Ricou (1991) argued that the Paikon unit of the central Vardar (Axios) zone represented a window into a regionally extensive Pelagonian platform, over which the Vardar (i.e. Almopias units) were thrust in the Early Tertiary. Alternatively, Bonneau *et al.* (1994) envisaged the Paikon unit as a stack of thrust sheets, also emplaced from the NE in the Early Tertiary. The implication of the latter hypothesis is that the entire Vardar (Axios) is allochthonous and derived from NE of the Serbo-Macedonian zone, as in Model 2B. However, recent detailed mapping now shows conclusively that the Paikon unit is, in fact, a single structurally coherent tectono-stratigraphic unit that originated and was emplaced entirely within the Vardar (Axios) zone (i.e. SE of the Serbo-Macedonian zone; Sharp & Robertson in press; Brown & Robertson in press). The Paikon unit cannot be correlated with the Pelagonian zone. The thrusting in the Paikon unit referred to by Bonneau *et al.* (1994) is the result of only localized post-emplacement compression of the Paikon unit within the Vardar (Axios) zone in the Early Tertiary.

Ongoing discussions revolve around the location and timing of closure and collisional deformation of different units in different areas (e.g. of the Kırşehir massif; e.g. Görür *et al.* 1984; Şengör *et al.* 1985), which is beyond the scope of this brief summary (see Robertson & Grasso 1995). In general, northward subduction beneath the Eurasian margin during the later stages of Tethyan closure gave rise to Andean-type magmatism (i.e. eastern Pontide arc) and rifting of the Black Sea marginal basin (Görür 1988). Neotethyan units were progressively accreted to Eurasia (e.g. Cretaceous Ankara Melange; Koçyiğit 1991). The deformation front migrated generally southward with time, towards Africa.

By the Late Miocene the accretionary collage in the east had collided with the Arabian promontory (Arabian sub-plate), and was followed by activation of the present convergent margin in the Eastern Mediterranean.

Palaeomagnetic data have provided much of the key evidence in unravelling the latest (Neogene to Recent) stages of the collision. These data have been extensively described elsewhere (e.g. Kissel & Laj 1988; Plazman *et al.* 1995; Kondopoulou *et al.* this volume; Piper *et al.* this volume; and others). The palaeomagnetic database on Neogene rotational deformation in the eastern Mediterranean has now reached the stage where the palaeogeography can be restored with reasonable accuracy to a pre-rotational framework. This should lead to improved tectonic models for the earlier Mesozoic to early Tertiary history.

Conclusions

Three principal alternative plate tectonic models have been discussed for the Late Palaeozoic-Early Mesozoic Tethyan history of the Eastern Mediterranean region, together with numerous more local interpretations.

Model 1 (Robertson & Dixon 1984) inferred a long-lived Tethyan ocean evolving from the Late Palaeozoic through the Mesozoic into the Tertiary, under the dominant influence of northward subduction and rifting of continental fragments from Gondwana. Tectonic units along the Eurasian margin were dominantly arc-accretionary complexes related to northward subduction. The Mesozoic palaeogeography of south Tethys was marked by interconnected oceanic strands interspersed with microcontinents that were rifted from Gondwana. Ophiolites mainly formed above subduction zones and were emplaced when crustal units (e.g. microcontinents) collided with subduction trenches. Based on the more recent evidence, this model remains broadly applicable, in that it successfully reconicles recent field evidence with a viable kinematic scenario.

Model 2A (Dercourt *et al.* 1986) envisaged a single Mesozoic Tethyan ocean basin with dominantly northward subduction beneath the Eurasian margin, as in Model 1. However, the ophiolites were seen as having formed at mid ocean ridges located within this single Mesozoic Tethys (Neotethys). Some ophiolites reached their present positions by very long-distance thrusting (i.e. hundreds of kilometres). Model 2B (Dercourt *et al.* 1993) envisaged a more complex palaeogeography, particularly in the Cretaceous. An arm of Permian-aged ocean

crust was seen as swinging through the south Aegean (Pindos ocean) to connect with rift basins further west. However, models 2A and 2B do not take into account the critical role of rifting and spreading in the Late Palaeozoic-Early Mesozoic in the south Aegean region. Also, the genesis of many large ophiolites above subduction zones and the derivation of ophiolites from several Mesozoic oceanic basins goes unrecognized.

Model 3 (Şengör *et al.* 1984) postulated southward-dipping subduction of Late Palaeozoic Tethys (Palaeo-Tethys) beneath the northern margin of Gondwana. A Triassic back-arc basin opened, rifting a continental fragment from Gondwana, that then drifted northwards to collide with a passive Eurasian margin in latest Triassic time, partially closing a marginal basin (Karakaya). This was followed by renewed rifting to form a Mesozoic (Neotethyan) ocean basin from the Early Jurassic onwards. More recent work, however, has not confirmed dominantly southward subduction of Tethys in the Late Palaeozoic-Early Mesozoic.

Our preferred model is now one in which the southern margin of Eurasia was active, undergoing northward subduction throughout much, or all, of Late Palaeozoic-Early Tertiary time. The Gondwana margin was passive during the Palaeozoic onwards. Continental fragments rifted, drifted across Tethys and were amalgamated to Eurasia. Important rift/drift events took place along the north margin of Gondwana in the Late Permian, Early-Mid-Triassic and Early Cretaceous. Rapid opening of Early Mesozoic basins in the south was partly accommodated by southward subduction of Late Palaeozoic Tethys, at least in NW Turkey. Large ophiolites formed above subduction zones mainly in the Late Permian-Triassic, Early-Mid-Jurassic and Late Cretaceous, while oceanic crust formed at 'normal' ocean ridges is mainly preserved as deformed and metamorphosed fragments in subduction-accretion complexes. These ophiolites were rooted in a number of both northerly and southerly oceanic stands and were mainly emplaced by trench-margin collisions. Tethys remained open into the Early Tertiary, when it finally closed by diachronous collisions of microcontinents, leaving a remnant only in the present easternmost Mediterranean Sea area adjacent to the Levant.

Palaeomagnetic data have helped define the regional framework of Africa and Eurasia. However, palaeomagnetic studies of the individual components of the orogen have so far been limited by the geological complexity and common problems of remagnetization. One exception is units in the south (e.g. Cyprus; see Morris, this volume) that were located in the most peripheral (i.e. external) areas of the orogenic belt and did therefore not experience tectonic thickening resulting from collision.

Helpful comments in the manuscript were received from G. Jones, P. Degnan and T. Danelian. D. Batty assisted with drafting the figures.

References

AKTA, Ş. G. & ROBERTSON, A. H. F. 1990. Late Cretaceous-Early Tertiary fore-arc tectonics and sedimentation: Maden Complex, S. E. Turkey. International Earth Sciences Congress on Aegean Regions, Izmir, Turkey. In: SAVASCIN, M. Y. & ERONAT, A. H. (eds) IESCA 1990 Proceedings, 2. Dokuz Eylul University, Izmir, 271–276.

AUBOUIN, J., BONNEAU, M., CELET, P., CHARVET, J., CLEMENT, B., DEGARDIN, J. M., MALLOT, H., MANIA, J., MANSY, J. L., TERRY, J., THIEBAULT, P., TSOFLIAS, P. & VERRIEUX, J. J. 1970. Contribution à la géologie des Héllenides: Le Gavrovo, le Pinde et la Zone Ophiolitique Subpélagonien. Annales de la Societé Géologique du Nord, 90, 277–306.

BAROZ, F., BEBIEN, J. & IKENNE, M. 1987. An example of high-pressure low-temperature metamorphic rocks from an island-arc: the Paikon Series (Innermost Hellenides, Greece). Journal of Metamorphic Geology, 5, 509–527.

BAUD, A., JENNY, C., PAPANIKOLAOU, D., SIDERIS, C. & STAMPFLI, G. 1991. New observations on the Permian stratigraphy in Greece and geodynamic interpretation. Bulletin of the Geological Society of Greece, 225, 187–206.

BAUMGARTNER, P. O. 1985. Jurassic sedimentary evolution and nappe emplacement in the Argolis Peninsula (Peloponnesus, Greece). Mémoire de la Société Helvetique pour la Science Naturelle.

BEBIEN, J., PLATEVOET, B. & MERCIER, J. 1994. Geodynamic significance of the Paikon Massif in the Hellenides: Contributions of the volcanic rock studies. 7th Congress of the Geological Society of Greece, Abstract volume.

——, OHNENSTETTER, D., OHNENSTETTER, M. & VERGELY, P. 1980. Diversity of Greek ophiolites; birth of ocean basins in transform systems, Ofioliti, 2, 129–197.

——, BAROZ, J., CAPEDRI, S. & VENTURELLI, G. 1987. Magmatisme basique associés a l'ouverture d'un basin marginal dans les Héllenides internes au Jurassic. Ofioliti, 12, 53–70.

BINGÖL, A. E., AKYÜREK, B. & KORKMAZER, G. 1973. ßiga yarımadasının jeolojisi ve Karakaya Formasyonunun bazı özellikleri Cumhuriyetin 50. Yılı Yerbilimleri Kongresi, Maden Tetkik ve Arama Erstitüsü, Ankara, 70–76.

BONNEAU, M., GODFRIAUX, I., MULAS, I., FOURCADE, E. & MASSE, J. P. 1994. Imbricate structure of the Paikon window (Macedonia, Greece). New biostratigraphical data. 7th Congress of the Geological Society of Greece, Abstract volume, 43.

BROWN, S. & ROBERTSON, A. H. F. a New structural evidence from the Mesozoic-Early Tertiary Paikon unit, north-eastern Greece. Bulletin of the Geological Society of Greece, in press.

—— & —— b Role of the Paikon Unit of the tectonic evolution of Neotethys, NE Greece. European Union of Geosciences, in press.

BURG, J.-P., IVANOV, Z., RICOU, L.-E., DIMOR, D. & KLAN, L. 1990. Implications of shear-sense criteria for the tectonic evolution of the Central Rhodope Massif, southern Bulgaria, Geology, 18, 445–454.

BUROLET, P. F., MUGNIOT, J. M. & SWENEY, P. 1978. The geology of the Pelagian block: the margins and basins of southern Tunisia and Tripolitania. In: NAIRN, A. E. M. & KAINES, W H. (eds) The ocean basins and their margins. Plenum, NY, 331–359.

CHANNELL, J. E. T. 1996. Palaeomagnetism and palaeogeography of Adria. This volume.

——, D'ARGENIO, B. & HORVATH, F. 1979. Adria, the African promontory, in Mesozoic Mediterranean palaeogeography. Earth Science Reviews, 15, 213–292.

CLIFT, P. D. 1990. Mesozoic/Cenozoic sedimentation and tectonics of the southern Greek Neotethys (Argolis Peninsula). PhD Thesis, University of Edinburgh.

—— 1992. The collision tectonics of the southern Greek Neotethys. Geologische Rundschau, 81, 669–679.

—— & ROBERTSON, A. H. F. 1990. A Late Cretaceous carbonate margin of the southern Greek Neotethys. Geological Magazine, 127, 825–836.

COLLINS, A. & ROBERTSON, A. H. F. New structural and tectono-stratigraphic evidence for the origin of the Lycian nappes, SW Turkey. European Union of Geosciences, Abstracts, in press.

DEGNAN, P. J. 1992. Tectono-sedimentary evolution of a passive margin: the Pindos Zone of the NW Peloponnese Greece. PhD thesis, University of Edinburgh.

—— & ROBERTSON, A. H. F. 1990. Tectonic and sedimentary evolution of the Western Pindos, Greece, In: 5th Congress of the Geological Society of Greece, Thessaloniki, May 1990, 38–39.

—— & —— Early Tertiary melange in the Peloponnese (S. Greece) formed by subduction-accretion processes. Bulletin of the Geological Society of Greece, in press.

DELAUNE-MAYERE, M. 1984. Evolution of a Mesozoic passive continental margin: Baër-Bassit (NW Syria). In: DIXON, J. E. & ROBERTSON, A. H. F. (eds) The Geological Evolution of the Eastern Mediterranean. Geological Society, London, Special Publications, 17, 151–160.

DEMRTAŞLI, E. 1984. Stratigraphic evidence of Variscan and early Alpine tectonics in southern Turkey. In: DIXON, J. E. & ROBERTSON, A. H. F. (eds) The Geological Evolution of the Eastern Mediterranean. Geological Society, London, Special Publications, 17, 129–146.

DERCOURT, J., RICOU, L. E. & VRIELYNCK, B. (eds) 1993. Atlas Tethys Palaeoenvironmental Maps. Beicip-Franlab, 1993.

——, ZONENSHAIN, L. P., RICOU, L. E., KAZMIN, V. G., LE PICHON, X., KNIPPER, A. L., GRANDJAC-

QUET, C., SBORTSHHIKOV, I. M., GEYSSANT, J., LEPVRIER, C., PERCHERSKY, D. H., BOULIN, J., SIBUET, J.-C., SAVOSTIN, L. A., SOROKHTIN, O., WESTPHAL, M., BAZHRNOV, M. L., LAUER, J.-P. & BIJU-DUVAL, B. 1986. Geological evolution of the Tethys belt from the Atlantic to the Pamirs since the Lias. *Tectonophysics*, **123**, 241–315.

DE WET, A. P. 1989. *Geology of part of Chalkidiki Peninsula, northern Greece*. PhD thesis, University of Cambridge.

DILEK, Y. & ROWLAND, J. C. 1993. Evolution of a conjugate passive margin pair in the Mesozoic southern Turkey. *Tectonics*, **11**, 12, 954–970.

——, THY, P., MOORES, E. M. & RAMSDEN, T. W. 1990. Tectonic evolution of the Troodos ophiolite within the Tethyan framework. *Tectonics*, **9**, 811–823.

DIMITRIADIS, S. & ASVESTA, A. 1993. Sedimentation and magmatism related to the Triassic rifting and later events in the Vardar-Axios Zone. *Bulletin of the Geological Society of Greece*, **28**(2), 149–168.

DIXON, J. E. & DIMITRIADIS, S. 1984. Metamorphosed ophiolitic rocks from the Serbo-Macedonian Zone, near Lake Volvi, north-east Greece. *In*: DIXON, J. E. & ROBERTSON, A. H. F. (eds) *The Geological Evolution of the Eastern Mediterranean*. Geological Society, London, Special Publications, **17**, 603–619.

ERSOY, Ş. 1990. Similarities of the western Taurus Belt with the external Hellenides. International Earth Sciences Congress on Aegean Regions, October 1990, Izmir, Turkey. *In*: SAVAŞÇIN, M. Y. & ERONAT, A. H. (eds) *IESCA 1990 Proceedings*, 129–142.

FERRIERE, J. 1982. Paleogéographie et téctoniques superposées dans les Héllenides internes: les massifs de l'Othrys et du Pelion (Grèce séptentrional). *Annales de la Societé Géologique du Nord*, **8**, 1970.

—— & STAIS, A. 1994. Le (ou les) bassin(s) Téthysien(s) Vardarien(s). *7th Congress of the Geological Society of Greece. Abstract volume*, 52.

FLEURY, J. J. 1980. Evolution d'une platform d'un bassin dans leur cadre alpin: Les zones de Gavrovo-Olonos (Grèce Continentale) et du Pinde-Olonos. *Annales de la Societé Géologique du Nord*, **4**, 1–651.

FOURCADE, E., DERCOURT, J., GUNAY, Y., AZEMA, J., KOZLU, H., BELLIER, J. P., CORDEY, F., CROS, P., DE WEVVER, P., ENAY, R., HERNANDEZ, J., LAUER, J. P. & VRIELYNCK, B. 1991. Stratigraphie et palaéographie de la marge séptentrionale de la plate-forme arabe au Mésozoique (Turquie de Sud-Est). *Bulletin de la Societé Géologique de France*, **161**, 27–41.

GALLET, Y., BESSE, J., KRYSTYN, L., MARCOUX, J., TÉVENIAUT, H. 1992. Magnetostratigraphy of the late Triassic Bolüceski Tepe section (southwestern Turkey): implications for changes in magnetic reversal frequency. *Physics of the Earth and Planetary Interiors*, **93**, 273–282.

——, ——, ——, THÉVENIAUT, H. & MARCOUX, J. 1993. Magnetostratigraphy of the Kavur Tepe section (southwestern Turkey): A magnetic polarity time scale for the Norian. *Earth and Planetary Science Letters*, **117**, 443–456.

GARFUNKEL, Z. & DERIN, B. 1984. Permian-early Mesozoic tectonism and continental margin formation in Israel and its implications for the history of the Eastern Mediterranean. *In*: DIXON, J. E. & ROBERTSON, A. H. F. (eds) *The Geological Evolution of the Eastern Mediterranean*. Geological Society, London, Special Publications, **17**, 187–202.

GASS, I. H. 1990. Ophiolites and oceanic lithosphere. *In*: MALPAS, J., MOORES, E. M., PANAYIOTOU, A. & XENOPHONTOS, C. (eds) *Ophiolites oceanic crustal analogues*. Proceedings of Symposium 'Troodos 1987', Cyprus Geological Survey Dept. 1–12.

GODFRIAUX, I. & RICOU, L.-E. 1991. Le Paikon, une fênetre tectonique dans les Héllenides Internes (Macedoine, Greece). *Comptes Rendus de l'Academie des Sciences, Paris*, **313**, Serie 11, 1479–1484.

GÖRÜR, N. 1988. Timing of opening of the Black Sea basin. *Tectonophysics*, **147**, 247–262.

——, OKTAY, F. Y., SEYMEN, I. & ŞENGÖR, A. M. C. 1984. Palaeotectonic evolution of the Tuzgölü basin complex, Central Turkey: sedimentary records of a Neo-Tethyan closure. *In*: DIXON, J. E. & ROBERTSON, A. H. F. (eds) *The Geological Evolution of the Eastern Mediterranean*. Geological Society, London, Special Publications, **17**, 467–482.

——, ŞENGÖR, A. M. C. & AKKÖK, R. 1985. *Mesozoic-Cainozoic Geology between Istanbul and Bursa*. Guide book for excursion to northwest Turkish Tethyan suture zones, Istanbul Technical University Faculty of Mines, 45–55.

GVIRTZMAN, G. & WEISSBROD, T. 1984. The Hercynian geanticline of Helez and the Late Palaeozoic history of the Levant. *In*: DIXON, J. E. & ROBERTSON, A. H. F. (eds) *The Geological Evolution of the Eastern Mediterranean*. Geological Society, London, Special Publications, **17**, 177–186.

HAENEL-REMY, S. & BEBIEN, J. 1985. The Oreokastro Ophiolite (Greek Macedonia): an important component of the innermost Hellenic ophiolite belt. *Ofioliti*, **10**, 279–296.

HARBURY, N. A. & HALL, R. 1988. Mesozoic extensional history of the southern Tethyan continental margin in the SE Aegean. *Journal of the Geological Society, London*, **145**, 283–301.

HIRSCH, F. 1984. The Arabian sub-plate during the Mesozoic. *In*: DIXON, J. E. & ROBERTSON, A. H. F. (eds) *The Geological Evolution of the Eastern Mediterranean*. Geological Society, London, Special Publications, **17**, 217–235.

——, FLEXER, A., ROSENFELD, A. & YELLIN-DROR, A. 1995. Palinspastic and crustal setting of the Eastern Mediterranean. *Journal of Petroleum Geology*, **18**, 149–170.

JACOBSHAGEN, V. & WALLBRECHER, E. 1984. Pre-Neogene nappe structure and metamorphism of the North Sporades and the southern Pelion Peninsula. *In*: DIXON, J. E. & ROBERTSON, A. H. F. (eds) *The Geological Evolution of the Eastern Mediterranean*. Geological Society, London, Special Publications, **17**, 591–602.

JONES, G. & ROBERTSON, A. H. F. 1991. Tectono-stratigraphy and evolution of the Mesozoic Pindos ophiolite and related units, northwestern Greece. *Journal of the Geological Society, London*, **148**, 267–288.

—— & —— Rift-drift-subduction and emplacement history of the Early Mesozoic Pindos ocean: evidence from the Avdella Melange, Northern Greece. *Bulletin of the Geological Society of Greece*, in press.

——, —— & CANN, J. R. 1991. Supra-subduction zone origin of the Pindos ophiolite, Northwestern Greece. *In*: PETERS, T. *et al.* (eds) *Ophiolites and Ophiolitic Lithosphere*. Proceeding of International Conference, Muscat, 1990, 771–800.

KARAMATA, S. 1988. 'The Diabase-Chert Formation' some genetic aspects. *Bulletin of the Serbian Academy of Science and Arts, Mathematics and Natural Sciences*, **28**, 1–11.

——, KRISTIC B. & STAJANOV, R. 1992. Terranes from the Adriatic to the Moesian Massif in the central part of the Balkan Peninsula. *Terra Nova*, **4**, Abstract Supplement, **2**, 78.

KISSEL, C. & LAJ, C. 1988. The Tertiary geodynamical evolution of the Aegean arc: a palaeomagnetic reconstruction. *Tectonophysics*, **146**, 183–201.

KOÇYIĞIT, A. 1991. An example of an accretionary forearc basin from north Central Anatolia and its implications for the history of subduction of Neo-Tethys in Turkey. *Geological Society of America Bulletin*, **103**, 22–36.

KONDOPOULOU, D., PAVLIDES, S. & ATZEMOGLOU, A. 1996. Palaeomagnetism as a tool in testing geodynamic models in the North Aegean: Convergencies, controversies and a further hypothesis. *This volume*.

KOZUR, H. 1991. The evolution of the Meliata-Hallstatt ocean and its significance for the early evolution of the Eastern Alps and Western Carpathians. *Palaeogeography, Palaeoclimatology, Palaeoecology*, **87**, 109–136.

KRAHL, J. 1992. The young Paleozoic and Triassic Tethyan rocks in the external Hellenides on Crete. *5th Congress of the Geological Society of Greece*, Abstracts, 59–60.

LAUER, J. P. 1984. Geodynamic evolution of Turkey and Cyprus based on palaeomagnetic data. *In*: DIXON, J. E. & ROBERTSON, A. H. F. (eds) *The Geological Evolution of the Eastern Mediterranean*. Geological Society, London, Special Publications, **177**, 483–492.

LIVERMORE, R. A. & SMITH, A.G. 1984. Some boundary conditions for the evolution of the Mediterranean region. *In*: STANLEY, D. J. & WEZEL, F.-C. (eds) *Geological Evolution of the Mediterranean Basin*. Springer-Verlag, 83–100.

——, SMITH, A. G. & VINE, F. J. 1986. Late Palaeozoic to early Mesozoic evolution of Pangea. *Nature*, **322**, 162–165.

M'RABET, A., BEN-ISMAIL, M., SOUSSI, M. & TURKI, M. 1989. Jurassic rifting and drifting of the North African margin and their sedimentary responses in Tunisia. *In*: *Abstracts of the 28th International Congress, Washington, D.C.*, **3**, 237.

MART, Y. 1987. Superpositional tectonic patterns along the continental margin of the Southeastern Mediterranean: a review. *Tectonophysics*, **140**, 213–232.

MAY, P. R. 1991. The eastern Mediterranean Mesozoic Basin: Evolution and oil habitat. *American Association of Petroleum Geologists Bulletin*, **75**, 1215–1232.

MERCIER, J. 1968. *Etude géologique des zones internes des Héllenides et Macedoine Centrale (Grèce)*. Annales Géologique de Pays Hélleniques, **20**.

—— & VERGELY, P. 1994. Is the Paikon massif a tectonic window in the Vardar zone? (Internal Hellenides, Macedonia, Greece). *7th Congress of the Geological Society of Greece, Abstract volume*, 60.

MORRIS, A. 1995. Rotational deformation during Palaeogene thrusting and basin closure in eastern Central Greece: Palaeomagnetic evidence from Mesozoic carbonates. *Geophysical Journal International*, **121**, 827–847.

—— 1996. A review of palaeomagnetic research in the Troodos ophiolite, Cyprus. *This volume*.

—— & ROBERTSON, A. H. F. 1993. Miocene remagnetisation of carbonate platform and Antalya Complex units within the Isparta Angle, SW Turkey. *Tectonophysics*, **220**, 243–266.

MOUNTRAKIS, D. 1984. Structural evolution of the Pelagonian Zone in northwestern Macedonia, Greece. *In*: DIXON, J. E. & ROBERTSON, A. H.F. (EDS) *The Geological Evolution of the Eastern Mediterranean*. Geological Society, London, Special Publications, **17**, 569–581.

MUSSALLAM, K. 1991. Geology, geochemistry and the evolution of an oceanic lithosphere rift at Sithonia NE Greece. *In*: PETERS, T., NICOLAS, A. & COLEMAN, R. G. (eds) *Ophiolite genesis and evolution of the oceanic lithosphere*. Kluwer Academic Publishers, London, 685–704.

OKAY, A. I. 1989. Geology of the Menderes Massif and the Lycien Nappes south of Denizli, western Taurides. *Mineral Research Exploration Bulletin (Ankara)*, **109**, 37–51.

—— & SIYAKO, M. 1993. The new position of the İzmir–Ankara Neo-Tethyan suture between İzmir and Balıkesir. *In*: TURGUT, S. (ed.) *Tectonics and hydrocarbon potential of Anatolia*. Proceedings of the Ozan Sungurlu Symposium, Ankara, 1993, 333–355 [in Turkish with an English abstract].

——, SIYAKO, M. & BURKAN, K. A. 1991. Geology and tectonic evolution of the Biga Peninsula, Northwest Turkey. *Bulletin of the Technical University of İstanbul*, **44**, 191–256.

OZKAYA, I. 1990. The origin of allochthons in the Lycien belt, southwestern Turkey. *Tectonophysics*, **17**, 367–379.

PAPANIKOLAOU, D. 1989. Are the medial crystalline massifs of the Eastern Mediterranean drifted Gondwanian fragments? *In*: PAPANIKOLAOU, D. & SASSI, F. P. (eds) *IGCP Project, 276, Newsletter Number 1*. Geological Society of Greece Special Publications, **1**, 63–90.

—— & SIDERIS, C. 1983. Le Paleozoique de l'autochthone de Chios: une formation à bloc de type wldflysch d'age permien (pro parte). *Comptes*

Rendus de l'Academie des Sciences, Paris, **297**, 603–606.

PEARCE, J. A., LIPPARD, S. J. & ROBERTS, S. 1984. Characteristics and tectonic significance of supra-subduction zone ophiolites. *In*: KOKELAAR, B.P. & HOWELLS, M. F. (eds) *Marginal Basin Geology*. Geological Society, London, Special Publications, **16**, 77–89.

PE-PIPER, G. & PIPER, D. W. J. 1990. Early oceanic subduction-related volcanic rock, Pindos Basin, Greece. *Tectonophysics*, **192**, 273–292.

PICKETT, E. A. 1994. *Tectonic evolution of the Palaeotethys ocean in NW Turkey*. PhD Thesis, University of Edinburgh.

——, ROBERTSON, A H. F. & DIXON, J. E. The Karakaya Complex, N.W. Turkey: a Palaeotethyan accretionary complex. *In*: *Geology of the Black Sea Region*. MTA, Ankara, in press.

PIPER, J. D. A., MOORE, J. M., TATAR, O., GURSÖY, H. & PARK, R. G. 1996. Palaeomagnetic study of crustal deformation across an intracontinental transform: The North Anatolian Fault Zone in Northern Turkey. *This volume*.

PLATZMAN, E. S., PLATT, J. P., TAPIRDAMAZ, C., SANVER, M. & RUNDLE, C. C. 1994. Why are there no clockwise rotations along the North Anatolian Fault Zone? *Journal of Geophysical Research*, **99**, 21705–21715.

POISSON, A. M. 1984. The extension of the Ionian trough into southwestern Turkey. *In*: DIXON, J. E. & ROBERTSON, A. H. F. (eds) *The Geological Evolution of the Eastern Mediterranean*. Geological Society, London, Special Publications, **17**, 241–249.

—— 1990. Neogene thrust belts in Western Taurides. The imbricate systems of thrust sheets along a NNW-SSE transect. *In*: SAVASCIN, M. Y. & ERONAT, A. H. (eds) *International Earth Sciences Congress on Aegean Regions* (IESCA) 1990, Proceedings, **2**. Dokus Eylul University, Izmir, 224–235.

REMY, P. 1984. Mise en evidence d'une metamorphisme dynamothermal dans les sediments au contact des ophiolites d'Oreokastro (Macedoine, greque). *Comptes Rendus de l'Academie des Sciences, Paris*, (Serie 11), 27–30.

—— & SARP, H. 1985. La zone de Kızılıca-Çorak göl. Un example de sillon intra plate-forme a la marge externe du massif de Menderes. *In*: IZDAR, I. & NOKAMAN, E. (eds) *6th Colloquium of the Geology of the Aegean region, Izmir, 1977*. Piri Reis, International Contribution Series, **2**, Dokuz Eylül University, İzmir, 555–564.

RICOU, L.-E., MARCOUX, J. & WHITECHURCH, H. 1984. The Mesozoic organisation of the Taurides: one or several oceanic basins. *In*: DIXON, J. E. & ROBERTSON, A. H. F. *The Geological Evolution of the Eastern Mediterranean*. Geological Society, London, Special Publications, **17**, 349–360.

——, —— & POISSON, A. 1979. L'allochthonie des Bey Dağları orientaux, Reconstructon palinspastique des Taurides occidentales. *Bulletin de la Societé Géologique de France*, **21**(7), 125–134.

ROBERTSON, A. H. F. 1993. Mesozoic-Tertiary sedimentary and tectonic evolution of Neotethyan carbonate platforms, margins and small ocean basins in the Antalya complex, SW Turkey. *In*: FROSTICK, L. E. & STEEL, R. (eds) *Sedimentation, Tectonics and Eustasy: sea-level changes at active margins*. Special Publication of the International Association of Sedimentologists, **20**, 415–465.

—— 1994. Tectonic Facies Concept and its application to Tethys in the Eastern Mediterranean region. *Earth and Planetary Science Reviews*, **37**, 139–213.

—— & DIXON, J. E. 1984. Introduction: aspects of the geological evolution of the Eastern Mediterranean. *In*: DIXON, J. E. & ROBERTSON, A. H. F. (eds) *The Geological Evolution of the Eastern Mediterranean*. Geological Society, London, Special Publications, **17**, 1–74.

—— & GRASSO, M. 1995. Overview of the Late Tertiary–Recent development of the Mediterranean region. *Terra Research*, in press.

—— & KARAMATA, S. 1994. The role of subduction–accretion processes in the tectonic evolution of the Mesozoic Tethys in Serbia. *Tectonophysics*, **234**, 73–94.

—— & WALDRON, J. W. F. 1990. Geochemistry and tectonic setting of Late Triassic and Late Jurassic–Early Cretaceous basaltic extrusives from the Antalya Complex, SW Turkey. *In*: SAVASCIN, M. Y. & ERONAT, A. H. (eds) *International Earth Sciences Congress on Aegean Regions, 1990, Proceedings*, **2**, 279–299.

—— & XENOPHONTOS, C. 1993. Development of concepts concerning the Troodos ophiolite and adjacent units in Cyprus. *In*: PRICHARD, H. M., ALABASTER, T., HARRIS, N. B. W. & NEARY, C. R. (eds) *Magmatic Processes and Plate Tectonics*. Geological Society, London, Special Publications, **76**, 85–119.

——, CLIFT, P. D., DEGNAN, P. & JONES, G. 1991. Palaeogeographic and palaeotectonic evolution of the Eastern Mediterranean Neotethys. *Palaeogeography, Palaeoclimatology, Palaeoecology*, **87**, 289–344.

ROBINSON, P. T. & MALPAS, J. 1990. The Troodos ophiolite of Cyprus: new perspectives on its origin and emplacement. *In*: MALPAS, J., MOORES, E. M., PANAYIOTOU, A. & XENOPHONTOS, C. (eds) *Ophiolites: Oceanic Crustal Analogues*. Proceedings of Troodos '87. Cyprus Geological Survey Department, 13–36.

SANDULESCU, M. 1989. Structure and tectonic history of the northern margin of Tethys between the Alps and the Caucasus. *In*: RAKUS, M., DERCOURT, J. & NAIRN, A. E. M. (eds) *Evolution of the northern margin of Tethys*. Mémoire de la Societé Géologique de France, **154**, III, 91–100.

SARIBUDAK, M., SANVER, M. & PONAT, E. 1989. Location of the western Pontides, NW Turkey, during Triassic time: preliminary palaeomagnetic results. *Geophysical Journal*, **96**, 43–50.

ŞENGÖR, A. M. C. 1992. The Palaeo-Tethyan suture: a line of demarcation between two fundamentally different architectural styles in the structure of Asia. *The Island Arc*, **1**, 78–91.

—— & YILMAZ, Y. 1981. Tethyan evolution of Turkey: a plate tectonic approach. *Tectonophysics*, **75**, 181–241.

——, YILMAZ, Y. & SÜNGÜRLU, O. 1984. Tectonics of the Mediterranean Cimmerides: nature and evolution of the western termination of Palaeo-Tethys. *In*: DIXON, J. E. & ROBERTSON, A. H. F. (eds) *The Geological Evolution of the Eastern Mediterranean*. Geological Society, London, Special Publications, **17**, 77–112.

——, CIN, A., ROWLEY, D. B. & SHANGYOU, N. 1991. Magmatic evolution of the Tethysides: a guide to reconstructoin of collage history. *Palaeogeography, Palaeoclimatologym, Palaeoecology*, **87**, 411–440.

——, GÖRÖR, N. & ŞAROĞLU, F. 1985. Strike-slip faulting and related basin formation in zones of tectonic escape. Society of Economic Mineralogists and Paleontologists, Special Publications, **37**, 227–264.

SHALLO, M., KODRA, A. & GJATA, K. 1990. Geotectonics of the Albanian ophiolites. *In*: MALPAS, J., MOORES, E. M., PANAYIOTOU, A. & XENOPHONTOS, C. (eds) *Ophiolites: Oceanic Crustal Analogues*. Proceedings of Symposium 'Troodos '87'. Geological Survey Department, Cyprus, 265–270.

SHARP, I. 1994. *The Mesozoic–Tertiary tectonic-sedimentary evolution of the Almopias zone, NW Greece*. PhD thesis, University of Edinburgh.

—— & ROBERTSON, A. H. F. *a* Evidence for Turonian rift related extensional subsidence and Tertiary backthrusting: The Almopias and Paikon isopic zones, Northern Greece. *Bulletin of the Geological Society of Greece*, in press.

—— & —— *b* Late Jurassic–Lower Cretaceous oceanic crust and sediments of the Eastern Almopias Zone, N.W. Greece; implications for the evolution of the Eastern 'internal' Hellenides. *Bulletin of the Geological Society of Greece*, in press.

——, —— & DIXON, J. E. 1991. Tectonic and sedimentary history of the Eastern margin of the Pelagonian Zone, NE Greece. *In*: EUG VI, *Terra Abstracts*, **3**. European Union of Geosciences, 302–310.

SIDHIROPOULOS, N. & DIMITRIADIS, S. 1989. Extension and melting of the continental crust during intracontinental emplacement of the Guevgueli ophiolite, eastern Vardar zone. *Terra Abstracts*, **1**, 57.

SMITH, A. G. 1993. Tectonic significance of the Hellenic-Dinaric ophiolites. *In*: PRICHARD, H. M., ALABASTER, T., HARRIS, N. B. W. & NEARY, C. R. (eds) *Magmatic Processes and Plate Tectonics*. Geological Society, London, Special Publications, **76**, 213–244.

——, HYNES, A. J., MENZIES, A. J., NISBET, E. G., PRICE, I., WELLAND, M. J. P. & FERRIERE, J. 1975. The stratigraphy of the Othris Mountains, eastern central Greece. *Eclogae Geolgicae Helvetiae*, **68**, 463–481.

STAIS, A. & FERRIERE, J. 1991. Nouvelles données sur la paleogéographie mésozoique du domaine vardarien: les bassins d'Almopias et de Peonais (Macedoine, Hellenides internes septentrionales). *Bulletin de la Societé Géologique de France*, **25**, 491–507.

STAMPFLI, G., MARCOUX, J. & BAUD, A. 1991. Tethyan margins in space and time. *Palaeogeography, Palaeoclimatology, Palaeoecology*, **87**, 373–410.

TEKELİ, O. 1981. Subduction complex of pre-Jurassic age, northern Anatolia, Turkey. *Geology*, **9**, 68–72.

TÜYSÜZ, O. 1990. Tectonic evolution of a part of the Tethyside orogenic collage: The Kargı Massif, Northern Turkey. *Tectonics*, **9**, 141–160.

USTAÖMER, T. 1993. *Pre-Late Jurassic sedimentary evolution of N. Tethys, Turkey*. PhD thesis, University of Edinburgh.

—— & ROBERTSON, A. H. F. 1993. A Late Palaeozoic-Early Mesozoic marginal basin along the active southern continental margin of Eurasia: evidence from the Central Pontides (Turkey) and adjacent regions. *Geological Journal*, **28**, 219–238.

—— & —— 1994. Late Palaeozoic marginal basin and subduction-accretion: evidence from the Palaeotethyan Küre Complex, Central Pontides, N. Turkey. *Journal of the Geological Society, London*, **151**, 291–305.

—— & —— A Pre-Late Jurassic tectonic evolution of the Central Pontides. *In*: *International Symposium on the Geology of the Black Sea Region, Ankara*. MTA, Ankara, in press.

WALDRON, J. W. F. 1984. Structural history of the Antalya Complex in the 'Isparta angle', Southwest Turkey. *In*: DIXON, J. E. & ROBERTSON, A. H. F. (eds) *The Geological Evolution of the Eastern Mediterranean*. Geological Society, London, Special Publications, **17**, 273–286.

WEIDMANN, J., KOZUR, H. & KAYA, O. 1992. Faunas and age significance of the pre-Jurassic turbidite-olistostrome unit in the western parts of Turkey. *Newsletter of Stratigraphy, Stuttgart*, **26**, 133–144.

YAZGAN, E. 1984. Geodynamic evolution of the Eastern Taurus region. *In*: TEKELİ, O. & GÖNCÜOĞLU, M. C. (eds) *Geology of the Taurus Belt*-MTA, Ankara, 199–208.

YILMAZ, Y. 1991. Allochthonous terranes in the Tethyan Middle East: Anatolia and the surrounding regions. *In*: DEWEY, J. F., GASS, I. G., CURRY, G. B., HARRIS, N. B. W. & ŞENGÖR, A. M. C. (eds) *Allochthonous terranes*. Cambridge University Press: 155–167.

ZIEGLER, P. A. 1990. *Geological atlas of western and central Europe*. Shell International Petroleum Company.

Palaeomagnetic investigations in Northern Albania and their significance for the geodynamic evolution of the Adriatic–Aegean realm

H. J. MAURITSCH[1], R. SCHOLGER[1], S. L. BUSHATI[2] & A. XHOMO[3]

[1] Geophysical Institute, Montanuniversität, Franz-Josef-Str. 18, A-8700 Leoben, Austria
[2] Geophysical Centre Tirana, Blloku 'Vasil Shanto' L. 9, Tirana, Albania
[3] Geological Centre Tirana, Blloku 'Vasil Shanto' L. 9, Tirana, Albania

Abstract: Palaeomagnetic results from the Inner and Outer Albanides south of the Shkoder–Pec line have established a clockwise rotation of about 45° for all tectonic units investigated so far. In Northern Albania, different tectonic zones surrounding the Shkoder–Pec line were sampled at 56 sites consisting of about 400 samples. Less than 50% of the samples had sufficiently high NRM intensities for thermal demagnetization and magnetization vector analysis. Fold and reversal tests support the results. The palaeomagnetic directions demonstrate a strong tectonic disturbance in the central part of the study area. The northernmost sites in the Albanian Alps show counterclockwise rotation (with respect to present north) in agreement with results from the Southern Dinarides. The pattern favours the idea that the Shkoder–Pec line is the transition zone between the counterclockwise rotation in the north and the clockwise rotation in the south.

One of the most significant problems in Mediterranean palaeomagnetism is the documentation of the timing, extent and mechanism of the clockwise rotation observed in Northern Greece, since Apulia to the west and the Dinarides to the north show either counterclockwise or no rotation. Márton (1987) first speculated about the possible importance of the Shkoder–Pec line with respect to the different rotation senses. Palaeomagnetic investigations in Albania identified a clockwise rotation from Northern Greece to the border of former Yugoslavia. Observed counterclockwise or no rotations on the island of Vis and near Split on one side (Márton & Milicevic 1993), and a clockwise rotation in Albania on the other side restrict the transition zone to the area between Split and the Shkoder–Pec line (Speranza et al. 1992; Mauritsch et al. 1994). Palaeomagnetic results from both sides of the Shkoder–Pec line, from the Albanian Alps in the north and the Mirdita zone in the south as well as the central Cukali subzone are presented in this study (Fig. 1).

Geological setting and sampling

The Albanides represent a key area for understanding the geological development of the Adriatic–Aegean realm. The Shkoder–Pec transversal divides the Albanides into a northern part, which constitutes the southeastern continuation of the Dinarides, and a southern part, which passes into the Hellenides. In order to study the role of the Shkoder–Pec line, we collected samples for palaeomagnetic analyses from the northernmost tectonic zones of the Albanides. Corresponding tectonic zones in the Dinarides and Hellenides can be distinguished based on the palaeogeographic development as follows: the Albanian Alps correspond to the High Karst zone in the Dinarides and the Parnassos zone in the Hellenides. The Cukali–Krasta zone is equivalent to the Budva zone in the Dinarides as well as the Pindos zone in the Hellenides. Finally, the Mirdita zone corresponds to the Serbian zone and the Sub-Pelagonian zone in the Dinarides and Hellenides, respectively (Fig. 1).

The movements along the Shkoder–Pec transversal dominate the tectonic zones in Northern Albania. Complex fault systems dominate Northern Albania, in contrast to the simple structures observed in Southern Albania (Mauritsch et al. 1995). The Albanian Alps overthrust the sediments of the Cukali–Krasta zone in northern Albania and are in turn overthrust by the Mirdita zone. Mesozoic carbonate sequences were deposited in two subzones of the Albanian Alps, each with an individual palaeogeographic evolution. Sediments were sampled in the Malesi E Madhe subzone, which occupies

From Morris, A. & Tarling, D. H. (eds), 1996, *Palaeomagnetism and Tectonics of the Mediterranean Region*, Geological Society Special Publication No. 105, pp. 265–275.

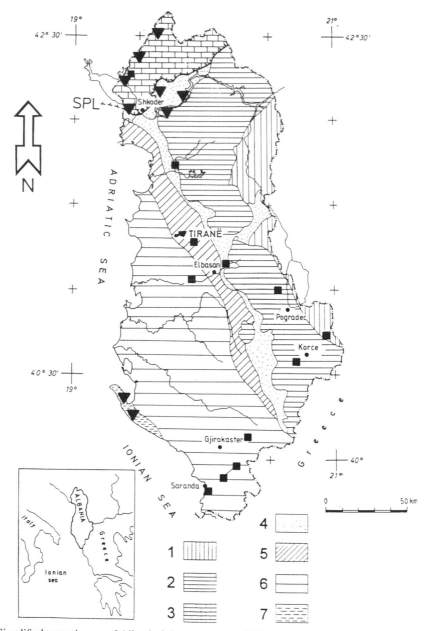

Fig. 1. Simplified tectonic map of Albania (after Pumo *et al.* 1982). Squares show localities sampled in 1989 and 1991 (Mauritsch *et al.* 1995), triangles for this study (see Fig. 2 for details). (1) Korab zone, (2) Mirdita zone, (3) Albanian Alps, (4) Cukali–Krasta zone, (5) Kruja zone, (6) Ionian zone, (7) Sazani zone. (SPL) Shkoder–Pec line.

the western part of the Albanian Alps. The subzone is characterized by neritic and reefal facies with biomicrites and biosparites. Most samples from this part of the study area showed magnetizations less than $10^{-5}\,\mathrm{A\,m^{-1}}$. Even with very careful laboratory treatment it was impossible to determine characteristic remanent magnetization directions at more than 70% of the sampling sites (Fig. 2). Figure 3 shows the stratigraphy of the Malesi E Madhe subzone,

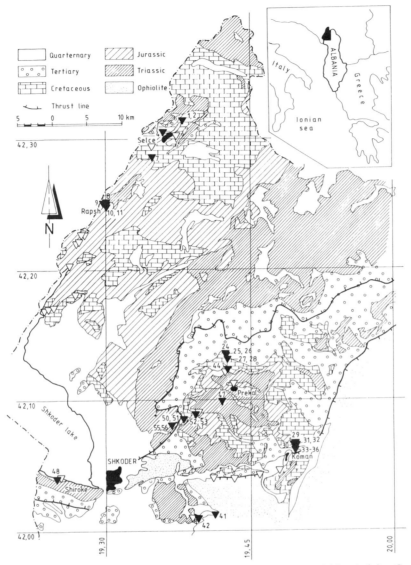

Fig. 2. Simplified geological map of the sampling area, the northernmost part of Albania (after Geologic Map of PSR of Albania, 1 : 200 000, Tirane 1983). Triangles mark the sampling sites of this study. Full symbols represent sites with successful demagnetization.

together with the stratigraphic levels of the sampling sites.

Biomicrites and schists dominate the Cukali–Krasta zone. Samples were collected in the half tectonic window of Cukali, to the NE of Shkoder. The window structure belongs to the Cukali subzone, which has a stratigraphy very similar to the Budva zone in the Dinarides (Fig. 3).

The Mirdita zone overlies and covers both the Cukali–Krasta zone and the easternmost part of the Albanian Alps. We collected samples from the NW of the Mirdita zone, which is dominated by Jurassic ophiolitic dunites, harzburgites, etc. (Fig. 2).

A total of 56 sites were collected in Northern Albania, consisting of 398 oriented cores for palaeomagnetic analyses and accompanying hand samples for biostratigraphy. The geographic coordinates of the sampling sites together with their stratigraphic ages are given in Table 1, in which only sites where demagnetization was

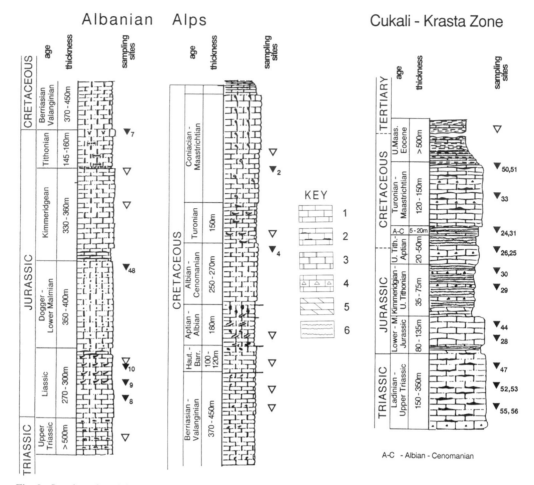

Fig. 3. Stratigraphy of the Albanian Alps and the Cukali–Krasta Zone. Triangles mark the sampling sites of this study. Full symbols with numbers represent sites with successful demagnetization. (1) Limestone; (2) limestone with chert lenses; (3) dolomite; (4) turbidite; (5) marl; (6) radiolarite.

successful are listed. Additionally, samples were collected from 23 Cretaceous sites in the Sazani zone in Southern Albania (Fig. 1), which is believed to constitute part of the Pre-Apulian foreland (Aubouin 1976). Since only one single site produced a statistically significant palaeomagnetic direction, the Sazani zone is not discussed further.

Palaeomagnetic results

All samples were progressively thermally demagnetized in order to study the stability of different components of the natural remanent magnetization (NRM). Additionally, we observed the saturation behaviour (IRM) and the anisotropy of magnetic susceptibility (AMS) in

selected samples. All measurements were carried out in the Palaeomagnetic Laboratory Gams of the Mining University Leoben using a 2G three-axis cryogenic magnetometer (measuring limit: $0.01\,\mathrm{mA\,m^{-1}}$) and a KLY-2 magnetic susceptibility bridge. Less than 50% of samples were found to be suitable for palaeomagnetic analysis.

The demagnetization paths of samples from seven sites in the Albanian Alps showed a single stable characteristic remanence vector after a few cleaning steps, whereas 18 remaining sites with NRM intensities typically less than $0.05\,\mathrm{mA\,m^{-1}}$ did not produce a stable component at all. Figure 4 shows representative demagnetization diagrams of samples from the Albanian Alps.

Table 1. *Palaeomagnetic results for successful northern Albanian sites*

Site	n	N/R	D_{bc}	I_{bc}	α_{95}	D_{ac}	I_{ac}	α_{95}	Stratigraphic age	Ma	Lat°N	Long°N
Albanian Alps												
Selce												
2	4	4/0	312	54	24.1	324	24	22.7	Coniacian–Campanian	74.0–88.5	42.518	19.629
4	4	3/1	327	49	16.1	327	39	8.5	Albian–Cenomanian	90.4–112.0	42.508	19.597
7	5	5/0	336	58	8.9	339	28	8.9	Tithonian	145.6–152.1	42.476	19.581
Rapsh												
8–10	9	6/3	47	6	15.2	49	44	14.4	Lower Jurassic	178.0–208.0	42.418	19.502
											42.415	19.498
											42.414	19.502
Shiroke												
48	4	0/4	29	–63	15.8	323	46	15.8	M. Jurassic–Oxfordian	154.7–178.0	42.067	19.471
Cukali–Krasta Zone												
Prekal												
50,51	10	4/6	56	26	19.9	53	46	17.0	Turonian–Maastrichtian	65.0–90.4	42.143	19.633
24	6	6/0	268	–41	6.3	264	25	11.9	Albian–Cenomanian	90.4–112.0	42.226	19.705
25,26	9	8/1	301	1	17.6	313	28	18.9	Tithonian	145.6–152.1	42.223	19.706
28	10	7/3	263	12	37.4	316	31	9.4	Lower–Middle Jurassic	157.1–208.0	41.218	19.709
44	8	0/8	79	38	19.7	62	19	11.4	Lower–Middle Jurassic	157.1–208.0	42.207	19.708
47	9	0/9	355	–6	55.3	3	31	20.0	Ladinian–Upper Triassic	208.0–239.5	42.165	19.700
52,53	7	7/0	57	49	11.4	8	25	11.4	Ladinian–Upper Triassic	208.0–239.5	42.149	19.655
55,56	6	0/6	16	12	29.4	11	25	28.0	Triassic	208.0–245.0	42.135	19.611
Koman												
33	4	4/0	46	38	15.6	140	23	12.6	Turonian–Maastrichtian	65.0–90.4	42.092	19.818
31	5	5/0	70	40	16.8	124	39	24.1	Albian–Cenomanian	90.4–112.0	42.109	19.826
29	6	6/0	40	38	15.0	79	50	15.0	Kimmeridgian–Tithonian	145.6–154.7	42.112	19.828
30	6	6/0	53	39	7.1	97	43	7.1	Kimmeridgian–Tithonian	145.6–154.7	42.112	19.828
Mirdita Zone												
42	5	5/0	13	19	14.6	13	19	14.6	Tithonian–Cenomanian	90.4–152.1	42.014	19.658
41	4	4/0	41	33	15.7	41	33	15.7	Middle–Upper Jurassic	145.6–178.0	42.020	19.689

n = number of samples; N/R = number of samples of normal/reverse polarity; D_{bc}, I_{bc} = declination, inclination of characteristic remanence direction before tilt correction; D_{ac}, I_{ac} = declination, inclination after tilt correction; α_{95} = semi-angle of 95% cone of confidence. Ages taken from Harland *et al.* (1990).

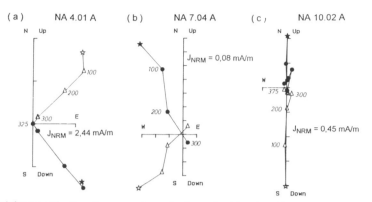

Fig. 4. Thermal demagnetization diagrams of samples from the Albanian Alps in orthogonal projection. Tilt corrected. Circles and triangles represent projection on the horizontal and vertical planes. NRM plotted with pentagrams. Numbers in diagrams denote demagnetization steps in °C. (**a**) Albian to Cenomanian from Selce. (**b**) Tithonian from Selce. (**c**) Lower Jurassic from Rapsh.

Samples from the Cukali–Krasta zone showed successful thermal demagnetization behaviour in 16 out of 25 sampling sites. Two-component magnetization was typical for samples from the Koman locality in the south of the Cukali window structure (Fig. 5a–c). The first component was aligned parallel to the recent magnetic field before tilt correction and was removed after heating up to 300–325°C. The second component was stable until the end of the demagnetization, which was limited by the measuring range of the magnetometer.

Samples from Prekal on the northern flank of the window structure yielded varying demagnetization behaviour with up to three magnetization components (Fig. 5d–i). The first component, comparable to the one in the locality Koman, was removed after heating at 300°C. The second component, stable up to 425–475°C, represented the characteristic magnetization in the majority of the samples (Fig. 5d–f,i). An additional high-temperature component was observed only in sites 24 to 28, the northernmost sampling sites in the Cukali–Krasta zone (Fig. 5g–h).

Extremely high NRM intensities in the range of $1\,A\,m^{-1}$ characterized the Jurassic ophiolites of the Mirdita zone. In some cases, stable high temperature components were randomly distributed within a site, while other samples were strongly affected by magnetic viscosity. Subsequent alternating field demagnetization after each heating step improved the results. Nevertheless, only two sites gave reliable palaeomagnetic results (Fig. 6). The Tithonian to Cenomanian samples yielded three magnetization components, which were separated at temperatures of 200°C and 475°C, respectively (Fig. 6a). Middle to Upper Jurassic samples

(Fig. 6b) showed two interacting vectors producing a great circle distribution in the temperature range between 300°C and 475°C, followed by a stable high-temperature component.

Stepwise acquisition of isothermal remanent magnetization (IRM) gave evidence for a magnetite-dominated magnetic phase in the majority of the samples, while the presence of iron-hydroxide is also evident. In the case of the Rapsh locality in the Albanian Alps, the presence of hematite is proved as well (Fig. 7a). This hematite could be of superparamagnetic particle size, since it is noticeable in the IRM acquisition curve, but not in the thermal demagnetization curve. Another piece of evidence for the presence of hematite is the light pink colour of the rocks. For samples from Prekali, one can notice a depressing influence on the acquisition curve, which could be due to iron sulphides (Fig. 7b, c). The micrites from Koman demonstrate a strong influence of iron-hydroxide, causing a delayed saturation at about 0.7–0.9 Tesla (Fig. 7d, e). This interpretation is strongly supported by the thermal demagnetization behaviour, where at least 50% of the NRM is lost in the temperature range up to 150°C.

The ultrabasic samples from the Mirdita zone (Fig. 7f) showed homogeneous magnetic behaviour. Saturation at about 0.2 Tesla and unblocking temperatures up to about 550°C clearly establishes magnetite to be the magnetic mineral. The demagnetization curve demonstrates a wide range of unblocking temperatures, which are due to grain-size effects.

Measurements of low field susceptibility showed a weak, predominantly diamagnetic bulk susceptibility for the majority of the samples from the Albanian Alps and weak

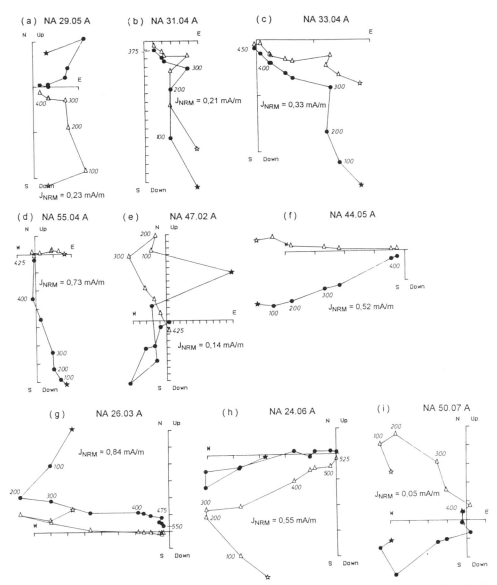

Fig. 5. Thermal demagnetization diagrams of samples from the Cukali–Krasta zone in orthogonal projection. Tilt corrected. Circles and triangles represent projection on the horizontal and vertical planes. NRM plotted with pentagrams. Numbers in diagrams denote demagnetization steps in °C. (**a–c**) samples from Koman: (**a**) Upper Jurassic, (**b**) Albian to Cenomanian, (**c**) Upper Cretaceous. (**d–i**) samples from Prekal: (**d**) and (**e**) Triassic, (**f**) Lower to Middle Jurassic, (**g**) Tithonian, (**h**) Albian to Cenomanian, (**i**) Upper Cretaceous.

paramagnetic susceptibility for the Cukali–Krasta zone, while ophiolites of the Mirdita zone reached values of 0.03 SI. Low-field anisotropy of magnetic susceptibility of the sediments was typically insignificant, except for some isolated samples with poorly defined magnetic foliation planes, oriented parallel to the sedimentary bedding planes. The ophiolites showed up to eight per cent total anisotropy without preferred orientation of the magnetic fabric.

Characteristic remanent magnetization directions (ChRM) for single samples were determined using principle component analysis (Kirschvink 1980). Mean values for sites (Table 1) were calculated following Fisher (1953).

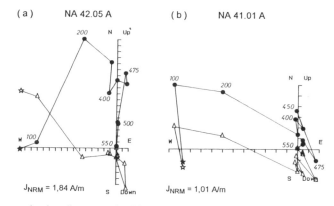

Fig. 6. Thermal demagnetization diagrams of ophiolite samples from the Mirdita zone in orthogonal projection. Tilt corrected. Circles and triangles represent projection on the horizontal and vertical planes. NRM plotted with pentagrams. Numbers in diagrams denote demagnetization steps in °C. (**a**) Tithonian to Cenomanian. (**b**) Middle to Upper Jurassic.

Discussion and conclusions

The present investigation was carried out on carbonates and ophiolites of Middle Triassic to Upper Cretaceous age on both sides of the Shkoder–Pec line. The palaeomagnetic results (Table 1) are shown in Fig. 8 together with the compiled results from Albania, taken from Mauritsch *et al.* (1995). In order to identify individual rotations of different tectonic units, we discuss the results with reference to predicted palaeodirections calculated from the African and Eurasian apparent polar wander paths of Besse & Courtillot (1991).

Selce, the northernmost locality under investigation, belongs to the Albanian Alps and comprises Upper Jurassic to Upper Cretaceous carbonates. A mean counterclockwise rotation of 30° with respect to present north is observed at this locality. Further to the SW, near Rapsh, a Lower Jurassic locality yielded a clockwise rotation of about 50°, which has to be interpreted as a local rotation around a vertical axis. This rotation is probably accommodated by observed steep faults, which cut the area into numerous blocks. The last locality in the Albanian Alps was along the southern shoreline of the Shkoder lake. Ophiolites and young sediments separate the Mesozoic sequence, believed to represent the southwestern continuation of the Albanian Alps, from the main tectonic unit. The palaeomagnetic direction is in good agreement with those from Selce. This may be an indication that they belong to the same tectonic unit. The results from both areas are not significantly different from the predicted African palaeodirections, although mean declinations from three of the four sites are displaced

in a counterclockwise sense from the African curve. This is consistent with results observed by Speranza *et al.* (1995) and Kissel *et al.* (1995) for the southernmost part of the Dinarides in Yugoslavia.

Triassic to Maastrichtian sediments of the Cukali–Krasta zone were sampled along the road from Shkoder towards the NE on the northern flank of the Cukali window structure. The Triassic and Jurassic yielded reliable palaeomagnetic data, confirmed by positive within-site fold tests. The Triassic results show a mean clockwise rotation of 27° with respect to predicted African palaeodirections from Irving & Irving (1982).

In contrast, in the Jurassic we can observe two completely different results. Sites 25–28 showed counterclockwise rotation of about 45°, whereas clockwise rotation of about 60° was found less than two km further to the south at site 44. Both results are clearly pre-folding magnetization directions as demonstrated by positive within-site fold tests (McElhinny 1964) at site 28 ($n = 10$, $K_1/K_2 = 10.53$, $F(18,18)$ 5% = 2.2) and site 44 ($n = 8$, $K_1/K_2 = 2.76$, $F(14,14)$ 5% = 2.48). This result favours the idea that the whole area is strongly affected by tectonic movements along the Shkoder–Pec line.

The Cretaceous biomicrites are partly remagnetized, which may be due to the dolomitization. For the subsequent interpretation, sites with proven remagnetization were discarded and only reliable sites taken into account.

In the Koman locality on the southern flank of the window structure, near the Shkoder–Pec line, sediments from Triassic through Cretaceous were sampled. The mean palaeodirections are much more consistent before tilt

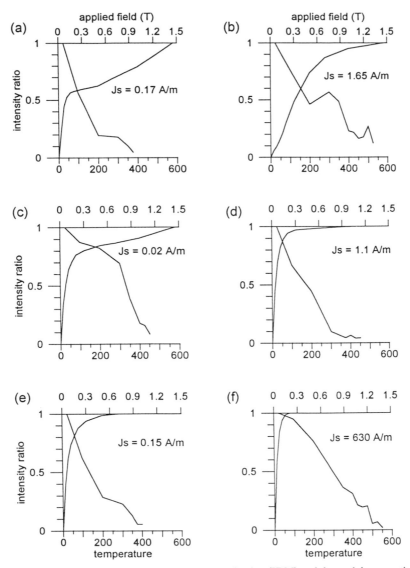

Fig. 7. Stepwise acquisition of isothermal remanent magnetization (IRM) and thermal demagnetization of NRM. Temperatures are given in °C. (**a**) Albanian Alps, site 10. (**b–e**) Cukali–Krasta zone, sites 24, 55, 33, 30. (**f**) Mirdita zone, site 41.

correction than afterwards. This can be explained by remagnetization as observed in site 31 with a negative within-site fold test and/or by the use of simple tilt corrections, which could be inappropriate (e.g. for site 33 with vertical bedding planes).

The results from the ophiolites from the Mirdita zone show a clockwise rotation of 47° (site 42) and 63° (site 41) with respect to Africa.

Summarizing the results from Northern Albania, we found strongly disturbed directions, disturbed by local tectonic phenomena. Particularly in the central part of the area in discussion, in the northern vicinity of the Shkoder–Pec line, we can notice individual movements within a few kilometres. Whereas the results from north of the Albanian Alps suggest a slight (not statistically significant) counterclockwise rotation with respect to predicted palaeodeclinations for Africa, similar to that observed in the Dinarides, the ophiolites of south of the Shkoder–Pec line suggest clockwise rotations, as observed

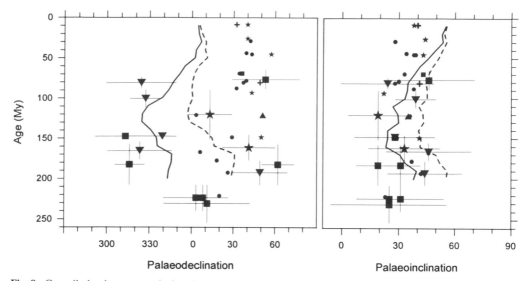

Fig. 8. Compiled palaeomagnetic data from Albania and predicted palaeodirections for Africa (solid curve) and Eurasia (dashed curve), calculated from master apparent polar wander paths (Besse & Courtillot 1991) for a location 40°N, 20°E. Small symbols for data taken from Mauritsch *et al.* (1995), larger symbols for this study (Table 1). Associated 95% confidence limits for declinations are projected onto the horizontal plane. Symbols indicate the different tectonic zones: Albanian Alps (triangles), Cukali–Krasta zone (squares), Mirdita zone (stars), Kruja zone (crosses), Ionian zone (circles).

throughout Albania (Speranza *et al.* 1992, 1995; Mauritsch *et al.* 1995). This leads to the conclusion that the investigated area along the geologically well-defined Shkoder–Pec line represents the transition zone between the counterclockwise rotation in the Dinarides and the clockwise rotation in the Albanides and Hellenides.

This conclusion contradicts that drawn by Mauritsch *et al.* (1995). The previous interpretation was based on Lower Cretaceous outcrops near Rapsh. The present study shows that this area has suffered a local clockwise rotation, which had not been noticed in the previous study.

With the Shkoder–Pec line as the transition zone, two models for the geodynamic evolution can be discussed. In the first model, the Greek mainland together with the adjacent areas and islands have undergone the post-Tortonian rotation as a single more or less rigid block around a rotation pole near the Shkoder–Pec line. In the second model, the NNW–SSE-striking Vardar Struma fault system and the E–W-striking North Anatolian fault system cut the whole area into small blocks (Zagorcev 1992). These blocks moved individually, but were driven by the same source and rotated in the same sense. The available palaeomagnetic results (Kondopoulou 1986; Pavlides *et al.* 1988;

Edel *et al.* 1990, 1992) favour the second model, since the palaeoinclinations from Albania to Thrace are the same within confidence limits. This means that the relative geographical positions of the areas in discussion have remained more or less stable since the Early Tertiary. If the first model were correct, a clockwise rotation of about 40–50° would be expected to cause a change in the latitude of the Peloponnesos and Crete of nearly 10°. Such a trend cannot be seen in the palaeomagnetic results, although it may be obscured by high scatter of the inclinations.

This study is part of a co-operative project between the Mining University Leoben and Geophysical Centre Tirana. Financial support was granted by the Austrian Research Fund (FWF-project P09815-PHY). We particularly acknowledge the technical support during fieldwork given by Director E. Zajimi and the help of R. Spitzer and R. Gurtner, as well as the improvement to the English by M. Pink. Finally, we acknowledge with many thanks the editing and critical comments of T. Morris, as well as the critical comments of an anonymous reviewer.

References

Aubouin, J. 1976. Alpine tectonics and plate tectonics: Thoughts about the Eastern Mediterranean. *In*: Ager, D. V. & Brooks, M. (eds) *Europe from Crust to Core*. Wiley, London, 143–158.

BESSE, J. & COURTILLOT, V. 1991. Revised and synthetic apparent polar wander path of the African, Eurasian, North American and Indian Plates, and true polar wander since 200 Ma. *Journal of Geophysical Research,* **96**, 4029–4050.

EDEL, J. B., KONDOPOULOU, D., PAVLIDES, S. & WESTPHAL, M. 1990. Multiphase paleomagnetic evolution of the Chalkidiki ophiolithic belt (Greece). Geotectonic implications. *Bulletin of the Geological Society of Greece,* **25**, 370–392.

——, ——, —— & —— 1992. Paleomagnetic evidence for a large counterclockwise rotation of Northern Greece prior to the Tertiary clockwise rotation. *Geodynamica Acta,* **5**, 245–259.

FISHER, R. A. 1953. Dispersion on a sphere. *Proceedings of the Royal Society of London,* **A217**, 295–305.

HARLAND, W. B., ARMSTRONG, R. L., COX, A. V., CRAIG, L. E., SMITH, A. G. & SMITH, D. G. 1990. *A geologic time scale 1989.* Cambridge University Press.

IRVING, E. & IRVING, G. A. 1982. Apparent polar wander paths Carboniferous through Cenozoic and the assembly of Gonwana. *Geophysical Surveying,* **5**, 141–188.

KISSEL, C. & SPERANZA, F. 1995. Paleomagnetism of external southern and central Dinarides and northern Albanides: implications for the Cenozoic activity of the Scutari-Pec transverse zone. *Journal of Geophysical Research,* **100**, 14 999–15 007.

KIRSCHVINK, J. L. 1980. The least-squares line and plane and the analysis of paleomagnetic data. *Geophysical Journal of the Royal Astronomical Society,* **62**, 699–718.

KONDOPOULOU, D. P. 1986. Tertiary rotational deformations in the Greek Serbo-macedonian massif. *Bulgarian Geophysical Journal,* **12**, 71–80.

MÁRTON, E. 1987. Paleomagnetism and tectonics in the Mediterranean region. *Journal of Geodynamics,* **7**, 33–57.

—— & MILICEVIC, V. 1993. Paleomagnetic investigations in the Dinarides between Zadar and Split. *ELGI Annual Report 1992.*

MAURITSCH, H. J., SCHOLGER, R., BUSHATI, S. L. & RAMIZ, H. 1995. Paleomagnetic results from Southern Albania and their tectonic significance for the geodynamic evolution of the Dinarides, Albanides and Hellenides. *Tectonophysics,* **242**, 5–18.

McELHINNY, M. W. 1964. The statistical significance of the fold test in palaeomagnetism. *Geophysical Journal of the Royal Astronomical Society,* **8**, 338–340.

PAVLIDES, S. B., KONDOPOULOU, D. P., KILIAS, A. A. & WESTPHAL, M. 1988. Complex rotational deformations in the Serbo-Macedonian massif (north Greece): structural and paleomagnetic evidence. *Tectonophysics,* **145**, 329–335.

PUMO, E., MELO, V. & Ostrosi, B. 1982. Albania. *In:* DUNNING, F. W., MYKURA, W. & SLATER, D. (eds) *Mineral deposits of Europe. Vol.2: Southeast Europe.* Spottiswoode & Ballantyne, London, 203–214.

SPERANZA, F., KISSEL, C., ISLAMI, I., HYSENI, A. & LAJ, C. 1992. First paleomagnetic evidence for rotation of the Ionian zone of Albania. *Geophysical. Research Letters,* **19**, 697–700.

——, ISLAMI, I., KISSEL, C. & LAJ, C. 1995. Paleomagetic evidence for Cenozoic clockwise rotation of the external Albanides. *Earth and Planetary Science Letters,* **129**, 121–134.

ZAGORCEV, I. S. 1992. Neotectonic development of the Struma (Kraistid) Lineament, southwest Bulgaria and northern Greece. *Geological Magazine,* **129**, 197–222.

Palaeomagnetism as a tool for testing geodynamic models in the North Aegean: convergences, controversies and a further hypothesis

D. KONDOPOULOU[1], A. ATZEMOGLOU[2] & S. PAVLIDES[3]

[1] *Department of Geophysics, Aristotle University of Thessaloniki, Greece*
[2] *Institute of Geology and Mineral Exploration, Thessaloniki, Greece*
[3] *Department of Geology and Physical Geography, Aristotle University of Thessaloniki, Greece*

Abstract: The north Aegean region is a geologically complicated area, in which numerous studies have focused during the last decade. Palaeomagnetism can give a decisive contribution in selecting possible models. In the present study all available palaeomagnetic data from the north Aegean and surrounding area (39.5°N to 42°N and 20°E to 26°E) are compiled, classified and evaluated. Their significance in terms of the rotational behaviour is given as follows.

In various plutonic, volcanic (Eocene–Oligocene) and sedimentary (Mio–Pliocene) formations, systematic easterly declinations are observed. The rare exceptions are caused either by remagnetized material or by very local phenomena. The declination gradually increases from recent to older periods, varying between approximately 20° and 40°. This distribution is seen in the area between south Albania and the western Greek Rhodope. In contrast, the eastern Greek Rhodope appears to be unrotated. The inclinations diverge systematically from the expected values. This could be due to either an alternative pole position for the Eocene (Westphal 1993) or possible tilting of blocks during the Oligocene–Miocene, associated with a NE–SW brittle–ductile extensional stress field. The data partially confirm geodynamic models proposed for the area by Taymaz *et al.* (1991), Sokoutis *et al.* (1993) and Jolivet (1993). The model proposed by Dinter & Royden (1992) does not predict rotations and is not supported by the palaeomagnetic data.

When the published volcanological, structural, seismic and palaeomagnetic data are combined they suggest that the north Aegean region during the Early Tertiary was tectonically similar to the present day active south Aegean subduction zone.

The North Aegean is a particularly promising research area where both extensional and strike-slip structures associated with active deformation are observed alongside reactivated older structures, block rotations and strong seismic activity. The area may be considered typically continental and the recent and active deformation as intracontinental for the following reasons: (1) the active subduction is restricted to the south Aegean arc; (2) the crust is essentially continental, in contrast to other marginal seas underlaid by oceanic crust, and the north Aegean crust is considerably thinner (25–30 km) than in most young orogens; (3) seismicity is comparable to other continental areas of the world and does not resemble that of plate boundaries (Tsapanos 1990); (4) the seismic slip rate is almost a factor of two greater than rates in the south Aegean fore-arc zone and back-arc area (Papadopoulos 1989). Along the North Anatolian–North Aegean Trough slip rates vary from 20 mm to 6 mm per year (C. B. Papazachos *et al.* 1992); (5) the maximum fault-segment length of normal and strike-slip neotectonic faults in the North Aegean are characteristic of active faults on continents; and (6) the North Aegean is not a region of typical back-arc extension as strike-slip movements are important (Pavlides & Caputo 1994).

Pavlides *et al.* (1990) have suggested that the loading conditions required by recently active deformation of the whole North Aegean area are similar to those required for right-lateral transtension. If the strike-slip faults along the North Aegean Trough represent the continuation of the North Anatolian Fault System, it is likely that movement on the faults was initiated between the late Miocene and Early Pliocene (Mercier *et al.* 1989). However, a late Pliocene or younger age for initiation of strike-slip deformation along the North Aegean cannot be discounted.

An important reorganization of the Aegean occurred during the mid-Miocene (Mercier *et al.* 1989) after the last Alpine compressive phase in the Langhian (16 Ma). The Ionian zone was folded and began to be thrust on to the

From Morris, A. & Tarling, D. H. (eds), 1996, *Palaeomagnetism and Tectonics of the Mediterranean Region*, Geological Society Special Publication No. 105, pp. 277–288.

278   D. KONDOPOULOU *ET AL.*

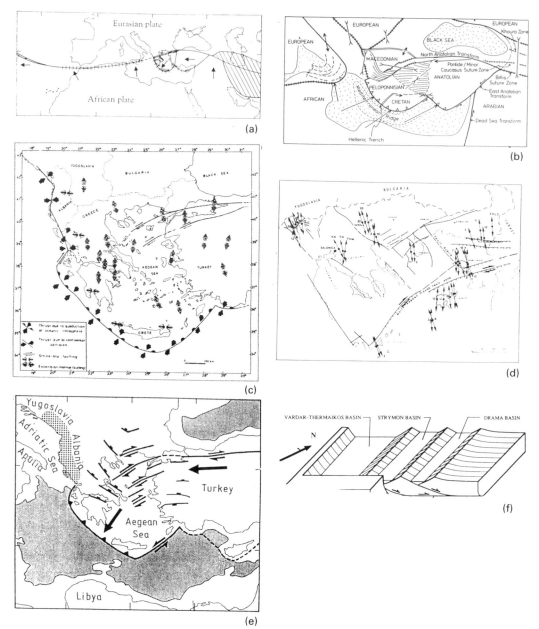

Fig. 1. Some of the proposed models for the tectonics of the Aegean and Eastern Mediterranean region from: (**a**) McKenzie (1970) reprinted by permission of Macmillan Magazines Ltd; (**b**) Dewey & Şengör (1979) reprinted by permission of Geological Society of America; (**c**) Papazachos *et al.* (1986) reprinted by permission of the author; (**d**) Mercier *et al.* (1989) reprinted by permission of Blackwell Science Ltd; (**e**) Taymaz *et al.* (1991) reprinted by permission of Blackwell Science Ltd; (**f**) Dinter & Royden (1991) reprinted by permission of Geological Society of America.

Preapulian zone, forming the convergent limit of an Ionian arc. The marine molassic Mesohellenic Trough vanished and new basins formed in the Central Aegean, while in the Northern Aegean and Bulgaria basins formed which connected either to the Mediterranean Sea or to the Parathethys. Several models have been suggested for the evolution of the whole Aegean (e.g. McKenzie 1970; Dewey & Şengör 1979). They generally sub-divide the region into a

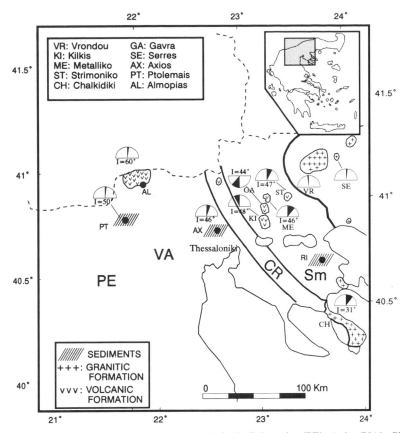

Fig. 2. Schematic configuration of palaeomagnetic results in the Pelagonian (PE), Axios (VA), Circum Rhodope (ER) and Serbomacedonian (Sm) geological zones. Thick lines: Rhodope limits and Strymon fault.

number of rigid microplates (e.g. Aegean, Anatolian, Macedonian) (Fig. 1).

Kinematics of minor and major faults in the North Aegean demonstrate three successive tensional directions from late Miocene to present day (Mercier et al. 1989). These directional changes are attributed to corresponding changes in the regional stress pattern, although rotations cannot be excluded for some areas. For example, Mercier et al. (1989) use structural data to suggest that the Lemnos block has rotated by an average of 15° clockwise relative to the neighbouring Samothraki–Thrace–Thassos block. Furthermore they propose that two major Hellenic arcs have been active in the Aegean since the Mid-Eocene: The Pelagonian-Pindic and the Aegean arc. Recent seismic tomography studies (C. B. Papazachos et al. 1994) suggest the presence of an old subduction zone dipping beneath the North Aegean Trough, which is distinctly different from the present day active subduction zone to the south.

Using improved earthquake focal mechan-

isms, Taymaz et al. (1991) describe the connection between the westward motion of Turkey relative to Europe and the extension in the Aegean Sea and surroundings. In their model, right-lateral strike-slip motion is distributed on NE- to ENE-trending faults in the central and eastern Aegean and NW Turkey. They suggest a deformation geometry consisting of a system of rotating broken slats attached to margins. This model is able to reproduce quantitatively a number of features in the Aegean, in particular senses and rates of rotation.

For the specific area of the north Aegean, Dinter & Royden (1993) suggest the NE–SW extension as responsible for the strike-slip motion along the North Aegean Trough from the Oligocene, while they completely ignore the role of E–W-trending neotectonic normal faults and the associated N–S Pleistocene extension. Furthermore, their model does not predict any block rotation.

According to Sokoutis et al. (1993), NE–SW-directed extension (initiated during the Oligo-

Table 1. *Published palaeomagnetic results for the study area (Eocene to Pliocene)*

Location (°N/°E)	Age	Rock type	N (sites)	D	I	α95	Remarks	References
Ptolemais (P) 40.4/21.5	U. Mioc.–Plioc.[†]	Sediments	6	4	50	15	Tilt corrected, negative fold test	Westphal et al. (1991)
Almopias (V) 41.2/22.0	2–4 Ma.[*]	Volcanics	6	195	-66.5	10		Bobier (1968)
Almopias (V) 41.2/22.0	2–4 Ma.[*]	Volcanics	4	17	54	16		Kondopoulou & Lauer (1984)
Axios (V) 40.6/22.5	Late Miocene[†]	Sediments	4	20	46	17	Positive reversal tests	Kondopoulou (1994)
Chalkidiki (CR) 40.3/24.0	35–45 Ma.[*]	Plutonics	10	37	31	9		Kondopoulou & Westphal (1986)
Kilkis (SM) 41.0/23.6	23–28 Ma.[*]	Volcanics	5	329	48	18		Pavlides et al. (1988)
Gavra (SM) 41.2/23.0	25 Ma.[*]	Volcanics	1	224.5	-44	6	Positive reversal test	Atzemoglou et al. (1994)
Metalliko (SM) 40.8/23.5	Eocene[†]	Volcanics	5	39	33	19	Tilt corrected	Westphal et al. (1991)
W. Greek Rhodope (Rh) 41.2/24.3	18–33 Ma.[*]	Plut.+Volc.	26	24	49	8	Positive reversal test	Atzemoglou et al. (1994)
Thrace (CR) 40.9/26.0	25–33 Ma.[*]	Plut.+Volc.	15	7	47	8	Positive reversal test/tilt corrected	Kissel et al. (1986)
Kirki–Essimi (CR) 40.9/26.0	Eocene[†]	Sediments	7	10	63	8		Spais (1987)
Essimi (CR) 40.9/26.0	30 Ma.[*]	Plut.+Volc.	4	2	52	6		Spais (1987)
Leptokarya (CR) 40.9/26.0	28 Ma.[*]	Plutonics	13	203	-46	5		Spais (1987)
Lemnos (Rh) 39.8/25.2	17–21 Ma.[*]	Volcanics	7	34	48	15		Westphal & Kondopoulou (1993)

P, Pelagonian Zone; V, Axios Zone; CR, Circum-Rhodope; SM, Serbomacedonian; Rh, Rhodope

[*] Radiometric

[†] Geological or biostratigraphical.

cene) is mainly responsible for the development of the Miocene–Quaternary sedimentary basins, with gravity collapse occurring along a detachment surface within a previously thickened lithosphere. This interpretation assumes a contemporaneous ductile–brittle rheology. It desagrees with the previous models, which attribute Tertiary ductile deformation to Alpine thrusting, and brittle extensional neotectonic deformation to back-arc extension above a subduction zone. By providing quantitative information on deformation, palaeomagnetism can be a very useful tool for testing geodynamic models. Attempts have already been made to combine palaeomagnetic and kinematic data in the central Aegean (McKenzie & Jackson 1986) and in central and north Aegean (Taymaz et al. 1991).

However, a considerable amount of new palaeomagnetic data have been presented for the Rhodope (Atzemoglou et al. 1994), enabling us to make a more complete analysis of the geodynamics of the north Aegean region than has been previously possible.

Palaeomagnetic data in the broader area

The palaeomagnetic method has proved very useful in documenting the geodynamic history of the Aegean region. The patterns of deformation established by palaeomagnetism are more clear in the external (convergent) zones than in the internal (extensional) zones where the picture is rather uncertain. In the last fifteen years an important palaeomagnetic dataset has been obtained in northern continental Greece and the north Aegean. These data come from both Mesozoic and Cenozoic formations. In the present paper we will concentrate on data from Eocene to Pliocene rocks. All the available data for this time span are compiled in Table 1, together with information about the sampled formations (ages, rock type, location). Schematic representations are given in Figs 2 and 3. The sampled units cover a range of rock types (sediments, volcanics, plutonics) distributed through the major geotectonic zones (Pelagonian, Axios, Circum Rhodope, Serbomacedonian, Rhodope), and have been obtained by researchers from various laboratories. It is considered that any systematic patterns arising from these data are reliable.

The Thessaloniki research group have made substantial contributions to palaemagnetic and structural studies in this region. Approximately 900 cores were collected (the majority drilled in the field) and subjected to standard rock magnetic and palaeomagnetic techniques, including: AF and thermal cleaning; IRM acqui-

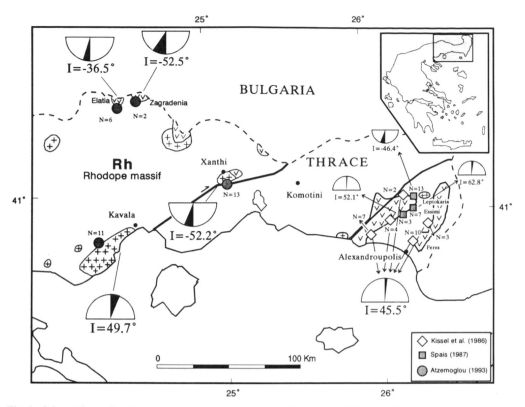

Fig. 3. Schematic configuration of palaeomagnetic results in Rhodope and Thrace. Thick lines: Kavala–Xanthi–Komotini fault and Circum-Rhodope limit.

sition and thermal demagnetization; Lowrie–Fuller tests; and thermomagnetic analyses. Additionally, for a number of igneous rocks collected in Thrace, reflected light microscope studies, RRM (rotational remanent magnetization) measurements, magnetic separation and XRD analyses were used to obtain estimates of the average size and concentration of the magnetic particles present (Spais 1987). Representative palaeomagnetic and rock magnetic results are shown in Fig. 4.

Fold tests have been possible in a few cases, using the method of McElhinny (1964) or, in ambiguous cases, McFadden (1990). In most of the volcanics and plutonics no tilt correction has been applied. This problem has been partially bypassed by considering the consistency between site mean directions over broad areas (40000 km^2), as well as the similarity of these directions with those obtained in sediments of neighboring parts of NW Greece (Kissel & Laj 1988).

Discussion of results

Declination values

Considering an W–E traverse from the Pelagonian zone of the West Greek Rhodope, declinations are predominantly directed eastwards, varying from 35° to 20° (Fig. 5). However, the following exceptions are noted: (i) N–S directions are observed in recent sediments in the Pelagonian zone. These sediments show evidence of a post-depositional magnetization, possibly associated with sulphide mineralization; and (ii) westerly declinations are observed in five sites in the Greek Serbomacedonian massif (Kilkis area). These directions are rather scattered ($\alpha_{95} = 18°$), but the implied counterclockwise rotation is supported by structural data which suggest clockwise rotation of the stress field (Pavlides et al. 1988). However in neighbouring formations of the same age only easterly declinations have

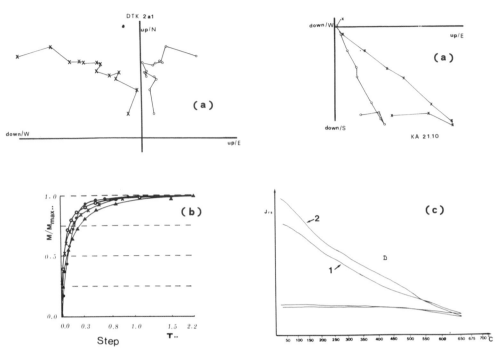

Fig. 4. (a) Examples of thermal demagnetization. Circles, horizontal plane; crosses, vertical N–S plane. (b) Examples of the IRM acquisition curves from representative samples. (c) Examples of thermomagnetic analysis by study of the decrease of saturation magnetization as a function of temperature. Vertical axis on arbitrary scale. 1, First heating curve; 2, second heating curve.

been observed. It is clear that additional detailed sampling is necessary in this area.

In Table 2, we have calculated expected declinations and inclinations for the region using the Eurasian and African poles given by Westphal *et al.* (1986). In almost all cases declinations diverge by 10–30° from the expected values, thus even small rotations seem to be well established.

Inclination values

It is clear from comparison of Tables 1 and 2 that inclinations are systematically lower than the expected values, which confirms observations made by various authors when considering data from the whole Aegean and continental Greece (Kissel & Laj 1988; Surmont 1989; Van der Voo 1993). Low palaeolatitudes with respect to those expected from the reference palaeomagnetic poles for stable Eurasia and Africa have already been reported for the whole Tethys belt (Westphal *et al.* 1986). Attempts to interpret these widespread inclination anomalies have been made by Westphal (1993), who after rejecting various alternative possibilities, suggests an alternative Eocene–Oligocene pole for Eurasia at 69°N, 215°E. In Table 2 we have also

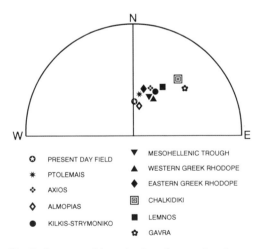

⊙ PRESENT DAY FIELD	▼ MESOHELLENIC TROUGH
✳ PTOLEMAIS	▲ WESTERN GREEK RHODOPE
✧ AXIOS	◆ EASTERN GREEK RHODOPE
◇ ALMOPIAS	▣ CHALKIDIKI
● KILKIS-STRYMONIKO	■ LEMNOS
	✿ GAVRA

Fig. 5. Stereographic projection of mean directions obtained in lavas and sediments of the broader area, for a period extending from Pliocene to Eocene. Numerical values from Table 1.

calculated the expected *D* and *I* values using this pole for Eocene and Oligocene formations. Rotations angles increase considerably with values diverging by 25–40° from the expected.

Table 2. *Expected D, I values on the basis of Eurasian, African or alternative VGP, for the areas and ages mentioned in Table 1*

Locality	Eurasian VGP* (°N, °E)	D_{exp}	I_{exp}	African VGP (°N, °E)*	D_{exp}	I_{exp}	Alternative VGP for Eurasia[+]	D_{exp}	I_{exp}
Ptolemais	83/142	8	56	84/161	5	55			
Axios	83/141	8	56	83/163	5	55			
Chalkidiki	81/146	9	55	80/185	4	50	69/215	356	36
Kilkis	83/136	8	57	84/170	4	55			
Gavra	83/136	8	58	84/170	4	56	69/215	355	37
Metalliko	78/156	10	52	76/199	1	46	69/215	356	37
W. Greek Rhodope	83/138	8	58	84/166	5	60	69/215	356	37
Thrace	83/136	8	58	84/170	4	56	69/215	357	36
Kirki-Essimi	78/156	11	52	76/199	2	45			
Lemnos	83/141	8	56	83/163	6	54			

* Westphal *et al*. (1986).
[+] Westphal (1993).

The compiled inclination values, however, are higher than those expected from this pole with two exceptions where a satisfactory convergence is observed: in Chalkidiki and Metalliko inclinations diverge by only 4–6° from the expected. In the other cases, possible explanations are imprecise age determinations or tilting of magnetic vectors in the extensional regime (Beck 1992). This latter possibility has been suggested by Atzemoglou *et al*. (1994) for the west Greek Rhodope data.

An alternative interpretation of Aegean inclination anomalies is given by Beck & Schermer (1994). These authors only consider the Aegean region, and examine and reject various alternatives before concluding that non-tectonic explanations are not valid or well established. They favour northward displacement of tectonic blocks, though this does not seem to be supported by the available geological evidence.

It is clear that both interpretations fail to solve the problem entirely. The alternative Eocene–Oligocene pole of Westphal (1993) cannot be applied to much of the discordant Aegean data which are of Miocene age, and it is difficult to accommodate more than 3° of northward displacement of Greece since the Eocene as the gap between Eurasia and Africa is not sufficiently large. Therefore, the problem of discordant inclination values remains open.

Other data sources

Palaeomagnetic data in the North Aegean show a quite uniform distribution of easterly declinations. This suggests movement of a lithospheric unit of regional extent rather than rotation of a series of discrete blocks. This observation is very important as the area is dissected by major and minor faults. For this reason it is essential to obtain information from independent data before drawing definite conclusions. Such data can be geophysical, seismological, petrological and volcanological.

Geophysical and seismological data

The Rhodope is a key to understanding North Aegean geodynamics. It has proved to be a mobile part of the Alpine belt rather than a stable microcontinent (Burg *et al*. 1990; Kilias & Mountrakis 1990; Kolokotroni & Dixon 1991; Atzemoglou *et al*. 1994). Within the Rhodope, the Komotini–Xanthi basin is a typical example of a Tertiary sedimentary basin, controlled principally by the Komotini–Xanthi–Kavala fault (Fig. 3). This basin is associated with a negative gravity anomaly. Modelling of this anomaly and of the aeromagnetic anomaly of the Xanthi granite suggests that: (i) the granite extends under the basin to the south; and (ii) the intrusion is very similar with the Leptokarya intrusion in petrological composition, age and magnetic susceptibility values (Maltezou & Brooks 1989). From the palaeomagnetic study it can be seen that both intrusions yield similar magnetization directions ($D = 203°, I = -46°$ for Leptokarya and $D = 208°, I = -53.5°$ for Xanthi; Spais 1987; Atzemoglou *et al.* 1994). The maximum depth of the bodies is estimated to be about 1.2 km and 6.5 km for the Xanthi and Leptokarya bodies respectively. The shallow depth and the general features of the Leptokarya body led Spais & Maltezou (1993) to interpret the palaeomagnetic direction in terms of rotation and tilting of the body due to its emplacement by thrusting. This interpretation seems plausible but cannot hold for the numerous formations (both plutonic and volcanic) which gave similar directions throughout the whole western Rhodope (Table 1).

A thorough study of all geophysical parameters in the Bulgarian and Greek Rhodope has been performed recently (Sideris *et al.* pers. comm.). The analysis of observed and transformed geophysical fields give evidence for a heterogeneous structure of the Rhodope Massif, which can be separated into two parts: western and southeastern. An increase of present dynamics to the south seems plausible. Furthermore, Sideris *et al.* (pers. comm.) propose three 'collage blocks': a Rhodopean, a Thrakian and a Serbomacedonian.

Shanov *et al.* (1992) used various geophysical fields (gravity, magnetotellurics) to support the view of a palaeosubduction zone beneath the Rhodope. These authors claimed that the intense seismic activity of southern Bulgaria is a strong indication of prolonged subduction which has not yet attenuated. The presence of an earlier subduction regime has been suggested by other workers (Boccaletti *et al.* 1974; Papazachos 1976; Spakman *et al.* 1988; Mercier *et al.* 1989) and is also supported by recent seismic tomography studies (C. B. Papazachos *et al.* 1994). Furthermore, the geophysical image obtained by tomography in the North Aegean is similar to that found in the southern Aegean, with a very thin crust (<25 km) and comparable heat flow maxima.

Petrological and volcanological data

Many detailed studies on the characteristics and emplacement mechanisms of the Tertiary granites of Northern Greece have been published in the last few years. Koukouvelas & Pe-Piper (1991) suggest a subduction-related origin for the Xanthi pluton and a different magmatic signature for the Kavala one, which post-dates the main deformation of the Rhodope. Additionally the authors see no evidence for subsequent strike-slip motion on the Kavala–Xanthi–Komotini fault (separating the Rhodope massif from the circum-Rhodope zone) since the intrusion of the Xanthi pluton. These observations reinforce the possibility of a regional lithospheric rotation rather than rotation of distinct fault blocks.

Considering both the Rhodope granitoids and associated volcanics, Jones *et al.* (1992) suggested that their geochemical characteristics are shared by other northern Greece granitoids (Chalkidiki, Thrace). They observed that the emplacement of granites in northern Greece was greatly influenced by pre-existing northerly dipping thrusts, and shows a N–S progressive younging from Serres to Xanthi to Kavala (*c.* 40 Ma in southern Bulgaria to *c.* 15 Ma in Kavala). The same age transition is evident in the intrusions east of Xanthi. In a recent study, Liati & Seidel (1995) state that the maximum burial depth, degree of overprint, temperature and age of metamorphism decrease from north to south. This southward younging is related by Del Moro *et al.* (1988) to subduction of the African plate beneath Eurasia, thereby favouring the model of Fytikas *et al.* (1984) which suggests a progressive increase in the angle of dip of the down-going slab. More recently, Innocenti *et al.* (1994) proposed that a north Aegean volcanic arc extended from Bulgaria and continental Greece to west Turkey in Tertiary times. Their model involves a sudden shift of subduction, in contrast to the model of Pe-Piper & Piper (1989) who proposed continuous subduction along the middle Aegean convergence zone.

Discussion and conclusions

By obtaining new, accurate palaeomagnetic data of mostly Oligo-Miocene age in the northern Serbomacedonian and western Greek Rhodope zones, it has been possible to extend the region of observed clockwise rotations in NW Greece further to the east. Along a W–E traverse from the Pelagonian zone to the western Greek Rhodope clockwise rotations, with one exception, prevail, but no clear systematic change can be observed from west to east. From Eocene to Pliocene a decrease of the measured rotations (from 38° to practically zero) is observed when different weight is given to the results, according

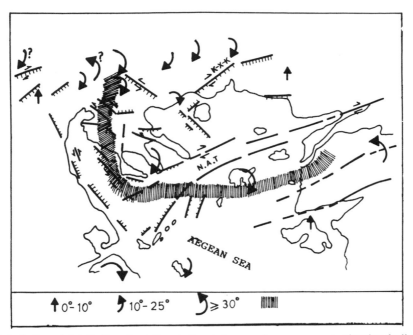

Fig. 6. Main faults and distribution of rotations in the study area. Arrows are proportional to declination values. K–X–K, Kavala–Xanthi–Komotini fault; N.A.T., North Aegean Trough. Location of a possible palaeo-subduction zone when previous models and palaeomagnetic data are taken into account is shaded.

to their reliability (number of sites and uncertainties in age). This pattern does not hold for Thrace, where the sampled formations belong to the eastern termination of the Circum Rhodope belt (Fig. 5). In this area, the Oligocene and Oligo–Miocene results yield almost N–S directions, with the exception of Leptokarya, discussed above. This regional rotation can be amplified in some cases by dextral movement along the length of the North Aegean Trough (e.g. Lemnos island).

The continuous dextral movement along the North Aegean Trough can be used to date the rotations in the vicinity of the fault in the following way: the angle of rotation (25–30°) observed in Chalkidiki (30–45 Ma) and Lemnos–Skyros (14–21 Ma), which reflect the North Aegean Trough movement, is very similar. Assuming that the rotation of Chalkidiki is at least partly related to this movement, then the clockwise rotation recorded along the North Aegean Trough could be younger than 14 Ma.

Further to the north, the Kavala–Komotini–Xanthi fault, is sub-parallel to the North Aegean Trough system and shares several of its characteristics, including a similar distribution of rotations along both faults. We suggest that this Kavala–Komotini–Xanthi fault is a key to the geodynamic evolution of the area, as it appears

to separate the rotated western Greek Rhodope from the unrotated part of northern Greece (Thrace). In contrast, the important Strymon 'thrust fault' is crossed without significant changes in declination. This observation reinforces the suggestion of Jones *et al.* (1992) that petrological and structural characteristics on both sides of this thrust are identical and that its role should be reconsidered. A similar suggestion is made by Burg *et al.* (1995) based on map continuity, similarities of lithologies, structures and strain regime between the Vertiskos–Kerdillion Massifs (Serbomacedonian) and the Rhodope. The authors integrate both massifs with the Rhodope into a single, major element of the Tethyan orogenic system and this is supported by the palaeomagnetic data.

In addition to these palaeomagnetic data and their implications, abundant independent studies suggest the existence of a subduction zone in the area during the Early Tertiary. It seems reasonable, therefore, to suggest that the tectonics of the central–northern Aegean sector during the Early Tertiary were similar to those of the southern Aegean during the initial stages of Alpide convergence (Le Pichon & Angelier 1979). In this case one would expect anticlockwise rotations further to the east, once the 'stable' part of Thrace is crossed. Such late

Miocene rotations (*c.* 30°) have been recently reported for two transects across the North Anatolian Fault Zone (Platzman *et al.* 1994). Additionally, unpublished data from Miocene formations in the Biga Peninsula, close to the north Aegean islands, yield a 20° westward declination (Platzman pers. comm.).

The above compiled data favour a regional lithospheric rotation rather than rotation of fault bounded blocks. A possible configuration of implied rotations is shown in Fig. 6. It is clear that geodynamic models which do not predict rotations in the area are not supported by these data. The Taymaz *et al.* (1991) model of broken slats predicts increasing rotations from north to south of opposite senses to the east and west. It could account, at least partially, for the observed deformational pattern. The question arises of the location of the stable margin to this deforming zone. We suggest that it should be found further to the north or northeast, beyond the limit of the extensional regime at 43°N. Therefore, new accurate data from the Bulgarian Rhodope should be obtained urgently.

Acknowledgements

We wish to thank G. Sideris and S. Nicolaou (IGME, Athens) for providing a preliminary version of their study on the Rhodope deep structure (in collaboration with A. Angelopoulos and D. Josifov). E. McClelland helped substantially with critical remarks on the first version of this paper. A. Morris contributed to the present version by a constructive review and numerous language corrections which greatly improved the presentation of the paper. Finally we would like to acknowledge for copyright permission Macmillan magazines Ltd, Blackwell Science Ltd, GSA and Papazachos *et al.* (1986).

References

ATZEMOGLOU, A., KONDOPOULOU, D., PAPAMARINO-POULOS, S. & DIMITRIADIS, S. 1994. Palaeomagnetic evidence for block rotations in the Western Greek Rhodope. *Geophysical Journal International,* **118**, 221–230.

BECK, M. 1992. Some thermal and paleomagnetic consequences of tilting a batholith. *Tectonics,* **11**, 297–302.

—— & SCHERMER, E. R. 1994. Aegean paléomagnétic inclination anomalies. Is there a tectonic explanation? *Tectonophysics,* **231**, 281–292.

BOBIER, C. 1968. Etude paléomagnétique de quelques formations du complexe volcanique d'Almopias. *Comptes Rendus de l'Académie des Sciences, Paris,* **267**, 1091–1094.

BOCCALETTI, M., MANETTI, P. & PECCERILLO, A. 1974. The Balkanides as an instance of a back-arc thrust belt: possible relation with the Hellenides. *Geological Society of America Bulletin,* **85**, 1077–1084.

BURG, J. P., IVANOV, Z., RICOU, L. E., DIMOR, D. & KLAIN, L. 1990. Implications of shear-sense criteria for the tectonic evolution of Central Rhodope massif, Southern Bulgaria. *Geology,* **18**, 451–454.

——, GODFRIAUX, I. & RICOU, L. E. 1995. Extension of the Mesozoic Rhodope thrust units in the Vertiskos–Kerdilion Massifs (Northern Greece). *Comptes Rendus de l'Academie des Sciences, Paris,* **320**, **IIa**, 889–896.

DEL MORO, A., INNOCENTI, F., KYRIAKOPOULOS, C., MANETTI, P. & PAPADOPOULOS, P. 1988. Tertiary granitoids from Thrace (Northern Greece): Sr isotopic and petrochemical data. *Neues Jahrbuch für Mineralogie Abhandlungen,* **159(2)**, 113–115.

DEWEY, J. F. & SENGOR, A. M. C. 1979. The Aegean and surrounding regions: Complex multiplate and continuum tectonics in a convergent zone. *Geological Society of America Bulletin,* **90**, 84–92.

DINTER, D. A. & ROYDEN, L. 1993. Late Cenozoic extension in northeastern Greece: Strymon Valley detachment system and Rhodope metamorphic core complex. *Geology,* **21**, 45–48.

FYTIKAS, M., INNOCENTI, F., MANETTI, P., MAZZUOLI, R., PECCERILLO, A. & VILLARI, L. 1984. Tertiary to Quaternary evolution of volcanism in the Aegean region. *In*: DIXON, J. E., ROBERTSON, A. H. F. (eds) *The Geological Evolution of the Eastern Mediterranean.* Geological Society, London, Special Publications, **17**, 687–699.

INNOCENTI, F., MANETTI, P., MAZZUOLI, R., PERTUASI, P., FYTIKAS, M. & KOLIOS, N. 1994. The geology and geodynamic significance of the island of Lemnos, North Aegean Sea, Greece. *Neues Jahrbuch für Mineralogie Abhandlungen,* **H11**, 661–691.

JOLIVET, L. 1993. Extension of thickened continental crust, from brittle to ductile deformation: examples from Alpine Corsica and Aegean Sea. *Annali di Geofisica,* **34**, 139–153.

JONES, C., TARNEY, J., BAKER, J. H. & GEROUKI, F. 1992. Tertiary granitoids of Rhodope, northern Greece: magmatism related to extensional collapse of the Hellenic Orogen?, *Tectonophysics,* **210**, 295–314.

KILIAS, A. & MOUNTRAKIS, D. 1990. Kinematics of the crystalline sequences in the Western Rhodope massif. *Geologica Rhodopica,* **2**, 100–116.

KISSEL, C. & LAJ, C. 1988. The Tertiary geodynamic evolution of the Aegean Arc: a palaeomagnetic reconstruction. *Tectonophysics,* **146**, 183–201.

——, KONDOPOULOU, D., LAJ, C. & PAPADOPOULOS, P. 1986. New palaeomagnetic data from Oligocene formations of Northern Aegea. *Geophysical Research Letters,* **13(10)**, 1039–1042.

KOLOKOTRONI, C. & DIXON, J. E. 1991. The origin and emplacement of the Vrondou granite, Serres, N.E. Greece. *Bulletin of the Geological Society of Greece,* **25**, 469–483.

KONDOPOULOU, D. 1994. Some constraints on the origin and timing of the magnetization for Mio-Pliocene sediments from N. Greece. *Proceedings of the VII Congress of the Geological Society of Greece,* in press.

—— & LAUER, J. P. 1984. Palaeomagnetic data from

Tertiary units of the North Aegean Zone. *In*: DIXON, J. E. & ROBERTSON, A. H. F. (eds). *The Geological Evolution of the Eastern Mediterranean*. Geological Society of London, Special Publications, **17**, 681–686.

—— & WESTPHAL, M. 1986. Palaeomagnetism of the Tertiary intrusives from Chalkidiki (N. Greece), *Journal of Geophysics*, **59**, 62–66.

KOUKOUVELAS, I. & PE-PIPER, G. 1991. The Oligocene Xanthi pluton, northern Greece: a granodiorite emplaced during regional extension. *Journal of the Geological Society, London*, **148**, 749–758.

LE PICHON, X. & ANGELIER, J. 1979. The Hellenic arc and trench system: a key to the neotectonic evolution of the Eastern Mediterranean region. *Tectonophysics*, **36**, 339–346.

LIATI, A. & SEIDEL, E. 1995. PT-paths of overprinted high-pressure rocks in Central Rhodope, N. Greece: implications for the regional tectonic evolution. *Terra Abstracts*, **7**, 45.

MALTEZOU, F. & BROOKS, M. 1989. A geophysical investigation of the post-Alpine granites and Tertiary sedimentary basins in Northern Greece. *Journal of the Geological Society, London*, **146**, 53–59.

McELHINNY, M. W. 1964. Statistical significance of the fold test in palaeomagnetism. *Geophysical Journal of the Royal Astronomical Society*, **8**, 338–340.

McFADDEN, P. L. 1990. A new fold test for palaeomagnetic studies. *Geophysical Journal International*, **103**, 163–169.

McKENZIE, D. P. 1970. Plate tectonics of the Mediterranean region. *Nature*, **226**, 239–243.

—— & JACKSON, J. 1986. A block model of distributed deformation by faulting. *Journal of the Geological Society, London*, **143**, 349–353.

MERCIER, J. L., SOREL, D., VERGELY, P. & SIMEAKIS, K. 1989. Extensional tectonic regimes in the Aegean basins during the Cenozoic. *Basin Research*, **2**, 49–71.

PAPAZACHOS, B. C. 1976. Seismotectonics of the Northern Aegean Area. *Tectonophysics*, **33**, 199–209.

——, KIRATZI, A. A., HATZIDIMITRIOU, P. M. & KARAKOSTAS, B. G. 1986. Seismotectonic Properties of the Aegean Area that Restrict Valid Geodynamic models. *2nd Wegener Conference*, Athens, 1–6.

PAPAZACHOS, C. B., KIRATZI, A. A. & PAPADIMITRIOU, E. E. 1992. Orientation of active faulting in the Aegean and surrounding area. *6th Congress of the Geological Society of Greece*, in press.

——, HATZIDIMITRIOU, P. M., PANAGIOTOPOULOS, D. G. & TSOKAS, G. N. 1995. Tomography of the crust and upper Mantle in S. E. Europe. *Journal of Geophysical Research*, **100**, **B7**, 12 405–12 422.

PAPADOPOULOS, G. A. 1989. Seismic and volcanic activities and aseismic movements as plate motion components in the Aegean area. *Tectonophysics*, **167**, 31–39.

PAVLIDES, S. & CAPUTO, R. 1994. The north Aegean region: a tectonic paradox? *Terra Nova*, **6**, 37–44.

——, KONDOPOULOU, D., KILIAS, A. & WESTPHAL, M. 1988. Complex rotational deformations in the

Serbo-Macedonian Massif (N. Greece): structural and palaeomagnetic evidence. *Tectonophysics*, **145**, 329–335.

——, MOUNRAKIS, D., KILIAS, A. & TRANOS, M. 1990. The role of strike-slip movements in the extensional area of Northern Aegean (Greece). A case of transtensional tectonics. *Annales Tectonicae*, **4**, 196–211.

PE-PIPER, G. & PIPER, D. J. W. 1989. Spatial and temporal variation in Late Cenozoic back-arc volcanic rocks, Aegean Sea region. *Tectonophysics*, **169**, 113–134.

PLATZMAN, E. S., PLATT, J. P., TAPIRDAMAZ, C., SANVER, M. & RUNDLE, C. C. 1994. Why are there no clockwise rotations along the North Anatolian Fault Zone? *Journal of Geophysical Research*, **99**, 21705–21715.

SHANOV, S., SPASSOV, E. & GEORGIEV, T. 1992. Evidence for the existence of a palaeosubduction zone beneath the Rhodopean massif (Central Balkans). *Tectonophysics*, **206**, 307–314.

SOKOUTIS, D., BRUN, J. P., VAN DEN DRIESSCHE, & PAVLIDES, S. 1993. A major Oligo-Miocene detachment in southern Rhodope controlling north Aegean extension. *Journal of the Geological Society, London*, **150**, 243–246.

SPAIS, C. 1987. *Palaeomagnetic and magnetic fabric investigations of Tertiary rocks from the Alexandroupolis area, N.E. Greece*. PhD thesis, University of Southampton.

—— & MALTEZOU, F. 1993. A combined palaeomagnetic and magnetic modelling study of the Leptokarya intrusion, Eastern Rhodope. *Proceedings of the 2nd Congress of the Hellenic Geophysical Union*, Florina, **2**, 307–319.

SPAKMAN, W., WORTEL, M. J. R. & VLAAR, N. J. 1988. The Hellenic subduction zone: a tomographic image and its geodynamic implications. *Geophysical Research Letters*, **15**, 60–63.

SURMONT, J. 1989. Paleomagnetisme dans les Hellenides internes: analyse des aimantations superposees par la methode des cercles de reaimantation. *Canadian Journal of Earth Sciences*, **26**, 2479–2494.

TAYMAZ, T., JACKSON, J. & McKENZIE, D. 1991. Active tectonics of the North and Central Aegean Sea. *Geophysical Journal International*. **106**, 433–490.

TSAPANOS, T. 1990. Relations between seismicity and tectonics in a global scale. *Bulletin of the Geological Society of Greece*, **25**, 3, 271–283.

VAN DER VOO, R. 1993. *Palaeomagnetism of the Atlantic, Tethys and Iapetus Oceans*. Cambridge University Press.

WESTPHAL, M. 1993. Did a large departure from the geocentric axial dipole hypothesis occur during the Eocene? Evidence from the magnetic polar wander path of Eurasia. *Earth and Planetary Science Letters*, **117**, 15–29.

—— & KONDOPOULOU, D. 1993. Palaeomagnetism of the Miocene volcanics from Lemnos island: implications for block rotations in the vicinity of the North Aegean Trough. *Annales Tectonicae*, **VII**, 2, 142–149.

——, BAZHENOV, M., LAUER, J. P., PECHERSKY, M. &

Sıbuet, J. P. 1986. Palaeomagnetic implications of the evolution of the Tethys belt from the Atlantic ocean to the Pamirs since the Triassic. *Tectonophysics,* **123**, 37–82.

——, Kondopoulou, D., Edel, J. B. & Pavlides, S. 1991. Palaeomagnetism of late Tertiary and Plio-Pleistocene formations from N. Greece. *Bulletin of the Geological Society of Greece,* **25**, 239–250.

Implications of ophiolite palaeomagnetism for the interpretation of the geodynamics of Northern Greece

H. FEINBERG[1], B. EDEL[2], D. KONDOPOULOU[3] & A. MICHARD[1]

[1] *Laboratoire de Géologie, CNRS UA 1316, ENS, 24 rue Lhomond, 75231 Paris Cedex, France*

[2] *Laboratoire de Paléomagnétisme, CNRS UA 323, IPG, 5 rue Descartes, 67084 Strasbourg, France*

[3] *Aristotle University of Thessaloniki, Department of Geology, 540 06 Thessaloniki, Greece.*

Abstract: Independent studies undertaken in Northern Greece (Axios–Chalkidiki belt and Vourinos massif), mainly on ophiolites but also on granites and metasediments, have shown that despite the occurrence of widespread remagnetizations it is possible to find a consistent arrangement of palaeomagnetic directions. Application of the fold test to four sites in the Kassandra peninsula, comprising pillow lavas and overlying metaflysch deposits demonstrates that the main regional magnetic component ($D = 47.7°$, $I = 47.7°$, $\alpha_{95} = 8.8°$) is a post-folding remagnetization. A comparison with directions found in the Sithonia granite shows that this component was probably acquired during the late-metamorphic emplacement of this granite (in the Eocene). An older component ($D = 325.4°$, $I = 34.1°$, $\alpha_{95} = 18°$) defined at a site in the sediments of the Kassandra area is also present in the ophiolites. This direction has probably been acquired during Late Jurassic–Early Cretaceous times. Further comparisons with the directions found in the Axios–Chalkidiki–Vourinos ophiolites and Florina granite show that the Mesozoic component is consistently found over a broad area.

The Cenozoic clockwise rotation of about 45° shown by the Eocene remagnetisation has been widely described by various authors, but the Mesozoic rotation indicated by the oldest component has been less investigated. This counterclockwise rotation is in the same sense as the rotation of the African plate during the Mesozoic, but is of larger amplitude. This suggests that: (i) Northern Greece was part of the African foreland before the Alpine collision; and (ii) oceanic spreading south of the Pelagonian zone during Cretaceous times would have increased the rotational movement of the studied area with respect to Africa.

Among the main geological issues in Northern Greece, the problem of the origin of the ophiolites has occupied an important place for many years (Smith *et al.* 1975; Mountrakis 1984; Vergely 1984; Ferrière 1985; Ricou *et al.* 1986; Robertson *et al.* 1991). Do the eastern, Vardar–Axios ophiolites and the western, Vourinos–Pindos–Othrys ophiolites originate in a single Jurassic ocean, formerly located to the NE of the Pelagonian continental block, or do they come from distinct oceanic basins located on the NE and SW sides of the Pelagonian domain, respectively? Palaeomagnetic data can be useful in resolving such geodynamic issues. Several palaeomagnetic results are now available for the Pelagonian and Axios zones, either in the southern part (Turnell 1988; Surmont 1989; Surmont *et al.* 1991; Morris 1995) or in the northern one (Edel *et al.* 1990, 1991/1992). In their study of the Chalkidiki ophiolites (Axios zone), Edel *et al.* (1991/1992) demonstrated that

the ophiolites of Northern Greece record several superimposed directions of magnetization which can be related to Mesozoic and Cenozoic events. An important Cenozoic remagnetization often hampers unravelling the orientation of the Mesozoic characteristic components. Taking advantage of these previous works, we have undertaken a study of selected sites in the ophiolitic massifs of northern Greece or in their associated sediments, identifying in every case the palaeohorizontal in order to perform field tests. In the present paper, we discuss first the Cenozoic magnetisation then we attempt to interpret the significance of the Mesozoic component.

Geological setting

The studied sites encompass the two main ophiolitic belts of Northern Greece, namely the Vardar–Axios zone to the east of the Pelagonian

From Morris, A. & Tarling, D. H. (eds), 1996, *Palaeomagnetism and Tectonics of the Mediterranean Region*, Geological Society Special Publication No. 105, pp. 289–298.

289

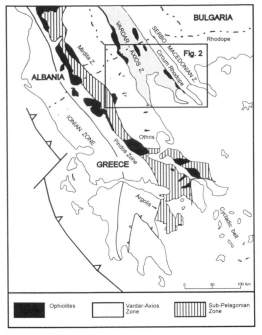

Fig. 1. Structural sketch map and location of studied area.

zone, and the Sub-Pelagonian zone to the west (Fig. 1). In the Vardar–Axios belt, samples were collected from four areas located in the eastern part of the belt, here named the Axios–Chalkidiki ophiolitic zone, which roughly corresponds to the Peonias zone of Mercier (1968) (Fig. 2). To the north of this zone the Gevgueli complex (four sites) offers a complete, although dismembered, ophiolitic suite, intruded by the Tithonian–Early Cretaceous Fanos granite (Bebien 1982, 1991; Spray *et al.* 1984). More southward, the Oreokastro massif (one site), which belongs either to the Axios–Chalkidiki zone or to the External Circum-Rhodope zone (Kaufmann *et al.* 1976; Kockel & Mollat 1977; Michard *et al.* 1994), corresponds to a solid gabbroic sliver vertically tilted together with its Late Tithonian cover of conglomeratic limestones (Mercier 1968; Kockel & Mollat 1977). In the Gerakini massif (Gauthier 1984; Edel *et al.* 1991/92), the Sonia site comprises layered gabbros close to the palaeo-Moho discontinuity. The Paliouri massif crops out at the southern tip of the Kassandra peninsula. On top of the spilitic pillow lavas of this small, and up to now neglected massif, two distinct sedimentary formations are observed (Fig. 2, insert). The lower formation, here named the Paliouri flysch, consists of rhythmic, volcano-clastic and pelagic

layers, which were apparently deposited directly after eruption of the pillow lavas in a deep-water environment. This volcano-clastic flysch is unconformably overlain by shallow-water deposits, consisting of reefal limestones and conglomeratic calcarenites with pebbles from the underlying formations (diabases and flysch). We interpret these shallow-water sediments as post-obduction deposits, equivalent to the Late Tithonian–Early Cretaceous Oreokastro limy conglomerates. The Paliouri pillow lavas and volcano-clastic flysch show greenschist-facies mineral assemblages, while the overlying Tithonian conglomerates are only affected by a very low grade metamorphism. Our interpretation contrasts with that of Kockel & Mollat (1977) who considered the volcano-clastic deposits east of Paliouri as younger than the Tithonian limestones.

In the Sub-Pelagonian zone, samples were drilled at four sites in the Vourinos massif (Fig. 2). The ophiolitic suite is complete and almost undisturbed in the Vourinos from the harzburgitic tectonites with chromite pods (Aetoraches site) up to the volcanic sequence (Krapa Hills dykes) (Moores 1969; Rassios *et al.* 1983). The volcanics are in turn overlain by the Krapa Hills limestones which include Calpionellids-bearing limestones at their base, followed by Early Cretaceous to Cenomanian rudist-bearing layers (Mavrides 1980). Amphibolite from the infraophiolitic sole was dated at 170 Ma (Spray & Roddick 1980), giving an upper limit to the age of ophiolite accretion and a lower limit to that of its obduction.

Two main orogenic periods can be recognized in the studied area: the Eo-Hellenic and the Meso-Hellenic phases. In broad terms, the 'Eo-Hellenic phase' (cf. Jacobshagen 1986) includes: (i) the Late Jurassic obduction of the Pelagonian/Sub-Pelagonian ophiolites; (ii) coeval metamorphic recrystallizations (cf. Paliouri area), the grade of which would increase eastward (i.e. Circum-Rhodope; Michard *et al.* 1994); and (iii) the emplacement of some granite massifs (Fanos, Monopigadhon, and more to the east, Arnea) at about 158–150 Ma (Spray *et al.* 1984, De Wet *et al.* 1989).

The 'Meso-Hellenic phase' post-dates the onlap of Tithonian–Cretaceous–Palaeocene sediments onto the Pelagonian ophiolites and their substrate. This Eocene phase caused westerly-directed thrust emplacement in the Vardar–Axios–Pelagonian–Sub-Pelagonian zones and was accompanied and followed by a metamorphic event involving a low-grade, greenschist-facies alteration in the studied massifs. Several Eocene–Oligocene granitoids were

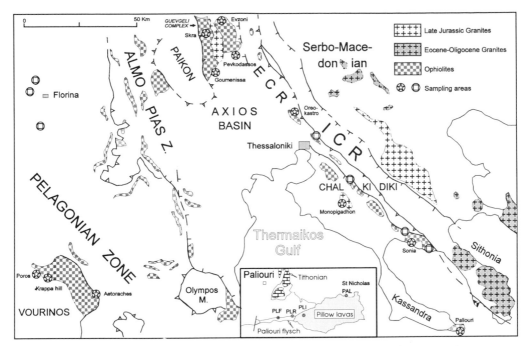

Fig. 2. Geological sketch map (ECR, External Circum Rhodope; ICR, Internal Circum Rhodope). Sampling sites: circled squares, sites of Edel *et al.* (1992); circled stars, present study and Kondoupoulou *et al.* 1994. Insert corresponds to the Paliouri area.

emplaced in the internal Circum-Rhodope zone and the juxtaposed Serbo-Macedonian zone, immediately to the east of the Chalkidiki ophiolites (Kondopoulou & Westphal 1986; De Wet *et al.* 1989).

Sampling, and laboratory measurements

A constant sampling strategy was applied for this study. All possible indications of the palaeohorizontal were searched for and their orientation carefully measured. The most simple case was that of oceanic sediments associated with ophiolites. In general, it was also possible to get values of dip and strike directly from the pillow lava flattening planes, which are related to their initial emplacement. In the gabbros we tentatively used the cumulate layering which is considered as almost horizontal when far from the top of the magmatic chamber (Nicolas 1989). When it was possible to assume an originally vertical setting (sheeted-dyke complex) the dyke orientations were also taken into consideration.

Palaeomagnetic measurements were conducted on 165 cored samples, oriented using both magnetic and sun compasses. Predominantly thermal and in some cases alternating field (AF) demagnetizations were performed.

Natural remanent magnetization (NRM) was measured at each step with either cryogenic or spinner magnetometers, depending on the intensity of NRM. Bulk susceptibility was measured on pilot samples after each demagnetization step to monitor possible mineralogical transformation during heating. Directions were analysed on orthogonal plots and selected with a least-squares routine. Isothermal remanent magnetization (IRM) acquisition and demagnetization experiments were performed on specimens representative of the different types of sampled rocks to investigate the magnetic mineralogy.

Results from the Paliouri area

Four sites have been sampled at the tip of the Kassandra peninsula near the village of Paliouri (Fig. 2, insert). Two sites, PLI (strike = 60°, dip = 30°) and PAL (strike = 25°, dip = 30°) consist of relatively well-preserved pillow lavas. The other two correspond to the 'Paliouri flysch', defined in the preceeding section. Site PLR (strike = 90°, dip = 30°) was sampled in a pyroclastic rock in stratigraphic contact with the pillows, while Site PLF (strike = 70°, dip = 15°) belongs to the true Paliouri flysch sequence of fine-grained sandstones and slumps.

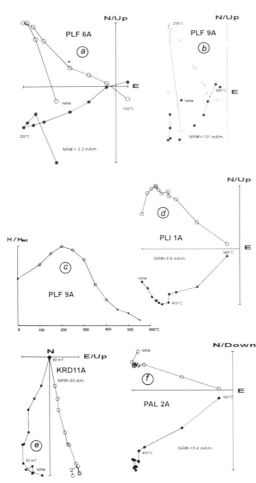

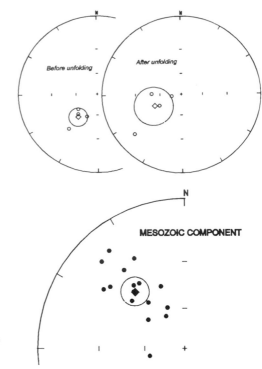

Fig. 4. (**a**) Stereographic projections of the intermediate direction found at the four sites of Paliouri, before and after tilt correction. (**b**) Stereographic projection of the Mesozoic component.

Fig. 3. Vector demagnetisation diagrams. Full (open) circles correspond to projections onto the horizontal (vertical) plane; (**a–e**), samples thermally cleaned; (**f**) alternating field cleaning.

Samples from different sites show different behaviour during demagnetization. The pillow lavas (after the destruction of a viscous component) display only one direction of magnetization between 200° and 580°C (Fig. 3f), whereas the sediments and especially those from site PLF exhibit two or three components as follows: (i) a viscous component destroyed before heating at 220°C; (ii) an intermediate component with a negative inclination and a southwestward declination, completely unblocked below 500–550°C; and (iii) a high temperature normal component visible up to 520°C (Fig. 3a–d). It was not possible in general to perform complete thermal cleaning of the sediments as mineralogical changes occurred during heating up to 570°C. The artificial production of magnetic minerals is indicated by a rapid increase of the susceptibility and of the remanent magnetization. On the other hand, it was impossible to successfully use AF cleaning on these samples. However, AF treatment gave good results with the pillow lavas.

The main magnetic mineral in the effusive rocks from the Paliouri area is likely to be magnetite or titano-magnetite, as the remanence shows a low coercivity and is completely destroyed at 580°C. Accordingly thin sections observed under reflected light show the common occurence of magnetite or more likely titano-magnetite in the pillow lavas. The magnetic mineralogy of the sediments from the same locality comprises a magnetic carrier with a high coercivity and intermediate to high unblocking temperatures, which might be hematite with different grain-size populations.

The intermediate unblocking temperature component present in the sediments has a similar direction to that of the single component observed in the pillow lavas (Table 1). The four site mean directions fail a fold test (Table 1; Fig.

Table 1. *Intermediate unblocking temperature components*

Site	Location	N	D_g	I_g	D_s	I_s	K	α_{95}
Paliouri area								
PLF	Paliouri flysch	15	223.6	−51.3	243.2	−55.7	79	4.1
PLR	Paliouri flysch	3	204.9	−55.1	260.7	−75.9	64	12.4
PAL	Paliouri pillows	14	219.6	−30.6	229.7	−14.3	29	6.9
PLI	Paliouri pillows	9	232.0	−55.9	271.9	−49.1	17	10.7
Mean		4	220.0	−48.6			35	15.8
					247.2	−50.4	8	34.2
Fold test: $K_1/K_2 = 0.24$, $F(6,6)$ 5% = 4.28								
Chalkidiki and Vourinos areas								
SKA	Skra layered gabbros	8	51.1	43.7	62.1	−6.9	8	17.5
SKB	Skra layered gabbros	18	42.0	44.0	66.3	−11.9	8	11.8
ORE	Oreokastro gabbros	9	29.5	39.2	188.6	49.8	9	15.5
GO	Goumenissa layered gabbros	4	78.7	46.4	89.3	49.2	25	14.1
AET	Aetoraches schlieren	8	46.3	57.6	39.8	28.3	57	6.5
KRC	Krapa hills limestone	8	85.9	44.2	151.1	56.5	36	8.3
KRD	Krapa hills dykes	4	35.0	40.3	42.8	−17.0	8	28.3
MO	Monopygadon granite	10	56.5	25.2			9	16
Mean (excluding MO data):		7	52.0	46.9			25	12.4
					63.2	13.3	4	36.3
Fold test: $K_1/K_2 = 0.15$, $F(12,12)$ 5% = 2.69								
Overall mean for intermediate temp component (excluding MO data):		11	47.7	47.7			28	8.8
					64.4	28.1	4	25.7
Fold test: $K_1/K_2 = 0.15$, $F(20,20)$ 5% = 2.12								

4; McElhinny 1964), thus the intermediate direction represents a post-folding remagnetization. The overprinted character of this component can also be seen on some vector diagrams for sediment samples where the intermediate direction does not appear to be directed towards the origin during demagnetization (Fig. 3a & b).

It has not been possible to successfully isolate the high temperature component in the majority of sediment samples as mineralogical changes occur during heating. However at Site PLF, 9 samples out of 15 allow a mean direction to be calculated for the high temperature component (Table 2). This older component is not seen in the Paliouri pillow lavas. The remagnetization which occurs in magnetite-bearing samples has not been so complete in the flysch-type sediments, probably due to the different magnetic mineralogy.

The intermediate temperature unblocking component is characterized by a southwest declination and a reversed inclination. We propose an Eocene age for this 'E' component. This assumption is based on a comparison with the direction of magnetization found in the Eocene (50–42 Ma) Sithonia granite. The direction measured by Kondopoulou & Westphal (1986) in this rock is close to the intermediate direction at Paliouri (see discussion). The Sithonia massif intrudes the Chalkidiki ophiolitic belt and the thermal event related to the regional metamorphism and to the emplacement of the granitic rocks appears to be the best candidate for the origin of the phase of remagnetization. Additionally, the emplacement of the intrusion might explain the observed tilting of the volcano-sedimentary formations.

Following the datation of the 'E' component, an age older than Eocene can be attributed to the high temperature or 'JC' component, but it remains difficult to decide whether this 'JC' magnetization has been acquired during the sedimentation of the Paliouri beds (Mid–Late Jurassic) or during the Late Jurassic–Early Cretaceous phase of obduction of ophiolites onto the Pelagonian domain. A remagnetization during this so-called Eo-Hellenic phase (which also produced granites and HP-greenschist facies regional metamorphism; see above) is, however, the most plausible hypothesis.

E and JC components in Axios–Vardar–Chalkidiki sites and in the Vourinos

The Eocene overprinted ('E') component is well defined near the village of Skra (Vardar region).

Table 2. *High unblocking temperature components*

Site	Location	N	D_g	I_g	D_s	I_s	K	α_{95}
Paliouri area								
PLF	Paliouri flysch	9	325.4	34.1	321.0	50.4	7	18
Chalkidiki and Vourinos areas								
KRD	Krapa hills dykes	5	257.6	63.1	64.8	51.6	12	18.2
SO	Sonia gabbros	3	318.2	35.7	316.0	−13.3	10	25.0
ORE	Oreokastro gabbros	4	338.8	42.1	236.0	34.0	68	8.5
EV	Evzoni sheeted dykes	3	329.8	62.6	339.6	28.6	88	8.6
Data previously obtained by Edel et al. *(1992)*								
Ka 9–10		9	129.5	−24			33	10.5
Ka 57		13	313.0	53			24	8.0
Ka 59		5	157.0	−57.0			88	8.0
Ka 60		11	322.0	11.0			48	11.0
Ka 17		11	306.0	22.0			30	8.0
Ka 58		10	319.0	41.0			45	7.0
Gr 22		6	321.0	33.0			43	10.0
Gr 43		8	312.0	39.0			177	4.0
Gr 44		7	322.0	23.0			50	8.5
Gr 47–48		7	316.0	12.5			23	19.0
Gr 50		6	150.5	−21.0			33	12.0
Overall mean for high unblocking temperature component		16	318.7	36.5			16	9.5

Two sites were cored in layered gabbros (magmatic bedding: strike = 10°, dip = 70°). A total of 26 samples have been demagnetized and measured. The stable intermediate temperature component shows a direction very close in geographic cordinates (*in situ*) to the Eocene direction observed at Paliouri (Fig. 4), but the application of tilt corrections using the magmatic bedding gives unreliable results (Table 1).

In the weakly magnetized Monopigadhon granitoid only the 'E' direction is observed. This may suggest an Eocene age for this granitoid by directional comparison with the Sithonia massif. However, geological observations (Kockel & Mollat, 1977; Gauthier, 1984) and preliminary isotopic dating (De Wet *et al.* 1989) suggest a Late Jurassic age for the Monopigadhon granite. The absence of an older 'JC' component in this unit would therefore reflect complete overprinting by the Eocene 'E' component.

In the Axios–Chalkidiki belt, the Eocene remagnetization component has also been found at Oreokastro in massive gabbros overlain by Titonian conglomerates (strike = 110°, dip = 90°) and at Goumenissa in layered gabbros (strike = 300°, dip = 30°). At Sonia, Evzoni and Pevkodassos the 'E' component is only seen as a weak overprint, and is not listed in Table 1.

In several sites, bedding correction (Table 1) results in a very shallow inclination. Taking into account the oldest possible age of magnetization

(Late Jurassic–Early Cretaceous) for these rocks, we think that these shallow inclinations are unrealistic as they indicate too low a palaeolatitude for both Africa and Eurasia. This again supports the interpretation of these components of magnetisation as the result of post-folding remagnetisation.

In the Vourinos area, the Eocene component was also found to be significant at Aetoraches in chromite layered schlieren (strike = 100°, dip = 10°), and in the Krapa Hills district both in sheeted dykes (strike = 340°, dip = 65°) and in overlying Jurassic-Cretaceous limestones (strike = 130°, dip = 49°). It also occurs as a weak overprint in the Poros limestone quarry.

The combined dataset from the four sites at Paliouri and seven of the sites in the Chalkidiki and Vourinos belts (excluding the data from Site MO, where a tilt correction could not be defined; Table 1) fails a fold test at the 95% confidence limit (McElhinny 1964). This negative area-wide fold test demonstrates the post-folding age of the 'E' component throughout the study area.

The Mesozoic 'JC' component or 'older direction' with westerly declination and normal inclination is more scarce and even when present it cannot be isolated in all specimens at a site. It is noticeable that when component 'JC' occurs in a site, the 'E' component is commonly also present as a light overprint. The rather scattered directions found for component 'JC' are listed in Table 2. This component is present in the Krapa Hills dykes (strike = 340°, dip = 65°), at Sonia in layered gabbros (strike = 215°, dip = 50°), and at Evzoni in a sheeted-dyke complex (strike = 260°, dip = 35°). Directions obtained after tilt correction are difficult to interpret. If the magmatic bedding gives often unrealistic tilt corrections for the 'E' component (see before), the case is not so clear for the 'JC' component.

Origin of the 'JC' component

In order to examine the origin and significance of the 'JC' component, we report in Table 2 the directions found previously by Edel *et al.* (1991/1992) in the Axios–Chalkidiki peridotites and in the Florina granite (NW Pelagonian zone). The combined 'JC' dataset gives densely distributed westward directions over the Chalkidiki–Axios–Vardar–Vourinos and Florina areas with the following *in situ* mean (Table 2; Fig. 4b):

$$N = 16, D = 318.7°, I = 36.5°, k = 16, \alpha_{95} = 9.5°.$$

This whole set of data displays a remarkable similarity of *in situ* mean directions over a broad area which can be considered as an area-wide fold test. In the following we will use the direction of the 'JC' magnetization in geographic coordinates. We consider this direction as a remagnetization which predates the Eocene one. The Eo-Hellenic phase of deformation is the best candidate for the regional thermal event which was probably responsible for the 'JC' remagnetization In this case, the moderate scatter of the 'JC' component would result from restricted tilting of the studied massifs, either before or during the Meso-Hellenic phase.

Discussion of geodynamic implications

The 'E' component: Cenozoic clockwise rotation

The palaeomagnetism of several batholiths of the Chalkidiki peninsula with ages ranging from 30 to 50 Ma, have been thoroughly studied by Kondopoulou & Westphal (1986). The best results have been obtained in the Sithonia granite, where a mean direction of $D = 37°, I = 31°, \alpha_{95} = 9°$ was obtained (after inverting through the origin, but directions were mostly of reversed polarity). This result, consistent with the overprint measured in the ophiolites of the present study, was interpreted by Kondopoulou & Westphal (1986) as indicating a 30° post-Eocene–Oligocene clockwise rotation of the area.

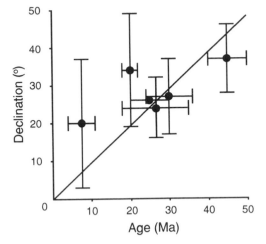

Fig. 5. Graph of declination against age for the Cenozoic data from Northern Greece given in Table 3.

We will now use a set of additional data obtained in a broader area of Northern Greece to better constrain this rotation (Table 3). The following observations can be made by comparing the ages and available declinations (Fig. 5) for the Cenozoic formations.

(1) The rotation angle increases almost always with the age of the magnetisation (Fig. 5). This observation supports the hypothesis of an homogeneous rotational pattern from the Ionian Islands to the Rhodope during the whole Oligo-Mio-Pliocene period, as previously suggested by Westphal *et al.* (1990).

(2) Some exceptions to this general pattern occur. For example, the rotation of the Lemnos Island volcanic formations reaches an unexpected value of 34° for a 18-22 Ma age. This can be explained by the additional effect of dextral movement along the North Anatolian Fault (Westphal & Kondopoulou 1993).

The 'JC' component : Mesozoic counterclockwise rotation

Evidence of Mesozoic westward directions of magnetization has already been found by Pucher *et al.* (1974), Turnell (1988), Surmont (1989), Lauer & Kondopoulou (1991) and Edel *et al.* (1991/1992), and the suggestion that these directions might correspond to a large counterclockwise rotation has been made by the latter authors.

In the present study the results obtained in the Paliouri area show that a component older than the Eocene remagnetization might be present in

Table 3. *Cenozoic directions of magnetization in Northern Greece*

Area	Rock type	Age (Ma)	D	I	K	α_{95}	Reference
Chalkidiki	Batholith	40–50	37	31	28	9	Kondopoulou & Westphal (1986)
Mesohellenic Trough	Sediments	24–36	27	47	16	10	Kissel & Laj (1988)
Greek Rhodope	Plutonics & volcanics	18–35	24	49	80	8	Atzemoglou *et al.* (1994)
Strymon	Volcanics	20–30	26	47			Westphal *et al.* (1990)
Lemnos	Volcanics	18–22	34	48	18	15	Westphal & Kondopoulou (1993)
Axios	Sediments	4–11	20	46	30	17	Kondopoulou (1994)

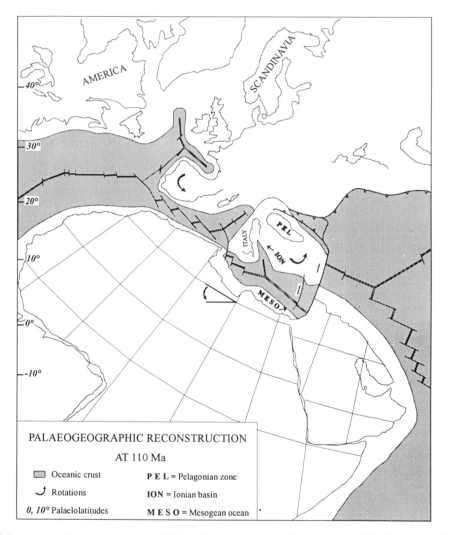

Fig. 6. Palaeogeographic reconstruction at 110 Ma. (References used : Dercourt *et al.* 1993; Theveniaut 1993).

the metasedimentary cover of the spilitic pillow lavas and that this direction is also found with consistency in the ophiolites of the Axios–Chalkidiki belt and Vourinos massif. The lump-ing of these data together with those of Edel *et al.* (1991/1992) gives a homogenous pattern of westward directions over the Vardar–Axios–Chalkidiki–Pelagonian and Sub-pelagonian areas. We conclude that the 'JC' component probably represents an Eo-Hellenic remagneti-sation, which has only been weakly affected (in many areas) by the subsequent Meso-Hellenic tectonics. Accordingly it seems possible to use the 'JC' declination to discuss block rotations at a regional scale (Northern Greece) during the Mesozoic.

The whole rotation shown by the 'JC' component is the sum of c. 40° deduced from the westward declination of 'JC' and of a 30–40° clockwise rotation that occurred during the Cenozoic (see above), giving a total Mesozoic counterclockwise rotation of c. 70–80°. This large rotation has not yet received an explanation. Two scales of lithospheric mechanism can be considered (Fig. 6). The first concerns the displacements of the Eurasian and African plates. Before the Alpine collision these two plates show quite different behaviours. While the Eurasian plate is fairly stable with only a moderate northward drift, the African plate has undergone a more important northward drift combined with a large counterclockwise rotation. The directions expected at 140 Ma (Jurassic–Cretaceous boundary) in the Chalkidiki are $D = 332°, I = 22°$ using the African apparent polar wander path (APWP) of Besse & Courtillot (1991) and $D = 337°, I = 24°$ using the APWP of Westphal et al. (1986). Thus 30° of counterclockwise rotation can be inferred from the African plate rotation if Northern Greece was attached to that plate in the Mesozoic.

A separate mechanism is required to explain the remaining 40–50° of counterclockwise rotation. At a smaller scale than that of the movement of the African plate, the postulated opening of the so-called Mesogean Ocean (Dercourt et al. 1993) occurred to the south of the Pelagonian–Tauric blocks during the Cretaceous and can explain this additional counterclockwise rotation of Northern Greece with respect to Africa. This is probably related to the scissors opening of the Mesogean ocean along transform faults accomodated by subductions north of the Pelagonian block. The stability of Apulia with respect to Africa (see Channell, this volume) indicates that the postulated rotation has not affected Italy, which was separated from Pelagonia by the Ionian basin. The mechanism proposed here is very similar to that acting during the coeval rotation of Iberia at the other end of the Mediterranean Sea.

We are greatly indebted to A. Morris for many improvements to our manuscript. Field work was supported by the French CNRS/Greek EIE cooperation program.

References

ATZEMOGLOU, A., KONDOPOULOU, D., PAPAMARINO-POULOS, S. & DIMITRIADIS, S. 1994. Paleomagnetic evidence for block rotations in the Western Greek Rhodope. *Geophysical Journal International*, **118**, 221–230.

BEBIEN, J. 1982. *L'association ignée de Guevgueli (Macédoine grecque): expression d'un magma-tisme ophiolitique dans une déchirure continentale.* Thèse Sci. Univ. Nancy.

—— 1991. Enclaves in lagiogranites of the Guevgueli ophiolitic complex, Macedonia, Greece. *In*: DI-DIER, J. & BARBARIN, B. (eds) *Enclaves and Granite petrology*, Developments in Petrology, **13**, Elsevier, 205–219.

BESSE, J. & COURTILLOT, V. 1991. Revised and synthetic Apparent Polar Wander Paths of the African, Eurasian, North American and Indian plates and true Polar Wander since 200 Ma. *Journal of Geophysical Research*, **96**, 4029–4050.

CHANNEL, J. E. T. 1996. *Palaeomagnetism and Palaeogeography of Adria*. This volume.

DE WET, A. P., MILLER, J. A., BICKLE, M. J. & CHAPMAN, H. J. 1989. Geology, geochronology of the Arnea, Sithonia and Ouranopolis intrusions, Chalkidiki peninsula, Northern Greece. *Tectono-physics*, **161**, 65–69.

DERCOURT, J., RICOU, L. E. & VRIELINCK, B. (eds) 1993. *Atlas of Tethys palaeoenvironmental maps*. Gauthier-Villars Paris.

EDEL, J. B., KONDOPOULOU, D., PAVLIDES, S. & WESTPHAL, M. 1990. Multiphase paleomagnetic evolution of the Chalkidiki ophiolitic belt (Greece). *Bulletin of the Geological Society of Greece*, **25**, 381–392.

EDEL, J. B., KONDOPOULOU, D., PAVLIDES, S. & WESTPHAL, M. 1991/1992. Paleomagnetic evidence for a large counterclockwise rotation of Northern Greece prior to the Tertiary clockwise rotation. *Geodinamica Acta*, **5**, 245–259.

FERRIERE, J. 1985. Nature et développement des ophiolites helleniques du secteur Othrys-Pelion. *Ofioliti*, **10**, 255–278.

GAUTHIER, A. 1984. *La ceinture ophiolitique de Chalcidique (Grèce du Nord)*. Thesis, Univer. Nancy.

JACOBSHAGEN, V. 1986. *Geologie von Griechenland*. Borntraeger, Berlin, 1–363.

KAUFMANN, G., KOCKEL, F. & MOLLAT, H. 1976. Notes on the stratigraphic and paleogeographic position of the Svoula formation in the innermost zone of the Hellenides (Northern Greece). *Bulletin de la Société Géologique de France*, **18**, 225–230.

KISSEL, C. & LAJ, C. 1988. The Tertiary geodynamic evolution of the Aegean Arc: a paleomagnetic reconstruction. *Tectonophysics*, **146**, 183–201.

KOCKEL, F. & MOLLAT, H. 1977. *Erläuterungen zur geologischen Karte der Kalkidiki und angren-zender Gebiete 1/100 000 (Nordgrichenland)*. Bundesanstalt für Geowissenschaften und Roh-stoffe, Hannover, 1–119.

KONDOPOULOU, D. 1994. Some constraints on the origin and timing of magnetization for Mio-pliocene sediments from N. Greece. *Proceedings of the 7th Congress of Geolological Society of Greece*, in press.

—— & WESTPHAL, M. 1986. Paleomagnetism of the Tertiary intrusives from Chalkidiki (Northern Greece). *Journal of Geophysics, 59*, 62–66.

——, FEINBERG, H., MICHARD, A. & MOUNDRAKIS, D. 1994. Contribution of ophiolites palaeomagnetism to the interpretation of N. Greece geodynamics. *Proceedings of the 7th Congress of Geological Society of Greece*, in press.

LAUER, J. P. & KONDOPOULOU, D. 1991. Paleomagnetism of the Nea Santa Rhyolites and comparisons with the Pelagonian Permo-Triassic. *Bulletin of the Geological Society of Greece, 25*, 369–379.

MCELHINNY, M. W. 1964. Statistical significance of the fold test in paleomagnetism. *Geophysical Journal of the Royal Astronomical Society, 8*, 338–340.

MAVRIDIS, A. 1980. A propos de l'âge de la mise en place tectonique du cortège ophiolitique du Vourinos (Grèce). *In*: PANAYIOTOU, A. (ed.) *Ophiolites*. Proceedings International Ophiolites Symposium, Cyprus, 1979. Geological Survey of the Republic of Cyprus, 349–350.

MERCIER, J. 1968. Etudes géologiques des zones internes des Hellenides en Macédoine centrale. *Annales Géologique des Pays Helleniques, 20*, 1–792.

MICHARD, A., GOFFE, B., LIATI, A. & MOUNTRAKIS, D. 1994. Découverte du faciès schiste bleu dans les nappes du Circum-Rhodope: un élément d'une ceinture HP-BT Eo-Hellénique en Grèce septentrionale. *Comptes Rendus de l'Academie des Sciences, Paris, 318*, 2, 1535–1542.

MOORES, E. 1969. Petrology and structure of the Vourinos ophiolotic complex of the Northern Greece. *Geological Society of America, Special Publications, 118*, 1–74.

MORRIS, A. 1995. Rotational deformation during Palaeogene thrusting and basin closure in eastern central Greece: palaeomagnetic evidence from Mesozoic carbonates. *Geophysical Journal International, 121*, 827–847.

MOUNTRAKIS, D. 1984. Structural evolution of the Pelagonian zone in NW Macedonia, Greece. *In*: DIXON, J. E. & ROBERTSON, A. H. F. (eds) *The Geological Evolution of the Eastern Mediterranean*. Geological Society, London, Special Publication, 17, 581–590.

—— 1986. The Pelagonian zone in Greece: a polyphase deformed fragment of the Cimmerian continent and its role in the geotectonic evolution of the Eastern Mediterranean. *Journal of Geology, 94*, 335–347.

NICOLAS, A. 1989. *Structures in ophiolites and dynamics of oceanic lithosphere*. Kluwer Academic Press, Dordrecht.

PUCHER, R., BANNERT, D. & FROMM, K. 1974. Paleomagnetism in Greece: indications for relative block movement. *Tectonophysics, 22*, 31–39.

RASSIOS, A., MOORES, E. & GREEN, H. 1983. Magmatic structure and stratigraphy of the Vourinos ophiolitic complex, Northern Greece. *Ofioliti, 8*, 377–410.

RICOU, L. E., DERCOURT, J., GEYSSANT, J., GRANDJACQUET, C., LEPVRIER, C. & BIJU-DUVAL, B. 1986. Geological constraints on the Alpine evolution of the Mediterranean Tethys. *Tectonophysics, 123*, 83–122.

ROBERTSON, A. H. F., CLIFT, P. D., DEGNAN, P. J. & JONES, G. 1991. Palaeogeographical and palaeotectonic evolution of the Eastern Mediterranean Neotethys. *Palaeogeography, Palaeoclimatology, Palaeoecology, 89*, 289–343.

SMITH, A. G., HYNES, A. J., MENZIES, M., NISBET, E. G., PRICE, I., WELLAND, M. J. & FERRIERE, J. 1975. The Stratigraphy of the Othris mountains, eastern and central Greece: a deformed continental margin sequence. *Eclogae Geologicae Helvetiae, 68*, 463–481.

SPRAY, J. G. & RODDICK, J. C. 1980. Petrology and $^{40}Ar/^{39}Ar$ geochronology of some Hellenic subophiolitic metamorphic rocks. *Contributions to Mineralogy and Petrology, 72*, 43–55.

——, BEBIEN, J., REX, D. C. & RODDICK, J. C. 1984. Age constraints on igneous and metamorphic evolution of the Hellenic–Dinaric ophiolites. *In*: DIXON, J. E. & ROBERTSON, A. H. F. (eds) *The Geological Evolution of the Eastern Mediterranean*. Geological Society of London, Special Publications, 17, 619–627.

SURMONT, J. 1989. Paleomagnétisme dans les zones internes: analyse des aimantations superposées par la méthode des cercles de réaimantation. *Canadian Journal of Earth Sciences, 26*, 2479–2494.

——, VRIELYNCK, B., FERRIERE, J., DECONINCK, J. F., AZEMA, J., STAIS, A., BAUDIN, F. & MOUTERDE, R. 1991. Paléogéographies du Toarcien et de la limite Jurassique-Crétacé dans les Hellenides entre le Pinde et le Vardar. *Bulletin de la Societé Géologique de France, 162*, 43–56.

THEVENIAUT, H. 1993. *Evolution de la Téthys occidentale et de la Pangée au Trias*. Thesis, Univer. of Paris 7.

TURNELL, H. B. 1988. Mesozoic evolution of Greek micro-plates from palaeomagnetic measurements. *Tectonophysics, 155*, 307–316.

VERGELY, P. 1984. *Tectonique des ophiolites dans les Hellenides internes; conséquences sur l'évolution des régions téthysiennes occidentales*. Thèse Sci., Univ. Paris-Sud, Orsay.

WESTPHAL, M. & KONDOPOULOU, D. 1993. Paleomagnetism and rock magnetism of Miocene volcanics from Lemnos island (N. Aegean): Implications for block rotations in the vicinity of the N. Aegean Trough. *Annales Tectonicae, 7*, 142–149.

——, BAZHENOV, M. L., LAUER, J. P., PECHERSKY, D. H. & SIBUET, J. C. 1986. Paleomagnetic implications on the Evolution of the Tethys belt from the Atlantic Ocean to the Pamirs since the Triassic. *Tectonophysics, 137*, 37–82.

——, KONDOPOULOU, D., EDEL, J. B. & PAVLIDES, S. 1990. Paleomagnetism of Late Tertiary and Plio-Pleistocene formations of Northern Greece. *Bulletin of the Geological Society of Greece, 25*, 3, 239–250.

Palaeomagnetic study of crustal deformation across an intracontinental transform: the North Anatolian Fault Zone in Northern Turkey

J. D. A. PIPER[1], JOANNA M. MOORE,[1], O. TATAR[2], H. GURSOY[2] & R. G. PARK[3]

[1]Geomagnetism Laboratory Department of Earth Sciences, University of Liverpool, Liverpool L69 3BX, UK

[2]Department of Geology, Cumhuriyet University, 58140 Sivas, Turkey

[3]Department of Geology, University of Keele, Keele, Staffordshire ST5 5BG, UK

Abstract: Eocene volcanic rocks spanning the North Anatolian Fault Zone in north central Turkey have a common reversed polarity and appear to record a short term volcanic episode useful for identifying subsequent tectonic rotations. Although regional differences are present, no distributed clockwise rotations caused by dextral motion across the fault zone since mid-Miocene times are found. Instead variable anticlockwise block rotations demonstrate that this fault system does not obey theoretical models for crustal behaviour across continental transforms. Deformation is found to be highly inhomogeneous with a narrow zone of intense clockwise rotation recognised within blocks bounded by strike-slip faults above, and parallel to, the fundamental lineament. Further from the lineament no systematic rotations with respect to the major bounding plates are detected. A zone of c. 30° anticlockwise rotation in the east may be either a consequence of emplacement of the Pontides or an ongoing consequence of continental collision. Slightly larger rotations south of the fault probably record block rotations into Anatolia as this region is being extruded westwards by continuing impingement of Afro-Arabia into the Eurasian Plate.

Turkey incorporates three major tectonic divisions (Fig. 1, inset). In the north the Pontides comprise at least two major continental fragments (Rhodope-Pontide and Sakarya; see Şengör & Yılmaz 1981) of disputed origins (Şengör et al. 1984; Robertson & Dixon 1984; Usumezsoy 1987). In Jurassic times this region occupied a passive south-facing continental margin to the Neo-Tethyan Ocean; it later converted into an active magmatic arc as the ocean basin was consumed between Aptian and Eocene times. The Anatolides (Fig. 1) are an amalgamation of terranes which have their origins as rifted fragments from the margin of Gondwana as the Neo-Tethys opened during the earlier part of Mesozoic times (Şengör et al. 1984). Collision with the Pontides commenced in Eocene times and terminated in mid-Miocene times during impingement of Arabia along the Bitlis Suture Zone (BSZ; Fig. 1); a consequence of this collision was the emplacement of the Tauride Orogen in southern Turkey (Fig. 1, inset).

The North Anatolian Fault Zone (NAFZ) is an intracontinental transform with dextral motion defining the tectonic boundary between the Eurasian Plate to the north and the Anatolian Block to the south. It is sited close to a former north-dipping subduction zone active between Late Cretaceous and mid-Eocene times. Subduction was finally terminated here by continental collision during mid–late Miocene times (Şengör & Yılmaz 1981). Thus a *palaeotectonic* history of ocean closure and subduction has been succeeded by a *neotectonic* history as impingement of the Afro-Arabian Plate has continued into the Eurasian Plate along the BSZ; this interaction is extruding Anatolia westwards towards the Aegean. The resulting dextral shear in the vicinity of the NAFZ remains the dominant motivation of present day tectonics in northern Turkey. The Anatolian Block bounded by these lineaments was recognised as a tectonic entity by Ketin (1948) from field studies following the 1930 Erzincan and 1942 Erbaa earthquakes; it was first defined in a plate tectonic context by McKenzie (1970) as a rigid plate rotating about a Euler pole at 19°N, 35°E.

Marine successions in the Pontides and Anatolides date from the Tethyan and palaeotectonic history of the orogen; they are emplaced as nappes and intensely deformed by

From Morris, A. & Tarling, D. H. (eds), 1996, *Palaeomagnetism and Tectonics of the Mediterranean Region*, Geological Society Special Publication No. 105, pp. 299–310.

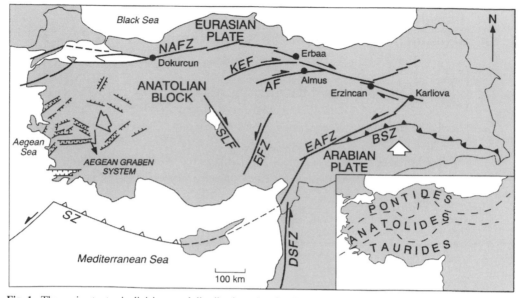

Fig. 1. The major tectonic divisions and distribution of major lineaments in Anatolia and adjacent regions. The large open arrows show relative motions of the plates and the smaller half arrows are directions of movement on major strike-slip faults. The abbreviations not given in the text are: SLF, Salt Lake Fault Zone; KEF, Kirikkale–Erbaa Fault Zone; AF, Almus Fault Zone; EFZ, Ecemis Fault Zone; SZ, East Mediterranean subduction Zone and DSFZ, Dead Sea Fault Zone. Some place names referred to in the text and the political boundary of Turkey are also shown. The inset shows the major terrane divisions in Turkey.

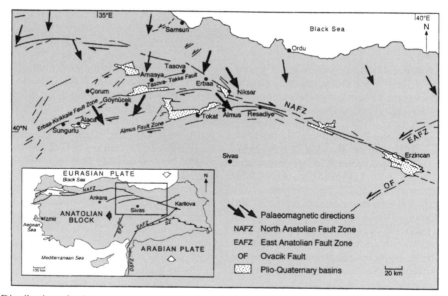

Fig. 2. Distribution of palaeomagnetic sampling sites of this study in the central part of the North Anatolian Fault Zone; the distribution of major faults in this region is also shown. The locations of previous palaeomagnetic studies shown by the stars are referred to in the text. The inset shows the regional location within the tectonic framework of the Eastern Mediterranean.

Alpine deformation. These rocks are typically succeeded by fluviatile and lacustrine deposits of Late Miocene–Pliocene age; the latter have been variably deformed by ongoing tectonic activity which has also reactivated older faults, generally altering their geometries (Koçyiğit

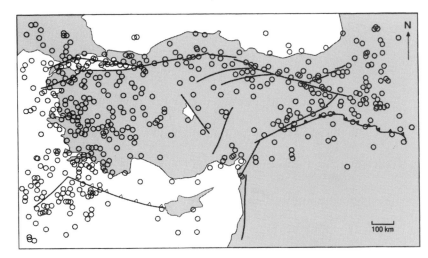

Fig. 3. The distribution of earthquakes occurring between AD 11 and 1967 in Anatolia and surrounding regions. The epicentres shown originated at depths of less than 70 km and had magnitudes of 4 or more. Simplified after Ergin *et al.* (1967). Major tectonic lineaments (see Fig. 1) are also shown.

1991). Magmatic activity in compressive and strike-slip regimes has produced sporadic but widespread volcanic activity mostly of calc-alkaline affinity (Yılmaz 1990). The products of this activity are the most obvious focus for palaeomagnetic investigations aiming to isolating tectonic deformations. For this study we have investigated occurrences of Lower Tertiary (probable Eocene) volcanics distributed along either side of the NAFZ in central Turkey between Çorum and Niksar (Fig. 2).

Deformation across the NAFZ

Recent deformation along the NAFZ is expressed by strike-slip motions which during the present century have ruptured the fault break from west to east producing major earthquakes between 1939 and 1961 (Ambraseys 1970), with the latest earthquake occurring at Erzincan in the east (Fig. 1) in March 1992. Focal mechanism solutions for these earthquakes indicate predominantly dextral movements on fault planes parallel to the major fault (McKenzie 1970; Platzman *et al.* 1994). In contrast motion along the BSZ (Fig. 1) is mainly compressional with a small component of sinistral strike-slip. The Eastern Anatolian Fault Zone (EAFZ) merges with this feature and extends NE to join the NAFZ at a triple junction near Karliova. The total displacement across the EAFZ has been estimated at *c.* 25 km (Rotstein 1984).

The signature of contemporary seismic activity (Fig. 3) demonstrates that the Anatolian

Block defined by the major lineaments is not a rigid plate. In addition to the prominent zone of distributed extensional faulting in the Aegean Graben System (Fig. 1), seismic activity is also present over a broad zone about the NAFZ with apparent contrast between distributions in the stronger Eurasian Plate and the weaker Anatolian Block. Partly because of this asymmetry it is unclear to what extent theoretical models for deformation across intracontinental transforms (England & McKenzie 1982; Sonder *et al.* 1986) can be applied to the NAFZ. Such models predict that a distributed zone of clockwise rotations should be observed across a broad zone about the fundamental lineament. At the surface however, most deformation is evident close to the lineament where a narrow zone a few kilometres in width comprises numerous subparallel faults. These segment the ground into depressions where the bounding faults diverge and pressure ridges where the bounding faults converge. Intermontane sedimentation in small strike-slip sedimentary basins with axes parallel to the principal direction of fault motion is also a feature of this zone (Koçyiğit 1990).

In addition, major arcuate fault systems splay at intervals from the dominant lineament into the weaker Block (Figs 1 and 3). The contribution of these faults to deformation of the Anatolian Block cannot be assessed from recent earthquake activity, but according to Rotstein (1984) net offset on the NAFZ diminishes from *c.* 85 km in eastern Anatolia to *c.* 20 km in the western sector, probably because motions are

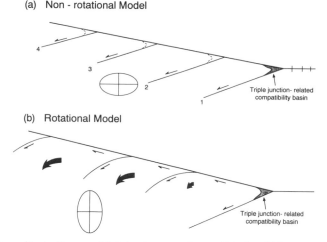

(a) Non - rotational Model

4

3

2

1

Triple junction- related
compatibility basin

(b) Rotational Model

Triple junction- related
compatibility basin

Fig. 4. Cartoon diagram illustrating possible lateral escape of crustal blocks within an extruding minor plate separated from major plates by two intracontinental transforms and joining at a triple junction analogous to the pattern in Turkey (see Fig. 1). A and B show geometries with and without constraint to escape; the ellipses show orientations of regional strain within the extruded block (after Şengör & Barka 1992).

progressively taken up along these side splays. Other assessments of slip along the NAFZ differ considerably from this view, although geological indicators in the Golova Basin define a minimum offset of 37 km near Erzincan in the east (Koçyiğit 1990; Fig. 1).

If westward extrusion of the Anatolian Block has been partly accommodated by movement along these side splays the fault blocks which they delineate may be sites of detectable rotation. To investigate this possibility we have studied the north-central segment of the NAFZ in northern Turkey where these faults are well known from their surface expression (Fig. 2). In each case formation of sedimentary basins has been a feature of the fault growth and throughout the region kinematic behaviour of the fault-bounded crustal blocks has controlled development of sedimentary basins (Koçyiğit 1991). Earlier compression has resulted in formation of E–W-trending molasse basins bounded by thrusts. Where compression has been replaced by strike slip, continuing basin development has produced younger *strike-slip basins*. These are typically smaller than the *hybrid basins* which have a pre-Pliocene compressive history succeeded by a neotectonic history of continuing evolution in the strike-slip regime (Aydin & Nur 1982).

The palaeomagnetism of these volcanic rocks has been studied in the vicinity of the town of Niksar by Van der Voo (1968), Saribudak (1989), Platzman *et al.* (1994) and Tatar *et al.* (1995). Platzman *et al.* (1994) also report K–Ar whole rock ages ranging from 45.3 to 41.8 Ma; these dates define a mid-Eocene age and support stratigraphic assessments described by Blumenthal (1950) and in unpublished reports of the Turkish Geological Survey (MTA). The present work extends sampling to the west (Fig. 2) where the volcanics have comparable stratigraphic relationships (Blumenthal 1950) although no specific age dates are available. Sample locations are sited within the Pontide Belt north of the NAFZ near a large bend in the main fault trace and within two blocks delimited by major side splays (Fig. 2).

Tectonic models

Sideways extrusion of the Anatolian Plate can be considered in terms of the escape of blocks bounded by the side splay faults. Analogous features are observed along other parts of the Alpine–Himalayan orogenic system and possible geometrical models are described by Şengör & Barka (1992) (Fig. 4) in terms of two limiting cases. Unconstrained lateral escape can take place along faults with trajectories which are essentially linear (the 'wedge in wedge' model of Şengör & Barka 1992). This results in elongation of the original escaping block in the direction of the extrusion. Ideally it could be achieved without rotation and no significant relative rotations would then be detected by the palaeomagnetic method (Fig. 4a). If sideways expulsion is restricted because the escaping block is being shortened, the lateral side splays will develop an arcuate form and a degree of rotation of the escaping blocks may then be detected (Fig. 4b).

Field and laboratory methods

Sampling of the early Tertiary volcanic suites was designed to cover at least three discrete tectonic blocks and extends a recent study made in the vicinity of the Niksar pull-apart basin by Tatar *et al.* (1995). Sites 1–10 come from tholeitic basalt and andesite lava flows north of the NAFZ in the Erbaa region (Fig. 4). Sites 11–22 come from lavas and tuffs at several locations south and west of the town of Amasya within the Anatolian Block and possibly within a single tectonic entity defined by side splays; this appears to be a westerly extension of the block impinging on the NAFZ near the town of Almus (Fig. 2) where Tatar *et al.* (1995) identify anticlockwise rotation of *c.* 35° with respect to the bounding plates. Sites 23–31 come from three locations within the next major fault-bounded block to the north and west (Fig. 2). Finally four additional sites, 32–35, are located to the west more than 20 km to the north of the NAFZ and well within the inferred limits of the Eurasian Plate.

At each location between seven and 12 cores were drilled with a portable motor and distributed across several metres of outcrop. Orientations were made with both Sun and magnetic compasses. Since volcanic units tend to acquire a nearly instantaneous record of the magnetic field dating from the time of initial cooling, it is essential to sample a number of different flows within each region to average effects of secular variation. Although outcrop in this region is typically inadequate to recognise the stratigraphic relationship of the sites with respect to each other, interbasaltic horizons, ropy lava surfaces and vesicle flow usually permit tilt of the volcanic units to be recognised and measured. Volcanoclastic rocks (sites 15–22 in this study) provide the clearest indications of bedding although as a consequence of diagenetic alteration their magnetic remanence is typically more complex than in the lava flows. Uncertainties in tectonic orientation are the most serious source of error in this kind of study because: (i) it is generally unclear whether the palaeohorizontal recorded at outcrop is representative of the unit as a whole; and (ii) the lavas and tuffs may have flowed down a pre-existing slope.

In the laboratory cores were prepared into 2.4 cm long cylinders. Magnetizations were measured by 'Minispin' magnetometers and progressively thermally demagnetized in steps of 50 or 100°C to 500°C and then in steps of 20°C to the Curie points of the remanence carriers. The components comprising the total natural remanent magnetization (NRM) were recognized by study of orthogonal projections and their directions calculated by principal component analysis. Detailed rock magnetic study was also conducted on core material from the first collection (sites 1–10) with objectives of recognising magnetic effects of low temperature alteration and determining whether this approach could be used to separate stable from unstable sites.

Rock magnetic study

Thermomagnetic determinations (saturation remanence, J_s, against temperature T) provide an initial diagnosis of the magnetic minerals present and their alteration with heating. Three of the six curve types distinguished by Mankinen *et al.* (1985) are recognized in these rocks but only type 2 is common. Type 1, corresponding to a titanium-rich titanomagnetite with a Curie point (T_c) below that of pure magnetite (580°C) is commonly observed in young basalts. It rarely survives in older lavas which tend to undergo subsolidus exsolution to form Ti-poor titanomagnetites and in this collection is found only at site 5 (Fig. 5). Type 2 is characterized by a single Curie point near to that of pure magnetite with variable reduction of saturation magnetization on heating, probably as a result of hematite formation; it is the predominant J_s–T expression in these lavas (e.g. site 1 in Fig. 5). Type 6 has an inflection only on the heating curve usually in the range 400–450°C which is the signature of inversion of Ti-poor maghemite to hematite; the latter has a low saturation remanence resulting in a drop in J_s; maghemite is identified in this way only at site 10 (Fig. 5a).

Low temperature susceptibility (KLT) analysis cools the rock to the temperature of liquid nitrogen (-196°C) and observes variations in magnetic susceptibility during subsequent return to room temperature. The susceptibility of multi-domain (MD) grains is controlled mainly by crystalline properties rather than shape (Uyeda *et al.* 1963) and magnetocrystalline anisotropy is strongly temperature dependent: in magnetite it changes from negative to positive at -150°C; in single domain (SD) grains susceptibility is largely dependent on shape anisotropy which is not greatly affected by temperature change. The curves (Fig. 5) are quantified in terms of the parameter RS which is the ratio of the susceptibility value at -196°C to the susceptibility at room temperature (Thomas

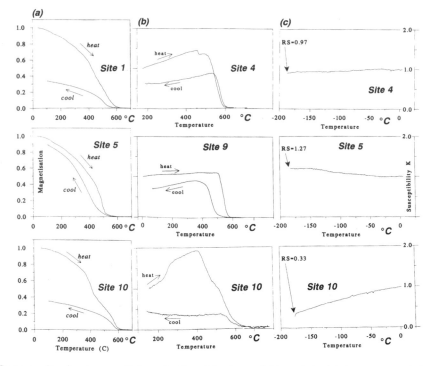

Fig. 5. Some results of rock magnetic investigation of the lavas of this study by (**a**) thermomagnetic (saturation magnetization, J_s, against temperature in °C), (**b**) high temperature susceptibility (KHT) analysis and (**c**) low temperature susceptibility (KLT) analysis. See text for an explanation of the curve shapes.

1992). RS values of >0.05 indicate that SD grains predominate whereas RS values <0.5 are typical of mixed assemblages including MD and/or cation-deficient (CD) grains. As is typical of volcanics subjected to sub-solidus exsolution, the lavas of this study show a dominance of SD assemblages: there is only weak expression of an MD/CD peak and RS values are >0.5 at all sites except 5.

High temperature susceptibility (KHT) analysis observes the variation in magnetic susceptibility as samples are heated to above their Curie temperatures. Dominant MD behaviour is characterized by a broad 'Hopkinson' peak and is a typical signature of MD behaviour in CD magnetite (maghemite); this was again observed only at site 10 (Fig. 5) confirming results of thermomagnetic analysis. As SD behaviour dominates the KHT spectrum the peak becomes suppressed and the maximum susceptibility is attained at higher temperatures. This is characteristic of the remaining sites (e.g. Fig. 5).

Thus rock magnetic analysis finds that low temperature hydrothermal alteration leading to development of CD magnetite is not prominent in these lavas and only detectable at site 10. Otherwise a dominance of SD remanence

carriers is evident. The rock magnetic tests did not, however, prove sensitive to palaeomagnetic stability. Site 10 with CD carriers yields stable vectors presumably resident in a SD fraction, whilst site 6 was dominated by Recent field components of low blocking temperature and yielded no stable components above 300°C; MD characteristics are developed at this site following KHT heating and this is a probable cause of instability during thermal demagnetization.

Palaeomagnetic study

Thermal demagnetization of the volcanic rocks mostly isolates two and three component structures with little or no overlap between the component spectra (e.g. Figs 6 and 8). When they are present, low blocking temperature (lbt) components in lava sites 1–10 from the Erbaa region are in the direction of the Recent field. Higher blocking temperature (hbt) components are largely or completely removed by the Curie point of magnetite and seven are of reversed polarity. Site 8 has an aberrant direction and site 5 is apparently of normal polarity although it remains significantly shallower than the Tertiary palaeofield both before and after tilt adjustment.

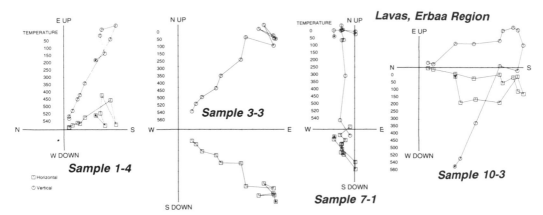

Fig. 6. Results of progressive thermal demagnetization of lava flows from the Erbaa region north of the NAFZ illustrated as orthogonal projections. The remanence vectors are projected onto the horizontal (square symbols) and vertical (circles) planes and sequential values are connected; the demagnetization steps in °C are listed. All directions are shown *in situ*.

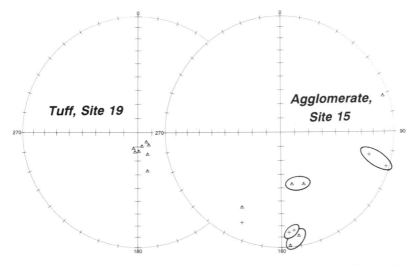

Fig. 7. Results of progressive thermal demagnetization of lavas and volcanoclastic units from the region west of Amasya. Symbols are as for Fig. 6 and the directions are shown *in situ*.

The seven reversed sites yield a mean direction $D/I = 195/-49°$ compatible with the early Tertiary palaeofield and the precision improves with adjustment for local tilts (Table 1), although this is not significant at the 95% confidence level.

In the region SW of Amasya no common overprint is present in the volcanics (Fig. 7). This is conclusively shown by study of clasts within agglomerates at sites 15–17: single component magnetizations often with a residual fraction resident in hematite (Fig. 7), are accordant within clasts and random between clasts (Fig. 8). The tuffs in this district have a more complex two and three component structure (Fig. 7) but the hbt components are accordant within sites and between sites 19–22.

The results from this region fall into two groups. Lavas at sites 11–14 from south of Çorum and close to the Erbaa–Kirikkale Fault Zone plus a lava at site 18 near Yenikoy are reversely magnetized and have S–SSE magnetizations following tilt adjustment. This adjustment yields an overall improvement in precision and suggests that the ChRM's predate deformation. Sites 19–22 in volcanoclastic rocks south of Amasya also have magnetizations of reversed polarity but they are directed S to SSW. Because dip is similar at most sites, tilt adjustment is not

Table 1. *Site and group mean palaeomagnetic results from Eocene Volcanics, North Anatolian Fault Zone, North-Central Turkey, after thermal cleaning*

Site No.	Tilt	N/n	R	D (in situ)	I (in situ)	D (tilt adjusted)	I (tilt adjusted)	k	α_{95}
Erabaa Region (North of NAFZ)									
1	34/071	8/8	7.94	149.9	−64.1	206.3	−52.5	118.6	5.1
2	29/090	7/4	3.93	154.1	−59.3	205.8	−59.3	42.7	14.2
3	43/048	7/7	6.80	149.7	−43.7	177.3	−24.0	30.6	11.1
4	25/260	7/4	3.98	236.0	−51.2	206.6	−71.5	154.8	7.4
5*	30/145	8/8	7.80	339.0	−7.0	340.0	22.1	35.3	9.4
6*	38/184	7/6	5.89	346.4	62.5	(PEF)		46.6	9.9
7	30/055	8/8	7.93	216.7	−75.3	228.4	−45.8	105.2	5.4
8*	33/183	7/7	6.97	100.7	−28.1	82.8	−27.4	213.6	4.1
9	29/224	8/8	7.81	196.3	−24.5	184.0	−48.9	37.5	9.2
10	14/148	7/6	5.95	191.0	−21.0	195.9	−30.9	99.2	6.8
Group mean result, 7 lavas:									
(*in situ*)			6.27	184.5	−52.5			8.2	22.5
(tilt adjusted)			6.64			194.6	−48.8	16.5	15.3
SW Amasya Block									
11	20/098	10/9	8.97	118.5	−61.8	143.9	−50.5	331.2	3.3
12	14/005	6/6	5.98	124.4	−35.4	131.9	−28.7	266.0	4.1
13	10/014	6/6	5.98	140.5	−42.2	146.8	−35.8	211.6	4.6
14	12/280	7/6	5.97	201.4	−45.8	188.8	−47.1	160.6	5.3
15	27/333	10/0	(Agglomerate)						
16	40/355	8/0	(Agglomerate)						
17	40/355	8/0	(Agglomerate)						
18	36/149	8/7	6.96	156.6	−15.4	160.4	−51.0	136.2	5.2
Group mean result, 5 lavas, Kalehisar–Yenikoy region:									
(*in situ*)			4.51	148.9	−43.9			8.2	28.4
(tilt adjusted)			4.79			152.6	−44.3	19.2	17.9
19	15/050	9/9	8.93	168.6	−76.0	199.6	−65.1	116.1	4.8
20	15/050	9/8	7.95	201.0	−67.3	211.7	−53.5	133.4	4.8
21	15/050	7/7	6.94	238.6	−66.9	235.5	−52.0	95.4	6.2
22	15/050	7/6	5.99	195.9	−54.1	204.2	−41.1	435.1	3.2
Group mean result, 4 tuffs, Karapinar region:									
(*in situ*)			3.92	203.1	−67.7			35.5	15.6
(tilt adjusted)			3.92			213.2	−53.7	35.7	15.6
NW Amasya Block									
24	28/190	8/6	5.79	186.4	−27.3	184.4	−55.2	23.3	14.2
29	56/200	7/7	6.76	1.7	55.1	(PEF)		24.6	12.4
30	28/300	7/5	4.80	163.8	−11.6	161.9	−9.8	19.7	17.7
SW Kastamonu Region (North of NAFZ)									
32	54/290	8/8	7.89	207.9	−50.4	157.8	−31.6	63.0	7.0
34	30/300	7/7	6.97	183.8	−44.4	165.8	−26.6	198.2	4.3
35	28/300	7/6	5.84	193.0	−50.6	169.1	−36.5	31.4	12.1
Group mean result, 3 andesite lavas:									
(*in situ*)			2.98	194.5	−48.9			84.8	13.5
(tilt adjusted)			2.99			164.2	−31.7	136.7	10.6

D and *I* are the mean declination and inclination derived from *n* sample components of magnetization out of a total site population of *N* cores. *R* is the magnitude of the resultant vector and *k* is the Fisher precision parameter ($=(n-1)/(n-R)$). α_{95} is the radius of the cone of 95% confidence about the mean direction.

* Sites excluded from mean calculations.

PEF, Present field component.

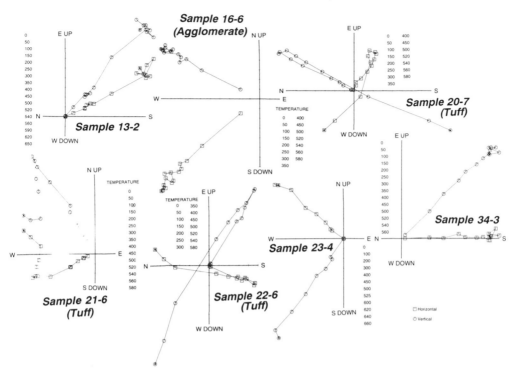

Fig. 8. Sample mean ChRM (*in situ*) directions from agglomerate (results from cores in common clasts enclosed) and tuff sites SW of Amasya. Lower hemisphere plots are shown as crosses and upper hemisphere plots are shown as triangles; equal area projection.

able to resolve the age of the remanence, although the inclination is closer to the predicted early Tertiary palaeofield after adjustment (Table 1). Because of the more complex nature of the remanence in these rocks it cannot be unequivocally assigned to the age of formation and is possibly younger than remanence in the lavas to the west.

Study of the block north west of Amasya (sites 23–31) was largely unsuccessful. Remanence in these lavas proved to be dominated by hematite components which, although stable within cores, were random within sites. Coherent ChRMs were either aberrant (23, 25) or present field overprints (24) and only two sites (24, 30) yielded reversed polarity magnetizations of value for tectonic analysis (Table 1). Three of four porphyritic andesite lava flows from a region north of the NAFZ and south west of the city of Kastamonu yield single component magnetizations of reversed polarity (Fig. 7). Because dips are similar, tilt adjustment is not diagnostic here but the in situ direction has an inclination closer to the predicted palaeofield than the tilt adjusted result; since these lavas show strong hydrothermal alteration it is possible that the remanence is ancient but post tilting.

Discussion and tectonic analysis

With one exception all lavas yielding coherent non-viscous magnetizations are of reversed polarity. This polarity is common to all lavas investigated in the Niksar and Almus regions to the west (Platzman *et al.* 1994; Tatar *et al.* 1995). Since the Eocene was characterized by frequent reversals of polarity (Harland *et al.* 1990) this suggests that the volcanic rocks throughout this region record only a brief pulse of activity. The longest period of constant reversed polarity during the Eocene appears to have lasted for about a million and a half years at *c.* 45 Ma (Harland *et al.* 1990), an age consistent with the K–Ar age dates obtained by Platzman *et al.* (1994) from the Niksar region.

Previous studies have resolved two scales of tectonic rotation in the vicinity of the NAFZ. Close to the fault zone (as defined at the surface by closely spaced strike-slip faults and pull-apart basins) large clockwise rotations (210–249°) are recognized in blocks up to 10 km across (Saribudak *et al.* 1990; Tatar *et al.* 1995). At Niksar this is apparent in volcanics which probably belong entirely to the Brunhes polarity chron (<700 000 years). Outside of the immediate fault zone (as defined by close parallel faulting)

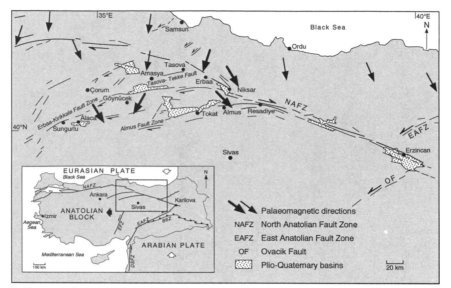

Fig. 9. Outline tectonic map of the NAFZ in north central Turkey showing the (reversed polarity) directions of magnetization in the Eocene volcanics from this and previous studies (Saribudak 1989; Tatar *et al*. 1995). The thick arrows are mean directions derived from substantial collections likely to have fully cancelled secular variation effects. The small arrows are from single sites which may incorporate secular variations; these must be regarded as ancilliary data only. Note that the westernmost result from the Kastamonou region, although based on three lavas, is plotted *in situ* as a small symbol because it is probably post-folding (see text).

only anticlockwise rotation is recognized both north and south of the NAFZ in the Niksar region (Platzman *et al*. 1994; Tatar *et al*. 1995).

Mean reversed field directions for this region of central Anatolia derived from the APW paths of Gondwana and Eurasia are:

	Eurasia	Gondwana
Recent	$180/-58°$	$180/-58°$
Lower Tertiary	$188/-47°$	$184/-50°$

Thus in the absence of later rotation a southerly-directed palaeofield direction of intermediate negative inclination would be expected in rocks of reversed magnetisation.

A common time of formation adds to the value of these results for tectonic analysis. However, since they are Eocene in age, rotations may have taken place between the studied locations during the palaeotectonic collisional history of this region; thus differences between tectonic blocks are not entirely attributable to the neotectonic history of post-Mid-Miocene times. Furthermore the sector of northern Turkey north of the NAFZ has been separated from the Eurasian Plate to the north by the back arc spreading which has produced the Black Sea. Robertson & Dixon (1984) place the Pontides at the Eurasian margin throughout Mesozoic times and Şengör *et al*. (1984) include

them within the Cimmerian continent. On these models the bulk of the back arc spreading was probably Cretaceous to Eocene in age and therefore mostly too old to be recorded by the Eocene and younger volcanics. Nevertheless it is clear from Fig. 9 that significant distortion of this southern orogenic margin of the Eurasian Plate is recorded by the Tertiary volcanics since the probable time of back arc spreading.

The second implication of the results plotted in Fig. 9 is that no general clockwise rotation of tectonic blocks has taken place across the NAFZ: declinations are either the same as those predicted from the adjoining Eurasian and Afro-Arabian Plates or are rotated anti-clockwise. This accords with the observations of Platzman *et al*. (1994) and Tatar *et al*. (1995) and implies that theoretical models treating the lithosphere as a thin viscous sheet obeying power law behaviour are inapplicable to this continental transform. Instead clockwise rotations are restricted to a narrow zone about the fault lineament (Saribudak *et al*. 1990; Tatar *et al*. 1994).

There is a general distinction, first noted by Saribudak *et al*. (1990), between southerly declinations compatible with the Eurasian Plate in the western part of the Pontide sector of Fig. 9 and SSE declinations in the eastern part of the

Pontides in this region. Results of the present study show that the mean remanence vector in the Erbaa lavas is not significantly rotated; however, comparable volcanics cropping out 40 km to the east and also located immediately north of the NAFZ are rotated anticlockwise by about 30° (Van der Voo 1968; Platzman et al. 1994; Tatar et al. 1995). This anticlockwise rotation of the eastern Pontides is also present in Upper Cretaceous results (Orbay & Bayburdi 1981). Thus the west to east change from unrotated to rotated declinations north of the NAFZ occurs rather abruptly along the northern Pontides and is accompanied by no change in strike direction of the fault; this implies an effect inherited from the emplacement history and unrelated to the neotectonic development of the NAFZ.

With the exception of the volcanoclastics in the Karapinar region at 35.5°E, 40.3°N which are rotated clockwise, anticlockwise rotations appear to be present along the whole length of the studied sector south of the NAFZ. The Eocene volcanic rocks south of the fault zone at Almus are rotated slightly more ($D = 144.1°$) than those immediately north of the NAFZ in this region at Niksar ($D = 152.4°$); this difference (which is embraced by the confidence limits) is the only possible effect so far detected of lateral extrusion along a side fault splay. In the lavas of the Corum area further to the west the anticlockwise declination $D = 152.6°$ is comparable to those observed in the Almus–Niksar region and contrasts with typical Eurasian declinations to the north side of the fault zone at this longitude (Fig. 9). Thus a larger rotation of the block constrained between the Erbaa–Kirikkale and Almus Faults may have occurred although more data would be required to confirm this. The anticlockwise rotation in this western region appears to be largely a consequence of the neotectonic history associated with deformation of the Anatolian Block because a smaller measure of anticlockwise rotation ($D = 164°$) is also identified by Platzman et al. (1994) in Miocene and younger volcanics of the Gerede region.

Conclusions

A coherent magnetic remanence is present in the volcanic rocks of north-central Anatolia containing an abundance of single domain magnetite carriers. The presence of a common reversed polarity implies that activity across this volcanic field occupied a short interval during Eocene times and late in the emplacement history of the Pontides. The presence of variable degrees of

anticlockwise block rotation on both sides of the NAFZ, which has been the site of a dextral intracontinental transform motion since Miocene times, shows that this fault does not accord with theoretical models based on thin skin tectonics. The predicted clockwise rotation appears to be confined to a zone a few kilometres wide where the transform breaks the surface. Regional variations in anticlockwise rotation in northern Anatolia are a consequence of back arc opening of the Black Sea and/or pre-Miocene deformation of the Pontide orogen. Westward extrusion of blocks within the Anatolian Plate is suggested by differential anticlockwise rotations on the south side of the NAFZ but is too small for significant definition.

We are grateful to the British Council for supporting an academic link between Sivas and Liverpool Universities permitting this cooperative study. We are also grateful to E. Platzman for providing a preprint of a related investigation of the NAFZ in Turkey.

References

AMBRASEYS, N. N. 1970. Some characteristic features of the Anatolian fault zone. *Tectonophysics*, **9**, 143–165.

AYDIN, A. & NUR, A. 1982. Evolution of pull-apart basins and their scale dependence. *Tectonics*, **1**, 91–105.

BLUMENTHAL, M. M. 1950. *Beitraege zur Geologie des Landschaften am Mittleren und Unteren Yesil Irmak (Tokat, Amasya, Havsa, Erbaa, Niksar).* Bulletin MTA, Ankara.

ENGLAND, P. & McKENZIE, D. P. 1982. A thin viscous sheet model for continental deformation. *Geophysical Journal of the Royal Astronomical Society*, **70**, 295–321.

ERGIN, K., GUGLU, U. & Uz, Z. 1967. *Turkie ve cirarinin deprem catalogu.* Istanbul Technical University, Maden Fak. Yayin, Istanbul.

HARLAND, W. B., COX, A. V., LLEWLLYN, P. G., PICKTON, C. A. G., SMITH, A. G. & WALTERS, R. 1990. *A Geologic Time Scale 1989.* Cambridge University Press.

KETIN, I. 1948. Uber die tektonisch-mechanischen Folgerungen aus den grossen anatolischen Erdbeden des letzten Dezenniums. *Geologische Rundschau*, **36**, 77–83.

KOÇYIĞIT, A. 1990. Tectonic setting of the Golova Basin: total offset of the North Anatolian Fault Zone, E. Pontide, Turkey. *Annales Tectonicae, Special Issue*, **4(2)**, 155–170.

—— 1991. An example of an accretionary forearc basin from northern Central Anatolia and its implications for the history of subduction of Neo-Tethys in Turkey. *Geological Society of America Bulletin*, **103**, 22–36.

MANKINEN, E. A., PRÉVOT, M. & GROMMÉ, C. S. 1985. The Steens Mountain (Orogen) geomagnetic polarity transition 1, directional history, duration of episodes and rock magnetism. *Journal of Geophysical Research*, **90**, 10 393–10 416.

McKenzie, D. P. 1970. Plate Tectonics of the Mediterranean region. *Nature*, **226**, 239–243.

Orbay, N. & Bayburdi, A. 1979. Palaeomagnetism of dykes and tuffs from the Mesudiye region and rotation of Turkey. *Geophysical Journal of the Royal Astronomical Society*, **59**, 437–444.

Platzman, E. S., Platt, J. P., Tapirdamaz, C., Sanver, M. & Rundle, C. C. 1994. Why are there no clockwise rotations along the North Anatolian Fault Zone? *Journal of Geophysical Research*, **99**, 21 705–21 716.

Robertson, A. H. F. & Dixon, J. E. 1984. Aspects of the geological evolution of the eastern Mediterranean. *In*: Dixon, J. E. & Robertson, A. H. F. (eds) *Geological Evolution of the Eastern Mediterranean*. Geological Society, London, Special Publications, **17**, 1–74.

Rotstein, Y. 1984. Counterclockwise rotation of the Anatolian Block. *Tectonophysics*, **108**, 71–91.

Saribudak, M. 1989. New results and a palaeomagnetic overview of the Pontides in northern Turkey. *Geophysical Journal International*, **99**, 521–631.

——, Sanver, M., Şengör, A. M. C. & Gorur, N. 1990. Palaeomagnetic evidence for substantial rotation of the Almacik flake within the North Anatolian Fault Zone, NW Turkey. *Geophysical Journal Interational*, **102**, 563–568.

Şengör, A. M. C. & Barka, A. A. 1992. Evolution of escape-related strike-slip systems: implications for disruption of collisional orogens (abstract), *29th International Geological Congress, Kyoto, Japan*, 232.

—— & Yílmaz, Y. 1981. Tethyan evolution of Turkey: a plate tectonic approach. *Tectonophysics*, **75**, 181–241.

——, —— & Sungurlu, O. 1984. Tectonics of the Mediterranean Cimmerides: Nature and evolution of the western termination of Palaeo-Tethys. *In*: Dixon, J. E. & Robertson, A. H. F. (eds) *Geological Evolution of the Eastern Mediterranean*, Geological Society, London, Special Publications, **17**, 77–112.

Sonder, L. J., England, P. C. & Houseman, G. A. 1986. Continuum calculations of continental deformation in transcurrent environments. *Journal of Geophysical Research*, **91**, 4797–4810.

Tatar, O., Piper, J. D. A., Park, R. G. & Gursöy, H. 1995. Palaeomagnetic study of block rotations in the Niksar overlap region of the North Anatolian Fault Zone, Central Turkey, *Tectonophysics*, **244**, 251–266.

Thomas, D. N. 1992. *Rock magnetic and palaeomagnetic investigation of the Gardar Lava Succession, South Greenland*. PhD thesis, University of Liverpool.

Usumezsöy, S. 1987. The NW Anatolian accretionary orogeny; western termination of Palaeo-Tethyan suture belt. *Geological Bulletin, Turkey*, **30**, 53–62.

Uyeda, S., Fuller, M. D., Belshé, J. C. & Girdler, R. W. 1963. Anisotropy of magnetic susceptibility of rocks and minerals. *Journal of Geophysical Research*, **68**, 271–291.

Van der Voo, R. 1968. Palaeomagnetism and the Alpine tectonics of Eurasia, Part 4, Jurassic, Cretaceous and Eocene pole positions from NE Turkey. *Tectonophysics*, **6**, 251–269.

Yilmaz, Y. 1990. Comparison of young volcanic associations of western and eastern Anatolia formed under a compressional regime: a review. *Journal of Volcanology and Geothermal Research*, **44**, 69–87.

A review of palaeomagnetic research in the Troodos ophiolite, Cyprus

ANTONY MORRIS

*Department of Geological Sciences, University of Plymouth, Drake Circus, Plymouth
PL4 8AA, UK*

Abstract: A summary of the most significant results of palaeomagnetic research within the well-known Troodos ophiolite of Cyprus is presented. The studies are broadly divided into those relating to microplate rotation and those which shed light on processes operating during crustal genesis. Early recognition that the Troodos Complex as a whole experienced a 90° anticlockwise rotation generated several attempts to constrain the precise timing of rotation. A series of palaeomagnetic data obtained within the *in situ* Late Cretaceous to Recent sedimentary cover of the ophiolite demonstrate that rotation began in the Campanian, soon after Turonian crustal genesis, and was complete by the Early Eocene. Rotation occurred at an approximately constant rate of *c.* $2°Ma^{-1}$. The most popular rotation mechanism is one involving a combination of oblique subduction to the south of Cyprus and collision of the trench with the Arabian continental margin to the east. A Troodos microplate became detached and rotated in response to the resultant anticlockwise torque exerted upon the supra-subduction zone crust. Detailed palaeomagnetic and structural studies have also revealed details of the Troodos spreading configuration, and specifically the relationships of horizontal and vertical axis rotations to ridge-parallel faulting and transform tectonism. Extension was achieved through a combination of magmatic accretion and amagmatic stretching, the latter producing significant rotations about ridge-parallel, sub-horizontal axes. Dykes and extrusives adjacent to the fossil Southern Troodos transform zone have been rotated clockwise in response to dextral shear along the transform, producing deviations away from the predominant N–S dyke trend within the Sheeted Dyke Complex of the ophiolite. The intersection of the transform with the principal Troodos spreading axis along the Solea graben has been identified, and transform-related rotations are now suggested to have occurred mainly in the active corner of the intersection. On-going palaeomagnetic research is now focused on the relationship between the Troodos Complex and the other major tectonostratigraphic terranes of Cyprus.

Ophiolitic terranes offer accessible, natural laboratories where important information on the structure and kinematics of constructive margins and transform faults can be gained. Ophiolites provide valuable constraints on regional tectonic settings. For example, geochemical studies of ophiolitic extrusives can distinguish crust generated at mid-ocean ridges from that formed by spreading above subduction zones (e.g. Pearce 1980). Such data lead to the development of improved palaeogeographical reconstructions.

The most extensively studied ophiolite within the Mediterranean region is the Troodos Complex of Cyprus. Unlike other Mediterranean ophiolites, the Troodos massif experienced little deformation during tectonic emplacement and has never been overthrust. Original sea-floor relationships have therefore been preserved. Palaeomagnetic studies have played a key role in interpreting the structure and history of the ophiolite. The contributions of palaeomagnetists to the Cyprus 'story' can be broadly subdivided into those relating to microplate tectonics and those dealing with tectonism and deformation during crustal genesis. This paper summarizes the most significant findings of these studies.

The paper is restricted to those aspects of the tectonic history of the Troodos which have been determined in part or in whole using palaeomagnetism. A comprehensive synthesis of the tectonic evolution of Cyprus is given by Robertson (1990), while for an extensive review of the history of geological research in Cyprus the reader is referred to Robertson & Xenophontos (1993).

Geological setting

The Mesozoic and Tertiary rocks of Cyprus can be assigned to three distinct terranes: Troodos; Mamonia; and Kyrenia. The Mamonia terrane is described briefly in the final section. The Kyrenia terrane is not discussed here. The Troodos terrane consists of a Late Cretaceous

From Morris, A. & Tarling, D. H. (eds), 1996, *Palaeomagnetism and Tectonics of the
Mediterranean Region*, Geological Society Special Publication No. 105, pp. 311–324.

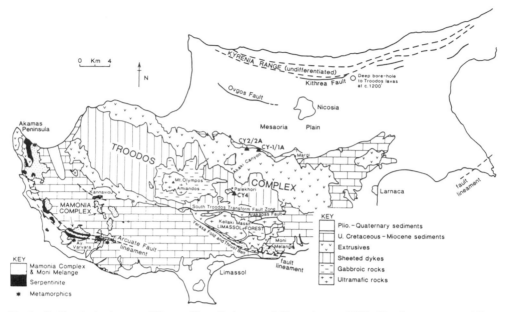

Fig. 1. Outline tectonic map of Cyprus (from Robertson & Xenophontos 1993). The three terranes of Cyprus are the Troodos Complex, the Mamonia Complex, and the Kyrenia Range. Palaeomagnetic studies have concentrated on the dykes and extrusives of the Troodos Complex and Southern Troodos transform zone, and on the sedimentary cover to the ophiolite. Palaeomagnetic research in the Mamonia Complex is on-going (Morris *et al.* in prep).

ophiolite and its *in situ* Late Cretaceous-Recent sedimentary cover (Gass 1968; Moores & Vine 1971). The ophiolite formed in a supra-subduction zone setting (Pearce 1975; Smewing *et al.* 1975; Schmincke *et al.* 1983; Robinson *et al.* 1983; Robertson 1990), and subsequently underwent a 90° anticlockwise rotation as a distinct microplate during the Late Cretaceous (Campanian) to Early Eocene (83.0–50.0 Ma) (Clube *et al.* 1985; Clube & Robertson 1986; see below). A complete ophiolite stratigraphy is preserved within the massif, with tectonised harzburgites, cumulate gabbros, sheeted dykes, mafic extrusives and a pelagic sedimentary cover (Anon. 1972; Moores & Vine 1971; Gass 1980). Recent rapid uplift has resulted in a concentric outcrop pattern with the deepest structural levels exposed in the centre (Fig. 1).

Along the southern margin of the main Troodos Complex lies the Arakapas Fault Belt (Simonian & Gass 1978). This major E–W structure forms the northern boundary of an area of anomalous ophiolitic crust, the Limassol Forest Complex, which is believed to have been generated within a leaky, transtensional oceanic transform zone (Murton 1986; MacLeod 1988, 1990). The transform tectonised sequences are collectively termed the Southern Troodos

Transform Fault Zone (MacLeod 1988; Fig. 1). To the south of the southern margin of the transform zone lies a separate crustal fragment, termed the Anti-Troodos plate (MacLeod 1988, 1990).

The main targets for palaeomagnetic research in the Troodos Complex and the Southern Troodos transform zone have been the sheeted dykes and overlying extrusives, and the pelagic sedimentary cover of the ophiolite.

The Sheeted Dyke Complex (SDC) and the extrusive series

The SDC lies between the plutonic complex and the extrusive series (Fig. 1). It is 1.0–1.5 km thick. The boundaries between the dykes and the gabbros below and extrusives above are gradational. The majority of individual dykes are less than 2.0 m wide, but can exceed 5.0 m. They consist principally of fine- to medium-grained, generally aphyric basalt and andesite with small amounts of olivine and clinopyroxene-phyric basalt (Xenophontos & Malpas 1987). More acid varieties are present but rare. In the western and central parts of the Troodos Complex dykes trend predominantly N–S, implying formation at a N–S-oriented spreading

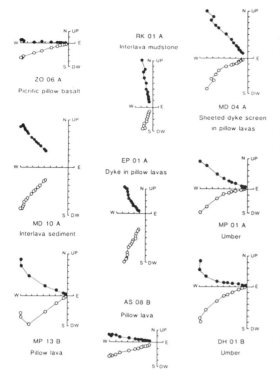

Fig. 2. Typical results of AF demagnetization of a range of Troodos lithologies (from Morris *et al.* 1990). Single stable magnetization components are easily isolated, following removal of minor north-dipping secondary components by fields of less than 20 mT.

due to such alteration were probably acquired at the spreading axis soon after TRM acquisition by cooling (Allerton & Vine 1987), however, and can be effectively considered as primary magnetizations. Thermal and AF demagnetization isolate stable components of magnetization in both zeolite and greenschist facies dykes. Localized alteration of the overlying extrusive series was restricted to hydrothermal pathways, and has had little effect on the TRM preserved in the lavas. Demagnetization of extrusive samples has recovered primary TRM vectors. These analyses have provided important clues to the nature of deformation associated with active spreading and transform tectonism. Examples of AF demagnetization behaviour of dyke and extrusive samples are shown in Fig. 2.

The pelagic sedimentary cover

The sedimentary cover of the Troodos ophiolite is continuous from the Late Cretaceous to Recent. The sequence consists of the Perapedhi, Kannaviou, Lefkara and Pakhna Formations. Palaeomagnetic studies of the cover sequence have established the timing of Troodos microplate rotation in detail (see below). Of particular palaeomagnetic interest are the umbers and radiolarites of the Perapedhi Formation and the pelagic chalks of the Lefkara Formation.

The Perapedhi Formation umbers are of Turonian-Campanian age and overlie the stratigraphically highest pillow lavas. Their distribution is sporadic, being confined to hollows in the lava surface, with a maximum thickness of 35 m (Robertson & Hudson 1974). In the field, umbers are pale, brown, or almost black, low density, fine grained mudstones which contain almost no calcium carbonate. The topmost umbers become increasingly clay-rich and change from dark brown to grey in colour (Robertson 1975). They are succeeded by pink, carbonate-free radiolarite and radiolarian mudstone.

Conformably overlying the Perapedhi and Kannaviou Formations is the Lefkara Formation, consisting of a sequence of pelagic chalks with subordinate cherts (Robertson & Hudson 1974). It is divided into Lower (Maastrichtian), Middle (Palaeocene to Eocene) and Upper (Oligocene) units on the basis of lithological and micropalaeontological criteria (Mantis 1970).

The sedimentary formations have NRM intensities which are approximately two orders of magnitude lower than those of the dykes and extrusives. Remanences are, however, easily measurable using cryogenic magnetometers.

system (present coordinates). However, significant deviations away from this azimuth are observed in the north and in the vicinity of the Southern Troodos transform zone (Simonian & Gass 1978).

The SDC passes up into a thick (1.0–1.5 km) extrusive succession (Fig. 1), dominated by pillowed and massive lava flows, with subordinate volcanic breccias and hyaloclastites (Schmincke *et al.* 1983). The series has been split into three divisions: Basal Group; Lower Pillow Lavas; and Upper Pillow Lavas. There is no strong evidence, however, for structural or metamorphic disconformities between these divisions (Gass & Smewing 1973), and geochemical studies suggest that the Troodos extrusive suite as a whole was co-magmatic (Smewing *et al.* 1975; Desmet 1976).

The sheeted dykes were initially considered unsuitable for palaeomagnetic study, specifically because of the possible destruction of titanomagnetite phases by hydrothermal alteration (Moores & Vine 1971). Chemical remanences

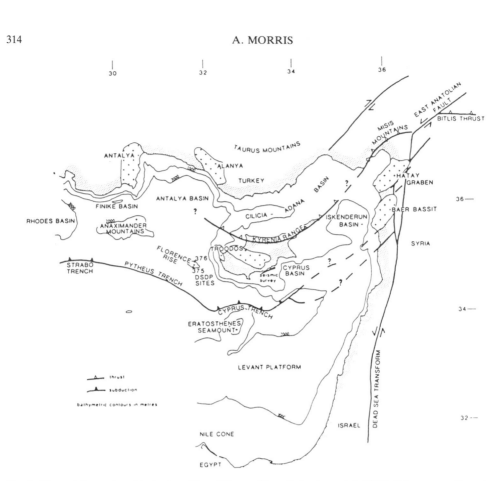

Fig. 3. Present day regional tectonic setting of Cyprus (from Robertson *et al.* 1991). Palaeomagnetic and geological data from surrounding areas suggest that the rotated Troodos microplate was small and located in the environs of Cyprus.

Stable, primary components of magnetization can be isolated by both AF and thermal demagnetization techniques. Examples of AF demagnetization behaviour of umber and inter-lava sediment samples are shown in Fig. 2.

The rotated Troodos microplate

Discovery and scale

Following the identification of the Troodos ophiolite as a fragment of oceanic crust (Gass 1968), the first palaeomagnetic study on Cyprus (Vine & Moores 1969) attempted to identify stripes of normally and reversely magnetized material within the Troodos complex. This would simultaneously confirm the Vine & Matthews (1963) hypothesis and provide compelling evidence that the massif was indeed formed by sea-floor spreading. Vine & Moores measured the NRM of over 900 hand specimens from 150 localities, using a portable fluxgate instrument in the field, and carried out a complementary laboratory study of 200 cores from a further 27 sites (Vine & Moores 1969; Moores & Vine 1971). These analyses did not reveal systematic linear magnetic anomalies, but instead identified a consistent magnetization direction of $D = 276°$, $I = 32°$ for the extrusives over the whole outcrop of the Troodos Complex and the two major inliers at Akamas and Troulli. The mean inclination indicated an origin for the Troodos Complex at a palaeolatitude of 17°N, consistent with a location between Gondwana and Eurasia during the Late Cretaceous. The westerly-directed mean declination recorded by the extrusives was completely unexpected, and was interpreted by Moores & Vine (1971) as the result of a 90° anticlockwise rotation of the whole Troodos Complex since its formation in the Late Cretaceous. This implied an original E–W spreading-axis orientation, more in line with the general Tethyan trend (Gass 1968).

A second extensive palaeomagnetic study of

the extrusive series was completed by Clube (1985). This documented uniform declinations across the strike of the ophiolite complex and confirmed the 90° anticlockwise rotation reported by Moores & Vine (1971).

The size of the rotated unit can be constrained by examination of palaeomagnetic and geological data from surrounding regions. Numerous palaeomagnetic studies in the Levant region to the east of Cyprus have identified significant rotations of fault-bounded blocks (Nur & Helsley 1971; Freund & Tarling 1979; Ron et al. 1984). Data from Late Cretaceous carbonates and basalts (i.e. coeval with Troodos genesis) define several rotational domains (Ron et al. 1984). These have experienced varying amounts of both clockwise and anticlockwise rotation, which are not related to the Troodos rotation but instead are linked to Neogene displacement along the Dead Sea Fault system (Fig. 3). Moreover, sheeted dykes in the Baër-Bassit ophiolite (60 km to the east of Cyprus; Fig. 3) are oriented E–W (Parrot 1977), but are believed to have formed within the same ocean basin as the Troodos. This suggests that Baër-Bassit did not undergo rotation along with the Troodos ophiolite. No palaeomagnetic data exist for the Kyrenia Range terrane in northern Cyprus (Fig. 1), but it follows the regional E–W structural trend. Recent data from southern Turkey indicate a 40° clockwise rotation of the Western Taurides belt during the Palaeogene (Kissel et al. 1993), and a 30° anticlockwise rotation of the Lycian Taurides during the Miocene (Kissel & Laj 1988; Morris & Robertson 1993). These data suggest that the boundaries of the rotated Troodos microplate lie to the west of the Levant and to the south of the Kyrenia Range (probably beneath the Pliocene sediments of the Mesaoria Plain; Fig. 1)). There is now a near consensus that the Troodos microplate was small and located in the environs of Cyprus (Robertson & Xenophontos 1993).

Timing of Troodos palaeorotation

The continuous Late Cretaceous to Recent *in situ* sedimentary cover to the ophiolite allows the precise timing of palaeorotation to be documented, since magnetic declinations within the sediments record rotation of the underlying ophiolitic basement. In the first investigation of this type, Shelton & Gass (1980) measured the NRM of a small suite of samples from the Perapedhi, Kannaviou and Lefkara Formations and from a sequence of Pliocene marls. They concluded that rotation was confined to a single Late Miocene event. This interpretation is unjustified, however, since it is based solely upon NRM directions of variable inclination and no detailed demagnetization experiments were performed.

The most thorough palaeomagnetic study of the circum-Troodos sedimentary cover was carried out by Clube (1985), who sampled extensively along the northern and eastern flanks of the ophiolite. AF and thermal demagnetization were used to isolate remanence components, and most samples were found to record single stable magnetizations following removal of low coercivity/unblocking temperature viscous components. Results were supported by positive fold and reversal tests. Clube (1985) found that remanence directions within the umbers and lowermost radiolarites of the Perapedhi Formation are indistinguishable from those in the underlying extrusives. Hence, rotation of the ophiolite complex must have been initiated after the deposition of these sediments. Data from the overlying Lefkara Formation pelagic chalks indicate that the Troodos crust rotated through at least 60° between the Late Campanian (c. 75 Ma) and the end of the Late Palaeocene (57 Ma), and that rotation was complete by the end of the Early Eocene (50 Ma) (Clube et al. 1985; Clube & Robertson 1986). Northerly remanences in Miocene sediments suggest that only minor rotation of the underlying ophiolitic basement took place after the end of the Early Eocene (Clube 1985).

A complementary investigation of the rotation history was carried out by Abrahamsen & Schönharting (1987). The majority of their samples were collected from two successions; an Upper Palaeocene–Lower Eocene section of Lefkara Formation pelagic chalks, and an Upper Eocene–Upper Miocene section of Lefkara chalks and reefal limestones of the Koronia Formation. The results were of variable quality due to low magnetization intensities (c. 10^{-4} Am^{-1}). However, demagnetized mean directions were found to cluster between north and northwest for the majority of sites, and the data obtained are thus in agreement with those of Clube (1985).

Further details of the timing of rotation were provided by Morris et al. (1990), as part of a study of fault block rotations associated with the Southern Troodos transform zone (see below). Sampling of continuous sections through the Perapedhi Formation at two localities showed a consistent difference in magnetization direction between the umbers and the stratigraphically overlying radiolarites. The umbers record the westerly-directed Troodos magnetization direction, demonstrating that the sampling sites have

Table 1. *Summary of well-constrained palaeomagnetic data relevant to the timing of Troodos microplate rotation*

Age	N	n	D	I	α_{95}	Reference
Pliocene	1	34	002	56	10.2	Clube (1985)
Upper Miocene	1	48	004	56	9.6	Clube (1985)
Lower Eocene	5	136	357	38	10.1	Clube & Robertson (1986)
Maastrichtian–Palaeocene						
Normal polarity	5	116	336	32	13.2	Clube & Robertson (1986)
Reversed polarity	6	101	152	−14	15.3	Clube & Robertson (1986)
Campanian	1	8	306	20	7.0	Morris *et al.* (1990)
	1	17	304	28	3.2	Morris *et al.* (1990)
Turonian	11	663	274	36	12.3	Clube & Robertson (1986)

N = number of sites; n = number of samples; α_{95} = semi-angle of 95% cone of confidence

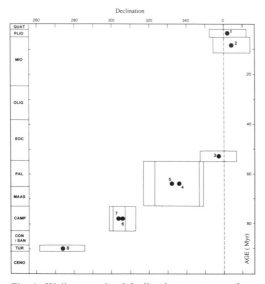

Fig. 4. Well-constrained declinations versus age for data relevant to the timing of Troodos microplate rotation. The linear trend of declinations from the Turonian to Early Eocene suggests a prolonged period of microplate rotation at a rate of *c.* $2°\,Ma^{-1}$, with no significant post-Early Eocene rotation. References: 1 & 2, Clube (1985); 3–5 & 8, Clube & Robertson (1986); 6 & 7, Morris *et al.* (1990).

is lower than those commonly observed in tectonically active regions of the Mediterranean (e.g. *c.* $5°\,Ma^{-1}$ in the Hellenic Arc; Kissel & Laj 1988). Successful tectonic models must be able to account for this well-defined, prolonged period of constant slow rotation.

Rotation mechanisms

Palaeorotation of a small oceanic microplate within a narrow Neotethyan ocean basin undergoing regional compression can be accommodated in several potential tectonic settings. The most popular models invoke rotation driven by either collision or oblique subduction.

Several authors have related rotation to inferred collision between a seamount or microcontinent and a subduction zone (Moores *et al.* 1984; Murton 1990; Robinson & Malpas 1990; Malpas *et al.* 1992; Fig. 5a). The Troodos terrane shows little evidence of regional compression or fault disruption, however, as expected from forceful collision and underthrusting, with partial or complete subduction of a seamount or microcontinent. Also, the palaeorotation of Troodos occurred during the Late Cretaceous to Early Eocene interval (a period of *c.* 30 Ma), and not during a short-lived collisional event (Robertson 1990).

Oblique subduction beneath the Troodos ophiolite was suggested as a driving mechanism for the palaeorotation by Clube *et al.* (1985). However, oblique subduction alone does not obviously provide the required torque. Clube & Robertson (1986) developed this idea further and proposed a revised model involving a combination of northward subduction beneath Troodos and collision outside Cyprus to drive the rotation (Fig. 5b). This is currently the most popular rotation mechanism.

Clube & Robertson (1986) noted the near coincidence of the timing of initiation of

not rotated with respect to the main Troodos ophiolite. More northerly directions recorded by the uppermost radiolarites indicate that 30° of Troodos rotation was complete before the end of Campanian radiolarite deposition.

Available well-constrained data ($\alpha_{95} \leqslant 15°$) on the timing of Troodos rotation are summarized in Table 1 and in Fig. 4, which shows mean declinations plotted against age. A linear declination trend implies that rotation took place at a uniform rate of *c.* $2°\,Ma^{-1}$, and was complete by the Early Eocene. The inferred rate of rotation

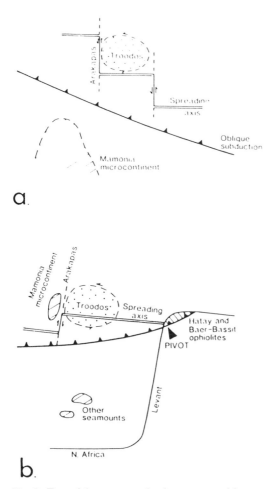

a.

b.

Fig. 5. Two of the many mechanisms proposed for the rotation of the Troodos microplate (from Robertson & Xenophontos 1993). (**a**) Collision, subduction erosion and underthrusting of a 'Mamonia microcontinent', essentially as envisaged by Moores *et al.* (1984), Murton (1990), Robinson & Malpas (1990) and Malpas *et al.* (1992). (**b**) Rotation related to oblique subduction and collision of a trench with the Arabian continental margin to the east, as in Clube & Robertson (1986) and Robertson (1990).

palaeorotation and the emplacement of the Hatay and Baër–Bassit ophiolites in the Campanian. The latter units represent fragments of supra-subduction zone crust which formed to the east of the Troodos spreading centre and which subsequently collided with the offset Levant continental margin. The emplaced Hatay ophiolite was transgressed by upper Campanian neritic carbonates (cf. the Campanian onset of rotation in Cyprus).

Collision to the east of Troodos pinned one segment of the supra-subduction zone ophiolite (Fig. 5b), while the Troodos spreading centre still remained above a downgoing slab. Continued northward movement of the African plate resulted in the trench pivoting about the intersection of the transform passive margin with the subduction front (Robertson 1990; Fig. 5b). Southwards 'roll-back' of the trench to the west of the pivot point exerted a pull on the overriding plate. As a consequence, an anti-clockwise rotational torque was developed in the forearc region. The Troodos microplate then became detached and began to rotate about an irregular ring of dextral strike-slip faults and associated relay faults (Clube & Robertson 1986; Robertson 1990). In this model, fragments of the northern continental margin (Mamonia Complex, Moni Melange) became attached to the rotating microplate along strike-slip lineaments (see below) and were then carried southward to their present position.

The inferred microplate boundary faults probably followed pre-existing zones of crustal weakness, particularly the westward extension of the Southern Troodos transform zone in SW Cyprus. The postulated curved microplate boundary must have cut across the pre-existing, more linear oceanic fracture zone (Robertson 1990). In southern Cyprus, microplate boundary faults are inferred to run close to the present coastline, south of the fossil transform zone (Clube & Robertson 1986).

This model for the palaeorotation is more consistent with the geology of Cyprus and the regional tectonic setting than models which invoke 'push' mechanisms involving microcontinental collision.

The Troodos spreading system

With the overall framework of the palaeorotation of Troodos established (Moores & Vine 1971; Clube *et al.* 1985; Clube 1986; Clube & Robertson 1986; Abrahamsen & Schönharting 1987; Morris *et al.* 1990), more recent palaeomagnetic work carried out in Cyprus has been aimed at specific, genetic problems associated with the structure and configuration of the Troodos spreading system. Results of these studies have been used by Allerton & Vine (1991) to produce an evolutionary model for formation of the Troodos ophiolite, involving periods of magmatic accretion, ridge-jumping and amagmatic stretching. Discussion here is limited to some of the more specific data incorporated in this model.

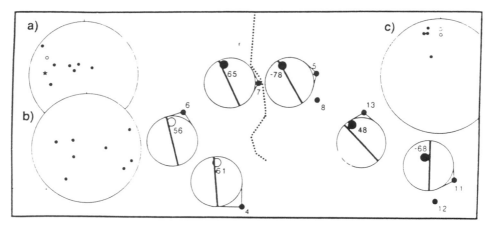

Fig. 6. Palaeomagnetic results obtained in the Solea graben by Allerton & Vine (1987). Net tectonic rotation axes (solid/open circles are lower/upper hemisphere), rotation angles (positive angles correspond to anticlockwise rotations by faulting) and initial dyke strikes (solid lines) are shown in the small stereonets for the preferred solutions. (**a**) mean site magnetization vectors (star is TMV). (**b**) present orientation of poles to dykes. (**c**) mean site rotation vectors. (from Allerton 1989*a*).

Rotations at spreading axes

Dykes in the SDC and extrusive series along the northern margin of the Troodos Complex deviate markedly from the predominant N–S dyke trend (Moores & Vine 1971). Clube (1985) discovered that magnetic remanence directions in these anomalous areas were not significantly different from those in areas of N-S trending dykes. Major discontinuities in dyke orientation must therefore by a primary feature of the Troodos oceanic crust and not the result of post-formation vertical axis rotations. In addition to variations in strike, Verosub & Moores (1981) noted significant differences in dyke dip within the SDC, which they suggested may result from rotational normal faulting at spreading axes. Verosub & Moores (1981) subdivided the ophiolite into a series of domains of similar dip. Building on this earlier work, Varga & Moores (1985) identified three structural grabens within the SDC, defined by listric and planar normal faults and rotated dykes that dip symmetrically toward graben axes. These are: the Solea graben, north of Mount Olympos; the Mitsero graben, north of the Southern Troodos transform zone; and the Larnaca graben, in the SE of the Troodos Complex. They suggested that rotation of originally vertical dykes about near horizontal axes was caused by extensional faulting near a ridge crest. The grabens were interpreted as fossil axial valleys produced by successive eastward jumps of an approximately north-trending (present coordinates), *slow-*

spreading ridge crest. This model, based primarily on observations of the present structural orientations of dykes, renewed interest in the palaeomagnetism of the SDC.

Allerton & Vine (1987) were the first to devise an independent palaeomagnetic test of the graben hypothesis. They developed a net tectonic rotation technique which allowed recovery of initial, pre-rotation dyke orientations together with the angles and corresponding poles of rotation, and hence reconstruction of spreading axis structure from palaeomagnetic data. A key assumption is that the angle, β, between the present pole to a dyke margin and the *in situ* magnetization vector remained constant during deformation. The analysis involves finding a single pole of rotation which simultaneously restores the *in situ* magnetization vector back to the Troodos magnetization vector (TMV; $D = 276°$, $I = 32°$; Moores & Vine 1971) *and* the present dyke pole as close to the horizontal as possible, while conserving the angle β. An anticlockwise rotation around this pole is described by a positive rotation angle. Two solutions are obtained for cases where the condition that dykes were intruded vertically is satisfied, with initial dyke poles symmetrically arranged either side of the TMV. The most probable solution is selected using external criteria. For example, restoring associated pillow flows to their initial orientations using both solutions frequently gives one reasonable and one steeply-inclined to overturned solution (Allerton 1989*a*). A strength of this net tectonic

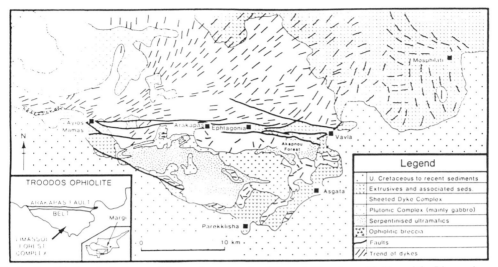

Fig. 7. Simplified geological map of the southern margin of the main Troodos ophiolite and the Limassol Forest Complex (after Simonian & Gass 1978), showing the progressive change in dyke trend into near parallelism with the Southern Troodos transform zone over a distance of 10–15 km (from Morris *et al*. 1990).

rotation technique is that it allows back-stripping of successive rotations where cross-cutting relationships are observed.

Palaeomagnetic research within the Solea graben by Allerton & Vine (1987) identified rotations of up to 78° around subhorizontal axes which are subparallel to the original dyke strikes (Fig. 6), consistent with an axial process for the development of the structure. Similar results were obtained by Allerton (1989*a*) in the Larnaca graben to the east, where a maximum rotation of 115° was found from currently flat-lying dykes. Initial dyke strikes were found to trend NW and to parallel the main normal faults in the area, implying that rotation occurred by slip on a set of ridge-parallel normal faults (Allerton 1989*a*). Further sampling within the Solea graben by Hurst *et al*. (1992) confirmed the importance of strike-parallel tilting on its western flank, but identified components of rotation around more steeply inclined axes on its eastern flank. Allerton & Vine (1987) suggested formation of the Solea graben by antithetic faulting on the west flank of an *intermediate- to fast-spreading* axis during a time of reduced magma supply. Subsequently, normal dyke intrusion resumed, generating the typical, simple structures observed over much of the complex.

Rotations associated with the Southern Troodos Transform Fault Zone

Much debate has centred on the sense of displacement along the fossil Southern Troodos transform zone in the Arakapas and Limassol Forest areas. A progressive change in dyke trend is observed within the SDC as the transform is approached from the north, from a predominant N–S orientation into eventual alignment with the transform lineament (Fig. 7). This has been interpreted as the result of either dyke injection into a sigmoidal stress field, implying that dykes are in their original orientations relative to the *sinistrally* slipping transform, or clockwise vertical axis fault block rotations in response to *dextral* slip (Simonian & Gass 1978; Fig. 8). These alternative models can be directly tested using palaeomagnetism, since tectonic rotations would have deviated remanences away from the regional Troodos magnetization direction. The first attempt at resolving the debate was made by Clube (Clube 1985; Clube & Robertson 1986), who sampled pillow lavas and interlava sediments along the Arakapas fault belt. Four out of five sites exhibited NW-directed remanences, in contrast to the westerly Troodos magnetization, and the results were taken as support for the block rotation model. More significant evidence for clockwise rotation of dykes within the SDC itself was provided by Bonhommet *et al*. (1988). Data from 13 sites were shown to cluster tightly with westerly declinations after correction of dykes back to the predominant N–S trend, thereby ruling out initial NE–SW trends. Further palaeomagnetic support for dextral shear along the transform zone came from the studies of Allerton & Vine (1990), MacLeod *et al*. (1990) and Morris *et al*. (1990). Importantly,

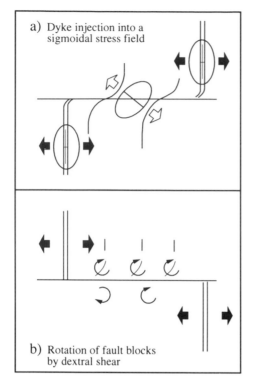

a) Dyke injection into a sigmoidal stress field

b) Rotation of fault blocks by dextral shear

Fig. 8. Possible alternative settings in which deviations in dyke trend could occur close to the Southern Troodos Transform fault Zone (from Morris *et al.* 1990). (**a**) dyke injection into a sigmoidal stress field operating across a sinistrally slipping transform between dextrally offset spreading axes. (**b**) rotation of fault blocks related to dextral slip along the active transform domain between sinistrally offset spreading axes. A modified form of this model (in which rotations are restricted to the ridge-transform corner) is now the accepted explanation for observed dyke deviations.

these authors applied the net tectonic rotation technique of Allerton & Vine (1987) to sets of cross-cutting dykes to demonstrate that clockwise block rotations were actively occurring during crustal genesis, rather than resulting from later reactivation of the fault zone.

The overwhelming palaeomagnetic evidence for clockwise fault block rotations associated with dextral slip along the Southern Troodos transform contrasts with structural data reported by Murton (1986) which suggests sinistral slip within the western Limassol Forest Complex (i.e. within the transform zone). This apparent conflict has been resolved by MacLeod & Murton (1995), however, who propose a model in which sinistral shear developed locally at

block boundaries within an overall dextral shear zone.

Detailed palaeomagnetic and structural analyses (MacLeod *et al.* 1990) have identified a western limit to the zone of transform-related rotations (*c.* 10 km west of the area shown in Fig. 7), beyond which only simple tilting of dykes has occurred. The changeover from rotated to unrotated dykes occurs across a complex zone about 2–5 km wide. This zone is interpreted by MacLeod *et al.* (1990) as a fossilised intersection between the Solea spreading axis and the Southern Troodos transform fault. They note that the radius of curvature of dyke swing remains approximately constant across the entire exposed width of the Troodos massif to the east of the ridge-transform intersection (see Fig. 7). MacLeod *et al.* (1990) conclude that transform-related rotations occurred within the active corner of the intersection itself rather than being accommodated progressively with increasing strike-slip displacement along the transform (as inferred in the model shown in Fig. 8). This supports a theoretical model of Allerton (1989b) for distortions within weak crust at ridge-transform intersections (Fig. 9).

The Mamonia Complex

The Mamonia Complex terrane of SW Cyprus (Fig. 1) consists of a series of Mesozoic deep-sea sedimentary and volcanic rocks representing remnants of a passive continental margin and oceanic crust formed in a small Neotethyan ocean basin (Robertson & Woodcock 1979). The Complex is separated from the main Troodos ophiolite to the east by arcuate high-angle fault lineaments marked by a series of slivers of Troodos-type crust and discontinuous strands of serpentinite (Fig. 1). The Mamonia and Troodos terranes are believed to have been juxtaposed by strike-slip faulting (Swarbrick 1979, 1980) and the terrane-bounding fault zones were transgressed by deep-water carbonates in the Early Tertiary.

Preliminary palaeomagnetic results from the Late Triassic Mamonia Complex lithologies have been reported by Clube (1985) and Clube & Robertson (1986). Bimodal distributions of magnetization directions were obtained in both the Mamonia pillow lavas (Dhiarizos Group) and the structurally overlying sediments (Vlambouros Formation of the Ayios Photios Group), implying the presence of normal and reversed polarities. Mean directions were poorly defined, however, possibly as a result of blanket demagnetization. No fold tests were possible and the age of magnetization is uncertain.

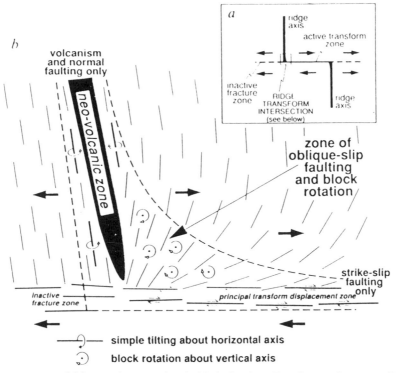

Fig. 9. The most recent model for rotations associated with the Southern Troodos transform zone (from MacLeod *et al.* 1990). (**a**) Schematic representation of an idealized ridge-transform system. (**b**) Relationships suggested by MacLeod *et al.* (1990) and Allerton (1989*b*) to explain the observed patterns of simple tilting and fault block rotations at the intersection between the Solea graben and Southern Troodos transform. Block rotations are essentially restricted in this model to the intersection itself.

Shallow inclinations and essentially N–S declinations suggest that these units were either magnetized close to the equator in the northern hemisphere and then rotated by *c.* 180°, or were magnetized close to the equator in the southern hemisphere and are unrotated. In either case, remanence directions are clearly unrelated to the westerly-directed Troodos magnetization.

The sense of displacement along the faults separating the Troodos and Mamonia terranes is currently under debate. Clube & Robertson (1986) suggested that Troodos and Mamonia crustal elements were interleaved as a direct result of Troodos microplate rotation (see above). This implies a dextral sense of shear along the terrane boundary, in agreement with evidence from dextral shear indicators in minor metamorphic rocks within the ophiolitic lineaments (Spray & Roddick 1981). More recently, Swarbrick (1993) has reported sinistral shear indicators within unmetamorphosed lithologies along the structural contacts, and has proposed a model in which anticlockwise rotation of the Troodos Complex occurred within a large-scale

sinistral strike-slip zone. This debate can potentially be settled by palaeomagnetism and is currently being addressed by Morris *et al.* (in prep.), who have sampled the Troodos-type slivers along the terrane-bounding faults in order to determine the sense of rotation of fault blocks with respect to the main Troodos Complex.

Summary

The Troodos Complex represents an uplifted fragment of Neotethyan oceanic crust (Gass 1968; Moores & Vine 1971) which formed above a northward dipping intra-oceanic subduction zone in the Late Cretaceous (Pearce 1975; Smewing *et al.* 1975; Schmincke *et al.* 1983; Robinson *et al.* 1983; Robertson 1990). Fabrics preserved within the ophiolite indicate that the Troodos system experienced non-steady state spreading (Varga & Moores 1985; Allerton & Vine 1987, 1991). Anomalous ophiolitic crust along the southern margin of the Troodos massif was generated within a transtensional Southern

Troodos transform fault zone, and a small fragment of 'Anti-Troodos' crust is also preserved (Murton 1986; MacLeod 1988, 1990).

Palaeomagnetic research has demonstrated that the ophiolite underwent a 90° anticlockwise rotation subsequent to its formation (Moores & Vine 1971). Detailed studies of the circum-Troodos sedimentary cover constrain the timing of this rotation to the Campanian to Early Eocene interval (Clube 1985; Clube et al. 1985; Clube & Robertson 1986; Abrahamsen & Schönharting 1987; Morris et al. 1990). The most favoured model for the palaeorotation (Clube & Robertson 1986; Robertson 1990) involves continental margin–trench collision and ophiolite emplacement to the east of Troodos generating an anticlockwise torque upon the suprasubduction zone Troodos crust. As a consequence, a small Cyprus-sized microplate detached and rotated around an irregular ring of dextral wrench faults.

Other palaeomagnetic and structural studies have documented tectonic processes operating during Troodos crustal genesis in considerable detail. Rotations of normal fault blocks about ridge-parallel, sub-horizontal axes occurred during sea-floor spreading (Allerton & Vine 1987; Allerton 1989a; Hurst et al. 1992). Vertical axis rotations adjacent to the Southern Troodos transform zone resulted from dextral shear along the transform (Clube & Robertson 1986; Bonhommet et al. 1988, Allerton & Vine 1990; Morris et al. 1990), but were probably generated within the active corner of the intersection between the Solea spreading axis and the transform zone (MacLeod et al. 1990). These palaeomagnetic inputs to the Cyprus 'story' provide a framework for successful models for the tectonic history of the ophiolite (e.g. Robertson 1990; Allerton & Vine 1991). Future palaeomagnetic studies, particularly concerning the relationship of the Troodos Complex to the other Cyprus terranes, can only further improve our understanding of the regional tectonics of the eastern Mediterranean.

I thank S. Allerton, K. Creer, C. MacLeod, B. Murton, A. Robertson, and F. Vine for enlightening discussions during the course of my own earlier research in Cyprus. These discussions greatly eased production of this review. My grateful thanks to the Vasilopoulos family of Athens for their generous hospitality during production of the final version of the paper, and especially Efrosini Vasilopoulou for use of her PC.

References

ABRAHAMSEN, N. & SCHÖNHARTING, G. 1987. Palaeomagnetic timing of the rotation and translation of Cyprus. Earth and Planetary Science Letters, 81, 409–418.

ALLERTON, S. 1989a. Fault block rotations in ophiolites: results of palaeomagnetic studies in the Troodos Complex, Cyprus. In: KISSEL, C. & LAJ, C. (eds) Palaeomagnetic Rotations and Continental Deformation. NATO ASI Series C. Volume 254, 393–410.

—— 1989b. Distortions, rotations and crustal thinning at ridge-transform intersections. Nature, 340, 626–628.

—— & VINE, F. J. 1987. Spreading structure of the Troodos ophiolite, Cyprus: some palaeomagnetic constraints. Geology, 15, 593–597.

—— & —— 1990. Palaeomagnetic and structural studies of the southeastern part of the Troodos complex. In: MALPAS, J., MOORES, E. M., PANAYIOTOU, A. & XENOPHONTOS, C. (eds) Ophiolites: Oceanic Crustal Analogues. Cyprus Geological Survey Department, 99–111.

—— & —— 1991. Spreading evolution of the Troodos ophiolite, Cyprus Geology, 19, 637–640.

ANON. 1972. Penrose field conference, Ophiolites. Geotimes, 24–25.

CLUBE, T. M. M. 1985. The palaeorotation of the Troodos microplate. PhD thesis, University of Edinburgh.

—— & ROBERTSON, A. H. F. 1986. The palaeorotation of the Troodos microplate, Cyprus, in the Late Mesozoic–Early Cenozoic plate tectonic framework of the Eastern Mediterranean. Surveys in Geophysics, 8, 375–434.

——, CREER, K. M. & ROBERTSON, A. H. F. 1985. The palaeorotation of the Troodos microplate. Nature, 317, 522–525.

DESMET, A. P. 1976. Evidence for co-genesis of the Troodos lavas, Cyprus. Geological Magazine, 113, 165–168.

FREUND, R. & TARLING, D. H. 1979. Preliminary Mesozoic palaeomagnetic results from Israel and inferences for a microplate structure in the Lebanon. Tectonophysics, 60, 189–205.

GASS, I. G. 1968. Is the Troodos massif of Cyprus a fragment of Mesozoic ocean floor? Nature, 220, 39–42.

—— 1980. The Troodos massif: Its role in the unravelling of the ophiolite problem and its significance in the understanding of constructive margin processes. In: PANAYIOTOU, A. (ed.) Ophiolites: Proceedings of the International Symposium, Cyprus, 1979. Cyprus Geological Survey Department, 23–35.

—— & SMEWING, J. D. 1973. Intrusion and metamorphism at constructive margins: evidence from the Troodos massif, Cyprus. Nature, 242, 26–29.

HURST, S. D., VEROSUB, K. L. & MOORES, E. M. 1992. Paleomagnetic constraints on the formation of the Solea graben, Troodos ophiolite, Cyprus. Tectonophysics, 208, 431–445.

KISSEL, C. & LAJ, C. 1988. The Tertiary geodynamical evolution of the Aegean arc: a palaeomagnetic reconstruction. Tectonophysics, 146, 183–201.

——, OVENBUCH, D., DE LAMOTTE, F., MONOD, O. & ALLERTON, S. 1993. First paleomagnetic evidence for a post-Eocene clockwise rotation of the

Western Taurides thrust belt east of the Isparta reentrant (Southwestern Turkey). *Earth and Planetary Science Letters*, **117**, 1–14.

MacLeod, C. J. 1988. *The tectonic evolution of the Eastern Limassol Forest Complex, Cyprus*. PhD thesis, Open University, Milton Keynes.

—— 1990. Role of the Southern Troodos Transform Fault in the rotation of the Cyprus microplate: evidence from the Eastern Limassol Forest. *In*: Malpas, J., Moores, E. M., Panayiotou, A. & Xenophontos, C. (eds) *Ophiolites: Oceanic Crustal Analogues*. Cyprus Geological Survey Department, 75–85.

—— & Murton, B. J. 1995. On the sense of slip of the Southern Troodos transform fault zone, Cyprus. *Geology*, **23**, 257–260.

——, Allerton, S., Gass, I. G. & Xenophontos, C. 1990. Structure of a fossil ridge-transform intersection in the Troodos ophiolite. *Nature*, **348**, 717–720.

Malpas, J., Xenophontos, C. & Williams, D. 1992. The Ayia Varvara Formation of SW Cyprus: a product of complex collisional tectonics. *Tectonophysics*, **212**, 193–241.

Mantis, M. 1970. Upper Cretaceous–Tertiary foraminiferal zones in Cyprus. *Epetiris*, **3**, 227–241.

Moores, E. M. & Vine, F. J. 1971. The Troodos Massif, Cyprus and other ophiolites as oceanic crust: evaluation and implications. *Philosophical Transactions Royal Society of London*, **A268**, 433–466.

——, Robinson, P. T., Malpas, J. & Xenophontos, C. 1984. A model for the origin of the Troodos Massif, Cyprus and other mideast ophiolites. *Geology*, **12**, 500–503.

Morris, A. & Robertson, A. H. F. 1993. Miocene remagnetization of carbonate platform and Antalya Complex units within the Isparta Angle, SW Turkey. *Tectonophysics*, **220**, 243–266.

——, Creer, K. M. & Robertson, A. H. F. 1990. Palaeomagnetic evidence for clockwise rotations related to dextral shear along the Southern Troodos Transform Fault, Cyprus. *Earth and Planetary Science Letters*, **99**, 250–262.

Murton, B. J. 1986. Anomalous oceanic lithosphere formed in a leaky transform fault: evidence from the western Limassol Forest Complex, Cyprus. *Journal of the Geological Society, London*, **143**, 845–854.

—— 1990. Was the Southern Troodos Transform Fault a victim of microplate rotation? *In*: Malpas, J., Moores, E. M., Panayiotou, A. & Xenophontos, C. (eds) *Ophiolites: Oceanic Crustal Analogues*. Cyprus Geological Survey Department, 87–98.

Nur, A. & Helsley, C. E. 1971. Paleomagnetism of Tertiary and Recent lavas of Israel. *Earth and Planetary Science Letters*, **10**, 376–379.

Parrot, J.-F. 1977. *Assemblage ophiolitique du Baër-Bassit et termes éffusives du volcano-sédimentaire*. Travaux et Documents de L'Orstrom, **72**.

Pearce, J. A. 1975. Basalt geochemistry used to investigate past tectonic environments in Cyprus. *Tectonophysics*, **25**, 41–67.

—— 1980. Geochemical evidence for the genesis and setting of lavas from Tethyan ophiolites. *In*: Panayiotou, A. (ed.) *Ophiolites: Proceedings of the International Symposium, Cyprus, 1979*. Cyprus Geological Survey Department, 261–272.

Robertson, A. H. F. 1975. Cyprus umbers: basalt-sediment relationships on a Mesozoic ocean ridge. *Journal of the Geological Society, London*, **131**, 511–531.

—— 1990. Tectonic evolution of Cyprus. *In*: Malpas, J., Moores, E. M., Panayiotou, A. & Xenophontos, C. (eds) *Ophiolites: Oceanic Crustal Analogues*. Cyprus Geological Survey Department, 235–252.

—— & Hudson, J. D. 1974. Pelagic sediments in the Cretaceous and Tertiary history of the Troodos Massif, Cyprus. *In*: Hsü, K. J. & Jenkyns, H. C. (eds) *Pelagic Sediments on Land and under the Sea*. International Association of Sedimentologists, Special Publication, **1**, 403–436.

—— & Woodcock, N. H. 1979. The Mamonia Complex, southwest Cyprus: the evolution and emplacement of a Mesozoic continental margin. *Geological Society of America Bulletin*, **90**, 651–665.

—— & Xenophontos, C. 1993. Development of concepts concerning the Troodos ophiolite and adjacent units in Cyprus. *In*: Prichard, H. M., Alabaster, T., Harris, N. B. W. & Neary, C. R. (eds) *Magmatic Processes and Plate Tectonics*. Geological Society, London, Special Publications, **76**, 85–119.

——, Eaton, S., Follows, E. J. & McCallum, J. E. 1991. The role of local tectonics versus global sea-level change in the Neogene evolution of the Cyprus active margin. *In*: MacDonald, D. I. M. (ed.) *Sedimentation, Tectonics and Eustasy – sea level changes at Active Margins*. International Association of Sedimentologists, Special Publications, **12**, 331–369.

Robinson, P. T. & Malpas, J. 1990. The Troodos ophiolite of Cyprus: new perspectives on its origin and emplacement. *In*: Malpas, J., Moores, E. M., Panayiotou, A. & Xenophontos, C. (eds) *Ophiolites: Oceanic Crustal Analogues*. Cyprus Geological Survey Department, 13–36.

——, Melson, W. G., O'Hearn, T. & Schmincke, H-U. 1983. Volcanic glass compositions of the Troodos ophiolite, Cyprus. *Geology*, **11**, 400–404.

Ron, H., Freund, R., Garfunkel, Z. & Nur, A. 1984. Block rotation of strike-slip faulting: structural and paleomagnetic evidence. *Journal of Geophysical Research*, **89**, 6256–6270.

Schmincke, H-U., Rautenschlein, M. Robinson, P. T. & Megehan, J. M. 1983. Troodos extrusive series of Cyprus: A comparison with oceanic crust. *Geology*, **11**, 405–409.

Shelton, A. W. & Gass, I. G. 1980. Rotation of the Troodos microplate. *In*: Panayiotou, A. (ed.) *Ophiolites: Proceedings of the International Symposium, Cyprus, 1979*. Cyprus Geological Survey Department, 61–65.

Simonian, K. O. & Gass, I. G. 1978. Arakapas fault belt, Cyprus: a fossil transform belt. *Geological Society of America Bulletin*, **89**, 1220–1230.

SMEWING, J. D., SIMONIAN, K. O. & GASS, I. G. 1975. Metabasalts from the Troodos Massif, Cyprus: genetic implications deduced from petrology and trace element compositions. *Contributions to Mineralogy and Petrology,* **51**, 49–64.

SPRAY, J. G. & RODDICK, J. C. 1981. Evidence for Upper Cretaceous transform metamorphism in West Cyprus. *Earth and Planetary Science Letters,* **55**, 273–291.

SWARBRICK, R. E. 1979. *The sedimentology and structure of SW Cyprus and its relationship to the Troodos Complex.* PhD thesis, University of Cambridge.

—— 1980. The Mamonia Complex of SW Cyprus and its relationship with the Troodos Complex. *In:* PANAYIOTOU, A. (ed.) *Ophiolites: Proceedings of the International Symposium, Cyprus, 1979.* Cyprus Geological Survey Department, 86–92.

—— 1993. Sinistral strike-slip and transpressional tectonics in an ancient oceanic setting: the Mamonia Complex, southwest Cyprus. *Journal of the Geological Society, London,* **150**, 381–392.

VARGA, R. J. & MOORES, E. M. 1985. Spreading structure of the Troodos ophiolite, Cyprus. *Geology,* **13**, 846–850.

VEROSUB, K. L. & MOORES, E. M. 1981. Tectonic rotations in extensional regimes and their paleomagnetic consequences for oceanic basalts. *Journal of Geophysical Research,* **86**, 6335–6349.

VINE, F. J. & MATTHEWS, D. H. 1963. Magnetic anomalies over oceanic ridges. *Nature,* **199**, 947–949.

—— & MOORES, E. M. 1969. Palaeomagnetic results from the Troodos Igneous Massif, Cyprus. *Transactions of the American Geophysical Union,* **50**, 131.

XENOPHONTOS, C. & MALPAS, J. G. (eds) 1987. *Field Excursion Guidebook. Troodos 87 – Ophiolites and Oceanic Lithosphere.* Cyprus Geological Survey Department.

Palaeomagnetism of some Cretaceous Nubian Sandstones, Northern Sinai, Egypt

A. M. KAFAFY[1], D. H. TARLING[2], M. M. EL GAMILI[3], H. H. HAMAMA[3] & E. H. IBRAHIM[2,3]

[1] *Department of Geology, Faculty of Science, University of Tanta, Egypt*
[2] *Department of Geological Sciences, University of Plymouth, PL4 8AA, UK*
[3] *Department of Geology, Faculty of Science, University of Mansoura, Egypt*

Abstract: A total of 23 sites have been sampled from the Nubian Sandstone of Gabal El-Minsherah and Gabal El-Halal, northern Sinai. Isothermal remanent magnetization acquisition showed that hematite is the predominant magnetic mineral carrying the remanence after the destruction of goethite at temperatures of 100–150°C. The *in situ* directions correspond to palaeomagnetic poles at 84°N 257°E (G. El Minsherah) and 78°N 288°E (G. El Halal). A fold-test shows that the magnetizations were acquired later than the Late Cretaceous (Laramide) tectonic disturbance. The presence of both normal and reversed polarities and *in situ* pole positions close to the expected Late Tertiary palaeomagnetic pole suggest that the magnetizations are of Miocene age. They are probably chemical magnetizations acquired during occasional fluid migrations through these permeable beds during arid Miocene climatic phases.

The Nubian Sandstone extends over a large region in northern Africa and Arabia. In the Gabal El-Minsherah and Gabal El-Halal areas, they overlie Jurassic strata and are topped by calcareous sediments of Upper Cretaceous age. The Nubian Sandstones and associated iron ores of Egypt have been studied palaeomagnetically by El-Shazly & Krs (1971, 1973), Hussain *et al.* (1976), Schult *et al.* (1978, 1981) and Hussain & Aziz (1983), but there are still many puzzles regarding their magnetization. This paper throws more light on the magnetization and geological evolution of the Nubian Sandstones of northern Egypt, particularly their post depositional development.

Geological background

Northern Sinai forms a distinct geomorphological and structural unit characterized by a large number of NE–SW-trending elliptical anticlines and synclines. These anticlines extend across Sinai, Palestine and Egypt where they appear at the surface in Gebel Shabrawet and Gebel Abu Rawash. At Qalb El Minsherah, in Sinai, the stratigraphic section starts with Upper Jurassic rocks which are overlain by unfossiliferous Nubian facies of Lower Cretaceous age which in turn are overlain by Cenomanian rocks. The base of the Cenomanian is distinct and conformable, but its upper limit is difficult to

Fig. 1. Location map of the sampling localities.

determine and has been taken arbitrarily to be the top of the 'Ammonite bed' (Shata 1960). Folding seems to have started as early as the Cenomanian and, on the basis of the sediment distribution, facies and thickness variations, continued intermittently until the uppermost Cretaceous (Said 1962). Said (1962) associated this compressive movement with the contemporaneous Laramide anticlinal structures in northern Sinai.

From Morris, A. & Tarling, D. H. (eds), 1996, *Palaeomagnetism and Tectonics of the Mediterranean Region*, Geological Society Special Publication No. 105, pp. 325–332.

325

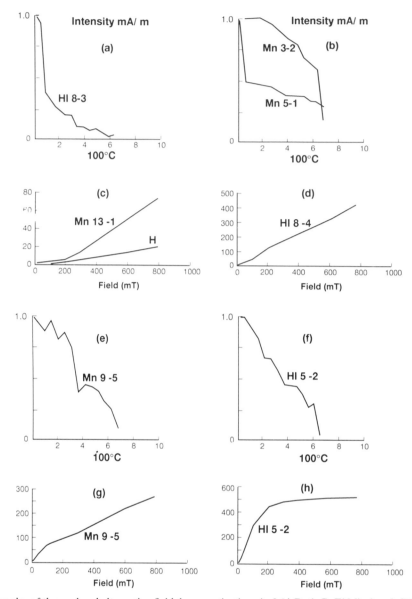

Fig. 2. Examples of thermal and alternating field demagnetization. (**a** & **b**) Both G. El Minsherah (Mn) and G. El Halal (Hl) samples commonly show a rapid decrease in intensity of NRM during thermal demagnetization which is interpreted as being due to the presence of goethite, but this is not present in all samples.
(**c** & **d**) The acquisition of IRM shows little or no indication of the presence of titanomagnetites, but could contain either goethite or hematite on the IRM properties alone.
(**e** & **f**) Intensity of NRM does not always drop rapidly, but the remanence becomes unmeasurable above 600°C. (**g** & **h**) The acquisition of IRM after heating to 150°C in order to destroy any goethite.

In northern Sinai, a total of 23 sites (165 samples) from the Nubian Sandstones were sampled in two sections (Fig. 1). The first section at G. El Minsherah was near Qalb El Minsherah (30.30°N, 33.60°E), 16 km SW of Bir El-Hasana, where 15 sites comprising 108 samples were collected. This section consists of about 200 m of variegated ferruginous sandstones of Nubian facies which become more marly and contain intercalations of oyster and marl beds in the uppermost parts. These rocks are overlain by Cenomanian and Turonian rocks. The second section was at G. El Halal, near Hadirat El Halal (30.60°N, 34.00°E), where the stratigraphic sequence is nearly the same at G. El Minsherah, but the base of the Nubian Sandstones is not exposed. Eight sites comprising 57 core samples were collected from this section. All samples were independently oriented using a sun compass, other than at midday or when overcast, when a magnetic compass was used. Each 2.5 cm diameter drill core was sliced into 2.1 cm high cylinders in the laboratory.

Methodology

The natural remanence magnetization (NRM) was measured using a Molspin MS2 magnetometer. Stepwise demagnetization was undertaken either with a Molspin MSA2 shielded alternating field (AF) demagnetizer which provides pure alternating magnetic fields up to 100 mT, or using a large Schonstedt-type furnace capable of heating up to 30 samples at temperatures up to 720°C. The low-field magnetic susceptibility along the z axis of each sample was measured using a bulk susceptibility bridge. The instrument has a sensitivity of 10^{-5} SI units. Isothermal remanence magnetization (IRM) studies were carried out in direct magnetic fields up to 0.9 T using a Molspin pulse magnetizer. All measurements were carried out at the Palaeomagnetic Laboratory of the Department of Geological Sciences, University of Plymouth, UK. Two pilot specimens from each site were selected for separate AF and thermal demagnetization. AF demagnetization produced virtually no changes in intensity or direction of remanence, even up to 100 mT, but thermal demagnetization enabled different components to be recognized within most samples On this basis, thermal cleaning was applied to the remaining samples, with low field magnetic susceptibility being measured after each temperature step to monitor any thermally induced changes to the magnetic mineralogy.

Rock magnetic observations

The colour of the rock was used to give some indication of the predominant magnetic mineralogy present, but this was supplemented by both IRM and blocking temperature studies. None of these can be considered to provide

unique identification of the magnetic mineralogy, but, when combined, enable grouping of the magnetic properties that can usually be related to specific magnetic minerals. The magnetic coercivity properties were determined by observing sample behaviour during progressive acquisition of IRM (Dunlop 1972). This method was supplemented by information on the blocking temperature spectra during thermal demagnetization as this enables separation of specific minerals. For example, the coercivities of hematite and goethite are often similar, but they can be distinguished readily by their different chemical stability to heating. Three groups could be distinguished on these forms of behaviour.

(1) Samples in the first group were characterized by a rapid drop in NRM intensity by 100–200°C (Fig. 2a) and by IRM acquisition curves that did not reach saturation even in fields as strong as 0.8 T (Fig. 2d). All these samples had a yellowish-orange colour. On the basis of these characteristics, the main magnetic mineral at room temperature was considered to be goethite. By 100–150°C, the goethite had been destroyed. At higher temperatures the intensity decreased slowly up to 650°C, consistent with the remaining magnetic mineral being hematite. Thermal demagnetization paths commonly showed a slight inflexion at about 600°C that could be due to the presence of either very small amounts of magnetite or large hematite grains. In order to distinguish between these possibilities, another sample from the same site was heated to 150°C, to destroy any goethite, and then its IRM acquisition was measured in steps up to 800 mT. As this IRM curve showed a rapid increase in intensity without any evidence for saturation until 800 mT, the inflexion is most likely to be due to hematite.

(2) In the second group, a high blocking temperature mineral was predominant which was quite stable up to 650°C (Fig. 2b). The IRM acquisition curves (Fig. 2c) showed a very slow increase until 100 mT, followed by a more rapid rise in isothermal moment, but with no evidence of saturation by 800 mT. These samples all had red staining (colour ranging from 5 R 8/2 to 10 R 3/4). The high blocking temperatures and coercivities indicate that the remanence is carried by very fine grained hematite, while the red coloration suggested this may be mainly present as a pigment.

(3) In the third group, all the samples were characterized by white to pink colours, suggestive of hematite. The IRM acquisition curves showed two types of behaviour. The first had a

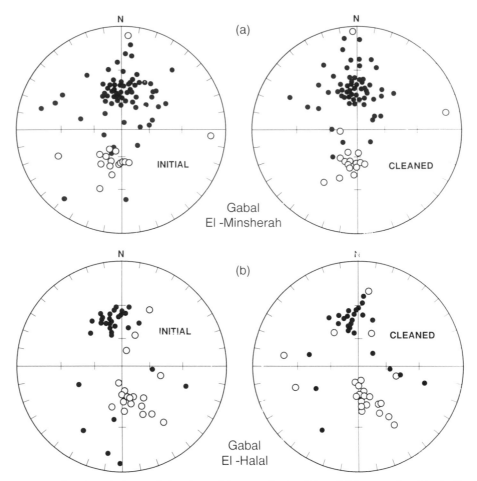

Fig. 3. Sample directions, before and after thermal demagnetization. The solid dots are of positive inclination and the hollow circles are of negative inclination. Equal-area stereographic projections.

weak shoulder, followed by a steady increase in remanence (Fig. 2g), while the second showed a sharp increase in intensity up to 300 mT followed by a slow increase at higher fields (Fig. 2h). During thermal demagnetization (see below and Fig. 2e & f), most of the remanence had been removed by just above 600°C suggesting that some hematite was present, although no consistent remanence could be determined between 620 and 680°C, so that magnetite could be indicated. To check whether the predominant high coercivity magnetic mineral was goethite or hematite, another sample was heated to 150°C to destroy any possible goethite. Its IRM behaviour remained the same as for the unheated sample, indicating that no goethite was present in either the unheated or heated samples. It

seems likely that most of the remanence is carried by hematite, but this is mostly not of single domain size, probably smaller and consequently exhibiting lower blocking temperatures than for single domain magnetite.

Palaeomagnetic results

Initial NRM intensities and susceptibilities showed wide ranges:

	Intensity (mA/m)	Susceptibility (m SI)
G. El-Minsherah	0.44–212.0	0.25–371.6
G. El-Halal	2.1–138.4	16.3–216.2

The colour of samples, using the Rock Colour Chart (Goddard 1984), varied from white to

Specimen Mn 6 -3

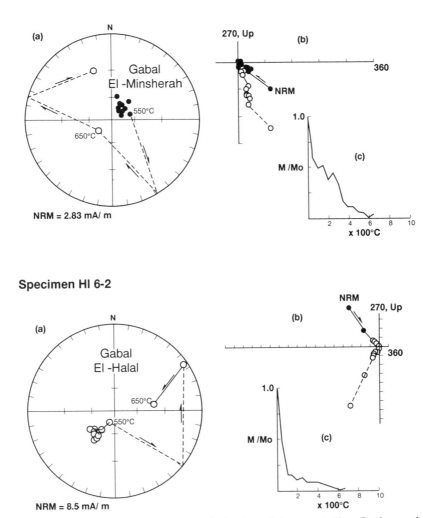

Fig. 4. Examples of directional and intensity changes during thermal demagnetization. On the equal-angle stereographic projections, the solid symbols are of positive inclination and the hollow symbols are of negative inclination. On the vector plots, the solid symbols represent the horizontal component and the hollow symbols are the vertical component.

pink and red to dark brown with some of grey colour. Generally, there was a correlation between the red colour of the samples and both the initial intensity of remanence and the volume susceptibility, i.e. the more red the sample, the higher the intensity and susceptibility. The initial directions of the samples (Fig. 3a) and their site mean directions (Table 1) were close to the present Earth's magnetic field or its antipole. Thermal demagnetization of pilot specimens

from G. El Minsherah showed a rapid decrease in initial intensity (0.23–3.0 mA m^{-1}) and most of the NRM intensity was lost by 600°C (Fig. 4a). The directional consistency indices ranged from 2.3 to 6.2. The initial susceptibilities were very low positive or negative values, indicative of the presence of paramagnetic minerals. No appreciable changes in susceptibility were observed during thermal demagnetization (Fig. 5a). Pilot specimens from G. El Halal, showed two main

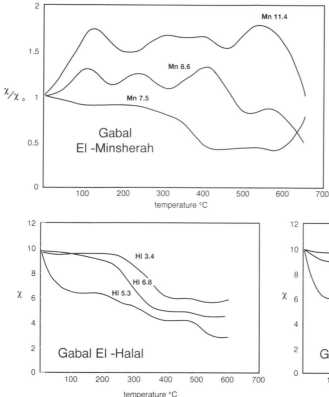

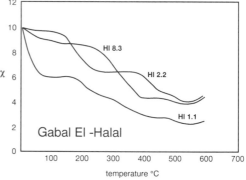

Fig. 5. Changes of low-field susceptibility during thermal demagnetization. (**a**) Many of the susceptibility values are close to the noise-level of measurement and the 'steps' in the curves may therefore be artificial. Although most susceptibility values decrease with increased heating, the values remain low in all cases.

types of behaviour during thermal treatment: (a) a sudden drop of the intensity by 100–150°C, followed by a slow decrease in intensity (e.g. Fig. 4b) accompanied by large changes in directions up to 550°C and a gradual decrease in low-field susceptibility (Fig. 5b). All samples directions were of normal polarity; b) directional consistency during heating to 650°C, but a gradual, steady decrease in susceptibility occurred up to 550°C, above which the susceptibility increased (Fig. 5b).

Using visual inspection of Cartesian plots, principal component analyses (Kirschvink 1980) for both linearity and planarity, and the directional consistency index (equivalent to the Stability Index of Tarling & Symons 1967) it was found that the characteristic component of remanence was isolated in most samples between

150° and 400°C. On this basis all remaining samples were heated to at least three temperatures within this range. Such a procedure meant that the linearity of the vector occurring within this range could be established. In most samples, the changes in direction were small (Fig. 3 a & b). About two third of the sites showed a reduction in the within-site scatter after this treatment; other sites showed only a small increase in within-site scatter. However, the within-site scatter at sites Mn 3, Mn 7, Hl 2 and Hl 8 remained unacceptably high ($\alpha_{95} > 20°$, Table 1) after thermal treatment and these sites were excluded from further statistical analysis. Giving unit weight to each of the remaining site directions and inverting the reversed directions to normal, the mean direction of all site mean directions, before and after bedding correction, was:

	in situ					Lat. (N°)	Long. (E°)	Bedding corrected	
	N	Dec.	Inc.	k	α_{95}			Dec.	Inc.
G. El Minsherah	13	359.7	44.5	55	5.7	85.9	217.3	11.3	60.3
G. El Halal	6	346.5	47.1	46	10.0	78.1	297.2	34.7	55.0

Table 1. *Site mean* in situ *directions for Nubian Sandstones from G. El Minsherah and G. El Halal*

Site	N	Initial				After cleaning			
		Dec.	Inc.	k	α95	Dec.	Inc.	k	α95
G. El Minsherah									
Mn 1	6	177	−53	523	3	180	−52	219	5
Mn 2	5	210	−30	7	32	204	−53	16	20
Mn 3	7	53	79	7	27	343	75	6	29
Mn 4	5	20	46	45	12	9	46	55	10
Mn 5	9	359	36	5	26	357	35	13	14
Mn 6	7	23	40	22	13	13	42	15	16
Mn 7	7	344	40	3	45	339	34	4	43
Mn 8	11	1	46	44	7	350	42	41	7
Mn 9	6	349	46	242	5	349	47	186	5
Mn 10	6	356	51	192	5	356	50	184	5
Mn 11	5	8	46	59	9	355	39	95	8
Mn 12	7	354	31	9	21	354	27	20	14
Mn 13	7	1	42	33	11	354	39	64	8
Mn 14	5	338	48	89	17	348	47	184	6
Mn 15	6	201	−50	85	7	196	−53	82	8
G. El Halal									
Hl 1	5	349	31	96	8	346	38	51	11
Hl 2	5	7	49	34	22	353	−15	2	89
Hl 3	5	159	−64	8	30	170	−58	54	13
Hl 4	5	167	−52	12	24	164	−53	15	20
Hl 5	13	351	37	55	6	354	39	70	5
Hl 6	5	144	−75	3	51	177	−57	78	9
Hl 7	5	150	−37	53	11	152	−36	70	9
Hl 8	3	346	44	156	5	351	35	25	26

It is clear that the bedding correction steepened the directions and increased the angular divergence between the site mean directions from 9.5° to 13.6°. However, these changes are small and only suggest that the remanence is probably post-folding in age. More significantly, the mean directions after tectonic correction are inconsistent with those for both Cretaceous and younger ages. Similarly, the presence of mixed polarities is inconsistent with the Aptian to Senonian age of the rocks as this period corresponds to that of the Cretaceous Normal Magnetochron. On this basis, it is considered that the remanence was acquired after folding and the appropriate palaeomagnetic poles are therefore those calculated prior to correction for bedding.

Discussion and conclusions

The normal and reversed directions (Fig. 3) are close to or antiparallel to the present Earth's magnetic field ($D = 2.4°$, $I = 44.7°$) in the sampled areas. This suggests that the magnetization could be of Recent age. However, the presence of mixed polarities, which have also been recorded in other Nubian Sandstone studies (e.g. Hussain *et al.* 1976; Schult *et al.* 1978, 1981; Hussain & Aziz 1983), clearly indicates ages older than 720 ka, hence the remanences are not associated with Recent weathering. The consistency of the polarity within each site indicates that each remagnetization was associated with a specific event during a single polarity zone. The remagnetization event for each site must therefore have had a duration of less than 1 Ma but was at different times in at least some of the sites as both normal and reversed polarities are involved. It is possible that the sites characterized by large internal directional scatter may have been magnetized during either a period of transitional polarity or during more than one polarity period as the magnetization of the individual samples seems to be stable. As these sandstones are porous and so are susceptible to remagnetization by oxidizing agents (e.g. Turner 1977), it seems most probable that the remagnetizations at both localities are caused by chemical remanences, probably during the formation of goethite and pigmentation. Palaeomagnetism therefore appears to date events associated with occasional flushing of fluids through these sediments during

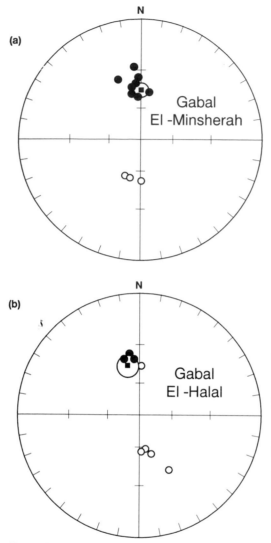

Fig. 6. Site mean directions. The site mean directions form two antiparallel groups in both localities. The circle marks the mean direction and circle of 95% confidence after the reversed site directions have been inverted to normal polarity. Equal-angle stereographic projection.

the Late Miocene and Pliocene when the region was predominantly arid.

Laboratory measurements were carried out at the Department of Geological Sciences, Plymouth University, UK, and we thank all the technical staff who gave assistance.

References

DUNLOP, D. J. 1972. Magnetic mineralogy of unheated and heated red sediments by coercivity spectra. *Geophysical Journal of the Royal Astronomical Society*, **27**, 37–55.

EL-SHAZLY, E. M. & KRS, M. 1971. Magnetism and palaeomagnetism of Oligocene basalts from Abu Zaabal and Qatrani, northern Egypt. *Geolysi Kalai Sbornik.*, **19**, 356, 261–270.

—— & KRS, M. 1973. Palaeogeography and palaeomagnetism of the Nubian Sandstone, Eastern Desert of Egypt. *Geologische Rundschau*, **62**, 212–225.

GODDARD, E. N. 1984. *Rock Colour Chart*. Geological Society of America.

HUSSAIN, A. G. & AZIZ, Y. 1983. Palaeomagnetism of Mesozoic and Tertiary rocks from east El Oweinat area, Egypt. *Journal of Geophysical Research*, **88**, 3523–3529.

——, SCHULT, A., SOFFEL, H. & FAHIM, M. 1976. Magnetization of the Nubian Sandstone in Aswan area, Qena-Safaga and Idfu-Mersa Alam District. *Helwan Observatory Bulletin*, 133–1.

KIRSCHVINK, J. L. 1980. The least-squares line and plane and the analysis of palaeomagnetic data. *Geophysical Journal of the Royal Astronomical Society*, **62**, 699–718.

SAID, R. 1962. *The geology of Egypt*. Elsevier Publ. Co., New York.

SCHULT, A., SOFFEL, H. C. & HUSSAIN, A. G. 1978. Paleomagnetism of Cretaceous Nubian Sandstone. *Journal of Geophysics*, **44**, 333–340.

——, HUSSAIN, A. G. & SOFFEL, H. C. 1981. Paleomagnetism of Upper Cretaceous volcanics and Nubian Sandstones of Wadi Natash, SE Egypt and implications for the Polar Wander Path of Africa in the Mesozoic. *Journal of Geophysics*, **50**, 16–22.

SHATA, A. 1960. The geology and geomorphology of El Qusaina area. *Bulletin de la Société Géologie et Geographie, Egypte*, **23**, 95–146.

TARLING, D. H. & SYMONS, D. T. A. 1967. A stability index of remanence in palaeomagnetism. *Geophysical Journal of the Royal Astronomical Society*, **12**, 443–448.

TURNER, P. 1977. Remanent magnetization of middle Old Red Sandstone lacustrine and fluviatile sediments from the Orcadian Basin, Scotland. *Journal of the Geological Society, London*, **133**, 37–50.

Palaeomagnetism of some Tertiary sedimentary rocks, southwest Sinai, Egypt, in the tectonic framework of the SE Mediterranean

A. L. ABDELDAYEM[1] & D. H. TARLING[2]

[1] *Department of Geology, Faculty of Science, Tanta University, Egypt*
[2] *Department of Geological Sciences, University of Plymouth, UK*

Abstract: Two hundred and fifteen limestone core samples of Palaeocene (nine sites), Lower Eocene (ten sites) and Miocene (nine sites) ages were collected from the Wadi Feiran area, southwest Sinai. The majority of samples contained magnetic vectors that were consistent in direction during thermal and alternating field demagnetization. Normal and reversed polarities were identified that passed a reversal test, but the bedding tilts were too small to enable an adequate fold test to be undertaken. When compared to coeval palaeomagnetic poles for stable Africa, the Sinai palaeomagnetic poles are consistently displaced by some 30°. This is unlikely to be due to crustal translation between Sinai and Africa, but is attributed to local tectonic rotations of fault blocks in a similar way as for other areas in the SE Mediterranean. In the Suez area, the rotation has been mainly about horizontal axes and is associated with the opening of the Gulf of Suez.

The Sinai peninsula has a very well exposed geology and consequently has been the object of many studies investigating its role as a microplate which separated from Africa due to Tertiary rifting in the Gulf of Suez (e.g. Kohn & Eyal 1981; Bentor 1985; Riva 1986; Bennett & Mosley 1987; Courtillot *et al.* 1987*a, b*; Garfunkel 1988). Nonetheless, few palaeomagnetic studies have been carried out (Wassif 1991; Heimann & Ron 1993; Kafafy & Abdeldayem 1993; Abdeldayem *et al.* 1994; Kafafy *et al.* this volume). The present study concerns the palaeomagnetic characteristics of some Palaeocene, Lower Eocene and Lower Miocene sedimentary rocks from the Wadi Feiran area in the south-west of the Sinai peninsula, i.e. on the eastern bank of the Gulf of Suez (Fig. 1). This area comprises a long NW-oriented homocline that dips uniformly to the NE at angles of less then 20° and is part of a discontinuous belt dissected into several fault blocks by NNE-trending strike-slip faults (Moustafa & Khalil 1987). These are the Gebel Nezzazat, Gebel Ekma, Gebel Abu Durba and Gebel Araba blocks, from north to south. Moustafa & Khalil (1987) studied the Durba–Araba boundary fault in detail and found left-lateral offsets of rocks of Precambrian to Early Miocene age. The north-eastern end of this fault terminates in three north-trending oblique-slip splay faults. Other workers (e.g. Chént *et al.* 1984, 1987; Jarrige *et al.* 1986; Riva 1986) have recognized NW–SE longitudinal and transverse normal and reverse faults associated with the extension during the Early Miocene opening of the Gulf of Suez.

Tilted blocks are common features along both coasts of the Gulf of Suez (Coffield & Smale 1987; Meshref 1990) and the area studied here is close to the Sidi Feiran area where Moustafa (1992) recognized three main sets of high-angle, predominantly normal faults bounding tilted blocks. Thus the Sinai peninsula is not as stable as it has often been thought, as now shown both geologically (e.g. Bartov *et al.* 1980; Chént *et al.* 1984, 1987; Jarrige *et al.* 1986; Moustafa & Khalil 1987, 1989; Evans 1988; Lyberis 1988; Moustafa 1992) and palaeomagnetically (Wassif 1991; Kafafy & Abdeldayem 1993). The results of the study presented here are discussed in the context of the local and regional tectonics of the Sinai, Israel, Lebanon and Syria.

Sampling and methodology

Standard 2.5 cm diameter samples were cored in the field, avoiding weathered and structurally complex exposures. Each sample was oriented using a sun compass and well-defined bedding plane orientations were measured at each site. All the samples were chalks or limestones. Nine sites (70 cores) were collected from the Palaeocene Esna Shales at Gebel Ekma where they extend as a narrow 20 m thick strip. Ten sites (79 cores) were collected at Gebel Nezzazat from the 180 m thick Lower Eocene Thebes Formation and nine sites (66 cores) in the 180 m thick Miocene Gharandal Group of Gebel Qabiliat. The natural remanent magnetization (NRM) of the specimens was measured using either Molspin MS2 or Jelinek JR4 magnetometers. AF

From Morris, A. & Tarling, D. H. (eds), 1996, *Palaeomagnetism and Tectonics of the Mediterranean Region*, Geological Society Special Publication No. 105, pp. 333–343.

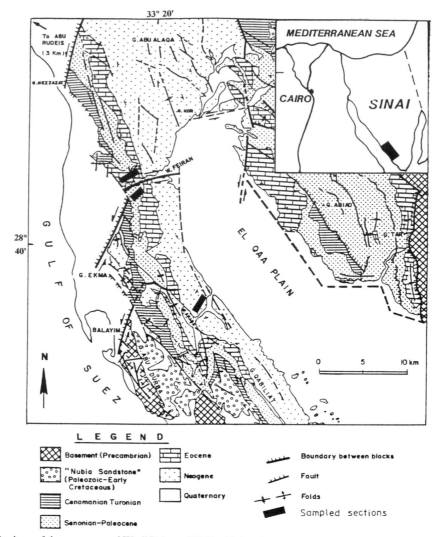

Fig. 1. Geology of the area around Wadi Feiran, SW Sinai (after Cherif *et al.* 1989). The sampled sections are shown in black.

demagnetization was carried out using a magnetically shielded Molspin MS A2 demagnetizer, with a peak field of 100 mT, within a Helmholtz coil system. Thermal demagnetization was conducted in a large shielded Schonstedt-type furnace, within a Helmholtz cage. Demagnetization data were analyzed visually, using stereographic and orthogonal plots, and statistically using the principal component analysis of Kirschvink (1980) for both linear and planar properties, while the directional consistency was established using Tarling & Symon's (1967) Stability Index (now known as Directional Consistency Index). All mean directions were computed using Fisher statistics (Fisher 1953). The properties of the

magnetic minerals were examined using isothermal remanence (IRM) acquisition experiments and subsequent thermal and AF demagnetization of IRM.

Magnetic mineralogy

Samples of different lithological varieties in each section were given an IRM in direct fields in steps from 10 mT to a maximum of 800 mT, using a Digico Pulse Magnetizer, to determine the rate of acquisition and saturation of IRM. They were then demagnetized in increasing opposite fields to determine their back-field coercivity. The samples were subsequently given an IRM along

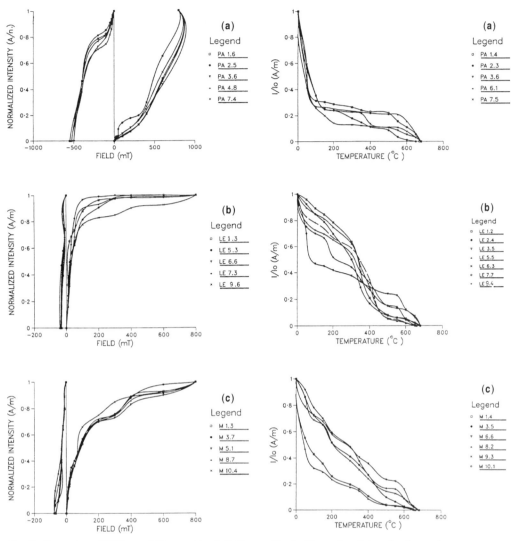

Fig. 2. Examples of IRM acquisition and back-field demagnetization curves. (**a**) Palaeocene; (**b**) Lower Eocene; and (**c**) Miocene samples. All curves are normalized to initial intensity of remanence.

Fig. 3. Examples of stepwise thermal demagnetization of IRM. (**a**) Palaeocene; (**b**) Lower Eocene; and (**c**) Miocene samples. The IRM used was the maximum achieved in an 800 mT field.

one axis, and this was then subjected to thermal demagnetization in order to determine the blocking temperature of IRM for the different lithologies.

The Palaeocene samples showed a slow rate of IRM acquisition and were not saturated by the peak field applied of 800 mT (Fig. 2a). On application of reversed direct fields, they slowly lost their IRM, which was completely destroyed by fields $\geqslant 500$ mT. Such behaviour demonstrates the predominance of high coercivity minerals. When the peak IRM was thermally demagnetized it showed an initial drop to <30%

of the original value at temperatures <100°C. At higher temperatures, the remaining IRM was steadily destroyed and was lost at 680°C (Fig. 3a), with a few samples showing kinks at 580°C. This indicates that one of the main high coercivity minerals present is goethite and that either hematite is also present, with very minor amounts of magnetite in samples exhibiting a kink at 580°C, or that the goethite converts to hematite on heating and that very minor quantities of titanomagnetite are occasionally present.

Most Lower Eocene samples showed an initial very rapid acquisition of IRM, with the majority

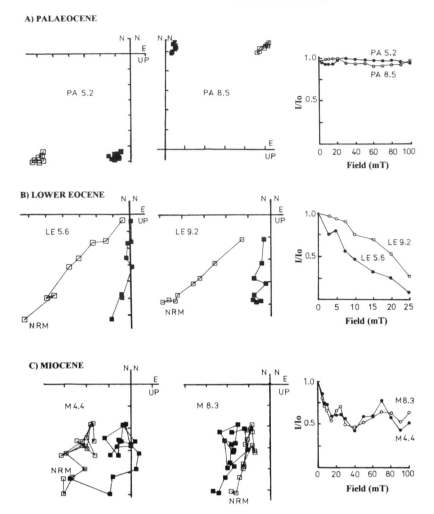

Fig. 4. Examples of AF demagnetization behaviour. (a) Palaeocene; (b) Lower Eocene; and (c) Miocene samples. The solid symbols on the Cartesian plots are the horizontal component and the hollow symbols represent the vertical component.

of samples acquiring >95% of their maximum IRM in fields of <100 mT, and becoming completely saturated in only slightly higher fields (Fig. 2b). The samples also lost their IRM in reversed fields of ≤50 mT. Such behaviour indicates that low coercivity magnetic minerals are dominant. In contrast, a few samples showed no saturation even in the maximum applied field. During thermal demagnetization, the majority of the Lower Eocene samples showed a gradual decay of IRM up to 350°C, when a sharp drop occurred followed by a slow decay and eventual destruction of IRM at 680°C. This behaviour indicates that maghemite and/or titanomaghemite are the predominant carriers

and that any hematite present is responsible for less than 15% to 30% of the total magnetic mineral content of the samples (Fig. 3b).

The Miocene samples showed an initial rapid IRM acquisition and acquired c. 60% of their total IRM at fields of <100 mT, but did not saturate by the maximum field applied (Fig. 2c). Demagnetization showed similar results with a 60% reduction in IRM by <50 mT and complete destruction by <100 mT. This indicates that both low and moderately high coercivity minerals are present. Thermal demagnetization of an axial IRM showed an initial drop of IRM in two samples, but most samples showed a continuous, relatively steady decay of IRM until

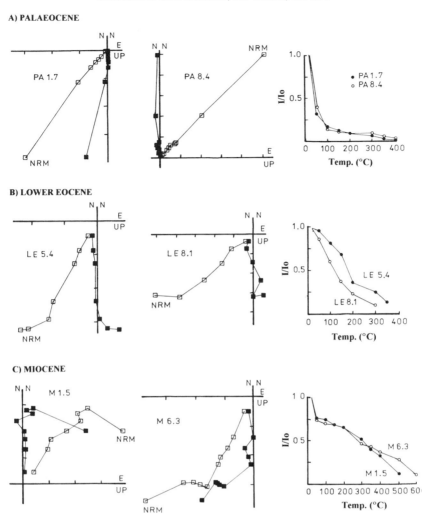

Fig. 5. Examples of thermal demagnetization behaviour. (**a**) Palaeocene; (**b**) Lower Eocene; and (**c**) Miocene samples. Symbols as for Fig. 4.

it was completely destroyed by 680°C. This behaviour is consistent with hematite being the principal magnetic carrier, but with goethite occasionally present.

Palaeomagnetic results and analysis

The initial intensity of remanence varied between the different age groups. The Palaeocene samples showed low NRM intensities of 0.62 to 6.28 mA m^{-1}, which was well above the noise of the Molspin magnetometer (0.02 mA m^{-1}). The NRM intensity of samples from both the Lower Eocene and Miocene sections were very low,

averaging 0.192 mA m^{-1} and 0.116 mA m^{-1} respectively, which were too weak for measurement using the Molspin magnetometer, but were above the noise level of the JR4 spinner (0.1 μAm^{-1}). Samples from sites 2, 3 and 10 from the Lower Eocene section were too weak to be measured on either magnetometer; these sites were excluded from further analysis.

Two groups of samples were selected for determining the optimum demagnetization levels that enabled the isolation of remanence components. At least one pilot specimen from each site was subjected to stepwise AF demagnetization and at least one other specimen to stepwise thermal demagnetization. The low field susceptibility of the thermally demagnetized samples was also measured after each temperature increment in order to monitor changes in the magnetic mineralogy. Both AF and thermally demagnetized behaviour were analysed to define characteristic components using both visual and statistical methods.

The Palaeocene samples, apart from a few initial fluctuations, showed no significant changes in either intensity or direction during AF demagnetization (Fig. 4a). In contrast, during thermal demagnetization they showed an initial 80% drop in intensity by temperatures of <200°C. At higher temperatures, the remanence decreased steadily towards the origin until it was completely destroyed at temperatures <500°C (Fig. 5a). The Lower Eocene limestone samples showed a continuous decay of intensity with little change in direction after the first two or three steps of both AF and thermal demagnetization. Both treatments therefore isolated a single linear vector which was destroyed in fields below 35 mT (Fig. 4b) or at temperatures above 300–400°C (Fig. 5b). The Miocene samples showed only a small initial drop in intensity during AF demagnetization (Fig. 4c). Thermal demagnetization also produced an initial small drop in intensity, but this was followed at higher temperatures by a more consistent decrease in intensity and fairly consistent directions, enabling one or two linear segments to be identified (Fig. 5c).

On the basis of the different reactions to AF and thermal demagnetization, it was decided to thermally treat the remaining Palaeocene samples at 200, 300, 350 and 400°C and the remaining Miocene samples at 400, 500, 550 and 600°C. These levels appeared to offer the optimum treatment at which individual vectors could be isolated. As the Lower Eocene pilot samples had behaved similarly to both AF and thermal treatment, it was decided to AF demagnetize the remaining samples at 10, 15, 20 and 25 mT.

Each of the remaining samples were then analysed for the linearity and consistency of vectors within and between different samples, and used to define site mean directions of magnetization (Table 1). Tectonic dips were low (<20°) and mostly in a similar direction; consequently bedding corrections could not be used to perform a fold test. Mean directions of magnetization were significantly different from the present Earth's magnetic field in the area (Dec = 2.5°, Inc = +41.5°).

All three age groups included antiparallel directions (Fig. 6), so a reversal test (McFadden & McElhinny 1990) was applied to the site mean

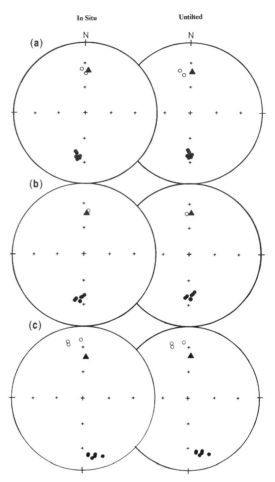

Fig. 6. The site mean characteristic directions before and after bedding correction. (a) Palaeocene; (b) Lower Eocene; and (c) Miocene sites. The plots are equal-area stereographic projections. Downward inclinations are shown with solid symbols. The triangle marks the present geomagnetic field direction at the sampling sites.

directions. The angle between the mean of the two polarity sets was 9.6°, with a critical angle of 11.4°, for the Palaeocene sites; 1.1°, with a critical angle of 10.2°, for the Lower Eocene sites; and 2.4°, with a critical angle of 10°, for the Miocene sites. Under the null hypothesis, directions from each of the three groups seem to have been drawn from populations with means that are 180° apart and their directions could be resolved to better than 10° but not better than 20° for both the Palaeocene and Lower Eocene and better than 10° but not better than 5° for the Miocene sites. The reversal classification is, therefore, positive in all of the three cases ('C' category for the Palaeocene and Lower Eocene

Table 1. *Site mean characteristic NRM directions and VGPs for the Palaeocene, Lower Eocene and Miocene rocks.*

Site	n	In situ		Corrected				VGP	
		Dec.	Inc.	Dec.	Inc.	k	α_{95}	Lat.	Long.
Miocene									
1	6	344.3	−23.0	341.3	−25.1	55	8.2	44.5	239.4
2	7	175.4	20.5	172.9	23.9	53	8.4	−48.4	43.9
3	6	174.8	21.2	172.2	24.5	76	7.0	−48.0	44.8
4	8	161.4	15.1	159.5	17.0	123	5.0	−47.8	64.4
5	7	344.4	−19.0	342.0	−21.1	138	4.7	46.8	239.7
6	7	171.4	16.3	169.4	19.3	45	9.1	-50.2	49.8
7	6	356.9	−19.5	354.9	−23.1	63	7.7	49.1	221.5
8	7	168.3	20.1	165.8	22.7	76	7.6	−47.4	54.2
9	7	168.7	18.1	166.4	20.8	292	5.8	−48.6	53.8
Mean	9	169.5	19.3	167.1	22.0	200	3.7	−48.2	52.6
Lower Eocene (sites 2, 3 & 10 scattered and unstable)									
1	8	184.4	40.9	175.7	44.4	30	9.5	−35.1	38.0
4	7	180.6	43.0	171.2	45.8	40	8.9	−33.5	42.7
5	8	189.7	37.8	182.1	42.2	53	7.1	−36.9	30.9
6	8	5.1	−39.1	357.0	−42.7	50	7.9	36.5	216.7
7	7	185.1	34.0	178.4	37.7	46	9.0	−40.1	35.3
8	6	191.4	34.7	184.7	39.4	45	9.1	−38.8	27.7
9	7	184.9	39.3	176.7	42.9	104	5.5	−36.3	37.0
Mean	7	186.0	38.5	178.1	42.2	360	3.2	−36.9	35.5
Palaeocene									
1	7	194.1	43.0	184.7	46.3	46	9.0	−33.5	28.3
2	7	187.1	36.4	179.6	38.7	53	8.3	−39.5	33.9
3	8	188.5	33.6	181.7	36.2	92	5.8	−41.2	31.2
4	9	185.7	38.3	177.7	40.3	183	3.8	−38.3	36.0
5	8	191.4	40.4	182.8	43.4	481	8.8	−35.9	30.2
6	8	192.1	35.8	184.8	39.0	123	5.0	−39.0	27.6
7	6	186.8	36.5	179.3	38.8	76	7.7	−39.4	34.1
8	6	0.5	−43.1	351.0	−44.1	136	5.8	34.8	223.2
9	7	354.5	−37.4	346.8	−37.5	79	6.8	38.7	229.2
Mean	9	186.8	38.4	178.7	40.6	198	3.7	−38.0	34.8

n, Number of samples in each site; Dec and Inc = *in situ* and structurally corrected magnetic declination and inclination; *k* and α_{95} = precision parameter and semi-angle cone of confidence (Fisher 1953).

sites and 'B' category for the Miocene sites). The overall mean directions, after bedding correction and giving unit weight to each site, are given in Table 1.

Discussion of the results from the Tertiary units of the Sinai

Characteristic directions of magnetization (ChRMs) were isolated in sites of all three ages. However, while the observed directions are different from (a) the present Earth's magnetic field and (b) the present axial dipole field direction in Sinai, the palaeomagnetic pole positions corresponding to the site ChRMs are not consistent with those of other African poles of the same age from recognized stable tectonic areas (e.g. Westphal *et al.* 1986; Besse & Courtillot 1991; Van der Voo 1993). Nonetheless, the occurrence of both normal and reversed

polarities in all three groups suggests that the magnetization is older than the last 700,000 years and also that they are not associated with axial dipole swings or short-lived dipole excursions (Westphal 1993). The differences in inclination could be interpreted as indicating that Sinai was some 30° further south than the African Plate. As such a difference is completely incompatible with the structural and lithological matches on both sides of the Gulf of Suez, the palaeomagnetic data were examined for the extent to which the observed directions could be attributed to local tectonic effects. This was done using the African reference poles given in Table 2. Using the methods outlined by Beck (1980), modified by Demerest (1983), the differences between observed and reference poles were described in terms of separate rotations about vertical and horizontal axes. The rotations were:

Palaeocene	$4.4° \pm 6.3°$	and	$-33.0° \pm 6.1°$	(vertical and horizontal axes respectively)
Lower Eocene	$4.2° \pm 6.1°$	and	$-34.2° \pm 5.9°$	
Miocene	$18.8° \pm 4.5°$	and	$-34.0° \pm 4.3°$	

It is noticeable that all three groups require almost identical rotations about horizontal axes, but also that the two older formations require little or no rotation about vertical axes, while the youngest requires a distinct clockwise rotation. It has long been known that rigid body rotation and fault translation take place to accommodate large strains. Rotations about horizontal and vertical axes are common rigid body rotations (Freund 1974; Ron et al. 1984; Wells & Coe 1985; Ron 1987). Rotations about inclined axes also occur but they are difficult to detect (Hudson & Geissman 1991). It has been found that in areas of small tectonic blocks where combined folding and faulting occur, rotation on inclined fault surfaces may take place (Piper 1987). Components of rotation about horizontal axes introduce changes in the inclination of magnetic remanences, and can result either from normal or reverse faulting (Ron et al. 1986) or as a consequence of rotations about inclined axes (Ron 1987).

In the region of study, NNE-trending strike-slip faults (Moustafa & Khalil 1987) and normal and transverse faults (Chént et al. 1984, 1987; Jarrige et al. 1986; Riva 1986) occur. The apparent southward pole translation identified in the present study (Table 2) seems to be due to the effect of normal faults associated with the extensional tectonics resulting from the opening of the Gulf of Suez in the Early Miocene. Apart from the resultant $18.8° \pm 4.5°$ clockwise rotation from the Miocene rocks, the strike-slip faults do not produce significant rotations around vertical axes. Perhaps this is because these faults terminate by splaying (Moustafa & Khalil 1987), thereby reducing the amount of horizontal displacement and converting it to dip-slip normal displacement (Ron 1987). The resultant clockwise rotation inferred from the Miocene results, on the other hand, is likely to be due to a rotation about a slightly inclined axis. The general agreement in horizontal rotation angles from the three studied units indicates that a uniform large-scale horizontal axis rotation may have affected the entire set of fault blocks under study. This rotation could also represent a composite of different rotations that took place as a result of the opening of the Gulf of Suez.

The question arising now is that if local tectonics are responsible for the anomalous magnetization directions in the Tertiary rocks, why have they not affected the results from the underlying Palaeozoic rocks obtained by Abdeldayem et al. (1994)? The answer could lie in one of the following possibilities.

(1) In the present area there is a difference in the mechanical behaviour of the longitudinal and transverse normal faults between the upper plastic sedimentary cover and the underlying more rigid Nubian Sandstone (Palaeozoic) and basement rocks (Riva 1986; Chént et al. 1987). In other words, the change in direction between the longitudinal and transverse faults is abrupt (i.e. brittle) in the Palaeozoic sediments and is more progressive (i.e. ductile) in the Tertiary sedimentary cover. This difference in mechanical behaviour between these contrasting units has led to a fault plane refraction and high deformation of the studied sediments (Riva 1986; Chént et al. 1987).

(2) The limited number of African palaeomagnetic poles for the Palaeozoic might prevent detection of such rotations.

(3) It is possible that there is an overestimation of the amount of rotation inferred from the present results, which would otherwise have appeared more clearly in the results from the underlying Palaeozoic units.

Palaeomagnetic evidence for local rotations in other tectonic elements of the SE Mediterranean

Rotations attributed to the effects of local tectonics have been widely reported throughout the SE Mediterranean region. Palaeomagnetic results from the following major tectonic features in the wider region are considered here: the Syrian Arc system (Kafafy & Abdeldayem 1993); the Gulf of Suez (Wassif 1991; Kafafy & Abdeldayem 1993); and the Gulf of Aqaba–Dead Sea transform (Freund & Tarling 1979; Ron et al. 1984; Ron 1987; Heimann & Ron 1993).

The Syrian arc system

This system forms a distinct geomorphological and structural unit characterized by NE–SW-trending elliptical anticlines and synclines that extend from NE Libya, across northern Egypt into Sinai, Israel and Syria (Krenkel 1925). Several models have been proposed for the tectonic evolution of this long chain of folds; the main differences being in the timing of initiation and development of the system (Jenkins 1990).

Table 2. *Formation and reference mean directions and pole positions*

Age	n	Direction			Pole			Area	
		Dec	Inc	α_{95}	Lat°	Long°	A_{95}	Lat°N	Long°E
Miocene	9	167.1	22.0	3.7	−48.2	52.6	3.5	28.5	33.4
Reference		185.9	−39.8		−82.0	350.0	4.0		
Lower Eocene	7	178.1	42.2	3.2	−36.9	35.5	3.4	28.7	33.3
Reference		182.4	−19.1		−71.0	26.0	6.5		
Palaeocene	9	178.7	40.6	3.7	−38.0	34.8	3.9	28.7	33.3
Reference		182.4	−19.1		−71.0	26.0	6.5		

Kafafy & Abdeldayem (1993) studied the palaeomagnetic properties of three of these arcs; two in north Sinai (Arif El-Naga and Maghara) and one in the north of the Eastern Desert of Egypt (Shabrawet). The latter is close to the northwestern end of the Gulf of Suez, and is considered as part of the Gulf of Suez Province. The two north Sinai arcs are known to have been folded during the Late Cretaceous–Early Tertiary, and they carry a secondary chemical magnetization that was probably acquired during recrystallization of magnetic minerals during the latest phase of folding. However, the magnetization directions are still deviant, and this has been attributed to the effect of rotations resulting from movement on thrusts bounding these folds along their southern margins. The results imply that these blocks were dragged along thrust planes during or soon after folding. The Arif El-Naga anticline was considered to have been dragged by about 40° to the SE while the Maghara anticline is dragged by about 30° in the same direction. These tectonic rotations are in addition to 22° of counterclockwise vertical axis rotation required for a complete restoration of the Maghara anticline to its assumed pre-thrusting position.

Gulf of Suez Province

The Lower Cretaceous carbonates and sandstones of the Shabrawet Syrian arc are characterized by a secondary magnetization which requires this anticline to have undergone a 28° clockwise vertical axis rotation (Kafafy & Abdeldayem 1993). This rotation was attributed to internal rigid deformation as a result of simultaneous strike-slip faulting and block rotation, the latter being associated with the opening of the Gulf of Suez and isolation of the Sinai microplate. Wassif (1991) obtained some peculiar Mesozoic palaeomagnetic results from two separate basaltic bodies at Gebel Faras El Azraq

(Middle Triassic) and Wadi Budra (Middle Jurassic) in the eastern part of the Gulf of Suez. These aberrant directions were attributed to either local tectonics or inadequate averaging of secular variation.

The Gulf of Aqaba–Dead Sea province

Anomalous palaeomagnetic directions have long been reported from areas along this major structural line (Van Dongen *et al.* 1967; Nur & Helsley 1971; Sallomy & Krs 1980; Gregor *et al.* 1974). These early results were initially interpreted in terms of rotation of coherent microplates (Van Dongen *et al.* 1967; Gregor *et al.* 1974). It was the study of Freund & Tarling (1979) which first suggested that these palaeomagnetic data (e.g. Gregor *et al.* 1974) could be more reasonably explained in terms of local zones of rigid block rotation along the plate margin in response to the dying out of the Dead Sea Fault system. These rotations, detected in Miocene volcanic rocks in particular, were found to be consistent with reconstructions of Miocene river systems. This model of rotation of individual blocks within a splayed strike-slip system was incorporated in the more recent studies of Ron *et al.* (1984), Ron (1987), Ron *et al.* (1990) and Heimann & Ron (1993). These later studies provided further palaeomagnetic data which gave detailed information about the amount and sense of individual block rotations.

Conclusions

The present study of Tertiary sedimentary rocks indicates rotations about horizontal axes due to local tectonics, mainly by movement on the normal faults which dissect the area. This faulting seems to be the result of extension in the Gulf of Suez which results in the formation of rotated tilted blocks (Meshref 1990). Similar local tectonic rotations appear to characterize many areas of the Sinai region, reflecting the considerable width and complexity of the Gulf of Suez and Aqaba–Dead Sea transform systems and the associated compressional and tensional stress fields. It can be stated with certainty that this area has been tectonically active throughout the Tertiary and should not be considered as a single, simple rigid plate.

All measurements were carried out at the Department of Geological Sciences, University of Plymouth. A.L.A. would like to thank P. Davies and all other technical staff for their help during his stay in Plymouth. The present data were analysed using PC programs developed by R. Enkin and by D. H. Tarling.

References

ABDELDAYEM, A. L., KAFAFY, A. M. & TARLING, D. H. 1994. Palaeomagnetic studies of some Palaeozoic sediments, southwest Sinai, Egypt. *Tectonophysics*, **234**, 217–225.

BARTOV, Y., LEWY, Z., STEINITZ, G. & ZAK, I. 1980. Mesozoic and Tertiary stratigraphy, paleogeography and structural history of the Gebel Arif en Naga area, eastern Sinai. *Israeli Journal of Earth Science*, **29**, 114–139.

BECK JR, M. E. 1980. Palaeomagnetic record of plate margin tectonic processes along the western edge of North America. *Journal of Geophysical Research*, **85**, 7115–7131.

BENNETT, J. D. & MOSLEY, P. N. 1987. Tieredtectonics and evolution, Eastern Desert and Sinai, Egypt. *In*: MATHEIS, G. & SCHANDELMEIER, H. (eds) *Current research in African earth sciences*. Balkema, Rotterdam, 79–82.

BENTOR, Y. K. 1985. The crustal evolution of the Arabo-Nubian massif with special reference to the Sinai peninsula. *Precambrian Research*, **28**, 1–74.

BESSE, J. & COURTILLOT, V. 1991. Revised and synthetic apparent polar wander paths of the African, Eurasian, North American and Indian plates, and true polar wander since 200 Ma. *Journal of Geophysical Research*, **96**, 4029–4050.

CHÉNT, P. Y., LETOUZEY, & ZAGHLOUL, E. A. 1984. Some observations on the rift tectonics in the eastern part of the Suez rift. *Proceedings of the 7th EGPC Exploration Seminar*, Cairo.

——, COLLETTA, B., LETOUZEY, J., DESFORGES, G., OUSSET, E. & ZAGHLOUL, E. A. 1987. Structures associated with extensional tectonics in the Suez rift. *In*: COWARD, M. P., DEWEY, J. F. & HANCOCK, P. L. (eds) *Continental Extensional Tectonics*, Geological Society, London, Special Publications, **28**, 551–558.

CHERIF, O. H., AL-RIFAIY, I. A., AL AFIFI, F. I. & ORABI, O. H. 1989. Foraminiferal bio-stratigraphy and paleoecology of some Cenomanian-Turonian exposures in west-central Sinai (Egypt). *Revue de Micropaleontologie*, **31**, 243–262.

COFFIELD, D. Q. & SMALE, J. L. 1987. Structural

geometry and synrift sedimentation in an accommodation zone, Gulf of Suez, Egypt. *Oil and Gas Journal*, **85**, 56–59.

COURTILLOT, V., ARMIJO, R. & TAPPONNIER, P. 1987a. Kinematics of the Sinai triple junction and a two phase model of Arabia–Africa rifting. *In*: COWARD, M. P., DEWEY, J. F. & HANCOCK, P. L. (eds) *Continental Extensional Tectonics*, Geological Society, London, Special Publications, **28**, 559–573.

——, —— & —— 1987b. The Sinai triple junction revisited. *Tectonophysics*, **141**, 181–190.

DEMAREST JR., H. H. 1983. Error analysis for the determination of tectonic rotation from paleomagnetic data. *Journal of Geophysical Research*, **88**, 4321–4328.

EVANS, A. L. 1988. Neogene tectonic and stratigraphic events in the Gulf of Suez rift area, Egypt. *Tectonophysics*, **153**, 235–247.

FISHER, R. A. 1953. Dispersion on a sphere. *Proceedings of the Royal Society, London*, **A27**, 295–305.

FREUND, R. 1974. Kinematics of transform and transcurrent faults. *Tectonophysics*, **21**, 93–134.

—— & TARLING, D. H. 1979. Preliminary Mesozoic palaeomagnetic results from Israel and inferences for a microplate structure in the Lebanon. *Tectonophysics*, **60**, 189–205.

GARFUNKEL, Z. 1988. Relation.between continental rifting and uplifting: evidence from the Suez rift and northern Red Sea. *Tectonophysics*, **150**, 33–49.

GREGOR, C. B., MERTZMAN, S., NAIRN, A. E. M. & NEGENDANK, J. 1974. The paleo-magnetism of some Mesozoic and Cenozoic volcanic rocks from the Lebanon. *Tectonophysics*, **21**, 375–395.

HEIMANN, A. & RON, H. 1993. Geometric changes of plate boundaries along part of the northern Dead Sea transform: geochronologic and paleomagnetic evidence. *Tectonics*, **12**, 477–491.

HUDSON, M. R. & GEISSMANN, J. W. 1991. Paleomagnetic evidence for the age and extent of Middle Tertiary counterclockwise rotation, Dixie Valley region, west central Nevada. *Journal of Geophysical Research*, **96**, 3979–4006.

JARRIGE, J. J., D'ESTEVOU, P. O., BUROLLET, P. F., THIRIET, J.-P., ICART, J. C., RICHERT, J. P., SEHANS, P., MONTENAT, C. & PRAT, P. 1986. Inherited discontinuities and Neogene structure: the Gulf of Suez and the northwestern edge of the Red Sea. *Philosophical Transactions of the Royal Society, London*, **A317**, 129–139.

JENKINS, D. A. 1990. North and central Sinai. *In*: SAID, R. (ed.) *The geology of Egypt*. Balkema, Rotterdam and Boston, 361–380.

KAFAFY, A. M. & ABDELDAYEM, A. L. 1993. Palaeomagnetism of some Syrian Arcs in north Sinai and Eastern Desert Egypt; tectonic implications. *Acta Mineralogica-Petrographica, Szeged*, **34**, 79–98.

KAFAFY, A. M., TARLING, D. H., EL GAMILI, M. M., HAMAMA, H. H. & IBRAHIM, E. H. 1996. Palaeomagnetism of some Cretaceous Nubian Sandstones, Northern Sinai, Egypt. *This volume*.

KIRSCHVINK, J. L. 1980. The least-squares line and plane and the analysis of palaeomagnetic data.

Geophysical Journal of the Royal Astronomical Society, **62**, 699–718.

KOHN, B. P. & EYAL, M. 1981. History of uplift of the crystalline basement of Sinai and its relation to opening of the Red Sea as revealed by fission track dating of apatites. *Earth and Planetary Science Letters*, **52**, 129–141.

KRENKEL, E. 1925. *Geologie Afrikeas*. Borntraeger, Berlin.

LYBERIS, N. 1988. Tectonic evolution of the Gulf of Suez and the Gulf of Aqaba. *Tectonophysics*, **153**, 209–220.

McFADDEN, P. L. & McELHINNY, M. W. 1990. Classification of the reversal test in palaeomagnetism. *Geophysical Journal International*, **103**, 725–729.

MESHREF, W. M. 1990. Tectonic framework. *In*: SAID, R. (ed.) *The geology of Egypt*. Balkema, Rotterdam and Boston, 113–115.

MOUSTAFA, A. R. 1992. Structural setting of the Sidri–Feiran area, eastern side of the Suez rift. M.E.R.C. *Ain Shams University, Earth Science Series*, **6**, 44–54.

—— & KHALIL, M. H. 1987. The Durba–Araba fault, southwest Sinai, Egypt. *Egypt Journal of Geology*, **31**, 15–32.

—— & —— 1989. Structural characteristics and tectonic evolution of north Sinai fold belts. *In*: SAID, R. (ed.) *The geology of Egypt*. Balkema, Rotterdam & Boston, 381–389.

NUR, A. & HELSLEY, C. E. 1971. Paleomagnetism of Tertiary and Recent lavas of Israel. *Earth and Planetary Science Letters*, **10**, 376–379.

PIPER, J. D. A. 1987. *Palaeomagnetism and continental crust*. Open University Press, Milton Keynes.

RIVA, E. T. 1986. Compressive features and wrench tectonics in western central Sinai. *Proceedings of the 8th EGPC Exploration Seminar*, Cairo.

RON, H. 1987. Deformation along the Yammuneh, the restraining bend of the Dead Sea transform: paleomagnetic data and kinematic implications. *Tectonics*, **6**, 653–666.

——, AYDIN, A. & NUR, A. 1986. Strike-slip faulting and block rotation in the Lake Mead fault system. *Geology*, **14**, 1020–1023.

——, FREUND, R., GARFUNKEL, Z. & NUR, A. 1984. Block rotation of strike-slip faulting: structural and paleomagnetic evidence. *Journal of Geophysical Research*, **89**, 6256–6270.

——, NUR, A. & EYAL, Y. 1990. Multiple strike-slip fault sets: a case study from the Dead Sea transform. *Tectonics*, **9**, 1421–1431.

SALLOMY, J. T. & KRS, M. 1980. A palaeomagnetic study of some igneous rocks from Jordan. *Bulletin of the Institute of Applied Geology*. (Jeddah), **3**, 155–164.

TARLING, D. H. & SYMONS, D. T. A. 1967. A stability index of remanence in palaeomagnetism. *Geophysical Journal of the Royal Astronomical Society*, **12**, 443–448.

VAN DER VOO, R. 1993. *Paleomagnetism of the Atlantic, Tethys and Iapetus Oceans*. Cambridge University Press.

VAN DONGEN, R. G., VAN DER VOO, R. & RAVEN, TH. 1967. Palaeomagnetic research in the central Lebanon mountains and in the Tartous area (Syria). *Tectonophysics*, **4**, 35–53.

WASSIF, N. A. 1991. Palaeomagnetism and opaque mineral oxides of some basalt from west central Sinai, Egypt. *Geophysical Journal International*, **104**, 319–330.

WELLS, R. E. & COE, R. S. 1985. Palemagnetism and geology of Eocene volcanic rocks of southwest Washington: Implications for mechanisms of rotation. *Journal of Geophysical Research*, **90**, 1925–1947.

WESTPHAL, M. 1993. Did a large departure from the geocentric axial dipole hypothesis occur during the Eocene? Evidence from the magnetic polar wander path of Eurasia. *Earth and Planetary Science Letters*, **117**, 15–28.

——, BAZHENOV, M. L., LAUER, J. P., PECHERSKY, D. M. & SIBUET, J. C. 1986. Palaeomagnetic implications on the evolution of the Tethys belt from the Atlantic Ocean to the Pamirs since the Triassic. *Tectonophysics*, **123**, 37–82.

ZIJDERVELD, J. D. A. 1967. A.C. demagnetization of rocks: analysis of results. *In*: COLLINSON, D. W. *et al.* (eds) *Methods in Palaeomagnetism*. Elsevier Publishing Co., Amsterdam, 254–286.

Emplacement temperatures of pyroclastic deposits on Santorini deduced from palaeomagnetic measurements: constraints on eruption mechanisms

LEON BARDOT[1], RICK THOMAS[2] & ELIZABETH McCLELLAND[1]

[1]*Department of Earth Sciences, Oxford University OX1 3PR, UK*
[2]*Cartograph Ltd, The Eden Centre, 47 City Road Cambridge, CB1 1DP, UK*

Abstract: In pyroclastic eruptions, explosive fragmentation of magma occurs, breaking up some of the surrounding volcanic edifice, and a clastic deposit is formed containing fragments of juvenile magma and other rock debris. The temperature at which such a deposit is emplaced places important constraints on eruption mechanisms. The palaeomagnetic technique applied to lithic clasts which had cooled and solidified prior to the eruption can provide a quantitative estimate of emplacement temperature if there has been no modification of the magnetic mineralogy of the lithics during emplacement; thermal demagnetization is used to determine the temperature which removes all of the uniformly directed overprint component acquired during cooling in the deposit.

The volcano of Santorini (or Thera) in the Aegean has experienced a series of cataclysmic eruptions over the past 200 ka. We have determined emplacement temperatures for many of these tephra, from plinian airfalls, lag breccias, ignimbrites and phreatomagmatic deposits. The palaeomagnetic technique will give erroneous results if the magnetic mineralogy of a lithic clast altered during the eruption. This would give a CRM overprint which might completely or partially replace the primary remanence in blocking temperature intervals which extend above the actual emplacement temperature. We have carried out palaeointensity experiments on overprinted lithics to assess the magnetic origin of the overprint (pTRM, CRM or a combination) and hence the reliability of the temperature estimates. Results from the co-ignimbrite lithic breccias show they were emplaced hot. The Middle Pumice and Minoan plinian airfalls display equilibrium temperatures that decrease with increasing distance from vent, and provide constraints on the predictions concerning the cooling of tephra during fallout from eruption columns. Emplacement temperatures for the Minoan Phases 2 (phreatomagmatic deposits) and 3 (massive structureless white deposits) are low considering the deposits are principally composed of juvenile material. For Phase 2 this supports the conventional view of such deposits being the products of lateral explosions caused when water entered the plinian vent. The juvenile material of Phase 3 probably cooled as a result of interaction with water in the vent system, and we interpret them as very low grade ignimbrites.

This paper is a review of studies carried out by the Oxford Palaeomagnetism group on the remanent magnetism of tephra erupted from Thera volcano. The island of Santorini (or Thera) in the Aegean (Fig. 1) has been the site of at least 12 pyroclastic eruptions over the past 200 ka (Druitt *et al.* 1989). Emplacement temperature estimates have been determined for many of these tephra including plinian pumice falls, phreatomagmatic deposits, ignimbrites and lag breccias. Temperature estimates are extremely useful for distinguishing hot-flow or fall deposits from flash flood or water saturated deposits, and place important constraints on the eruption mechanism.

Druitt *et al.* (1989) described the stratigraphy, volcanology and geochemistry of the deposits they termed the Thera Pyroclastic Formation. These tephra record two major cycles of mafic to silicic volcanism, the first cycle ending with the Lower Pumice 2 eruption about 100 ka, and the second cycle ending with the Minoan eruption some 3500 years ago. Five major eruptions are recognized within the first cycle, and seven in the second cycle (Fig. 2). Each eruption began with a plinian pumice-fall phase and most concluded with the emplacement of pyroclastic flows. The deposits are well exposed in the caldera walls, and in various now-disused quarries.

Palaeomagnetic results are presented from: (1) co-ignimbrite lithic breccias from the Lower Pumice 2 (erupted at the end of the first cycle;

From Morris, A. & Tarling, D. H. (eds), 1996, *Palaeomagnetism and Tectonics of the Mediterranean Region*, Geological Society Special Publication No. 105, pp. 345–357.

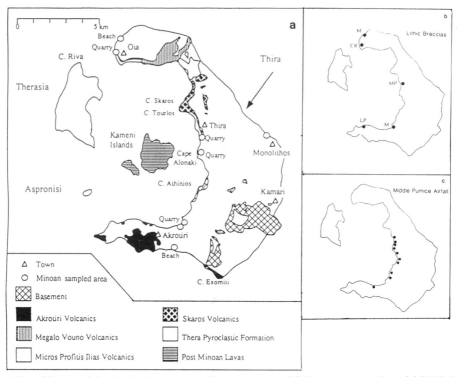

Fig. 1. Map of Santorini showing basic geology and sampled areas: (**a**) Minoan sample sites ; (**b**) Lithic breccia sample sites; M Minoan; CR Cape Riva; MP Middle Pumice; LP Lower Pumice 2; (**c**) Middle Pumice airfall sample sites.

approx 100 ka), Middle Pumice (erupted close to the start of the second cycle; approx 79 ka), Cape Riva (18 ka), and Minoan eruptions (3.5 ka); (2) a welded plinian airfall deposit from the Middle pumice eruption; and (3) the first four phases of the Minoan eruption. The co-ignimbrite lithic breccias were described by Druitt & Sparks (1982). These lithic breccias are believed to have formed by proximal sedimentation from pyroclastic flows. These breccias are clast- to matrix-supported, massive to bedded, contain lithic blocks up to 2 m in diameter, and may pass laterally or vertically into non-welded ignimbrite. The Middle Pumice airfall unit was described by Sparks & Wright (1979), who called it the Thera Tuff. This deposit mantles topography and is welded within about 1.5 km of the vent.

The Bronze Age eruption of Thera has attracted considerable attention over the last three decades. This has been due in part to speculation about its role in the destruction of the Minoan civilisation on Crete. The eruption was of great magnitude (between 16 and 39 km^3 eruption volume) and considerable rarity as the

average recurrence time on a world-wide basis for an eruption of this size has been estimated to be approximately 300 years (Decker 1990). On the island itself, a large settlement was covered by the tephra of the eruption. Because no human remains have been uncovered, arguments have been presented to claim that precursor seismic activity caused the inhabitants to abandon the town before the start of the eruption, but this is not conclusive. Comparison with Herculaneum where all the victims were found beside the harbour, suggests that until the harbour area of Minoan Santorini is found and excavated, this problem will not be solved.

The Minoan eruption can be separated into five successive phases (Sparks & Wilson 1990). Phase 1 produced a plinian airfall deposit. Phase 2 produced base surges and contemporaneous plinian deposits. Phase 3 tephra are chaotic, poorly sorted, massive units, containing abundant lithic blocks and appear to be low temperature primary pyroclastic flow deposits. Phase 4 deposits are dominated by non-welded ignimbrite, with intercalated lithic breccias, and are believed to be of hot origin (Sparks & Wilson

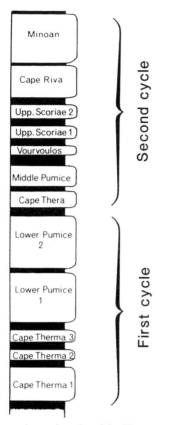

Fig. 2. Schematic stratigraphy of the Thera Pyroclastic Formation (after Druitt *et al*. 1989). The major, named eruptions are separated by minor tephra layers and palaeosols (shown in black).

1990). Phase 5 deposits are alluvial lithic breccias. The mechanics of such large eruptions are of great interest, and the origin of the massive phase three deposits, in particular, is still uncertain.

Principles of emplacement temperature estimation by palaeomagnetic methods

In pyroclastic eruptions explosive fragmentation of magma occurs, breaking up some of the surrounding volcanic edifice, forming a clastic deposit containing fragments of juvenile magma, lithics and crystals. The incorporated lithic clasts will have been originally magnetized prior to the eruption. If the pyroclastic deposits were emplaced above ambient temperature then the lithic clasts will have been heated during their incorporation into the deposit and will have cooled to ambient temperature in their present position. On heating, a portion of the original

magnetization with blocking temperatures less than or equal to the maximum temperature of the deposit, will have been demagnetized. It will have been replaced by a new partial thermoremanence (pTRM) on subsequent cooling. The original high blocking temperature remanence within each clast will have a random orientation, whereas the low blocking temperature remanence will have the same orientation in each clast, parallel to the local Earth's magnetic field at the time of cooling.

The emplacement temperature of a lithic clast can be determined by thermal demagnetization. The sample is heated and cooled in a magnetic field of less than 5 nT to increasing peak temperatures and the remanence is measured after each temperature step. This process demagnetizes the grains with blocking temperatures up to the peak heating temperature, and an increment of the total remanence is removed at each step. On laboratory heating to temperatures up to the emplacement temperature, only the low blocking temperature, Earth's present field component is demagnetized; above this temperature the original high blocking temperature component is removed. The estimate of emplacement temperature is the temperature interval between the highest temperature at which the low temperature Earth's present field magnetization component was present, and the subsequent temperature step. The temperature at which the magnetization of a given magnetic grain is unblocked is dependent on the timescale of the heating process (Dodson & McClelland Brown 1980). This means that a cooling rate correction should in principle be applied when temperatures are estimated from laboratory unblocking temperatures. However, in the case of pyroclastic deposits, where cooling is relatively rapid, the cooling rate correction is small.

This technique may give erroneous results if the magnetic mineralogy of a lithic clast altered during the eruption. This could give a CRM overprint which may partially or completely replace the primary remanence in blocking temperature intervals which extend above the emplacement temperatures.

The palaeomagnetic estimation of emplacement temperature requires that the low temperature component of magnetization is of thermal origin. Palaeointensity experiments were carried out to check whether the low temperature components fulfilled this criterion. In Thellier-type palaeointensity studies thermoremanent magnetization acquired by a sample in a known field in the laboratory is compared to the sample NRM. On the acquisition of TRM there is usually a good linear proportionality

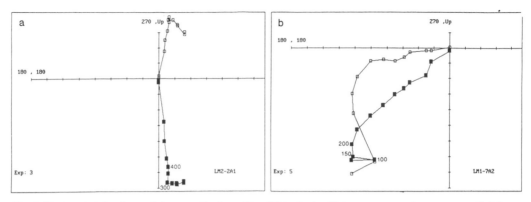

Fig. 3 Vector plots for thermal demagnetization of two lithic clasts with two component remanences. Solid squares give the magnetization vector of the clast, projected onto the horizontal plane, at different laboratory temperatures (given in °C by numbers besides the squares); open squares give the vector projected onto a vertical plane in a N–S direction at the same laboratory temperatures. (**a**) Two-component remanence of lithic clast from the Minoan breccia (ideal behaviour). Clast emplacement temperature $=300°C$. (**b**) Two-component remanence of lithic clast from the Minoan breccia, with curvature between the two components indicating chemical overprinting.

between the intensity of the remanence and the field applied during cooling, as long as the field is low. If a magnetization is of thermal origin, the ratio between its intensity (M) and the intensity acquired thermally in the laboratory (M_L) equals the ratio between the strength of the original magnetizing field (H) and the strength of the field applied in the laboratory (H_L), i.e. $M/M_L = H/H_L$. This ratio will be constant for all temperature intervals if all the magnetization was acquired in a cooling event. The palaeointensity of the field can be determined as it is the only unknown in the equation. It is generally accepted that consistent intensity estimates from a suite of samples, which pass reliability checks that test for alteration during heating, are indicative of TRM in the material studied. It is not certain whether the existence of CRM can lead to straight line segments on NRM/TRM plots, and this point has been hotly debated over the last decade. There is some evidence that suggests that in some cases it can, but that the intensity will not be correct, and will be variable between samples. In other cases, CRM may give non-linear NRM/TRM plots.

If a CRM is formed during incorporation into a pyroclastic deposit it is possible that the blocking temperature of the new or modified grains is greater than the formation temperature (and maybe greater than the emplacement temperature). If this is the case then it is possible to have two suites of grains with blocking temperatures greater than emplacement temperatures, one primary group carrying the earlier randomly oriented remanence direction,

and the other secondary group carry a remanence in the ambient field direction. One effect of such overlapping blocking temperature spectra is to cause curvature in the vector plots above emplacement temperature, e.g. Fig. 3b (cf. McClelland Brown 1982). The estimate of the high blocking temperature direction will therefore be biased towards the overprint direction.

Sampling and experiental procedure

The main procedure used for sampling was the same as that used by McClelland & Druitt (1989), in which oriented hand samples were taken. Lithic clasts were oriented by gluing rigid plastic plates onto the surface of the clast and marking the strike and dip of this plate. Pumice blocks were oriented either by this method or by cutting a flat surface on the block and marking the strike and dip directly onto the pumice surface. Standard (one inch diameter) cores were cut from the blocks in the laboratory. Magnetization of the samples was measured using either a CCL cryogenic magnetometer or a Molspin spinner magnetometer. Samples were demagnetized using a furnace with a residual field less than 5 nT in temperature steps ranging between 25–50°C until the remaining intensity was less than 5% of the NRM. The data were plotted on Zijderveld orthogonal demagnetization diagrams and the principal components of magnetization were analysed using LINEFIND (Kent *et al.* 1983).

A modified version of the Thellier method for

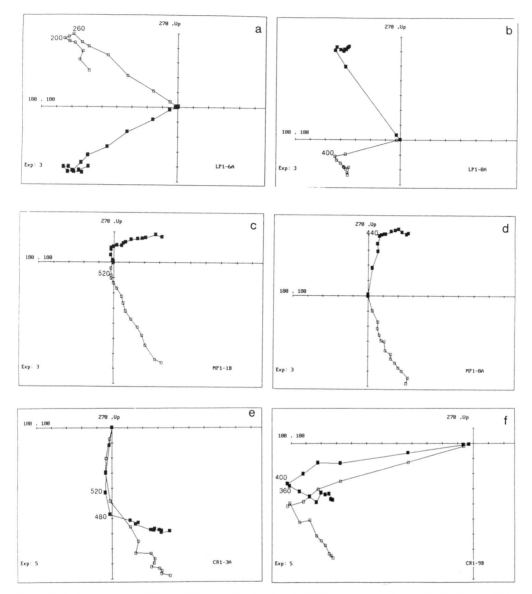

Fig. 4. Typical vector plots of thermal demagnetization data for lithic breccias; symbols as in Fig. 3. (**a** and **b**). Lithic clasts from Lower Pumice 2. Clast emplacement temperatures are 200–230°C and 400°C respectively. (**c** and **d**) Lithic clasts from Middle Pumice with emplacement temperatures of 480–520°C and 440°C respectively. (**e** and **f**). Lithic clasts from Cape Riva with emplacement temperatures of 480–520°C and 360–400°C respectively.

palaeointensity determination was used, involving three heatings at each step (Walton 1984); two demagnetizations (zero field) separated by one remagnetization (40 μT); with temperature steps of 50°C up to 450°C and 30°C at higher temperatures. Extensive pTRM checks were also carried out.

Palaeomagnetic results from lithic breccias

Most of the lithic clasts sampled (Fig. 1b) from the four different eruption units have a two component remanence. Figure 3 shows vector plots for two lithic clasts from Minoan lithic breccias. Sample LM2-2A1 shows an ideal

L. BARDOT *ET AL.*

Table 1 *Statistics for magnetization components of lithic breccias*

Eruption	Component	Dec.	Inc.	α_{95}	n	R	k	$3R^2/n$
Lower Pumice 2	Low T_b	346.5	42.8	17.1	7	6.6	13.3	18.38
	High T_b	127.7	−31.0	>90	6	1.9	1.2	1.86
Middle Pumice	Low T_b	346.0	60.0	7.7	7	6.9	61.9	20.42
	High T_b	340.1	52.7	85.4	7	2.9	1.5	3.58
Cape Riva	Low T_b	14.3	51.2	3.6	8	8.0	232.6	23.82
	High T_b	242.6	82.9	67.3	6	3.4	1.9	5.88
Minoan	Low T_b	355.3	63.2	8.6	15	14.3	20.7	41.03
	High T_b	60.5	31.8	47.6	14	6.2	1.7	8.25

two-component remanence where there is no overlap between the low-temperature overprint and the high temperature component, so there is no curvature between the two straight line segments. This sample gives an emplacement temperature of 300–350°C. Sample LM1-7A2 shows a less well behaved two-component remanence, where the recent Earth's field component is only seen below 150°C, but there is a considerable temperature range (150–300°C) over which both overprint and pre-Minoan components are demagnetized at the same temperature. This leads to curvature on the vector plot, and we interpret this to be due to CRM acquired in the Minoan heating, with blocking temperatures above the re-heating temperature. The maximum emplacement temperature estimate is therefore the lowest temperature at which the pre-Minoan component is seen in combination with the Minoan direction, i.e. 150–200°C for this sample. Two component vector plots from lithics from the Lower Pumice 2, Middle Pumice, and Cape Riva lithic breccias are shown in Fig. 4. These show little evidence of overlapping components.

Table 1 gives mean direction and component statistics for the four lithic breccia units. Rayleigh (1919) derived a test for random grouping of directions. At a level of 95% confidence a group of directions is randomly directed if the value of $3R^2/n$ is less than 7.81 (where R is the vector resultant of a set of n vectors). Statistically the low blocking temperature remanences for all the lithic breccias are significantly grouped and lie close to the Earth's field direction at the time of emplacement (approximately declination 360°, inclination +60°). This indicates the low blocking temperature component of the lithic clasts were acquired while cooling *in situ* in the deposit.

The high blocking temperature remanences of the Lower Pumice 2, Middle Pumice, and Cape Riva lithic breccias are randomly distributed (Table 1), indicating these components predate

emplacement and were randomized by the depositional process. The grouping of the high temperature components of the Minoan lithic breccias however is significant at the 95% confidence level (Table 1), although the grouping is not as tight as for the low temperature components of these clasts. The non-randomness of the high temperature remanences of the Minoan lithics indicates some CRM overprinting has occurred in grains with blocking temperatures above the emplacement temperature, biassing the distribution towards the Earth's field (see later for discussion).

This analysis of component directions therefore led us to suspect that determining temperatures from the Minoan deposits would be more problematical than from the other three lithic breccias, due to CRM overprinting. We therefore made palaeointensity determinations on a number of lithic blocks from the Minoan breccia, in order to attempt to discriminate between samples where the CRM overprinting has made it impossible to accurately determine an emplacement temperature, and those where the temperature is reliable. Other workers have estimated the intensity of the Earth's field at the time of the Minoan eruption (disputed to be either 1500 BC or 1630 BC). Data from Minoan kilns on Crete spanning the time of eruption suggest values of 50–60 μT (Liritzis & Thomas 1980), whilst unpublished data from Downey (1983) produced from hearth material in Akrotiri excavations on Santorini give an average value of 51 ± 4.9 μT with individual estimates from 40 to 70 μT. Work by Walton (1984), re-evaluating the effect of alteration during palaeointensity experiments, suggests that these values should be somewhat lower, perhaps in the region of 45–50 μT.

Representative NRM/TRM plots are shown in Fig. 5 for the two Minoan samples described above, whose vector plots are shown in Fig. 3. Sample LM2-2A1 has an ideal, two component remanence, with no overlap of components; the

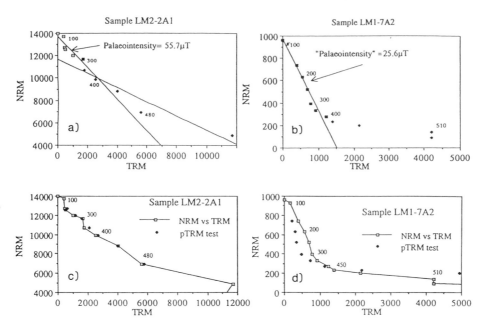

Fig. 5. NRM/TRM plots of (**a**) sample LM2-2A1, and (**b**) sample LM1-7A2. Numbers beside the points represent the temperature of demagnetization (and remagnetization) in °C. NRM/TRM plots with pTRM checks for (**c**) sample LM2-2A1, and (**d**) sample LM1-7A2. Open squares are first pTRM measurement, solid diamonds are repeat pTRM after higher temperature step.

lower temperature line extends up to 350°C. The NRM/TRM plot from this sample is characterized by two line segments: a steeper initial segment from 0 to 400°C and a shallower one from 350 to 570°C. The scatter in the higher temperature data means that the 350°C and 400°C data points lie on both lines. Our directional data indicate that the pTRM Minoan overprint does not extend above 350°C, and the palaeointensity data do not contradict that interpretation. Figure 5c shows pTRM checks which suggest that alteration does not occur until above 480°C. The low temperature line gives a palaeointensity of 55.7 μT (correlation coefficient, $R^2 = 0.906$), which falls in the range of other published values for this age. This reasonable palaeointensity value indicates that this component is of TRM origin, and that we can accept the temperature estimate.

The two remanence components from sample LM1-7A2 overlap over a considerable temperature range. The Minoan direction is only isolated below 150°C, and the pre-Minoan component above 300°C. The NRM/TRM plot from this sample gives a linear segment between 100 and 300°C with an apparent palaeointensity of 25.6 μT (correlation coefficient, $R^2 = 0.956$). However, pTRM checks (Fig. 5d) indicate that considerable alteration has occurred in this

range during the laboratory experiment, so the palaeointensity estimate is meaningless. We can see that the repeat pTRM check measurement is less than the first measurement at a given temperature, until a temperature of 400°C is reached. Above this temperature, the repeat pTRM becomes larger than the first pTRM measurement at a given temperature. This indicates that the capacity of the sample to acquire TRM is being reduced by the laboratory alteration, in the blocking temperature range up to 400°C, and increased in higher blocking temperature intervals. When we compare the direction of remanence after the first demagnetization step with that after the second demagnetization step (Fig. 6), we can see that a significant amount of the remanence acquired during cooling in an applied field is not removed by the second demagnetization, and therefore has higher blocking temperatures than its formation temperature.

What can be deduced about the natural remanence from the alteration behaviour in the laboratory? In the laboratory, some of the low blocking temperature fraction is being destroyed and the high blocking temperature fraction enhanced. This phenomenon is likely to have also occurred during the incorporation of this lithic clast into the Minoan breccia. Clearly, our

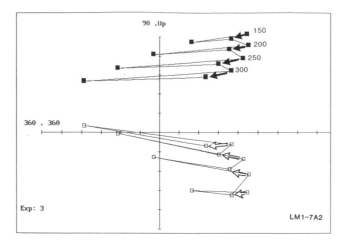

Fig. 6. Vector plot showing all three heatings during a Thellier experiment for four temperature steps(T) for sample LM1-7A2; symbols as in Fig.2. The arrow links the first thermal demagnetization with the second demagnetization step, indicating the CRM component created in the intervening remagnetization step. The symbols to the left represent the vector sum of pTRM acquired below T and the NRM remaining after thermal demagnetization below T. The pTRM is directed slightly west of zero and slightly upwards in the arbitrary reference frame of this figure.

interpretation of the curvature of the vector plots as being indicative of alteration is correct. Do we have any reason to believe that our estimate of the emplacement temperature is a significant over-estimate? This would be the case if the CRM that formed during the Minoan heating completely overprinted pre-Minoan remanence in higher blocking temperature fractions than the heating temperature (T_f). In this case, the heating temperature should be less than our estimate of 150°C. Our data suggest that this has not occurred, as the CRM acquired in the laboratory experiments that has $T_b > T_f$ is added vectorially onto the remaining NRM direction, and there is no indication of significant removal of an NRM component during the pTRM acquisition and subsequent second demagnetization.

Although the palaeointensity experiment on LM1-7A2 failed in the classic sense, in that a palaeointensity could not be determined due to alteration, the experiment carries very valuable information about the characteristics of the magnetic alteration that enable us to assess the reliability of our directional temperature estimates. In this example, we have no reason to believe that the maximum emplacement temperature estimate of 150°C is not reliable.

The Lower Pumice 2 lithic breccia shows a range of emplacement temperatures from 220 to ⩾580°C (Fig. 7), with all but one clast in the temperature range 220–450°C. The Middle

Pumice breccia also shows a range of emplacement temperatures from 200 to ⩾580°C. However this deposit has more high temperature clasts than Lower Pumice 2 (Fig. 7). Emplacement temperatures for the Cape Riva breccia range from 380 to ⩾580°C. The Minoan lithic breccia displays the widest range of emplacement temperatures from 120 to ⩾580°C with over 50% of the temperatures occurring in the range 200–350°C.

Emplacement temperatures show considerable variation at most localities. McClelland & Druitt (1989) considered the thermal histories of lithic clasts emplaced with a range of temperatures. As the deposit cools down *en masse*, individual clasts will heat up or cool down to some average temperature due to heat transfer. Under normal conditions lithics with an initial temperature greater than the average temperature would cool with time after emplacement and all the grains with blocking temperatures less than the initial temperature will be remagnetized. Clasts with initial temperatures less than the average temperature will initially heat up to the average temperature before cooling to ambient, so that remagnetization of all the grains with blocking temperatures less than the average temperature will occur. Lithic clasts in a pyroclastic deposit will therefore record emplacement temperature values ranging from the average temperature of the deposit to the maximum initial temperature of a clast, which may

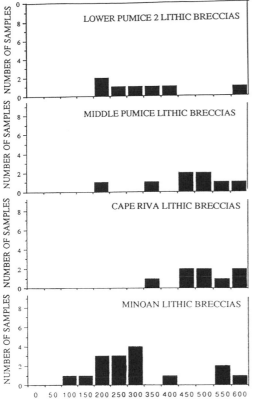

Fig. 7. Emplacement temperatures of co-ignimbrite lithic breccias from the Lower Pumice 2, Middle Pumice, Cape Riva, and Minoan eruptions.

from the airfall was sampled at ten localities downwind from the vent (Fig. 1c), from welded to non-welded deposits. At each locality between two and four lithic blocks, 5–20 cm in diameter were collected for thermal demagnetization. Emplacement temperature estimates as a function of distance from vent are plotted in Fig. 8. Three lithic clasts from the most proximal locality where the airfall is welded have single component magnetizations in the Earth's field direction consistent with emplacement above 580°C (the Curie point of magnetite). In the transition zone between welded and non-welded, temperatures range from 520 to 580°C. Lithics from the non-welded airfall give emplacement temperatures of 224–580°C.

The dashed line on Fig. 8 shows how the minimum emplacement temperatures (or equilibrium temperatures) fall off away from the vent. The expected magma temperature for these andesitic pumices would be 950°C or greater (Cas & Wright 1988). Our data suggest that little heat would have been lost by the juvenile magma during its ascent into the plinian column and fall-out close to the vent, but that a very significant loss of heat occurred for material deposited over 2 km from the vent.

Palaeomagnetic results from lithics from the Minoan eruption

Palaeomagnetic sampling has been carried out in all four phases, and in as many localities as possible (see Fig. 1a). The data are only summarized here, and will be described in more detail elsewhere.

Phase 1, plinian airfall

The Minoan plinian airfall consists of rhyodacite pumice with subsidiary lithic clasts. The low blocking temperature components are well grouped about the recent Earth's field direction, while the high blocking temperature components are randomly oriented between blocks (Fig. 9a). The emplacement temperatures estimated from lithic blocks within the Minoan airfall are considerably lower than those from the Middle Pumice airfall. Isopach maps from Druitt et al. (1989) show that the dispersal axis was to the south east from a vent close to the present town of Thera. Two sites close to the dispersal axis, one site in Thera quarry and one in Akrotiri quarry give emplacement temperatures of between 150 and 350°C (Fig. 9b, c). A site from Oia quarry, which was upwind of the eruption column, gives lower temperatures from zero to 250° (Fig. 9d). The lowest temperature

exceed the Curie temperature. Thus an average or equilibrium temperature of a deposit is most closely approximated by the lowest value of emplacement temperature at a particular location.

The Lower Pumice 2 and Middle Pumice lithic breccias both show equilibrium temperatures of approximately 200°C, while Cape Riva displays an equilibrium temperature of 380°C showing these lithic breccias were unambiguously emplaced hot. The Minoan, though emplaced hot, had a lower equilibrium temperature of 120°C. These results are consistent with the interpretation, based on field observations, that all these breccias formed by proximal sedimentation of lithic clasts from pyroclastic flows.

Palaeomagnetic results from lithics from the Middle Pumice Airfall

The Middle Pumice airfall consists of andesitic pumice with occasional lithic material. Material

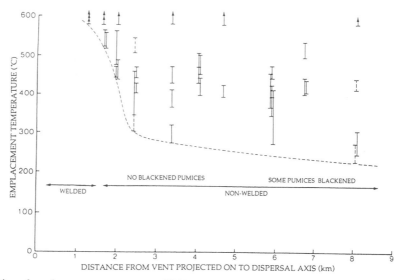

Fig. 8. Variation of emplacement temperature of lithic clasts from the Middle Pumice airfall with distance from vent (after McClelland & Druitt 1989).

was taken to be the equilibrium temperature of the deposit, and hence the distal equilibrium temperatures were less than 100°C. Thomas & Sparks (1992) calculated how heat is lost when material falls out of an eruption cloud during a plinian eruption. Large clasts (>25 cm diameter) lose very little heat; heat loss from clasts in the diameter range 1.5–25 cm increases with decreasing grain size and increasing fall height, clasts smaller than 1.5 cm are deposited cold. We have measured grain sizes of lithics from two localities to estimate the deposit equilibrium temperature by summing the heat contribution from each grain size fraction, assuming an initial magma temperature of 820°C (Thomas, 1993). The lithic content is 15% at Thera quarry. If this were initially cold, then the equilibrium temperature is calculated to be 180°C if lithics and pumices were of equal size. The equilibrium temperature is calculated to be 400°C if the pumices were twice as large as the lithics (pumice sizes cannot be measured directly, as they fragment on impact). The lithic content is 4% at Akrotiri quarry. Using the same assumptions as for Thera quarry, the calculated temperature range is 50°C to 160°C. These temperatures bracket the range estimated from the palaeomagnetic technique. It is possible that the range of temperatures estimated from lithic clasts arises both from locally variable grain size of the surrounding pumices, and from preheating of the lithic clasts before incorporation into the eruption column.

Phase 2, phreatomagmatic deposits

Sites were sampled in Thera, Oia and Akrotiri quarries. Minimum temperatures of 100–150°C were determined for lithics from Thera and Oia quarries, while one sample from Akrotiri quarry showed no overprinting at all, despite having a favourable blocking temperature spectrum. Some hot lithics were found. Cooling rate corrections were applied for the average deposit thickness of 10 m, and result in a maximum equilibrium temperature estimate of 100°C for this deposit. The low temperature of these base surge deposits supports the conventional interpretation of such deposits as the products of lateral explosions caused when water entered the plinian vent. Water vaporized in the eruption would have a major cooling effect on the high temperature (820°C) juvenile material. The proportion of lithics is low in this phase, and they will have had only a minor cooling effect.

Phase 3, massive structureless white deposits

Great thicknesses of up to 50 m of these deposits are found on the island. Their origin is uncertain. Small lithic blocks (<10 cm) were collected from 19 localities around the island, and long cores were drilled from the largest blocks available at several locations in order to investigate the effects of thermal disequilibrium. A range of emplacement temperatures was

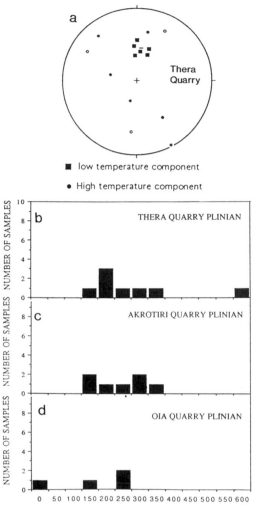

Fig. 9. Results from Minoan plinian air-fall. (**a**) Stereographic projection of directional data from Thera quarry; (**b, c** and **d**) show range of emplacement temperatures from three localities.

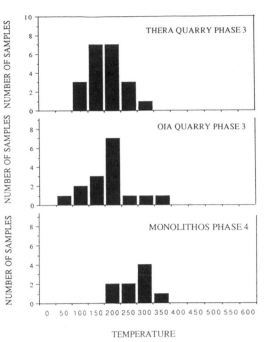

Fig. 10. Range of emplacement temperatures from Minoan Phase three and Phase 4 deposits.

found at each site (Fig. 10), but the lowest temperature, indicating the deposit equilibrium temperature, was between 0 and 100°C. Lithics account for up to 20% by volume of the Phase 3 deposits, and thus are not responsible for the cold nature of the tephra. The deposits were clearly deposited from quite energetic flows, so they are interpreted as very low grade ignimbrites laid down from pyroclastic flows. The juvenile material probably cooled as a result of substantial interaction with water in the vent system. No thermal disequilibrium was found in any of the long (20 cm) cores drilled from large

blocks. The maximum unblocking temperature of the overprint was found to be the same for the interiors of the blocks as for the outer surfaces. These findings contradict the conclusions of Downey (1983) that the Phase 3 deposits were hot (>350°C), based on maximum unblocking temperature of juvenile pumice material and the existence of a dominant recent Earth's field overprint at the surface of a large block, which had no evidence of an overprint in the interior. We believe his data should be discounted. The maximum unblocking temperature of remanence in pumices simply reflects the grain size distribution, as the remanence is predominantly CRM acquired during cooling in the deposit (see below). Moreover thermal disequilibrium in such a thick deposit is impossible to sustain, as the cooling time for the deposit as a whole is much longer than the time it would take a cold block incorporated in a hot matrix to heat up to the equilibrium temperature of the deposit.

Phase 4, lithic rich ignimbrite and lithic breccias

The fine-grained deposits are incontrovertibly ignimbrites (Sparks & Wilson 1990), but the lithic breccias may be either flood deposits intercalated with the ignimbrites, or lag breccias of pyroclastic origin. Classic non-welded ig-nimbrite facies from Monolithos beach give emplacement temperature estimates from 200 to 350°C (Fig. 10). Here the lithic content is up to 50% of the total material present, although the grain size is very fine. If the lithics were initially cold, then the bulk temperature of the source material, mixing cold lithics with juvenile magma at 820°, would have been close to 400°C. The deposit equilibrium temperatures of 200°C therefore do not necessitate the input of water, supporting their ignimbrite origin. Breccias deposited in the initial stages of Phase 4 of the eruption at Oia show a consistent variability in emplacement temperature related to clast size. Small blocks show relatively high emplacement temperatures of 150–300°C, while large blocks show essentially no overprint, except close to the surface. Our interpretation is that the breccia was rapidly transported in a hot flow, where the small blocks were able to reach temperatures up to 300°C, but large blocks were only superficially heated. Once the breccia was deposited, there was not enough heat stored in the small lithics to heat the greater mass of large, essentially cold blocks. Later lag breccias give higher equilib-rium temperatures of 200–300°C. Our data suggest that the lithic breccias of Phase 4 are all of pyroclastic origin.

Palaeomagnetic results from pumices

Pumices from the Middle Pumice airfall and Phases 1 to 4 of the Minoan eruption mostly have a single component remanence, acquired after deposition in the direction of the Earth's field. In all these deposits the maximum blocking tem-perature of the remanences is considerably higher than the emplacement temperature as estimated for the deposits. Downey & Tarling (1991) argue that the maximum blocking tem-perature of pumices gives an accurate em-placement temperature of the tephra. We disagree strongly with this assumption, and we believe nearly all their NRM is of CRM origin. Zlotniki *et al.* (1984) undertook a palaeoin-tensity study of pumices from recent pyroclastics from Guadeloupe. They showed that, although the pumices have a single component rem-anence, only a low blocking temperature frac-tion gives a sensible palaeointensity, while the high T_b fraction gives an intensity that is too low by a factor of two. They conclude that a significant proportion of pumice remanence is a CRM with blocking temperatures exceeding the emplacement temperature, acquired during cooling in the deposit, with the same direction as the TRM component (see McClelland & Druitt

1989 for further discussion). We therefore only use the temperatures obtained from lithic clasts to estimate the equilibrium temperature of the deposits.

Five pumice blocks from Phase 1 of the Minoan eruption have a two component rem-anence (Thomas 1993). The NRM's of these samples are deflected away from the Earth's field direction towards the direction of the high temperature component, indicating the high temperature magnetization was produced natur-ally. A number of mechanisms have been proposed for the origin of high temperature magnetizations in juvenile material: (a) cooling during transport (Downey 1983); (b) movement/ rotation of fragments after deposition, probably during compaction (Hoblitt *et al.* 1985); and (c) stationary cooling of fragments before incorpor-ation into the deposit. Thomas (1993) argues that (c) is most likely for these pumices as cooling in flight is unlikely to produce a coherent magnetization, and the amount of rotation needed to produce the directional difference between the low and high temperature com-ponents seems unlikely to have been compaction related.

Conclusions

(1) Thermal demagnetizaton is a valuable tool for determining emplacement temperatures of pyroclastic deposits. Two-component remanen-ces in lithic clasts provide a quantitative esti-mation of emplacement temperature of the lithic clasts, and the lowest temperature estimate at a particular location can be used to estimate the equilibrium temperature of the whole deposit.

(2) Palaeointensity experiments are a useful tool for determining if the low temperature component of magnetization is of thermal origin. If alteration has occurred, they also provide valuable information about the charac-teristics of magnetic alteration, thus allowing assessment of the reliability of temperature estimates in these cases.

(3) Emplacement temperature estimates from individual locations from all the deposits sampled show a wide range in temperatures (typically a few hundred degrees). One factor contributing to this range is chemical overprint-ing in lithics, which reduces the precision of the emplacement temperatures. However, some data sets show little sign of chemical overprint-ing indicating this is not the only contributing factor. At many locations samples were col-lected from all levels within the deposit, some of which are slightly compositionally zoned,and were probably zoned thermally during eruption.

It is therefore likely that some scatter in emplacement temperature is due to thermal zoning. However this does not account for neighbouring clasts displaying emplacement temperatures differing by over 100°C. Probably the most important factor for this variation is that clasts had a wide range of temperatures on emplacement. For example, clasts taken from deep within the conduit are likely to be significantly hotter than clasts taken from near the surface.

(4) Emplacement temperature estimates place important constraints on eruption mechanisms. Results from the co-ignimbrite lithic breccias show they were emplaced hot. The Middle Pumice and Minoan airfalls display equilibrium temperatures that decrease with increasing distance from vent, and provide constraints on the predictions of Thomas & Sparks (1992) concerning the cooling of tephra during fallout from eruption columns. Emplacement temperatures for the Minoan Phases 2 and 3 are low considering the deposits are principally composed of juvenile material. For Phase 2 this supports the conventional view of such deposits being the products of lateral explosions caused when water entered the plinian vent. The juvenile material of Phase 3 probably cooled as a result of interaction with water in the vent system.

Research funds for these studies have been provided by a Natural Environmental Research Council Studentships to L. B. and R. T. and by a research fellowship to E. M. from the Royal Society of London.

References

CAS, R. A. F. & WRIGHT, J. V. 1988. *Volcanic successions: Modern and Ancient.* Unwin Hyman, London.

DECKER, R. W. 1990. How often does a Minoan eruption occur? *In:* HARDY, D. (ed.) *Thera and the Aegean World III Vol. 2.* The Theran Foundation, London, 444–452.

DODSON, M. H. & McCLELLAND BROWN, E. 1980. Magnetic blocking temperatures during slow cooling. *Journal of Geophysical Research*, **85**, 2625–2637.

DOWNEY, W. S. 1983. *Magnetic studies of Santorini Minoan tephra and Cretan archaeological materials.* PhD thesis, Newcastle upon Tyne University.

——— & TARLING, D. H. 1991. Reworking characteristics of Quaternary pyroclastics, Thera (Greece), determined using magnetic properties. *Journal of Volcanology and Geothermal Research*, **46**, 143–155.

DRUITT, T. H. & SPARKS, R. S. J. 1982. A proximal ignimbrite breccia facies on Santorini, Greece. *Journal of Volcanology and Geothermal Research*, **13**, 147–151.

———, MELLORS, R. A., PYLE, D. M. & SPARKS, R. S. J. 1989. Explosive volcanism on Santorini, Greece. *Geological Magazine*, **126**(2), 95–126.

HOBLITT, R. P., REYNOLDS, R. L. & LARSON, E. E. 1985. Suitability of non-welded pyroclastic flow deposits for studies of magnetic secular variation: a test based on deposits emplaced at Mount St. Helens, Washington. 1980. *Geology*, **13**, 242–245.

KENT, J. T., BRIDEN, J. C. & MARDIA, K. V. 1983. Linear and planar structure in ordered multivariate data as applied to progressive demagnetization of palaeomagnetic remanence. *Geophysical Journal of the Royal Astronomical Society*, **62**, 699–718.

LIRITZIS, Y. & THOMAS, R. 1980. Palaeointensity and thermoluminescence measurements on Cretan kilns from 1300 to 2000 BC. *Nature*, **283**, 54–55.

McCLELLAND BROWN, E. 1982. Discrimination of TRM and CRM by blocking-temperature spectrum analysis. *Physics of the Earth and Planetary Interiors*, **30**, 405–414.

McCLELLAND & DRUITT, T. H. 1989. Palaeomagnetic estimation of emplacement temperatures of pyroclastic deposits on Santorini, Greece. *Bulletin of Volcanology*, **51**, 16–27.

RAYLEIGH, Lord. 1919. On a problem of vibrations, and of random flights in one, two and three dimensions. *Philosophical Magazine*, **37**(6), 321–347.

SPARKS, R. S. J. & WILSON, C. J. N. 1990. The Minoan deposits: a review of their characteristics and interpretation. *In:* HARDY, D. (ed.) *Thera and the Aegean World III.* The Theran Foundation, London, 89–98.

——— & WRIGHT, J. V. 1979. Welded air-fall tuffs. *Geological Society of America Bulletin*, **180**, 155–166.

THOMAS, R. M. E. 1993. *Determination of emplacement temperatures of pyroclastic deposits by theoretical and palaeomagnetic methods.* PhD thesis, Bristol University.

——— & SPARKS, R. S. J. 1992. Cooling of tephra during fallout from eruption columns. *Bulletin of Volcanology*, **54**, 542–553.

WALTON, D. 1984. Reevaluation of Greek archaeomagnetudes. *Nature*, **310**, 740–743.

ZLOTNIKI, J., POZZI, J. P., BOUDON, G. & MOREAU, M. G. 1984. A new method for the determination of the setting temperature of pyroclastic deposits (example of Guadeloupe: French West Indies). *Journal of Volcanology and Geothermal Research*, **21**, 297–312.

Palaeomagnetic controls on the emplacement of the Neapolitan Yellow Tuff (Campi Flegrei, Southern Italy)

MAURIZIO DE' GENNARO[1,4], PAOLA R. GIALANELLA[2], ALBERTO
INCORONATO[1], GIUSEPPE MASTROLORENZO[3] & DEBORA NAIMO[1]

[1]*Dipartimento di Scienze della Terra, Università degli Studi di Napoli 'Federico II', largo
S. Marcellino 10, I-80138 Napoli, Italy*
[2]*Institut für Geophysik, ETH Hönggerberg, CH-8093 Zürich, Switzerland*
[3]*Osservatorio Vesuviano, via Manzoni 249, I-80123 Napoli, Italy*
[4]*Istituto Policattreda di Scienze Geologiche e Mineralogiche, Università degli Studi di
Sassari, via Giovanni Maria Angioi 10, I-07100 Sassari, Italy*

Abstract: Two outcrops of Neapolitan Yellow Tuff at Torregaveta and Posillipo have been
investigated palaeomagnetically. Analyses of low blocking temperature components, in
association with mineralogical, volcanological and magnetic anisotropy data, suggest the
following sequence of events: (i) deposition of pyroclastics at temperatures not lower than
275°C, at least at Torregaveta; (ii) cooling associated with zeolitization and modification of
the original positions of the clasts; (iii) further dislocation of the clasts after cooling; and (iv)
lithification. Such a model reconciles otherwise conflicting volcanological and palaeomagnetic data.

The Neapolitan Yellow Tuff (NYT) is the major
pyroclastic tuff deposit in Campi Flegrei and
comprises pumices, obsidian fragments, crystals
and lithics embedded in an ashy matrix. The
pyroclastic deposit is inferred to cover at least
$500 km^2$ and to have a volume of about $40 km^3$
(Orsi *et al.* 1992). They are dated around 12 ka
(Alessio *et al.* 1971, 1973; Scandone *et al.* 1991)
and outcrop mainly in the steep relief bordering
the Campi Flegrei caldera (Fig. 1). In the caldera
depression, their thicknesses vary between a few
metres to more than 100 m, and they are mostly
covered by younger pyroclastic deposits. The
formation is mainly hydromagmatic and consists
of two main diagenetic facies; a yellow, lithified,
un-welded facies, rich in zeolites, and a grey,
non-lithified facies called 'Pozzolana' (Scherillo
1955). Both vertical and lateral transitions
between the two facies occur, even over a few
metres. The stratigraphic and sedimentological
analysis of the NYT (Cole & Scarpati 1993)
indicates the presence of a lower member,
consisting mainly of fall-out units and an upper
hydromagmatic one. The lower member is more
widespread than the upper and outcrops both
inside and outside the caldera with thicknesses
varying between about 10 m close to the caldera
rim to less than 1 m in distal locations. It exhibits
a thin stratification of phreatoplinian fall out and
surge layers with a prevailing fine-grained
fraction. In contrast, the upper member consists

of a thick, coarse-grained pyroclastic flow and
surge deposits with subsidiary fall-out units. The
eruptive mechanism and the location of the vent
been controversial. Two contrasting hypotheses
have been proposed: (a) the products were
emitted at different times from different eruptive centres (Parascandola 1936; Rittmann *et al.*
1950; Capaldi *et al.* 1987); and (b) the formation
was generated from a single huge eruption
(Scherillo & Franco 1967; Lirer & Munno 1975;
Di Girolamo *et al.* 1984; Lirer *et al.* 1987). The
latter is strongly supported by recent detailed
stratigraphical analyses (Orsi & Scarpati 1989;
Scarpati 1990; Orsi *et al.* 1991; Scarpati *et al.*
1993; Cole & Scarpati 1993). It is also supported
by preliminary anisotropy of magnetic susceptibility (AMS) studies (Incoronato 1982; Favara *et
al.* 1987) suggesting that the location of the vent
is in the eastern part of the caldera depression
(Fig. 1). In this work it is accepted that the NYT
resulted from a single eruption through a single
vent. The zeolitization of the glassy matrix of the
NYT is characterized by crystallisation of phillipsite and chabazite (de' Gennaro *et al.* 1987;
de' Gennaro & Franco 1988), which occurred
during cooling (de' Gennaro & Colella 1991).
This timing is also suggested by the lack of
mineralogical zoning expected in zeolites related
to an open hydrological system (de' Gennaro *et
al.* 1990) or with the presence of a water table
(de' Gennaro *et al.* 1994). The zeolites in the

From Morris, A. & Tarling, D. H. (eds), 1996, *Palaeomagnetism and Tectonics of the
Mediterranean Region*, Geological Society Special Publication No. 105, pp. 359–365.

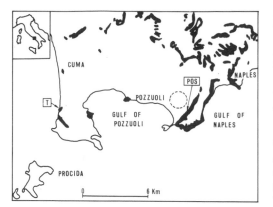

Fig. 1. Schematic map of Campi Flegrei volcanic area showing the main volcanic and volcano-tectonic elements (after Di Vito *et al.* 1985). The principal outcrops of the NYT are exposed mostly along the caldera rim. The internal part of the caldera is filled by recent volcanic deposits younger than 12 ka. Sampling sites: T, Torregaveta; POS, Posillipo; The presumed location of the vent (dashed circle) is also indicated on ground of preliminary AMS studies (data from: Incoronato 1982 (T); Favara *et al.* 1987).

NYT have no regular stratigraphic distribution, and the greater and lesser concentrations alternate randomly, which may be related to a sequence of flow units, each with different temperatures, cooling rates, and more or less rapid loss of volatile components (de' Gennaro *et al.* 1990).

Temperature estimates by thermal remanent magnetic investigations

Emplacement temperatures can be determined from Thermal Remanent Magnetization (TRM) analyses of oriented clasts (Hoblitt & Kellogg 1979). Emplacement temperatures above the maximum blocking temperature (type I deposit) will cause all previous remanences to be erased and TRMs, parallel to the local Earth's Magnetic Field (EMF), will be acquired as cooling proceeds. Emplacement temperature equal to the ambient temperature (type II deposit) will not affect any previous remanences of clasts which will thus be randomly oriented as a result of rotations during transport. Emplacement temperature below the maximum blocking but above the ambient one (type III deposit) will cause part of the existing remanence to unblock and, as cooling proceeds, these components will acquire a TRM parallel to the local EMF, while the existing TRMs that were not unblocked by the heating will retain their random orientation. As far as type III deposits are concerned, the

estimate of emplacement temperature equals the upper limit of temperature interval for which all low blocking temperature components are parallel to both one another and the direction of the local EMF.

Palaeomagnetic estimate of emplacement temperatures have been successfully carried out on lithic clasts from AD 79 Vesuvius (Kent *et al.* 1981) and Santorini (Tarling 1978; Downey 1983; McClelland & Druitt 1989; McClelland & Thomas 1990; Bardot *et al.* this volume) pyroclastic deposits, mostly using Principal Cmponent Analyses (PCA), based on the least squares method (Kirschvink 1980) to identify the components. Clasts from the NYT were similarly subjected to progressive thermal demagnetization (PTD). Only preliminary investigation had been carried out on lithic clasts at Torregaveta (Incoronato 1990). These had suggested emplacement temperatures of about 275°C, and this study of two other outcrops is part of a programme to systematically map the distribution of emplacement temperatures in the NYT.

Sampling, measurements and results

The sections of NYT sampled at Posillipo and Torregaveta (Fig. 1) were both part of the lithified upper member of the formation and exhibit similar general textural features to several other depositional units. The tuff sequence consists of a crudely stratified wavy to planar alternations of coarse-grained disorganised matrix-supported layers, thinly laminated discontinuous beds and massive, even fine ash layers. Each thick, coarse-grained bed is delimited by a lower and an upper massive, indurated fine ash bed. The section at Posillipo presents a more chaotic texture, possibly due to instability within the pyroclastic deposit on the steep slope formed by the rapid accumulation of the pyroclastic deposit during the eruption. This section is characterized by a generally lower sorting, it is rich in coarse lithic blocks, suggesting a near vent position, and at places, there was some evidence of syn-depositional reworking. In contrast, the tuff deposit of Torregaveta was better sorted and finer, and presented much less evidence of reworking. The Posillipo sampling site was on the SE internal border of the Campi Flegrei caldera, near to its rim, while the Torregaveta sampling site was on the western caldera rim. Extra care was taken to ensure that the sampled outcrops did not exhibit evidence of reworking. Sampling was carried out by the plastic disc method (Tarling 1983) and the specimens were oriented using a sun compass

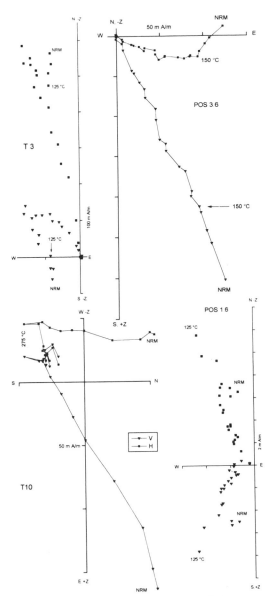

Fig. 2. Cartesian projections of some NYT clasts from Torregaveta (T) and Posillipo (POS) outcrops subjected to thermal demagnetization (for locations see the volcanological map). H (V), horizontal (vertical) projections.

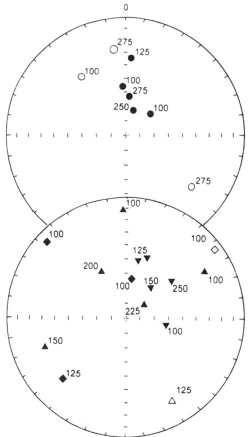

Fig. 3. Lambert equal-area stereographic projections of low blocking temperature components of clast specimens from Torregaveta (top) and Posillipo (bottom ▲, I flow unit; ◆, II flow unit, ▼, III flow unit) NYT outcrops (for locations see Fig. 1). Open (closed) symbols, positive (negative) inclinations. Numerals refer to the upper value of the temperature range within which the low blocking temperature components have been calculated (see Table 1) according to Kirschvink (1980).

prior to removal. At Posillipo, 30 lithic clasts from three different flow units (10 clasts for each unit) were sampled. It was also decided to reassess the emplacement temperature at Torregaveta but, because of alteration at the previous site (Incoronato 1990), the new sampling (10 lithic clasts) was from a stratigraphically higher flow unit. The diameter of the lithic clasts ranged from approximately 4 to 6 cm. The size of the oriented clasts was reduced in the laboratory to fit them into hollow non-magnetic cylinders (2.5 cm diameter, 2.2 cm high). During this process, four of the 30 Posillipo specimens were destroyed. The surviving clasts were subjected to PTD from 75°C, at 25°C intervals, using a Schonstedt thermal demagnetizer. The remanences were measured using a Molspin magnetometer. Low fields susceptibility was monitored after each heating step but no changes were detected during the treatment. The intensities of four Posillipo specimens dropped below the instrumental

Table 1. *Low blocking temperature components of specimens of NYT from Torregaveta (T) and Posillipo (POS) outcrops (for location refer to Fig. 1)*

Specimen	T (°C) range	Dec	Inc	da
T1	20–100	354.1	55.8	10.2
T2	20–100	48.0	67.1	4.3
T3	20–125	4.0	33.8	8.2
T4	20–275	128.2	−29.8	6.2
T6	75–275	352.1	−27.5	9.4
T7	20–100	327.6	−37.5	9.6
T8	20–250	14.4	71.6	7.2
T10	20–275	6.1	62.1	4.2
POS 1.1	20–100	358.0	10.9	7.4
POS 1.3	20–150	252.7	29.0	6.9
POS 1.4	20–100	58.9	24.0	3.8
POS 1.6	20–125	149.0	−23.0	6.9
POS 1.7	20–200	330.7	53.6	8.8
POS 1.10	75–225	51.4	72.8	5.7
POS 2.1	20–100	51.2	−6.9	2.0
POS 2.3	20–100	8.2	62.0	2.9
POS 2.6	75–125	226.2	28.3	4.2
POS 2.8	20–100	307.7	8.8	7.5
POS 3.1	75–250	51.1	49.7	9.7
POS 3.3	20–100	97.2	61.5	6.8
POS 3.6	20–150	38.5	62.6	2.5
POS 3.7	75–125	18.9	45.9	8.5
POS 3.10	75–125	12.1	48.9	4.2

T (°C) range, temperature interval for which the component has been identified; Dec, declination; Inc, inclination; da, diagonal angle (Kirschvink 1980).

noise level after a few steps and, therefore, no further analysis was carried out on them. Examples of demagnetization paths of specimens subjected to PTD are given in Fig. 2. As the primary aim was to determine emplacement temperatures, attention was focused on the magnetic components defined in the lower part of the blocking temperature spectrum. Where a component could be identified, its direction, with the associated temperature interval and diagonal angle (Kirschvink 1980) was listed (Table 1). In some specimens, no low blocking temperature components could be identified in the lower part of the blocking temperature spectrum as the directions of magnetization changed erratically. The low blocking temperature component directions for Torregaveta (Fig. 3) shows grouping of some remanences close to the direction of the axial geocentric geomagnetic field at the sampling site. The upper limit of the temperature range of these vectors varied between 100°C and 275°C and two directions were reversed. Directions from the Posillipo clasts were very scattered (Fig. 3) and two directions are reversed. The upper limit of the temperature range within which directions have been calculated varied between 100°C and 200°C.

Discussion and conclusions

The directional behaviour of magnetization components of the lower part of the blocking temperature spectrum (Fig. 3; Table 1) suggests that the investigated deposits are of type II (Hoblitt & Kellogg 1979). This suggests that the zeolitization process also took place at ambient temperature, but there is no evidence for such low temperature crystallization of zeolites in these deposits. The apparent conflict can be resolved if post-emplacement dislocation of clasts is taken into account. The grouping of the low blocking temperature directions (Fig. 3) suggest that such dislocations are of different importance in the two sites studied and that most of these dislocations were small and therefore not detectable in the field.

Preliminary Anisotropy of Magnetic Susceptibility (AMS) studies have been carried out on matrix specimens from Torregaveta (Incoronato 1982) and Posillipo (Favara *et al.* 1987). Curie temperature and isothermal remanent magnet-

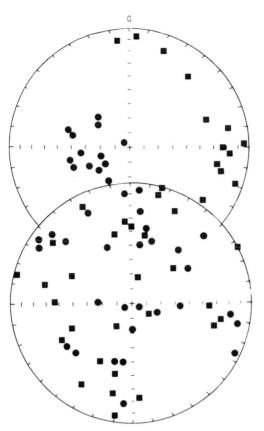

Fig. 4. Lambert equal-area stereographic projections of maximum (squares) and minimum (dots) susceptibility axes from NYT matrix specimens from Torregaveta (T) and Posillipo (POS) (preliminary data from: Incoronato 1982 (T); Favara *et al.* 1987).

ization determinations of matrix samples from Torregaveta (Incoronato 1982) and isothermal remanence studies of matrix samples from Posillipo (Favara *et al.* 1987) indicate that the magnetic carrier is magnetite. The magnetic fabric (Fig. 4) is poorly defined at Posillipo, whereas at Torregaveta it is imbricated and foliated, suggesting that post-emplacement dislocation of clasts is more prevalent at Posillipo than at Torregaveta. The different magnitude of post-emplacement clast dislocation can be related to the fact that Posillipo is close to the vent, so the deposit is thicker and more voluminous that at Torregaveta, which is relatively distal with lower thickness and shallower palaeodepositional slopes (Pescatore & Rolandi 1981) and so is less favourable to post-depositional clast dislocation. Post-emplacement mechanical compaction and vapour-phase crystallization (MacDonald & Palmer 1990; Sheridan & Ragan

1976) may be a possible cause of such post-emplacement dislocations.

It is possible to estimate the lower threshold of the emplacement temperature, at least at Torregaveta. Those directions close to the direction of the axial geocentric geomagnetic field at the sampling site (Table 1; Fig. 4b) indicate that emplacement temperatures were not lower than 275°C. Laboratory tests on zeolitization of trachytic pumices suggest that a critical temperature of about 300°C was reached (de' Gennaro & Colella 1991). In fact, crystallization of phillipsite and chabazite, which characterise the NYT, only occurs if the temperature does not exceed this value (otherwise analcime and k-feldspar crystallize which are absent from the NYT deposits). This suggests that the lower threshold temperature of 275°C, estimated palaeomagnetically, is quite close to the real emplacement temperature value. This value is also similar to the preliminary results obtained by Incoronato (1990) at Torregaveta, about 2 m below the flow unit investigated in this study. The different values for the threshold temperatures indicate that each clast changed its original emplacement position at somewhat different temperature to other clasts, resulting in the scattering of the directions of the low blocking temperature components, and that this process continued after the ambient temperature had been reached. This means that the zeolitization, which took place during cooling, could not have caused the lithification of the deposit, as previously suggested by Franco (1975). Such a conclusion is supported by recent studies which show no relationship between the zeolitic content of the NYT and its breaking load (de' Gennaro & Rippa, work in progress). It seems probable that a gel-like aluminosilicate phase may play an important role in the lithification process (de' Gennaro & Colella 1989) and that this can cause lithification later than previously supposed.

This study shows that if palaeomagnetic investigations of pyroclastic deposits are integrated with mineralogical and sedimentological assessments, then valuable insights can be obtained on emplacement temperatures, post-depositional modification and lithification of such deposits. Future attention will be focused on sections of the NYT where detailed vertical palaeomagnetic and mineralogical sampling, with associated sedimentological study, can be co-ordinated.

We thank the editors and anonymous referees for improvements to the manuscript. This work was supported through grants of GNV (Gruppo Nazionale per la Vulcanologia).

References

ALESSIO, M., BELLA, F., IMPROTA, S., BELLUOMINI, G., CORTESI, C. & TURI, B. 1971. University of Rome, Carbon-14 Dates. *IX Radiocarbon*, **13**, 395–411.
—, —, BELLUOMINI, G., CALDERONI, G., CORTESI, C. & TURI, B. 1973. University of Rome Carbon-14 dates. *X Radiocarbon*, **15**, 165–178.
BARDOT, L., THOMAS, R. & McCLELLAND, ? 1996. Emplacement temperatures of pyroclastic deposits on Santorini deduced from palaeomagnetic measurements: constraints on eruption mechanisms. *This volume.*
CAPALDI, G., CIVETTA, L., DI GIROLAMO, P., LANZARA, L., ORSI, G.,SCARPATI, C. 1987. Volcanological and geochemical constraints on the genesis of Yellow-Tuffs in the Neapolitan-Phlegraen area. *In*: DI GIROLAMO P. (ed.) *The Volcaniclastic Rocks of Campania (Southern Italy)*. Rendiconti dell'Accademia delle Scienze fisica e matematiche, Special issue, Napoli, 25–40.
COLE, P. & SCARPATI, C. 1993. A facies interpretation of the eruption and emplacement mechanisms of the upper part of the Neapolitan Yellow Tuff. C. F. (Southern Italy). *Bulletin Volcanologique*, **55**, 311–326.
DE' GENNARO, M. & COLELLA, C. 1989. Use of thermal analysis for the evaluation of zeolite content in mixtures of hydrated phases. *Thermochimica Acta*, **154**, 345–353.
—— & —— 1991. The role of temperature in the natural zeolitization of volcanic glass. *N. Jb. Min. Mh.*, **8**, 355–362.
—— & FRANCO, E. 1988. Mineralogy of Italian sedimentary phillipsite and chabazite. *In*: KALÒ D. & SHERRY H. S. (eds) *Occurrence, Properties and Utilization of Natural Zeolites*. Akademiai Kiadò, Budapest, 87–97.
——, ADABBO, M., LANGELLA, A. 1994. Hypothesis on the genesis of the zeolites in some European deposits. *In*: MOUPTAN, S. A. & MING, D. (eds) *Proceedings of Zeolite 93*, in press.
——, FRANCO, E., ROSSI, M., LANGELLA, A. & RONCA, A. 1987. Epigenetic minerals in the volcaniclastic deposits from central-southern Italy: A contribution to zeolite genesis. *In*: DI GIROLAMO P. (ed.) *The Volcaniclastic Rocks of Campania (Southern Italy)*. Rendiconti dell'Accademia delle Scienze fisica e matematiche, Special issue, Napoli, 107–131.
——, PETROSINO, P., CONTE, M. T., MUNNO, R. & COLELLA, C. 1990. Zeolite chemistry and distribution in a Neapolitan Yellow Tuff deposit. *European Journal of Mineralogy*, **2**, 779–786.
DI GIROLAMO, P., GHIARA, M. R., LIRER, L., MUNNO, R., ROLANDI, G. & STANZIONE, D. 1984. Vucanologia e petrologia dei Campi Flegrei. *Bollettino della Società Geologia Italiana*, **103**, 149–213.
DI VITO, N., LIRER, L., MASTROLORENZO, G., ROLANDI, G. & SCANDONE, R. 1985. Volcanological map of Phlegraen Fields. *Min. Prot. Civ.*, University of Naples.

DOWNEY, W. S. 1983. *Magnetic studies on Santorini tephra and Minoan Cretan archaeological materials*. PhD Thesis, University of Newcastle upon Tyne.
FAVARA, V., FLORIO, G. & OLIVETTA, L. 1987. *Studio di Anisotropia di Suscettivita' Magnetica sui depositi di Tufo Giallo Napoletano*. Laurea thesis, University of Naples.
FRANCO E. 1975. La zeolitizzazione naturale. *In*: *Zeoliti e zeolitizzazione*. Proceedings of a Meeting Convegni Accademia Nazionale Lincei, 33–60.
HOBLITT, R. P. & KELLOGG, K. S. 1979. Emplacement temperatures of unsorted and unstratified deposits of volcanic rock debris as determined by paleomagnetic techniques. *Geological Society of America Bulletin*, **9**, 633–642.
INCORONATO, A. 1982. *Palaeomagnetic studies in the Southern Apennines, Italy*. PhD thesis, University of Newcastle upon Tyne (UK).
—— 1990. Determinazione paleomagnetica della temperatura di messa in posto di un deposito di TGN (Torregaveta, C.F.). *Proceedings of the 2nd National Meeting of Geo-Elettro-Magnetismo*, Palermo 13–15 September, 8.
KENT, D. V., NINKOVICH, D., PESCATORE, T. & SPARKS, S. R. I. 1981. Paleomagnetic determination of emplacement temperatures of Vesuvius AD 79 pyroclastic deposits. *Nature*, **290**, 393–396.
KIRSCHVINK, J. I. 1980. The least squares line and plane and the analysis of palaeomagnetic data. *Geophysical Journal of the Royal Astronomical Society*, **62**, 699–718.
LIRER, L. & MUNNO, R. 1975. Il Tufo Giallo Napoletano. *Periodico Mineralogia*, **44**, 103–108.
——, MASTROLORENZO, G. & ROLANDI, G. 1987. Un evento pliniano nell'attivita' recente dei Campi Flegrei. *Bollettino della Società Geologia Italiana*, **106**, 461–473.
MacDONALD W. D. & PALMER H. C. 1990. Flow direction in ash-flow tuffs: a comparison of geological and magnetic susceptibility measurements, Shirege member (upper Bandelier Tuff), Valles caldera, New Mexico, USA. *Bulletin Volcanologique*, **53**, 45–59.
McCLELLAND, E. A. & DRUITT, T. H. 1989. Palaeomagnetic estimates of emplacement temperatures of pyroclastic deposits on Santorini, Greece. *Bulletin Volcanologique*, **51**, 16–27.
—— & THOMAS, R. 1990. *In*: *Thera and Aegean world. A Paleomagnetic Study of Minoan Age Tephra from Thera*. Thera Foundation, London, **3**, 129–138.
ORSI, G. & SCARPATI, C. 1989. Stratigrafia e dinamica eruttiva del Tufo Giallo Napoletano. *Bollettino GNV*, Roma 917–930.
——, D'ANTONIO, M., DE VITA, S. & GALLO, G. 1992. The Neapolitan yellow Tuff, a large-magnitude trachytic phreatoplinian eruption: eruptive dynamics, magma withdrawal and caldera collapse. *Journal of Volcanology and Geothermal Research*, **53**, 275–287.
——, CIVETTA, L., APRILE, A. D'ANTONIO, M., DE VITA, S., GALLO, G. & PIOCHI, N. 1991. The Neapolitan Yellow Tuff: Dynamics emplacement

mechanisms and magma evolution of a phrea-toplinian to plinian eruption. *IAVCEI CEV, Field guide*, **76115**.

PARASCANDOLA, A. 1936. I vulcani occidentali di Napoli. *Bollettino della Società Naturalisti Napoli*, **48**, 39–58.

PESCATORE, T. S. & ROLANDI, G. 1981. Osservazioni preliminari sulla stratigrafia dei depositi vulcano-clastici nel settore SO dei Campi Flegrei. *Bollettino della Società Geologica Italiana*, **100**, 233–254.

RITTMANN, A., VIGHI, L., VENTRIGLIA, V. & NICO-TERA, P. 1950. Rilievo geologico dei Campi Flegrei. *Bollettino della Società Geologica Italiana*, **69**, 117–362.

SCANDONE, R., BELLUCCI, F., LIRER, L., & ROLANDI, G. 1991. The structure of Campanian plain and the activity of the Neapolitan volcanoes (Italy). *Journal of Volcanology of Geothermal Research*, **48**, 1–31.

SCARPATI, C. 1990. *Stratigrafia, geochimica e dinamica eruttiva del Tufo Giallo Napoletano*. PhD thesis, University of Naples, 1–159.

SCARPATI, C., COLE, P. & PERROTTA, A. 1993. Neapolitan Yellow Tuff, a large volume multi-phase eruption from Phlegraen Fields (Southern Italy). *Bulletin Volcanologique*, **55**, 343–356.

SHERIDAN M. F. & RAGAN D. M. 1976. Compaction of ash-flow-tuffs. *In*: CHILLINGARIAN G. V. & WOLF K. H. (eds) *Developments in Sedimentology*. Elsevier. New York, **2**, 677–717

SCHERILLO, A. 1955. Petrografia chimica dei tufi flegrei. Tufo Giallo, Mappamonte, Pozzolana. *Rendiconti dell'Accademia delle Scienze Fisiche e Matematiche, Napoli*, **22**, 317–330.

SCHERILLO, A. & FRANCO, E. 1967. Introduzione alla carta stratigrafica del suolo di Napoli. *Atti Accademia Pontamiana, Napoli*, **6**, 27–37.

TARLING, D. H. 1978. Magnetic studies of the Santorini Tephra deposits. *Proceedings of the 2nd International Thera Conference*, 195–201.

—— 1983. *Palaeomagnetism*. Chapman and Hall.

Magnetic stratigraphy procedures in volcanic areas: the experience at Vesuvius

ALBERTO INCORONATO

Dipartimento di Scienze della Terra, Universita' degli Studi di Napoli 'Federico II', Largo S. Marcellino 10, I-80138 Napoli, Italy

Abstract: A procedure involving stringent criteria for the definition of a linear vector, combined with visual assessment, is demonstrated to be essential for magnetic dating of volcanic lavas during historical times and that the partial demagnetization process must be extended to where the entire spectrum has been destroyed. As a result, it is demonstrated that some magnetic dating of Medieval lavas on Vesuvius are incorrect. Five flows are shown to have identical directions of magnetization, $D = 13°$, $I = +64°$ ($\alpha_{95} = 0.9°$) which can be shown to date to the AD 1631 'big eruption', confirming that this eruption produced several lava flows, despite recent reports to the contrary. It is also shown, mainly using data from Etna, that five other flows on Vesuvius correspond to separate eruptions during a time span between the nineth and eleventh centuries.

Knowledge of the nature, frequency and distribution of past volcanic products plays a significant role in assessing the present-day hazard of any volcanic area. One of the more hazardous volcanic areas is that of Vesuvius, being situated in a densely populated area. However, several major aspects of the volcanic history of Vesuvius remain questionable or unresolved even for events occurring during the last few hundred years. In particular, the 'big eruption' of AD 1631 has been regarded by Le Hon (1865) and Johnston Lavis (1884) as a major explosive and effusive event and this has been supported by more recent studies by Burri & Di Girolamo (1975) and Rolandi & Russo (1991). In contrast, the same event is regarded by Rosi & Santacroce (1986) and Arno' *et al.* (1987) as only an explosive event with virtually no associated lavas, and they suggested that the lava flows ascribed to this eruption by other authors were actually much older and probably erupted within the AD 968–1037 time interval. Other lavas, such as those outcropping in the Villa Inglese quarry and some nearby areas (Fig. 1) were also ascribed to the same time interval (Arno *et al.* 1987). It is therefore important to date these different lavas in order to assess the nature of some of the past volcanic events associated with Vesuvius. On this basis, in 1990 a large scale magnetostratigraphic survey of Vesuvius was initiated in order to assess whether the problematic flows had acquired their magnetic remanences at the same or at different times and, where possible, to assign a magnetic age to the flows. Such techniques are long established (e.g. David 1904; Chevallier 1925; Hoye 1981) and depend

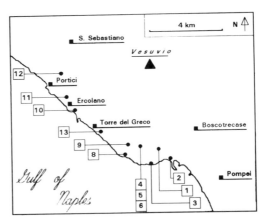

Fig. 1. Site locations on Mount Vesuvius. The lava sites sampled are shown as dots, with the corresponding site number in a square. Sites 1, 2, 8, 9, 10, 11, 12 are lava flows with emplacement ages which have recently been questioned (see text); other sites 3, 4, 5, 6, 13 refer to lava flows emplaced in unspecified Medieval times.

on observation of natural remanent magnetization (NRM). The NRM of a lava unit is the sum of partial magnetizations acquired over different cooling intervals (Thellier 1951), enabling the isolation of components of remanence that can be attributed to the direction of the geomagnetic field as the lavas cooled. More recent studies used principal component analyses (Kirschvink 1980) of vectors determined on specimens demagnetized up to 50 mT only, in order to determine the directions of the remanent vectors (Carracedo *et al.* 1993). This article shows that

From Morris, A. & Tarling, D. H. (eds), 1996, *Palaeomagnetism and Tectonics of the Mediterranean Region*, Geological Society Special Publication No. 105, pp. 367–371.

367

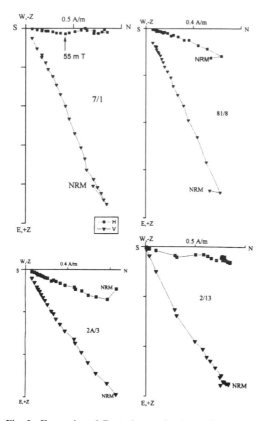

Fig. 2. Examples of Cartesian projections of separate lavas. The squares represent the horizontal component (N v. E) and the triangles refer to the vertical component (H v. Z). The top two examples illustrate the behaviour during AF demagnetization of samples from the putative AD 1631 eruption (sites 1 and 8) (after Gialanella *et al.* 1993). The arrow on the left diagram indicates the 55 mT value which separates the two components (see text). The lower diagrams illustrate results from two Medieval flows, with AF treatment on the left and thermal treatment on the right (sites 3 and 13 respectively) (after Gialanella *et al.* 1995).

magnetostratigraphic surveys in volcanic areas can only be successful when the remanent vectors are analysed using more stringent criteria (Gialanella *et al.* 1993, 1994; Sarno 1994) and more complete demagnetization than were used in the study by Carracedo *et al.* (1993). These new evaluations provide more realistic magnetic dating for the nature and age of some of the fifteenth century and Medieval activity of Vesuvius.

Sampling, treatment and analyses

Specimens were directly cored in the field, using a portable electric drill, and oriented with a solar

compass prior to removal. The cores were later sliced into standard specimen cores (2.5 cm diameter. 2.1 cm high). At each sampling site, 12 cores were collected (except site 1 (9 specimens); site 12 (22); and site 3 (19)). Initially six specimens from each site were subjected to either progressive alternating field demagnetization (PAFD) up to 100 mT, with a Molspin AF demagnetizer, or progressive thermal demagnetization (PTD) up to 600°C, with a Schonstedt thermal demagnetizer. In both cases, the remanences were measured using a Molspin magnetometer. All treatment and analyses were undertaken in the Laboratorio di Paleomagnetismo of the Dipartimento di Scienze della Terra, Universita' degli Studi di Napoli 'Federico II'.

Lavas attributed to the AD 1631 eruption

Attention was first focused (Gialanella *et al.* 1993) on the lava deposits attributed to the 1631 emplacement date, whose age has recently been questioned (Carracedo *et al.* 1993). Seven sites (sites 1, 2, 8, 9, 10, 11 and 12; Fig. 1) were sampled, including some sites previously reported by Carracedo *et al.* (1993). At a site level specimens could exhibit either single and multi-component magnetization (Fig. 2) and these components were isolated over different coercivity/blocking temperature ranges, even for specimens from the same site (Nigro 1992; Sarno 1994). It is thus evident that very careful, stringent analyses are required to assess the most meaningful components of remanence that can then be combined to determine the mean direction of the geomagnetic field at the time of emplacement of the lavas. This can be exemplified by specimen 7/1 (Fig. 2). Routine PCA, which considered remanent vectors to be acceptable if the diagonal angle (da) was less than 10°, only isolated one component for the entire coercivity spectrum with a direction $D = 1.3°, I = 70.9°$ (da = 3.1°). Visual inspection of the horizontal component in particular shows that at least two components can be inferred within the medium-high coercivity range. PCA carried out above 55 mT and between 35 mT and 55 mT identifies the following well-defined components: $D = 12.0°, I = 66.6°$ (da = 2.5°, range 55 mT to 100 mT) and $D = 355.7°, I = 71.4°$ (da = 2.5°, range 35 mT to 55 mT). Both components isolated in this way differ from that defined by the less stringent PCA corresponding essentially to the vector sum of the two components subsequently isolated. Assuming that the higher coercivity component is most likely to correspond to the primary magnetization, i.e. the component acquired when the lava was

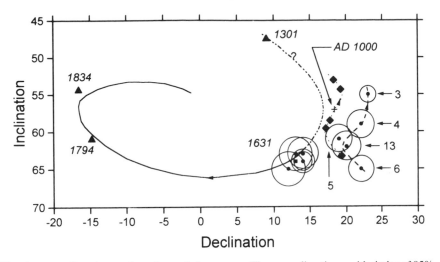

Fig. 3. The site mean directions and secular variation curves. The mean directions, with circles of 95% confidence, for the putative AD 1631 flows are unnumbered, being essentially all in the same direction (the square refers to their mean direction), while the Medieval sites are numbered. The secular variation curve for Vesuvius is shown as a solid line, but dashed for the Medieval period for which it is speculative (Hoye 1981). The secular variation curve for Etna during the period around AD 1000 (Tanguy *et al.* 1985), after transposition to Vesuvius (see text), is shown as a dotted curve. (Gialanella *et al.* 1995).

Table 1. *The site mean directions for Vesuvius lavas.*

Site	N_c/N_m	D	I	α_{95}	k
Putative 1631 Lavas					
1	(05/09)	11.8	65.1	2.2	1100
2	(06/08)	14.2	63.7	1.6	1675
8	(06/09)	13.7	64.1	1.2	3120
10	(06/09)	12.8	63.2	2.2	961
12	(09/11)	14.3	62.9	1.9	585
Mean (5 sites)		13.4	63.8	0.9	7703
Villa Inglese Lavas					
3	(06/08)	22.8	55.1	1.2	3037
4	(07/12)	22.1	58.8	1.9	1049
5	(05/19)	19.3	61.3	2.2	1063
6	(06/12)	22.2	64.8	1.8	1359
13	(06/06)	20.0	62.3	2.0	1098

For site location refer to Fig. 1. N_c/N_m = number of specimens from which the mean direction at site level has been computed over number of specimens measured for that site (see text). D = declination; I = inclination; α_{95} and k = statistical parameters as defined by Fisher (1953).

originally emplaced, then this component can be used to define the site mean direction corresponding to the geomagnetic field direction at the time that the lavas erupted and cooled.

The rapid acquisition of isothermal remanent magnetization (IRM) suggested that the main magnetic carriers consist of magnetite and that there was no evidence for the presence of

hematite or ilmenohematites (Gialanella *et al.* 1993; Sarno 1994). On this basis, it was considered that the high-temperature component could be attributed to the direction of remanence acquired during the original cooling. As the direction of this component was identical to the high coercivity component isolated in specimens from the same site, it was considered that the high coercivity component could similarly be related directly to the direction of the geomagnetic field at the time the lava originally cooled. The vector diagrams for the demagnetization, thermal or alternating field, for specimens from seven sites that may have been erupted in AD 1631 were then examined using the same visual inspection and rigorous application of stringent criteria for defining the specimen vectors. This resulted in the rejection of most of the specimens from sites 9 and 11 (Gialanella *et al.* 1993), therefore, no site mean directions for these sites are listed in Table 1. Fourteen specimens from the remaining sites were also rejected (Gialanella *et al.* 1993). The high coercivity/blocking temperature components of the remaining specimens were then combined to determine the site mean directions (Table 1 and Fig. 3). The directions for sites 1, 2, 8, 10 and 12 were statistically identical to each other, clearly indicating that they had acquired their remanence when the geomagnetic field was in an identical direction, i.e. they are coeval.

The Medieval sites

The magnetic vectors (Fig. 2) for specimens from five flows of uncertain Medieval age (sites 3, 4, 5, 6 & 13; Fig. 1) were analysed in an identical way to those for the putative AD 1631 flows described above. These sites included three successive lavas (4, 5 & 6 from top to bottom) within the Villa Inglese quarry. In these flows, almost half (27) of the 56 specimens had to be rejected but this resulted in site mean directions (Gialanella *et al.* 1994) that were defined by an α_{95} less than or equal to 2.2° (Table 1). All three overlying lavas had different directions, although one of them (site 5) had a statistically identical direction (at a 95% probability level) to that of site 13. Thus the flows at sites 5 and 13 may be coeval, but all others are associated with geomagnetic field directions associated with different times.

Comparison with other data

The site mean directions for the putative AD 1631 events (sites 1, 2, 8, 10 and 12) lie well away from those for the other Medieval sites (3, 4, 5, 6 and 13; Fig. 3). Stratigraphic data clearly support an AD 1631 age for site 9 which rests above AD 1631 pyroclastics (Rolandi & Russo 1989) and underlies a building which has been historically dated to AD 1698 (Rolandi & Russo 1987, 1989, 1993). Permission for this building was granted in order to encourage resumption of activities in an area severely affected by the AD 1631 eruption (Rolandi & Russo 1989). Similarly, sites 10 and 8 are now known to be the same flow (Rolandi & Russo 1987, 1989, 1993). Although some of the earliest palaeomagnetic measurements were undertaken on volcanic rocks from Vesuvius (Melloni 1853; David 1904), the observations for this century were summarized by Hoye (1981) and subsequent published studies have been largely concentrated on Etna (Tanguy *et al.* 1985). Hoye (1981) provided a secular variation curve for the last 700 years based on these observations on Vesuvius (Fig. 3), although the curve was necessarily speculative for the AD 1301–1631 period. In fact the AD 1631 eruption occurred after 500 years of quiescence and the AD 1301 direction of the geomagnetic field (Fig. 3) was derived from analysis of lavas from Ischia Island, in the Gulf of Naples. The direction reported by Hoye (1981) for the AD 1631 eruption (site unknown) was $D = 14.9°$, $I = +64.7°$ ($\alpha_{95} = 1.7°$), which is identical to that found for the five flows reported here, $D = 13.4°$, $I = +63.8°$ ($\alpha_{95} = 0.9°$). Assuming the regional geomagnetic field to be that of an inclined geocentric dipole (Tarling 1983; Noel & Batt 1990), it is possible to calculate the probable direction of the geomagnetic field at Vesuvius at about this time using the data for an AD 1646 flow on Etna (Angelino, work in progress). This gives a direction of $D = 5.6°$, $I = +63.3°$ ($\alpha_{95} = 1.9°$), which is slightly west of the Vesuvius value, probably reflecting the 15 year older age of the Etna observation. In conclusion, the consistency of all directions from these five sites confirms that these are correctly assigned to the AD 1631 'big eruption'. As stratigraphical data indicate that sites 9 and 11 are part of the flow of sites 8 and 10, respectively, it can be concluded also that sites 9 and 11 can be correctly assigned to the AD 1631 event. This volcanic event clearly did include significant effusive stages, in addition to the explosive events.

The other Medieval flows (sites 3, 4, 5, 6 and 13) are less well constrained than the AD 1631 flow (site 9). Site 3 4, 5 and 6 are in stratigraphic order (oldest to youngest), thereby suggesting that the secular variation curve remains easterly but was steepening during the time that these flows were erupted (Fig. 3) and suggesting that site 3 is likely to be the oldest. However, this flow must be older than the eleventh to twelfth century coastal watch tower built on it during the Angevin kingdom (Rolandi & Russo 1987, 1989). Comparison with the Etna data for this period (Fig. 3), corrected in the same way as the AD 1646 flow direction, shows that the new Vesuvius data lie some 5° east of the speculative secular variation path of Hoye (1981), whereas the extrapolated Etna directions (AD 850–1200 approximately; Tanguy 1981; Tanguy *et al.* 1985) lie between the speculative and observed curves. However, the same westerly motion is indicated for the youngest parts of both the Etna and Vesuvius curves, suggesting that the Medieval flows sampled on Vesuvius do belong to the nineth to eleventh centuries (Gialanella *et al.* 1995), with the flows at sites 5 and 13 being possibly of identical age. It is therefore demonstrated that such studies must be undertaken rigorously, using complete, fine-scale demagnetization spectra and strict component analyses together with careful visual inspection of the vector diagrams.

Palaeomagnetic investigations in progress on some Etna Medieval lavas (Angelino, work in progress), using the same stringent criteria as in the present study, have produced a SV curve which when relocated to Vesuvius (in the same way as the AD 1646 Etna data) overlaps almost entirely with the coeval Vesuvius SV curve of the present study.

The author thanks the editors for improvements to the manuscript. This work was supported through grants of GNV (Gruppo Nazionale per la Vulcanologia).

References

ARNO', V., PRINCIPE, C., ROSI, M., SANTACROCE, R., SBRANA, A. & SHERIDAN, M. F. 1987. Eruptive history. In: SANTACROCE, R. (ed.) Somma -Vesuvius. Quaderni Ric. Sc., CNR Roma, 114, 53–103.

BURRI, C. & DI GIROLAMO, P. 1975. Contributo alla conoscenza delle lave della grande eruzione del Vesuvio del 1631. Rendiconti della Società Itani Minerologia Petrologia, 30, 705–740.

CARRACEDO, J. C., PRINCIPE, C., ROSI, M. & SOLER, V. 1993. Time correlations by palaeomagnetism of the 1631 eruption of Mount Vesuvius. Volcanological and volcanic hazard implications. In: DE VIVO, B., SCANDONE, R. & TRIGILA, R. (eds) Mount Vesuvius. Journal of Volcanology and Geothermal Research, 58, 203–209.

CHEVALLIER, R. 1925. L'Aimantation des laves de l'Etna et l'orientation du champ terrestre en Sicile du XIIᵉ au XVIIᵉ siècle. Annales Physicae, 4, 5–162.

DAVID, P. 1904. Sur la stabilité de la direction d'aimantation dans quelques roches volcaniques. Comptes Rendus de l'Acadamie des Science, Paris, 138, 41–42.

FISHER, R. A. 1953. Dispersion on a sphere. Proceedings of the Royal Society, A217, 295–305.

GIALANELLA, P. R., INCORONATO, A., RUSSO, F. & NIGRO, G. 1993. Magnetic stratigraphy of Vesuvius products. I – 1631 lavas. In: DE VIVO, B., SCANDONE, R. & TRIGILA, R. (eds) Mount Vesuvius. Journal of Volcanology and Geothermal Research, 58, 211–215.

——, ——, ——, SARNO, P. & DI MARTINO, A. 1995. Magnetic stratigraphy of Vesuvius products. II – Medieval lavas. Quaternary International, in press.

HOYE, G. S. 1981. Archaeomagnetic secular variation record of Mount Vesuvius. Nature, 291, 216–218.

JOHNSTON LAVIS, H. J. 1884. The geology of Mt. Somma and Vesuvius: being a study of volcanology. Quarterly Journal of the Geological Society of London, 40, 135–139.

KIRSCHVINK, J. L. 1980. The least-squares line and plane and the analysis of palaeomagnetic data. Geophysical Journal of the Royal Astronomical Society, 62, 699–718.

LE HON, M. 1865. Histoire complete de la grande eruption du Vesuve de 1631. Bulletin de l'Acadamie des Sciences, Lettres et Beaux Arts, Belge, 20, 483–538.

MELLONI, M. 1853. Ricerche intorno al magnetismo della rocce. Memorie della Reale Accamia di Napoli, 1, 121–164.

NIGRO, G. 1992. Studio paleomagnetico dei prodotti lavici vesuviani del 1631. Laurea Thesis, Universita' degli Studi di Napoli 'Federico II'.

NOEL, M. & BATT, C. M. 1990. A method for correcting geographically separated remanence directions for the purpose of archeomagnetic dating. Geophysical Journal International, 102, 753–756.

ROLANDI, G. & RUSSO, F. 1987. Contributo alla conoscenza dell'attivita' storica del Vesuvio. La stratigrafia di Villa Inglese (Torre del Greco). Rendiconti della Accademaia delle Scienze Fisica e Matematiche, Napoli, 54, 123–157.

—— & —— 1989. Contributo alla conoscenza dell'attivita' storica del Vesuvio: dati stratigrafici e vulcanologici nel settore meridionale tra Torre del Greco, localita' Villa Inglese, e Torre Annunziata. Bollettino della Società Geologica Italiana, 108, 521–536.

—— & —— 1993. L'eruzione del Vesuvio del 1631. Bollettino della Società Geologica Italiana, 112, 315–332.

ROSI, M. & SANTACROCE, R. 1986. L'attivita' del Somma-Vesuvio precedente l'eruzione del 1631. Dati stratigrafici e vulcanologici. In: ALBORE LIVADIE, C. (ed.) Tremblement de terre, eruptions volcaniques et vie des hommes dans le Campanie antique. Biblioteque d'Institute Français Naples, 7, 15–33.

SARNO, P. 1994. Studio paleomagnetico e stratigrafico di lave vesuviane (post 472–1794 AD). Laurea Thesis, Universita' degli Studi di Napoli 'Federico II'.

TANGUY, J. C. 1981. Les éruption historiques de l'Etna: cronolgie et localisation. Bulletin Volcanologique, 44, 585–640.

——, BUCHER, I. & THOMPSON, F. C. 1985. Geomagnetic secular variation in Sicily and revised ages of historic lavas from Mount Etna. Nature, 318, 453–455.

TARLING, D. H. 1983. Palaeomagnetism: principles and applications in geology, geophysics and archaeology. Chapman & Hall, London.

THELLIER, E. 1951. Proprietes magnetiques des terres cuites et des roches. Journal de Physic et Radium, 12, 205–218.

Archaeomagnetic results from the Mediterranean region: an overview

M. E. EVANS

Institute of Geophysics, Meteorology, and Space Physics, University of Alberta, Edmonton, Canada T6G 2J1

Abstract: Palaeomagnetic methods were first applied to archaeological materials and very young geological formations in the Mediterranean region a century and a half ago. Although the pioneering work established the general validity of the method, the data remained sparse and permitted only limited conclusions. By the mid-1920s, thorough investigations of the lava flows of Mt. Etna demonstrated how geomagnetic secular variation could be traced back into pre-observatory epochs. In the following decade, systematic study of archaeological features commenced, but it was not until the 1960s that the database had grown to the point where useful geomagnetic inferences could be drawn. For the Mediterranean region, there are currently available two extensive data sets (one for France, the other for Bulgaria) containing a total of *c.* 400 directional results. For Italy and Greece, a useful start has been made (*c.* 100 results); for other countries (Morocco, Tunisia, Turkey) only a handful of results have been published. Collectively these data enable us to trace back the directional behaviour of the geomagnetic field with reasonable confidence for the last 2000 years, but with less certainty for earlier times. Nevertheless, a picture is emerging in which static flux bundles and flux spots undergoing zonal retrograde drift can be recognized. Since the determination of geomagnetic field strength does not require *in situ* material, the investigator is freed of the need for accurately oriented samples and the scope of archaeomagnetic research is therefore enlarged. Many results have been published, but their status remains questionable due to the on-going debate concerning the validity of the different experimental procedures. Intensity and directional results have potential for archaeological dating both in an absolute and in a relative sense. Some authors claim the possibility of achieving a precision approaching ±50 years, but this depends strongly on the region and time interval in question. One example of particular interest is the application of archaeomagnetic data to the demise of the Minoan civilization.

Significant investigations relevant to Mediterranean archaeomagnetism and palaeomagnetism were first reported by Macedonia Melloni (1853), who studied samples of lava flows from Vesuvius and the Phlegrian Fields. For this purpose he constructed a primitive astatic magnetometer. This *magnetoscope*, as he called it, pre-dates by a century Blackett's seminal instrument. Melloni also recognized the importance of thermoremanence in generating strong and stable magnetization in rocks. This promising start was not followed up until the work of Folgerhaiter (1894) on volcanic rocks in Latium, and subsequently on archaeological materials (1899). For the most part these were ceramic vases of various types: Etruscan, Corinthian, Attic, and Pompeiian, representing almost a millenium (*c.* 800 BC to AD 79). Assuming they were fired in an upright position, Folgerhaiter argued that the inclination of the geomagnetic field operating at the time of fabrication could be deduced. He summarized his findings in a graph of inclination vs time, from which he concluded that in the sixth century BC the magnetic equator passed through central Italy. Prior to that '...l'extrémité la plus basse d'une aiguille magnétique aurait été non le pôle nord, comme à présent, mais le pôle sud'; in modern terminology this would be called a polarity reversal. Of course, subsequent work has shown this to be incorrect; the most likely explanation being the invalidity of the assumption concerning the orientation of the vases during firing. However, the boldness of the idea generated enough debate to prompt further fruitful investigations, as will be seen shortly. Aware of the limitations of knowing inclination only, Folgerhaiter made a plea for future work to include measurements of declination and intensity. For this he drew attention to the promise of *in situ* kilns and buildings exposed to fire, and also proposed a palaeointensity method involving laboratory heating in a known field. For these inspired ideas, and for his ground-

From Morris, A. & Tarling, D. H. (eds), 1996, *Palaeomagnetism and Tectonics of the Mediterranean Region*, Geological Society Special Publication No. 105, pp. 373–384.

373

374   M. E. EVANS

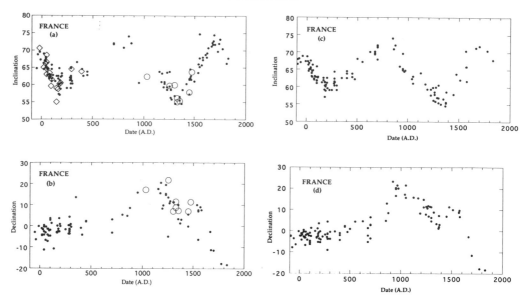

Fig. 1. (**a**) Inclination data for France spanning the last two millenia. Dots from Thellier (1981), circles from Bucur (1986), diamonds from Langouet *et al.* (1983). All values have been reduced to the latitude of Paris. (**b**) Corresponding declination data. There are no results from Langouet *et al.* (1983) because they studied only displaced objects. (**c**) Updated version of (a) derived from Bucur (1994). (**d**) Updated version of (b) derived from Bucur (1994).

breaking experimental work, Giuseppe Folgerhaiter must surely be regarded as the father
of archaeomagnetism.

Early this century, Bernard Brunhes and
Pierre David began their well-known studies in
France concerning lavas and their underlying
baked sediments. In his summary of their work,
Brunhes (1906) pays great attention to Folgerhaiter's findings, particularly the suggestion
that the older vases were made during an
interval of reversed polarity. He admitted that
this result was certainly surprising, and that
experts in terrestrial magnetism, notably
Carlheim-Gyllenskjold at the Stockholm Observatory, did not accept it. However, the
French investigations provided undeniable evidence of reversed polarity during the Miocene,
so that eventually the opponents were won over
to the reality of geomagnetic reversals, at least
in the geological past if not during recorded
history. To make sure, Brunhes took the
trouble to send a suitable sample to Carlheim-
Gyllenskjold, and makes a point of reporting
his conversion to 'notre manière de voir'.

Brunhes goes on to report the archaeomagnetic study made by his colleague David of
igneous paving slabs from the Temple of Mercury built by the Romans at the summit of the
Puy-de-Dôme in the first century BC. Apparently, these enormous slabs (typically 1.5 × 1.0

× 0.4 m) were quarried horizontally and thus
retain the correct inclination values (seven
samples yield values between 52° and 58°),
although some of them seem to have been
placed in position upside-down. These early
workers were aware of viscous overprinting by
the present field, and Brunhes therefore
stresses that the retention of the original remanence by the inverted slabs is strong evidence
for magnetic stability, at least for two millenia.
Here again, however, one notes the prescience
of Folgerhaiter (1899) who made the same
point in connection with fragments of pottery
buried in random orientations: the archaeological equivalent of the well-known 'conglomerate
test'. Brunhes describes how David wished to
locate the source of the paving slabs, since their
composition is distinctly different from the
rocks comprising the Puy-de-Dôme itself. To
this end, David determined the magnetization
of likely material from ancient quarries about
4 km away. Two samples had inclination values
in the range obtained from the paving slabs.
This is an early example of sourcing, paving the
way, as it were, for later attempts to determine
provenance of specific materials by archaeomagnetic means.

During the inter-war years, Mediterranean
archaeomagnetism was advanced by the initiation of systematic studies of young geological

formations (Chevallier 1925) and of archaeo-logical features (Thellier 1938). Chevallier's work concerns historic lava flows of Mt Etna which he attempted to use for defining a secular variation curve spanning the last seven centuries. Unfortunately, subsequent scrutiny has cast considerable doubt on the dates assigned to the various flows. So, although Chevallier's work stands as a classic early application of archeomagnetism to geophysical problems, the actual data involved have been completely superceded. In the case of Thellier's investigations of archaeological material, several decades were to pass before they came to fruition and a comprehensive discussion appeared. We turn now, therefore, to a summary of the main results, both geological and archaeological, as they stand at the present time.

Archaeodirections for the last two millenia

France

In large measure, the current state of archaeomagnetism derives from the work of Emile Thellier in Paris, whose *magnum opus* (Thellier 1981) provides a comprehensive summary of almost half a century of diligent work in France. For the reader used to grappling with geological material, it is perhaps worth pointing out that familiar problems, such as weakly magnetized samples, polyphase remanence, and tectonic complexities, rarely arise in archaeomagnetism. This generally leads to high precisions. Thellier's data, for example, have an average precision parameter (k) in excess of 1000 and corresponding α_{95} values almost always <3°. Consider first the inclination data. Thellier states that there are 142 data from *in situ* structures, although his Table IV contains only 137. Of these, 16 have no reported age control. In order to retain only the most reliable data, results were omitted whose reported date is constrained no better than a hundred years, as well as those based on less than five samples, and a single result for which α_{95} exceeds 5°. In addition to the results from *in situ* structures, there are 32 results from displaced objects (Thellier's Table III), all of which are well-dated and of high precision. To Thellier's data can be added the results of Langouet *et al* (1983) for the Gallo-Roman period ($N = 16$), and those of Bucur (1986) for the Middle Ages ($N = 8$). The inclination magnetogram for the last two millenia thus consists of 148 points (Fig. 1a). Coverage is adequate for the first half of the first millenium and from the twelfth century onwards, but there is a serious paucity of results for the intervening six hundred years for which

cultural remains are more elusive. Nevertheless, a clear pattern is observed; values range between 55° and 75° (mean = $63.5 \pm 0.4°$), with prominent minima in the second and fourteenth centuries. The maximum in the early eighteenth century is confirmed by the observatory record at Paris. Currently, the inclination in Paris is *c.* 64°, whereas a purely axial dipole field would yield a value of 66.5°. Declination can only be obtained from *in situ* structures. The values observed range from 20°E to 20°W, with a mean close to zero (Fig. 1b). During the first five centuries AD, the inclination changes noted above are accompanied by very little discernable declination change, whereas between the twelfth and nineteenth centuries there has been a more-or-less steady swing to westerly directions culminating in the historical extremum of 22°W recorded about 1820. Currently, the declination in Paris is *c.* 2°W and *decreasing* by *c.* 1.5° per decade.

Shortly after presenting this paper at the London meeting (and concurrently submitting the written version requested by the organizers) Ileana Bucur provided a reprint of her excellent new summary of the French archaeomagnetic data (Bucur 1994). It appears that some 90 new structures have been investigated since the publication of Thellier's (1981) paper. She includes in her compilation only *in situ* structures for which both declination and inclination are available. This provides an initial total of 184 results, but this drops to 119 when structures of 'uncertain or unknown archaeological date' ($N = 51$) and those yielding 'abnormal values' ($N = 14$) are excluded. The surviving results are all of high quality, with α_{95} never exceeding 2.8°. These data are presented here in Fig.1(c, d), re-plotted to facilitate comparison with Fig.1(a, b). The great similarity between the two compilations comes as no surprise since the two data sets are by no means independent. However, it is clear that there is now improved coverage in the first century BC and in the interval AD 500–1200. As Bucur herself points out, the secular variation curve for France for the last 21 centuries now 'seems fairly complete' and offers, under some circumstances, a dating tool with 'a precision of the order of a quarter century'.

Italy

One of the most important studies reported in the early decades of this century was that of Raymond Chevallier (1925), involving the lava flows of Mt Etna. This has now been brought completely up-to-date by Tanguy (1970, 1980). The most vexing problem has always been the

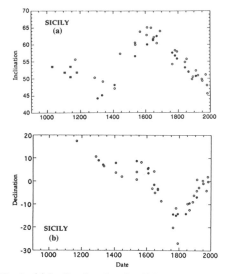

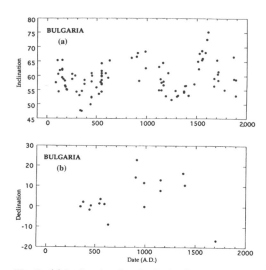

Fig. 3. (a) Inclination data for Bulgaria reported by Kovacheva (1992). (b) Corresponding declination results.

Fig. 2. (a) Inclination data for Sicily. Solid symbols from Tanguy *et al.* (1985), dots representing Mt Etna lava flows, squares representing bricks from eleventh and twelfth century Norman churches. Open circles are from the Mt Etna lava flows studied by Rolph & Shaw (1986). (b) Corresponding declination data. Tanguy *et al.* (1985) report no declination results from the Norman churches.

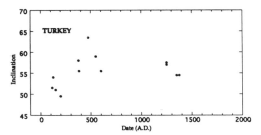

Fig. 4. Inclination results from Ephesus (Turkey) reported by Bammer (1964).

validation of the supposed historic dates of many of the flows. In the most recent summary (Tanguy *et al.* 1985), 12 units are regarded as reasonably well-dated, whereas 13 are of dubious age. Nevertheless, the results can be used to extend the record back beyond the earliest observatory recordings, to AD 1329. If the important result from the island of Ischia (Bay of Naples) is included, this can be pushed back to AD 1301. In addition, four earlier points are provided by inclination data from bricks of eleventh and twelfth century Norman buildings in Sicily (Fig. 2). Rolph & Shaw (1986) have independently investigated Etnean lava flows, and their results are included in Fig. 2. Minor differences between the two sets of results can be noted, but on the whole the agreement is satisfactory.

Declination data for Sicilian lavas are also summarized in Fig. 2, there being no corresponding values for the Norman churches. Again, the two sets of results, obtained quite independently by two separate laboratories, are in good agreement. It is noted that the late eighteenth century point reported by Rolph & Shaw seems to be rather too far to the West. Nevertheless, the archaeomagnetic results clearly reflect a westerly extremum at about 1800

which compares favourably with the historical observatory records.

Etna is undoubtedly the most thoroughly studied volcano in the Mediterranean region, but magnetic results have also been reported for Vesuvius (Hoye 1981). These involve 11 flows dated between 1631 and 1944 whose archaeomagnetic directions after cleaning are very well defined (mean α_{95} = 1.9°). They describe a clockwise elliptical loop consistent with the well-known European observatory data, but their limited age span does not provide any useful extention into pre-observatory times.

Bulgaria

Kovacheva (1992) summarizes her own extensive data for Bulgaria covering the last two millenia. Applying the same acceptance criteria used above for the French data, leads to an

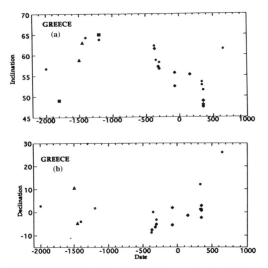

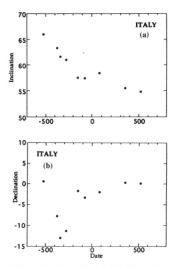

Fig. 5. (a) Inclination data for Greece (reduced to the latitude of Athens). Dots are from Belshé *et al.* (1963), diamonds are from Evans & Mareschal (1988) and Evans (1994), triangles are from Downey & Tarling (1984). Squares represent Becker's results from Turkey (cited by Jones 1986). The earlier of the two results of Downey & Tarling (1984) is that from Akrotiri on the island of Santorini, the other represents the mean of the well-grouped directions obtained from Minoan palace sites in central Crete. (b) Corresponding declination data for Greece.

Fig. 6. (a) Inclination data for Italy (reduced to latitude 40°N using the dipole assumption) (Evans & Mareschal (1989). (b) Corresponding declination data.

inclination magnetogram comprising 93 points, but only 17 declination values survive (Fig. 3). The inclination curve is broadly similar to that for France. Declinations, on the other hand, are too few to permit firm conclusions, although the general shape outlined agrees with the French data. The apparently rapid westerly shift implied by the isolated point in the seventh century needs to be verified by further work before it can be regarded as real.

Turkey

Inclination results from 14 sets of bricks and tiles from Ephesus ranging in age from the second to the fourteenth centuries have been reported by Bammer (1964). One of them is poorly constrained in time; the others are illustrated in Fig. 4. If the latitude gradient is allowed for, the observations are compatible with the Bulgarian data, and suggest that further studies would be worthwhile.

Other archaeodirectional results

Six sites in Morocco (ranging in age from the third century BC to the fifth century AD) have

been sampled by Kovacheva (1984), but only four of them were studied in sufficient detail to permit the calculation of statistical error limits. This hardly suffices to define the secular variation, but these few data do extend the geographical coverage of Mediterranean archaeomagnetism. For completeness, they are listed here (Dchar Jdid, third–second centuries BC, $D = 1.8°$, $I = 53.0°$, $\alpha_{95} = 5.1°$, $N = 10$; Al Kouass, first century BC, $D = 353.9°$, $I = 54.3°$, $\alpha_{95} = 3.3°$, $N = 5$; Volubilis (large furnace), third century AD, $D = 352.2°$, $I = 50.3°$, $\alpha_{95} = 3.6°$, $N = 10$; Volubilis (south furnace), fifth century AD, $D = 353.3°$, $I = 51.7°$, $\alpha_{95} = 2.5°$, $N = 5$).

Thellier (1981) reports three results from Carthage. Two are from Punic kilns which ceased operation at the time of the Roman invasion (146 BC), the third is Roman (older than AD 300). The two Punic kilns yield very similar results (no. 1, $D = 2°W$, $I = 57°$; no. 2, $D = 2°W$, $I = 58°$); the Roman kiln has a distinctly shallower magnetic vector ($D = 1°W$, $I = 51°$).

A preliminary archaeomagnetic survey of a variety of structures in Greece was published many years ago by Belshé *et al.* (1963). They investigated 31 features (mostly kilns and burnt walls) ranging in age from 2000 BC to AD 700. The results were rather heterogeneous and they therefore attempted a classification (partly subjective) aimed at eliminating inferior data. Only those classified as 'good' or 'fair' were considered acceptable; there are 20 such results, but 10 of them have α_{95} values >5°. The 10

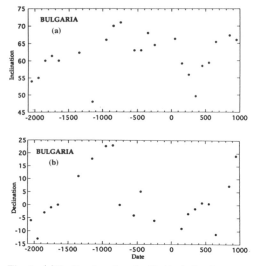

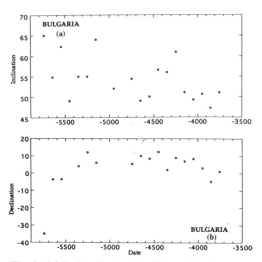

Fig. 7. (a) Inclination data for Bulgaria for the interval 2000 BC to AD 1000 (Kovacheva 1980). (b) Corresponding declination data.

Fig. 8. (a) Inclination data for Bulgaria for the interval 5800 BC to 3800 BC (Kovacheva 1980). (b) Corresponding declination data.

remaining points are shown on Fig. 5, which includes results reported by Downey & Tarling (1984), Evans & Mareschal (1988), and Evans (1994). There are still too few points to allow firm conclusions, but the low inclination values in the first few centuries AD are consistent with the French and Bulgarian results (Figs 1 and 3).

Evans & Mareschal (1989) report results from southern Italy spanning the interval from the sixth century BC to the sixth century AD, illustrated here in Fig. 6. The 11° decrease in inclination matches that found in Greece, but the observed westerly declinations are more pronounced in the Italian data.

In an earlier report, Kovacheva (1980) presented Bulgarian archaeomagnetic results (in the form of century means) reaching back as far as the seventh millenium BC. Those falling in the interval 2100 BC to AD 1000 are illustrated in Fig. 7 for comparison with the corresponding interval for Greece (Fig. 5). Agreement is reasonable, except for the Bulgarian inclination result for the twelfth century BC. Rapid changes like this cannot be ruled out, but since this point happens to fall in a 400-year gap it needs to be confirmed before it can be accepted as a genuine record of geomagnetic behaviour.

For dates prior to 2000 BC the only results available are those of Kovacheva (1980) for Bulgaria. This remarkable sequence extends back to 6300 BC, although there are only two century-mean values between the twenty-first and thirty-eighth centuries BC. There is also a

four-century gap prior to the fifty-eighth century BC, which rather isolates the few earliest points. Bearing this in mind, the results spanning the two millenia preceding the thirty-eighth century BC have been plotted (Fig. 8). Inclinations range between 65° and 47°, but the mean (54°) is significantly lower than the current value at Sofia (60°). Declinations are also systematically displaced, averaging about 8°E throughout almost the entire interval.

Archaeointensity results

Folgerhaiter's original suggestion (1899) that laboratory reheating in a known field could be used to determine the intensity of the ancient field was first seriously taken up by Thellier in the 1930s. Subsequently, Thellier & Thellier (1959) summarized their early work, based on results from France, Switzerland, and Tunisia. These already indicated that the magnetic field in the Mediterranean area had decreased by 50% over the last two and a half millenia. Within a decade, Cox (1968) was able to show that this extremely important observation applied to the entire globe, and was consistent with the known dipole decrease found from spherical harmonic analyses since the first determinations made by C. F. Gauss in the 1830s. More recent summaries (McElhinny & Senanayake 1982) confirm this finding. In the decades since the Thelliers' early work, a great deal of effort has been expended in determining archaeointensities, much of it involving sites in the Mediterranean

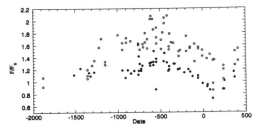

Fig. 9. Archaeointensity results reported by Walton (1979, 1990). Open circles represent 1979 results, dots represent 1990 results. All values are given in terms of the ratio of the ancient field strength (F) to the present-day value at the site (F_0).

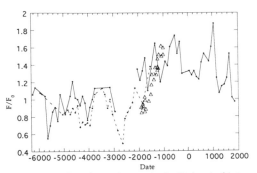

Fig. 10. Archaeointensity results for Bulgaria (dots and solid line; Kovacheva 1980), Greece (dots and dashed line; Thomas 1981), Egypt and western Asia (open triangles, Aitken *et al.* 1984).

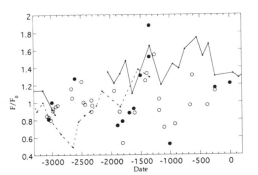

Fig. 11. Archaeointensity results from Egypt (Games 1980). Solid dots represent Thellier-type experiments, open circles represent non-thermal SRM data. Corresponding results from Bulgaria and Greece are shown for comparison (see Fig. 10).

area (Aitken *et al.* 1983, 1986; Games 1980; Kovacheva 1980, 1983; Thomas 1981, 1983; Walton 1979, 1984, 1987). The experimental difficulties encountered have been the source of much debate, summarized by Aitken *et al.* (1988) and Walton (1988). Some of the earlier data implied rapid fluctuations, exceeding $200\,nT\,a^{-1}$ in Greece (Walton 1979), but these were significantly reduced by improved experimental technique (Walton 1990). Furthermore, the newer results are systematically lower than the older ones (Fig. 9). During the first millenium BC, for example, the ancient field in Athens did not often exceed 1.3 times the present-day value (Walton 1990), whereas the older results imply that this ratio was rarely less than 1.5 (Walton 1979). Evidently, great caution is called for when assessing published archaeointensities. Nevertheless, certain broad features do emerge, and it would be hasty to discard all the older data.

Results for Bulgaria, Greece, Egypt and western Asia (Kovacheva 1980; Thomas 1981; Aitken *et al.* 1984) are summarized in Fig. 10. The Bulgarian samples indicate that the field was generally stronger after 2000 BC than before, and the Greek results are consistent with this. The data from Egypt and western Asia imply a relatively rapid rise broadly compatible with other contemporaneous data. A least squares line through the data of Aitken *et al.* (triangles in Fig. 10) fits quite well ($R = 0.966$) and implies a modest rise of only $38\,nT\,a^{-1}$. Bucha & Mellart (1967) report nine archaeointensity determinations from the important Neolithic settlement of Çatal Hüyük in Turkey. These relate to the interval 6500–5750 BC and yield ancient field intensities ranging from 1.11 to 1.44 times the present value at the site ($46\,\mu T$). Thomas (1981) provides a single point in this interval of time, and Kovacheva (1980) lists a further three results. These are in good agreement with each

other (Fig. 10), and are compatible with five of the nine Turkish values (1.11–1.19). The other four results from Çatal Hüyük are rather high (1.29–1.44) and imply two separate rapid oscillations in field strength. In view of the paucity of data for these very early times, these high values should be regarded as tentative rather than proven.

Results from Egypt for the first three millenia BC were obtained by Games (1980) using a novel technique involving sun-dried (adobe) bricks, checked by Thellier-type determinations on fired materials. The manufacture of adobe does not involve firing. Games (1980) refers to its magnetism as a shear remanent magnetization (SRM) acquired when the clay is thrown into the mould. A corresponding laboratory SRM is given by crushing the brick, mixing with water, and re-casting it 'in a way which simulates the original brick-making process'. There seems to

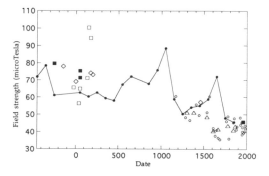

Fig. 12. Archaeointensity results from Mt.Etna (open circles, Rolph & Shaw 1986; open triangles, Tanguy 1975), French Samian ware (open squares, Shaw 1986), and Italian kilns (solid squares, Evans 1986). Kovacheva's (1980) Bulgarian results (dots and solid line) and the original results of Thellier & Thellier (1959: open diamonds) are shown for comparison. Currently the geomagnetic field strength at Paris and Sofia is 47 μT, while in southern Italy it is 45 μT. Rolph & Shaw (1986) obtained no result from the 1910 flow, but Shaw (1974) reports a value of 42 μT.

be no theoretical basis for the procedure, but the results obtained are generally compatible with Games's Thellier-method data from fired samples (Fig. 11). With one or two marked exceptions, Games's Thellier results are consistent with the Bulgarian and Greek results. The most notable discrepancy is the extremely low value near 1000 BC, and the smaller peak at 2600 BC. For the moment, these two results, and their associated SRM data, should be treated with caution.

For more recent times, there are results from several historic flows from Mt Etna (Tanguy 1975; Rolph & Shaw 1986). Between the twelfth and sixteenth centuries, and from 1750 to the present-day, these compare favourably with Kovacheva's Bulgarian results (Fig. 12). However, there is a serious disagreement in the seventeenth century. Since the Bulgarian data have been used in conjunction with Japanese results to support the persistence of the geomagnetic westward drift (Evans 1987), it is important that this disagreement be scrutinized further. The original results of Thellier & Thellier (1959) are also shown in Fig. 12, along with six results from French Samian ware (Shaw 1979) spanning the interval 50 BC to AD 200, and four results from Italian kilns (Evans 1986). Agreement amongst the various data sets is reasonable, except for two of the Samian ware results which yield very high field values urgently needing verification.

Vesuvius, AD 79

The eruption of Vesuvius which destroyed the towns of Pompeii and Herculaneum was witnessed by Pliny the Younger and is accurately dated; it began in the afternoon of 24 August AD 79. It therefore provides an opportunity to obtain precisely dated archaeomagnetic information. Hoye (1981) collected 20 samples from the deposits which buried Herculaneum, and reports a mean direction of $D = 351.9°$, $I = 57.0°$, $k = 212$, $\alpha_{95} = 2.2°$. Kent et al. (1981) obtained a similar result ($D = 352.5°$, $I = 55.8°$, $k = 188$, $\alpha_{95} = 6.7°$), but since it is based on only four samples, the error cone is rather large. Evans & Mareschal (1989) studied a kiln at Pompeii and report a mean direction of $D = 358.0°$, $I = 59.1°$ ($k = 952$, $\alpha_{95} = 1.7°$, $N = 9$). Evans (1991) used six of their samples for an archaeointensity investigation, and obtained a value of 61 ± 1 μT for the ancient field strength. Archaeomagnetism has also been applied to the problem of establishing the exact nature of the deposits which buried Herculaneum. These were interpreted as mudflows by Maiuri (1958), but other authors (Merrill 1918; Sparks & Walker 1973) consider them to be hot pyroclastic flows. Kent et al (1981) studied six lithic fragments associated with the four ash and pumaceous matrix samples discussed above. During step-wise thermal demagnetization, a well-grouped magnetic component was revealed ($D = 359.2°$, $I = 56.8°$, $k = 251$, $\alpha_{95} = 4.2°$). This is indistinguishable from the mean directions given by the matrix samples and by the kiln at Pompeii, and is interpreted as a partial TRM dating from the time of emplacement. It indicates that the deposits were hot pyroclastic flows, emplaced at c. 400°C.

Discussion and conclusions

Archaeological implications

From a purely archaeological point of view, the incentive for archaeomagnetic research has always been the goal of finding an additional means of providing chronological information. Thus, for example, Aitken et al. (1984) conclude that 'the geomagnetic intensity recorded in pottery, bricks and tiles can be useful in giving a more accurate date, approaching ±50 yr'. This can be appreciated by inspecting Fig. 10, but it must be remembered, as Aitken and his colleagues stress, that the objects involved must already be known to derive from the general area and broad time interval covered by their master reference curve. For other regions and other time intervals, the prognosis is not always so favourable. Furthermore, the doubt now being cast upon much of the older archaeointensity data adds to the difficulties. Nevertheless, Walton (1990) feels that dating between 900 BC and AD 400 is possible with precision between 100 and 200 years. For more recent centuries, Rolph *et al.* (1987) successfully used archaeointensity and archaeodirectional results to re-assess the eruption dates attributed to some of the lava flows of Mt Etna. In this way, two particular flows originally thought to have been erupted in 1329 and 1651 were re-assigned to the interval AD 800–1000. This closely follows the procedure previously used by Tanguy *et al.* (1985) in their work on Mt Etna. It relies on transferring the French secular variation curve (Fig. 1) to Sicily since there are no Italian results between the sixth and eleventh centuries.

Where the possibility of absolute magnetic dating is ruled out by the lack of a suitable master curve, it is still sometimes possible to provide relative dates. An important example concerns the demise of the Minoan civilization. The sudden destruction of many palaces and other important sites by volcanic activity of Santorini some three and a half millenia ago was first proposed by Marinatos (1939), and has been widely debated ever since (for a recent summary, see Manning 1990). Archeomagnetic investigations (Downey & Tarling 1984; Tarling & Downey 1989) clearly indicate that the Minoan sites on Crete were not all destroyed simultaneously, as would be expected for some great natural disaster. Sites in central Crete yield a tight group of archaeodirections ($D = 355.3°$, $I = 60.0°$, $\alpha_{95} = 1.4°$), but sites in eastern Crete yield a significantly shallower direction ($D = 355.7°$, $I = 55.6°$, $\alpha_{95} = 1.2°$). These suggest two short, but distinct, episodes of destruction. One can, of course, imagine *two* volcanic events, but such a suggestion is frustrated by there being a third, intermediate, direction ($D = 354.7°$, $I = 57.5°$, $\alpha_{95} = 1.4°$) obtained from a site (Makrygialos) located geographically between the central and eastern sites. These three archaeodirections can be combined with other data from Crete and Santorini to form a secular variation pattern reminiscent of the historical observatory record in Europe (Evans & Mareschal 1988). The natural progression observed could be attributed to a sequence of carefully timed volcanic eruptions, but such an explanation is rather contrived. A wave of destruction by enemy action is equally likely. Of course, this could have been brought on by economic and political disorder resulting from natural geohazards. The existence of distinct archaeodirections in central and

eastern Crete has been used by Tarling & Downey (1989) to 'date' a kiln at Knossos which yielded a mean direction ($D = 355.2°$, $I = 60.9°$, $\alpha_{95} = 1.7°$), indistinguishable from the central Cretan group. Since Knossos is located in central Crete this could have been presumed *a priori* but for the fact that habitation actually continued at the site long after the cessation of Minoan civilization.

Apart from chronological information, magnetic measurements have also proved useful for diagnostic characterization of Mediterranean archaeomaterials, as we saw with the sourcing of the paving slabs at the Temple of Mercury studied by Pierre David early this century. The early work simply involved inclination measurements, but more recent studies have applied the whole range of rock-magnetic and mineral-magnetic parameters. In this way, for example, McDougall *et al.* (1983) effectively discriminated between several sources of obsidian in the eastern Mediterranean on the basis of susceptibility and intensity of natural and laboratory remanences. Williams-Thorpe & Thorpe (1993) measured the susceptibility of hundreds of granite columns at several sites in Rome and its environs and identified their potential sources from quarries in Italy, Turkey and Egypt. Within the realm of provenancing and other technological aspects, mention should also be made of the application of magnetic measurements to numismatics (Tarling 1982; Hoye 1983). It appears that NRM directions, even after AF cleaning, are too scattered to offer any potential for archaeomagnetic dating. However, IRM acquisition curves offer a non-destructive means of discriminating between cast and struck coins. One final point worthy of note is that the direction of remanence of all struck coins implies that they were manufactured with the obverse die below (i.e. 'heads' down). As Sullivan (1981) remarks, it was obviously 'taboo to hit the emperor in the face'.

Geomagnetic implications

From a geophysical point of view, the central goal of archaeomagnetic work has always been the extension of the severely restricted time coverage provided by actual instrumental records. For the Mediterranean region the observatory record goes back no more than a few centuries, whereas the spectrum of geomagnetic variations has considerable power at time scales well beyond a millenium (Barton 1983). We have seen how Chevallier's early work demonstrated the feasibility of using Etnean lava flows to push the record back to the

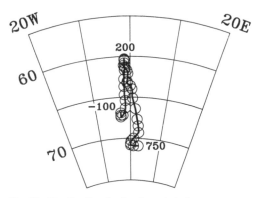

Fig. 13. Declination-inclination plot of archaeomagnetic results from France for the interval 100 BC to AD 750. Derived from Bucur (1994, Figs 5 and 6). (25 year intervals between successive circles).

thirteenth century, and how Thellier's studies of baked clays further extended this to Roman times. These important results were further enhanced by the demonstration that the strength of the field over the Mediterranean had decreased by 50% since the second century BC.

One of the most persistent notions concerning the temporal behaviour of the magnetic field is that of the westward drift; an idea suggested more than 300 years ago (Halley 1692; Evans 1988). The well-known secular variation pattern at London and Paris takes the form of an open loop which is readily explained in terms of a westerly drifting source in the outermost core. Zonal retrograde drift of this kind involving an eccentric dipole moving at a steady rate of $0.17°$ per year gives a satisfactory fit to the London data (Malin & Bullard 1981). For a more realistic picture one should substitute spots, or bundles, of flux (Bloxham & Gubbins 1985) for small drifting dipoles. In this context, it is instructive to compare the inclination curves for France and Sicily (Figs 1 and 2). Careful inspection indicates that the peak at c. AD 1610 in Sicily occurs about 70 years later at Paris implying a drift rate of $0.2°$ per year. For earlier epochs there are no convincing parallels. This could be due to practical limitations such as the scarcity, and dispersion, of much of the relevant data. It seems probable, however, that westward drift was not continuous throughout the entire time interval in question. For example, the French data indicate that the familiar looping pattern was not present prior to the sixteenth century. As pointed out above, declination changes very little between the first century BC and the nineth century AD, whereas inclination

undergoes significant variation. This is the signature of a pulsating stationary source located near the meridian of the site (Fig. 13). The archaeomagnetic record for the Mediterranean region thus provides examples of drifting flux spots *and* static flux bundles (Bloxham & Gubbins 1985). In the former, geomagnetic variations result primarily from spatial changes, in the latter it is temporal changes which are responsible.

I thank the Natural Sciences and Engineering Research Council of Canada for financial support, and Gerry Hoye for preparing the AUTOCAD plot (Fig. 13).

References

AITKEN, M. J., ALCOCK, P. A., BUSSELL, G. D. & SHAW, C. J. 1983. Palaeointensity studies on archaeological material from the near east. *In*: CREER, K. M., TUCHOLKA, P. & BARTON, C. E. (eds) *Geomagnetism of Baked Clays and Recent Sediments*. Elsevier, 122–127.

——, ALLSOP, A. L., BUSSELL, G. D. & WINTER, M. B. 1984. Geomagnetic intensity in Egypt and western Asia during the second millennium B.C., *Nature*, **310**, 306–308.

——, ——, —— & —— 1986. Palaeointensity determination using the Thellier technique: reliability criteria. *Journal of Geomagnetism and Geoelectricity*, **38**, 1353–1363.

——, ——, —— & —— 1988. Determination of the Intensity of the Earth's Magnetic Field during Archaeological times: reliability of the Thellier technique. *Reviews of Geophysics*, **26**, 3–12.

BAMMER, 1964. Die gebrannten Mauerziegel von Ephesos und ihre Datierung. *Jahrbuch der Österreichen Archäologische Institut*, **47**, 290–299.

BARTON, C. E. 1983. Analysis of palaeomagnetic time-series: techniques and applications. *Geophysical Surveys*, **5**, 335–368.

BELSHÉ, J. C., COOK, K. & COOK, R. M. 1963. Some archaeomagnetic results from Greece. *Annual of the British School at Athens*, **58**, 8–13.

BLOXHAM, J. & GUBBINS, D. 1985. The secular variation of the Earth's magnetic field. *Nature*, **317**, 777–781.

BRUHNES, B. 1906. Recherches sur le direction d'aimantation des roches volcaniques. *Journal de Physique*, **5**, 705–724.

BUCHA, V. & MELLART, J. 1967. Archaeomagnetic intensity measurement on some Neolithic samples from Çatal Hüyük, (Anatolia). *Archaeometry*, **10**, 23–25.

BUCUR, I. 1986. Fourteenth century archaeomagnetic field directions from widely distributed sites in France. *In*: *Proceedings of the 24th International Archaeometry Symposium*. Smithsonian Institution Press, Washington, DC, 449–459.

—— 1994. The direction of the terrestrial magnetic field in France, during the last 21 centuries.

Physics of the Earth and Planetary Interiors, **87**, 95–109.

CHEVALLIER, R. 1925. L'aimantation des laves de l'Etna et l'orientaton du champ terrestre en Sicile du XII au XVII siècle. *Annales Physicae*, **4**, 5–162.

COX, A. 1968. Lengths of geomagnetic polarity intervals. *Journal of Geophysical Research*, **73**, 3247–3260.

DOWNEY, W. & TARLING, D. 1984. Archaeomagnetic dating of Santorini volcanic eruptions and fired destruction levels of late Minoan civilization. *Nature*, **309**, 519–523.

EVANS, M. E. 1986. Palaeointensity estimates from Italian kilns. *Journal of Geomagnetism and Geoelectricity*, **38**, 1259–1267.

—— 1987. New Archaeomagnetic Evidence for the Persistence of the Geomagnetic Westward Drift. *Journal of Geomagnetism and Geoelectricity*, **37**, 769–772.

—— 1988. Edmond Halley, Geophysicist. *Physics Today*, **41**, 41–45.

—— 1991. An archaeointensity investigation of a kiln at Pompeii. *Journal of Geomagnetism and Geoelectricity*, **43**, 357–361.

—— 1994. Recent archaeomagnetic investigations in Greece and their geophysical significance. *In: Proceedings of the 2nd Congress Hellenic Geophysical Union, Florina*, 578–585.

—— & MARESCHAL, M. 1988. Secular variation and magnetic dating of fired structures in Greece. *In:* FARQUHAR, R. M., HANCOCK, R. G. & PAVLISH, L. A. (eds) *Proceedings of 26th International Archaeometry Symposium.* Archaeometry Laboratory, Univ. Toronto. 75–79.

—— & MARESCHAL, M. 1989. Secular variation and magnetic dating of fired structures in southern Italy, *In:* MANIATIS, Y. (ed.) *Archaeometry, Proceedings of the 25th International Symposium.* Elsevier, 59–68.

FOLGHERAITER, G. 1894. Orientation et intensité du magnétisme permanent dans les roches volcanique du Latium. *Rendiconti Reale Accademia Lincei*, 165.

—— 1899. Sur les variations séculaires de l'inclinaison magnétique dans l'antiquité. *Archives des Sciences Physiques et Naturelles*, **8**, 5–16.

GAMES, K. P. 1980. The magnitude of the archaeomagnetic field in Egypt between 3000 and 0 B.C. *Geophysical Journal of the Royal Astronomical Society*, **63**, 45–56.

HALLEY, E. 1692. An account of the cause of the change of the variation of the magnetical needle; with an hypothesis of the structure of the internal parts of the Earth. *Philosophical Transactions of the Royal Society, London*, **127**, 563–578.

HOYE, G. S. 1981. Archaeomagnetic secular variation record of Mount Vesuvius. *Nature*, **291**, 216–217.

—— 1983. Magnetic properties of ancient coins. *Journal of Archaeological Science*, **10**, 43–49.

JONES, R. E. 1986. *Greek and Cypriot Pottery.* British School at Athens, Fitch Laboratory Occasional Papers **1**.

KENT, D. V., NINKOVICH, D., PESCATORE, T. & SPARKS, S. R. J. 1981. Palaeomagnetic determi-

nation of emplacement temperature of Vesuvius AD79 pyroclastic deposits. *Nature*, **290**, 393–396.

KOVACHEVA, M. 1980. Summarized results of the archaeomagnetic investigation of the geomagnetic field variation for the last 8000 years in south-eastern Europe. *Geophysical Journal of the Royal Astronomical Society*, **61**, 57–64.

—— 1983. Archaeomagnetic data from Bulgaria and south-eastern Yugoslavia. *In:* CREER, K. M., TUCHOLKA, P. & BARTON, C. E. (eds) *Geomagnetism of Baked Clays and Recent Sediments.* Elsevier, 106–110.

—— 1984. Some archaeomagnetic conclusions from three archaeological localities in north-west Africa. *Comptes Rendus de l'Academie Bulgare des Sciences*, **37**, 171–174.

—— 1992. Updated archaeomagnetic results from Bulgaria: the last 2000 years. *Physics of the Earth and Planetary Interiors*, **70**, 219–223.

LANGOUET. L., BUCUR, I. & GOULPEAU, L. 1983. Les problemes de l'allure de la courbe de variation seculair du champ magnétique en France, Nouveaux results archaeomagnétiques. *Revue d'Archeometrie*, **70**, 37–43.

MAIURI, A. 1958. Pompeii. *Scientific America*, **198**, 68–78.

MALIN, S. R. C. & BULLARD, E. C. 1981. The direction of the earth's magnetic field at London, 1570–1975. *Philosophical Transitions of the Royal Society, London*, **A299**, 357–423.

MANNING, S. W. 1990. The Thera eruption: the third congress and the problem of the date. *Archaeometry*, **32**, 91–100.

MARINATOS, S. 1939. The volcanic destruction of Minoan Crete. *Antiquity*, **13**, 425–439.

McDOUGALL, J. M., TARLING, D. H. & WARREN, S. E. 1983. The magnetic sourcing of obsidian samples from Mediterranean and Near Eastern sources. *Journal of Archaeological Science*, **10**, 441–452.

McELHINNY, M. W. & SENANAYAKE, W. E. 1982. Variations in the Geomagnetic Dipole 1: The Past 50,000 years. *Journal of Geomagnetism and Geoelectricity*, **34**, 39–51.

MELLONI, M. 1853. Du Magnétisme des roches. *Comptes Rendus de l'Academie sciences de Paris*, **37**, 966–968.

MERRILL, E. T. 1918. Notes on the eruption of Vesuvius in 79 A.D. *American Journal of Archaeology*, **22**, 304–309.

ROLPH, T. C. & SHAW, J. 1986. Variations of the geomagnetic field in Sicily. *Journal of Geomagnetism and Geoelectricity*, **38**, 1269–1277.

——, —— & GUEST, J. E. 1987. Geomagnetic field variations as a dating tool: application to Sicilian lavas. *Journal of Archaeological Science*, **14**, 215–225.

SHAW, J. 1974. A new method for determining the magnitude of the palaeomagnetic field. *Geophysical Journal of the Royal Astronomical Society*, **39**, 133–141.

—— 1979. Rapid changes in the magnitude of the archaeomagnetic field. *Geophysical Journal of the Royal Astronomical Society*, **58**, 107–116.

SPARKS, R. S. J. & WALKER, G. P. L. 1973. The

Ground Surge Deposit: a Third Type of Pyroclastic Rock. *Nature*, **241**, 62–64.

SULLIVAN, W. 1981. Testing of Relics Results in Surprises. *New York Times, May 24*, 39.

TANGUY, J. 1970. An archaeomagnetic study of mount Etna: the magnetic direction recorded in lava flows subsequent to the twelfth century. *Archaeometry*, **12**, 115–128.

—— 1975. Intensity of the geomagnetic field from recent Italian lavas using a new palaeointensity method. *Earth and Planetary Science Letters*, **27**, 314–320.

—— 1980. *L'Etna*, PhD Thesis, Université Pierre et Marie Curie, Paris.

——, BUCUR, I. & THOMPSON, J. F. C. 1985. Geomagnetic secular variation in Sicily and revised ages of historic lavas from Mount Etna. *Nature*, **318**, 453–455.

TARLING, D. H. 1982. Archaeomagnetic properties of coins. *Archaeometry*, **24**, 76–79.

—— & DOWNEY, W. S. 1989. Archaeomagnetic study of the Late Minoan kiln 2, Stratigraphical Museum Extension, Knossos. *Annual of the British School at Athens*, **84**, 345–352.

THELLIER, E. 1938. Sur l'aimantation des terres cuites et ses application géophysiques. *Annales de l'Institute Physique dy Globe*, Paris, **16**, 157–302.

—— 1981. Sur la direction du champ magnetique terrestre, en France, durant les deux derniers millenaires. *Physics of the Earth and Planetary Interiors*, **24**, 89–132.

—— & THELLIER, O. 1959. Sur l'intensité de champ magnétique terrestre dans le passé historique et géologique, *Annales de Géophysique*, **15**, 385–376.

THOMAS, R. C. 1981. *Archaeomagnetism of Greek pottery and Cretan kilns*. PhD Thesis, Edinburgh University.

—— 1983. Summary of prehistoric archaeointensity data from Greece and eastern Europe. *In*: CREER, K. M., TUCHOLKA, P. & BARTON, C. E. (eds) *Geomagnetism of Baked Clays and Recent Sediments*. Elsevier, 117–122.

WALTON, D. 1979. Geomagnetic intensity in Athens between 200 B.C. and 400 A.D., *Nature*, **277**, 643–644.

—— 1984. Re-evaluation of Greek archaeomagnitudes. *Nature*, **310**, 740–743.

—— 1987. Improving the accuracy of geomagnetic intensity measurements. *Nature*, **328**, 789–791.

—— 1988. The lack of reproducibility in experimentally determined intensities of the Earth's magnetic field. *Reviews of Geophysics*, **26**, 15–22.

—— 1990. Changes in the intensity of the geomagnetic field. *Geophysical Research Letters*, **17**, 2085–2088.

WILLIAMS-THORPE, O. & THORPE, R. S. 1993. Magnetic susceptibility used in non-destructive provenacing of Roman Granite columns. *Archaeometry*, **35**, 185–195.

Additional bibliography

ABDELDAYEM, A., TARLING, D. H., MARTON, P., NARDI, G. & PIERATTINI, D. 1992. Archaeomagnetic study of some kilns and burnt walls in Selinunte archaeological township, Sicily. *Science and Technology for Cultural Heritage*, **1**, 129–141.

MÁRTON, P., ABDELDAYEM, A., TARLING, D. H., NARDI, G. & PIERATTINI, D. 1992. Archaeomagnetic study of two kilns at Segesta, Sicily. *Science and Technology for Cultural Heritage*, **1**, 123–127.

——, TARLING, D. H., NARDI, G. & PIERRATINI, D. 1993. An archaeomagnetic study of roof tiles fromTemple E, Selinunte, Sicily. *Science and Technology for Cultural Heritage*, **2**, 131–136.

NAJID, D. & TARLING, D. H. 1984. Archaeomagnetic results from Morocco (abs.). *Geophysical Journal of the Royal Astronomical Society*, **77**, 309.

SARIBUDAK, M. & TARLING, D. H. 1993. Archaeomagnetic studies of the Urartian civilization, eastern Turkey. *Antiquity*, **67**, 620–628.

TARLING, D. H. 1992. The Eruption of Santorini and the End of the Minoan Civilization in Crete. *PACT* **25**(6), 107–115.

—— & DOWNEY, W. S. 1990. Archaeomagnetic results from Late Minoan destruction levels on Crete and the 'Minoan' Tephra on Thera. *In*: HARDY, D. A. & RENFREW, A. C. (eds) *Thera and the Aegean World III*. The Thera Foundation, 146–159.

Archaeomagnetic directions: the Hungarian calibration curve

P. MÁRTON

Geophysics Department, L. Eötvös University, H-1083 Budapest, Ludovika tér 2, Hungary

Abstract: The first archaeomagnetic dating in Hungary was made using a reference inclination curve derived by interpolation between three inclination curves then available from continental Europe, namely from France, Bulgaria and the Ukraine. As the corresponding declination curves could not be interpolated with confidence, all three declination curves were used only for an estimation of the time interval to which the measured declinations might be assigned. As data accumulated it became feasible for dating to be made exclusively on the basis of archaeomagnetic and direct observational results for Hungary. The results now cover the last 2000 years with relatively short gaps (sixth and thirteenth centuries AD), for which data interpolation is plausible. The variation of inclination is fairly sinusoidal and well resolved. It exhibits two maxima (*c.* 70°) at approximately AD 800 and AD 1550 and three minima at about AD 300 (*c.* 58°), AD 1300 (*c.* 55°) and in the twentieth century (*c.* 62°). Although the variation of the declination is twice as large as that of the inclination, the declination record has inherently less resolution (dD = dI/cos$I$ > dI). Approximately zero declination values during the first half of the first millennium AD were followed by westerly values during the sixth to eighth centuries (*c.* 10°). It appears that by AD 900 the declination was already easterly and rapidly increasing until AD 1000 when it peaked with a value over 20°E. A newly discovered feature is a short oscillation beginning with 16° at AD 1300 and ending with 16.5° at AD 1600 during which the declination became westerly with a minimum of −10° in the first half of the fifteenth century. From AD 1600 on the declination rapidly decreased and described a negative half circle between about AD 1620–30 and 1950 with a minimum of −18° at AD 1800. The general pattern of the directional secular variation for Hungary is in agreement with that for France, Sicily, Britain, the Ukraine and the Balkans.

Secular variations are a few decade or longer changes of the geomagnetic field. For a given location, the secular variation is best studied by taking yearly averages of the local geomagnetic elements (e.g. declination, inclination and intensity) and analysing the resulting time series. Such studies, however, can be pursued back for only a very short time, say a hundred years or so, except for a few places where the record is longer. For example, directional data for London can be traced back to the end of the sixteenth century and suggest a quasi-periodic behaviour of the secular variation. During these 400 years the maximum variation of the declination was about 35° and that of the inclination about 8°. For older periods directions can only be inferred from the remanent magnetization of features dated by archaeological or other means. Fortunately, the accuracy with which these directions of remanent magnetization can be recovered is amply sufficient for changes with periods greater than a hundred years or so to be detected, but is not considered satisfactory for shorter periods.

Acquisition of archaeomagnetic data is a protracted and on-going process (for an earlier summary see Márton 1990). All features hitherto studied were fired material (mostly clay) possessing thermoremanent magnetizations (TRM). The direction of the TRM is believed to parallel the local magnetic field during cooling in antiquity and the latter is usually assumed to be identical to or at least basically not different from the local geomagnetic field. The intensity of the TRM is also proportional to the strength of the ambient field at the time of cooling but the intensity determinations (Burlatskaya *et al.* 1986) have proved to be inferior in resolution to the directional measurements and will not be dealt with in the present study.

This contribution describes the present state of knowledge of directional changes of the geomagnetic field for Hungary during the past two thousand years using geomagnetic and archaeomagnetic data currently available from the region. Following a brief exposition of the direct geomagnetic observations, the major part of the paper is devoted to archaeomagnetism and data analysis.

From Morris, A. & Tarling, D. H. (eds), 1996, *Palaeomagnetism and Tectonics of the Mediterranean Region*, Geological Society Special Publication No. 105, pp. 385–399.

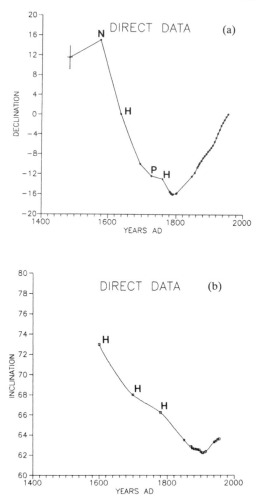

Fig. 1. Declination (**a**) and inclination (**b**) data from direct observations for Hungary. (**a**) Back to AD 1696 unlabelled data are from Schenzl (1881) and Barta (1957). These and the *D* value with standard deviation 2.5° for 1483 ± 7 years (Márton 1990) are originally for Budapest. Data from Hansteen (1819) are labelled with H and originally refer to Vienna. P is for Pozsony (see reference in Márton (1990) and finally, N corresponds to a declination mark on a portable sundial from Nagyszeben (Transylvania) from 1576. (**b**) Unlabelled data are from Schenzl (1881) and Barta (1957). H refers to readings by interpolation from I-charts of Hansteen (1819) for 1780, 1700 and 1600, but the earliest *I*-value is very uncertain. The best fit lines were generated by computer using cubic spline approximation with a tension factor of 1.

Direct geomagnetic and archaeomagnetic observations

The direct declination and inclination data (Fig. 1) pertaining to the directional secular variation for Hungary can be considered separately. Back to 1781, reference can be made to an essentially continuous set of declination observations (Schenzl 1881; Barta 1957). Individual declination values also are known from different sources for 1728 (1), 1696 (7), 1576 (1) and 1476–91 (7) (Márton 1990). The direct record of inclination can be traced back to 1848 (Schenzl 1881; Barta 1957) but can also be extended using readings interpolated from old inclination charts for 1780, 1700 and 1600 (Hansteen 1819). All these directional values refer originally to Budapest (47.5°N, 19.1°E) except the declination data with labels (Fig. 1) which have not been reduced to Budapest because their uncertainty is probably greater than the correction itself.

The sampled archaeological features are remnants of fired objects that have either remained completely unmoved since firing or where it is possible to fully reconstruct their original positions by simple tilt corrections. Dates are based on archaeological evidence (pottery, coin, documentary) supplied by the local archaeologists. In a few instances, radiocarbon dates are also available (see Table 1). Samples were obtained by the block sampling technique. Prior to removal, fiducial marks were drawn directly on the surface of the samples and orientation measurements were made using a sun compass and/or magnetic compass. After careful removal, the samples were packed and transported to the laboratory where they were cut into 2 cm cubic specimens (at least two from each sample) for magnetic measurements.

The primary magnetization of the archaeological material is a TRM which is strong enough to be easily measurable with commercially available magnetometers such as the JR-2, JR-3 and JR-4 spinners. The routine measurements applied to every specimen (subsample) included an incremental demagnetization procedure either by alternating magnetic field (AF) or thermal methods or both.

Low field susceptibility was monitored as part of the thermal treatment. The direction of the stable component of the magnetization, which is usually identified with the primary TRM, was defined by linearity analysis or just by the direction measured at the beginning of the linear segment of the demagnetization diagram. Either method gives, as expected, statistically identical site mean directions (Fisher 1953) and the one with the greater precision was adopted.

All the directional data acquired from 250 BC up to the eighteenth century AD are given in their original form (Table 1) as obtained after demagnetization and any tilt correction followed

Table 1. *Archaeomagnetic site mean directions from Hungary for historical ages*

Locality/feature	Lat/Long	N/n	D_{BP}	I_{BP}	k	α_{95}	Cleaning	Archaeological date
Sopron–Krautacker dűlő	47.68/16.62	6/34	346.8 / 347.7	67.8 / 67.4	188	4.9	460°C	Celtic 200 ± 50 BC
Two pottery kilns		4/24	359.8 / 359.7	68.0 / 67.9	645	3.6	200–460°C	Celtic 100 ± 50 BC
Nagyberki.	46.38/17.97	14/28	1.8 / 1.6	59.3 / 60.3	1143	1.2	400°C	Celtic 0 ± 50 BC
Two ovens, floors		14/28	7.3 / 7.6	60.6 / 61.6	1078	1.2	400°C	
Óbuda, Budapest. Oven(?), floor	47.50/19.10	4/4	353.2	64.9	191	6.7	20–50 mT after 130°C	Probably 1st century AD
Zalalövő.	46.85/16.58	7/17	355.1 / 355.6	57.6 / 58.0	347	3.2	40 mT	Roman AD 100 ± 15
Two hearths, floors		6/12	349.3 / 349.4	63.0 / 63.2	280	4.0	40 mT	
Petőháza Hearth, floor	47.58/16.87	5/15	6.8 / 7.0	62.1 / 62.2	270	4.7	(200°C) 30 mT	Roman AD 200 ± 100
Burnt earth, floor		5/10	358.5 / 358.8	61.3 / 61.2	586	3.2	(200°C) 30 mT	Roman AD 200 ± 100
Óbuda, Budapest.	47.50/19.10	6/6	356.8	58.3	340	3.6	30 mT after 130°C	Late 2nd–3rd century
Two burnt graves, walls		5/5	0.7	54.6	218	5.2	30 mT after 130°C	
Tác-Gorsium. Oven, floor	47.12/18.30	3/3	358.2 / 358.4	56.0 / 56.4	201	8.7	100–150°C	Roman AD 250 ± 50
Sopron. Potters kiln	47.68/16.62	10/11	355.4 / 355.8	60.6 / 60.3	416	2.4	250°C	Probably 3rd century AD 350
Fenékpuszta. Baking oven, floor	46.70/17.20	3/5	2.5 / 2.7	61.5 / 62.2	1359	2.7	300°C	AD 350

Table 1. *Continued*

Locality/feature	Lat/Long	N/n	D/D_{BP}	I/I_{BP}	k	α_{95}	Cleaning	Archaeological date
Beregsurány. Four ovens, wall	48.18/22.52	4/6	2.3 1.9	62.0 61.4	389	4.7	200–500°C	AD 400 ± 25
Sopron–Potzmann–dűlő. Oven, floor	47.68/16.62	8/8	3.7 3.8	64.3 64.2	250	3.5	20 mT after 130°C	Roman AD 375–500
Zalalövő. Fireplace, bricks	46.85/16.58	8/15	351.9 352.4	56.2 56.5	138	4.7	40 mT	Roman AD 350–500
Fenékpuszta. Two baking ovens, walls and floors	46.70/17.20	4/14	4.1 4.5	58.3 59.1	437	3.7	300°C	AD 456
Kölked–Fekete–kapu. Oven, floor	45.92/18.68	3/3 (6)	346.7 346.2	68.6 69.6	1142	3.6	200–250°C	Avar AD 600–800
First reconstruction		5/5 (10)	351.0 351.5	70.0 71.0	345	4.1	200–250°C	
Second reconstruction		5/5 (10)	357.1 356.4	72.5 73.4	558	3.2	200–250°C	
Three ovens, floors		7/13	351.3 350.1	67.9 69.0	311	3.3	200–350°C	
		10/10	352.9 352.5	69.4 70.4	823	1.7	250°C	
		6/12	3.5 3.5	72.2 73.2	335	3.7	250°C	
Zamárdi.	46.90/18.07							
Iron smelting furnace		10/28	346.6 346.1	72.0 72.3	673	1.9	400°C	Avar AD 660–780
Burnt earth		3/6	358.3 358.2	66.1 66.5	3223	2.2	400°C	
Iron smelting furnace		7/14	353.9 353.6	70.7 71.0	429	2.9	400°C	
Oven, floor		3/5	342.0 341.6	69.3 69.5	272	7.5	400°C	
Iron smelting furnace		4/7	357.5 357.1	72.0 72.4	1043	2.8	400°C and 30 mT	AD 600–800
Iron smelting furnace		5/10	357.9 357.7	69.1 69.5	414	3.8	400°C and 30 mT	

Table 1. *Continued*

Locality/feature	Lat/Long	N/n	D / D_BP	I / I_BP	k	α_{95}	Cleaning	Archaeological date
Dénesfa.	47.40/17.00	4/11	342.7 / 342.4	70.0 / 69.7	556	3.9	70 mT	AD 700–900
Two iron smelting furnaces, floors		4/21	2.9 / 2.3	71.7 / 71.7	480	4.2	70 mT	
Karos–Mókahomok.	48.30/21.78	9/15	353.0 / 353.6	70.4 / 70.0	401	2.6	130°C and 20 mT	9th century AD
Two ovens, floors		10/10	4.1 / 4.5	70.7 / 70.1	238	3.2	130–450°C	
Karos–Tobolyka.	48.30/21.78	8/8	351.0 / 351.8	71.6 / 71.3	243	3.6	250–450°C	8th–9th centuries AD
Six ovens, floors		6/6	343.9 / 344.5	69.8 / 69.6	795	2.4	130–450°C	
		6/6	351.5 / 352.6	73.0 / 72.7	1356	1.8	250–450°C	
		8/13	15.2 / 14.8	65.2 / 64.2	329	3.1	250–450°C	9th–10th centuries AD
		8/8	23.2 / 22.8	67.3 / 66.2	482	2.5	250–450°C	
		10/10	18.8 / 18.7	69.5 / 68.6	271	2.9	250–450°C	
Sopron–Potzmann–dűlő	47.68/16.62	4/4	9.2 / 8.8	69.7 / 69.8	248	5.8	130–320°C	10th century AD
Burnt pit, wall Two iron smelting furnaces		8/8	28.7 / 28.5	68.9 / 69.3	158	4.0	30 mT after 130°C	
		7/7	11.5 / 11.0	70.2 / 70.3	274	3.7	70 mT after 130°C	
Somogyvámos.	46.58/17.67	8/8	10.2 / 10.3	67.8 / 68.6	675	2.1	300°C	Possibly 9th–10th centuries AD
Three iron smelting furnaces		5/5	14.0 / 14.0	69.8 / 70.6	734	2.8	150°C	
		7/7	12.5 / 12.6	68.4 / 69.2	601	2.5	520°C	
Edelény. Oven, floor + wall	48.27/20.75	9/18	20.5 / 20.3	69.5 / 68.8	376	2.7	150–300°C and 20 mT	10th century AD

Table 1. *Continued*

Locality/feature	Lat/Long	N/n	D / D_{BP}	I / I_{BP}	k	α_{95}	Cleaning	Archaeological date
Burnt clay wall in stone building		6/12	7.2 / 7.1	67.0 / 66.3	923	2.2	300°C and 20 mT	
Drassburg, East Austria. Burnt dyke	47.73/16.52	8/32	22.8 / 22.6	68.2 / 68.5	497	2.5	300°C and 30 mT	AD 970 ± 75
Somogyfajsz Two iron smelting furnaces	46.50/17.62							
Wall		10/10	20.5 / 21.0	62.6 / 63.6	198	3.4	150–400°C	10th–11th centuries AD
Floor		8/8	21.3 / 21.7	66.1 / 67.1	201	3.9	150–400°C	
Szakony–Békástó–dülö	47.40/16.74	5/7	24.8 / 24.4	66.5 / 66.2	60	3.8	25 mT	Early 11th century AD
Seven iron smelting furnaces		4/8	32.3 / 32.5	67.1 / 67.8	135	7.9	25 mT	
		4/11	34.1 / 34.6	63.6 / 64.4	255	5.8	25 mT	
Four furnaces combined		4/14	28.9 / 29.0	67.1 / 67.7	58	12.1	25 mT	
Sopron.	47.68/16.62	5/10	17.2 / 17.3	64.4 / 64.7	184	5.6	400°C and 50 mT	11th century AD
Burnt dyke, three sampling sites		6/12	17.7 / 17.6	66.7 / 66.9	308	3.8	400°C and 50 mT	
		5/10	22.3 / 22.3	66.0 / 66.3	1642	1.9	400°C and 50 mT	
Röjtökmuzsaj.	47.58/16.86	9/27	18.9 / 18.6	69.4 / 69.7	246	3.3	460°C	11th century AD
Three iron smelting furnaces, floors		5/16	21.0 / 20.8	68.5 / 68.8	82	8.5	460°C	
		4/13	18.0 / 18.2	62.6 / 62.9	171	7.0	460°C	
Répcevis.	47.40/16.68	8/28	18.8 / 19.1	63.2 / 63.7	739	2.0	500°C and 40 mT	11th century AD
Two iron smelting furnaces		15/52	17.5 / 17.7	64.7 / 64.7	251	2.4	500°C and 40 mT	

Table 1. *Continued*

Locality/feature	Lat/Long	N/n	D_{BP}	I_{BP}	k	α_{95}	Cleaning	Archaeological date
Ópusztaszer.	46.50/20.08	11/11	22.9 23.2	63.8 64.3	442	2.2	250–540°C	AD 1100 ± 25
Two ovens, floors		6/10	16.3 16.4	61.5 62.1	347	3.6	400°C	AD 1125 ± 25
Óbuda, Budapest.	47.50/19.10	5/10	13.6	66.7	716	2.9	20–30 mT after 130°C	Early (?) 12th century AD
Three hearths and seven ovens, floors		5/5	9.6	59.9	613	3.1	10 mT after 130°C	
		4/4	16.2	60.7	450	4.3	10 mT after 130°C	
		5/5	6.9	61.3	484	3.5	10 mT after 130°C	
		6/6	15.6	62.7	471	3.1	10 mT after 130°C	
		5/5 5/10	13.3 12.9	61.0 58.6	757 1224	2.8 2.2	250°C 130°C and 10 mT	
		6/6	17.0	59.7	730	2.5	20 mT after 130°C	
		9/9	15.6	60.7	162	4.1	20 mT after 130°C	
		5/15	13.0	61.8	933	2.5	7.5–15 mT	
Hidegség. Rectangular hearth, walls and floor	47.63/16.74	12/71	12.2 12.4	62.7 62.9	433	2.1	20–70 mT (250°C)	Late 12th century AD
Oven, floor		8/12	18.0 18.3	58.8 58.9	296	3.2	20–40 mT	
Oven, floor		10/12	19.7 20.2	58.7 59.1	222	3.2	20–40 mT	
Óbuda, Budapest. Oven, floor	47.50/19.10	8/16	14.1	58.2	1608	1.4	30 mT after 130°C	Late 12th–early 13th century AD
Csongrád–Várhát. Brick kiln	46.72/20.20	24/53	16.3 16.2	55.9 56.4	253	1.9	280–400°C	About AD 1300
Óbuda, Budapest. Lime kiln, tuff	47.50/19.10	16/28	6.1	58.1	187	2.7	250–585°C and 20–60 mT	About AD 1350

Table 1. *Continued*

Locality/feature	Lat/Long	N/n	D / D_{BP}	I / I_{BP}	k	α_{95}	Cleaning	Archaeological date
Dömös	47.80/18.90	5/8	4.8 / 4.8	60.1 / 59.8	292	4.5	200–500°C	About AD 1400
Two brick kilns		10/15	3.4 / 3.4	62.4 / 62.2	458	2.3	200–500°C	
Szombathely.	47.23/16.63	5/10	0.5 / 0.8	60.8 / 61.1	1758	1.8	10–40 mT after 130°C	About AD 1410
Fireplace, floor three consecutive reconstructions		5/9	356.3 / 356.4	63.3 / 63.4	467	3.5	10–40 mT after 130°C	About AD 1420
		11/24	351.9 / 352.1	62.8 / 62.8	706	1.7	250°C	About AD 1430
Szentkirály. Fireplace, floor	46.92/19.92	11/11	16.5 / 16.7	66.8 / 67.1	430	2.2	150–450°C	AD 1591–1606(?)
Buda–Császárfürdő.	47.50/19.10	11/22	6.4	71.0	794	1.6	40 mT after 130°C	^{14}C data cal AD 1560 ± 80
Three cylindrical lime kilns. Limestone, walls		6/12	5.9	69.7	647	2.6	60 mT after 130°C	
		7/14	10.9	68.3	344	3.0	50 mT after 130°C	
Nagyrév. Hearth, floor	46.93/20.25	10/10	353.9 / 353.9	66.6 / 67.1	335	2.8	20–70 mT	AD 1650 ± 50
Structures with uncertain archaeological ages								
Sopron–Potzmann–dűlő.	47.68/16.62	8/8	5.0 / 5.0	65.6 / 65.6	1395	1.5	20 mT after 130°C	Early Medieval
Three ovens, floors		10/10	3.0 / 2.9	66.1 / 66.0	508	2.1	20 mT after 130°C	
		5/8	345.5 / 344.9	72.6 / 72.2	199	5.4	350–520°C	
Secondary use		5/9	358.3 / 357.8	70.3 / 70.1	439	3.7	130–250°C	
Örménykút.	46.82/20.73	9/14	17.4 / 17.3	59.5 / 59.8	201	3.6	400°C and 30 mT	Allegedly 9th–10th centuries AD

Table 1. *Continued*

Locality/feature	Lat/Long	N/n	D_{BP}	I_{BP}	k	α_{95}	Cleaning	Archaeological date
Four ovens, floors		10/10	31.8	68.0	129	4.3	30 mT	
			32.2	68.0				
		6/13	17.1	60.5	224	4.4	450°C and 30 mT	
			17.0	60.8				
		10/22	8.4	59.1	127	4.3	400°C and 30 mT	
			8.2	59.6				
Győr–Ménfőcsanak.	47.63/17.63	9/9	24.5	60.3	302	3.0	130–450°C	Possibly Medieval
			25.7	60.6				
Two ovens, floors		6/6	14.4	64.1	174	5.1	250–450°C	
			14.7	64.2				
Sopron. Oven, floor	47.68/16.62	8/8	15.8	55.1	594	2.3	200–500°C	Medieval
			16.4	55.4				
Sopron–Bánfalva. Lime kiln	47.68/16.62	12/72	349.3	62.7	111	4.1	500°C and 30 mT	Probably last century
			349.6	62.3				

Site means are based on the number of the independently oriented samples. Lat, geographical latitude (°); Long, geographical longitude (°); N, number of independently oriented samples; n, number of specimens measured; D, site mean declination (°); D_{BP}, same reduced to Budapest; I, site mean inclination (°); I_{BP}, same reduced to Budapest; k, Fisher's precision parameter; α_{95}, 95% confidence limit; Cleaning: optimum demagnetization step/interval. Archaeological dates are given by the site archaeologists and based on pottery, coin, documentary or radiocarbon evidence.

by normalization to Budapest. The latter correction was made on the assumption that the regional field can be largely simulated by an inclined geocentric dipole whose pole is defined by the observed site direction (e.g. Tarling 1988). The first 89 directions were from features well-dated archaeologically or by radiocarbon while the last 12 are, for the time being, poorly dated or of unknown age.

Discussion of errors

In archaeomagnetic work, sampling and measuring errors are considered negligible in comparison to the errors which can result from various effects causing magnetic distortions (see Tarling 1988 for discussion). With directional measurements in fired materials the main sources of deviation are believed to be inhomogeneity, shape effects and anisotropy.

Inhomogeneity

Within the same structure, certain directions can sometimes differ by as much as 10–15°. These deviations are thought to be associated with the inhomogeneity of the magnetizing field, which could be produced by strongly magnetic objects (e.g. iron, iron slag, bricks) within the structure during its last cooling. The effect of inhomogeneity is difficult to quantify but, given a sufficient number of samples, the resulting distortions can be averaged out and are not considered appreciable. In cases of fewer samples, as well as in general, averaging of sites of the same age range will probably also reduce the effect to an insignificant level.

Shape effect

The most exact treatment of the effect of shape anisotropy on the direction of the TRM was given by Coe (1978). Qualitatively, the effect is to deflect the TRM away from the shortest dimension of the body. The effect is larger for less equidimensionally-shaped cooling objects and for more strongly magnetic materials. For planar bodies with a high Q value, such as the floors and walls of various archaeological features, theory predicts significant deflection only if the TRM susceptibility surpasses 0.12–0.13 SI units (Coe 1978). To estimate the role of shape anisotropy, the strongest sample from each site was selected and its TRM susceptibility and Q value were computed as follows (Coe 1978):

$$\text{TRM susceptibility} = 0.68 \frac{\text{Initial magnetization}}{\text{Magnetic field strength}}$$

$$Q = \frac{\text{Initial magnetization}}{\text{Induced susceptibility} \times \text{Magnetic field strength}}$$

A value of 40 mT was estimated as the magnetic field strength for the relevant period. While the Q values were always much greater than 1 (to over 100), the TRM susceptibilities varied between 0.0014 and 0.17 SI units. Fortunately, most features (including the flat ones) turned out to have TRM susceptibilities smaller than 0.12 SI units. Hence these must be free from deviations due to the shape effect. In fact, there are only five structures with TRM susceptibilities greater than 0.13 SI units: the Csongrád and Óbuda kilns and one of the Dömös kilns (0.14–0.15), the Zalalövő fireplace (0.16) and one site of the burnt dyke of Sopron (0.17) (Table 1). However, the last two of these represent rather equidimensional features, so the shape effect is not likely. In the case of the burnt dyke of Sopron the site mean direction of magnetization is in particularly good agreement with other two (Table 1), each of which represents a site with a TRM susceptibility of only 0.07 SI units. The remaining three kiln structures also are believed to possess TRMs not appreciably deviating from the ancient geomagnetic field. Proof of the lack of deviations due to shape anisotropy is provided by the fact that the individual sample directions show no correlation with the geometrical positions of the samples within either of these structures.

Magnetic anisotropy

Magnetic anisotropy due to particle alignment may also contribute to the deflection of the magnetization from the ambient field vector when the TRM is acquired. The effect is conveniently described by the orientation and magnitude of the anisotropy ellipsoid of the TRM relative to the sample. If the TRM ellipsoid is known, the deflection of magnetization can be corrected. Stephenson et al. (1986) showed experimentally that the TRM and low field IRM (isothermal remanent magnetization) ellipsoids are of identical shape and the latter can be determined without changing the chemistry of the samples. Therefore they suggested using the IRM ellipsoid to deduce the direction of the ambient field which produced the TRM. Experiments to apply these ideas to the archaeomagnetic material have been in progress for some time. Planar structures (floors) have a low field (1–5 mT) IRM that is usually anisotropic by

Table 2. *Archaeomagnetic directional data obtained from those in Table 1 after combining site mean directions with the same archaeological age ranges*

Age range*	D	I	k	α₉₅
250–150 BC	13.3	67.4	188	4.9
150–50 BC	−.3	67.9	645	3.6
50 BC– AD 50	4.7	61.0	824	0.9
0–100	−6.8	64.9	191	6.7
85–117	−7.0	60.6	226	2.8
100–300	2.8	61.8	324	2.7
150–300	−1.3	56.6	247	2.9
200–300	−1.9	59.5	344	2.2
340–360	2.7	62.2	1359	3.3
375–425	1.9	61.4	389	4.7
350–500	−7.6	56.5	138	4.7
375–500	3.8	64.2	250	3.5
440–460	4.5	59.1	437	4.7
600–800	−5.9	71.1	1092	1.7
660–780	−9.8	69.9	500	4.1
700–900	−9.6	71.1	789	2.7
800–900	−1.3	70.6	267	2.1
800–1000	13.1	69.3	4634	1.4
900–1000	16.5	68.1	543	2.4
895–1045	22.6	68.5	497	2.5
900–1100	21.4	65.2	192	2.5
1000–1050	30.2	66.6	1134	2.7
1000–1100	18.9	66.0	1040	1.7
1075–1125	23.2	64.3	442	2.2
1100–1150	13.6	61.4	982	1.5
1170–1200	17.1	60.3	732	4.6
1140–1240	14.1	58.2	1608	1.4
1280–1320	16.2	56.4	253	1.9
1340–1360	6.1	58.1	187	2.7
1375–1425	3.9	61.5	384	2.0
1390–1430	−3.5	62.5	1186	3.6
1591–1606	16.7	67.1	430	2.2
1530–1690	7.8	69.7	2372	2.5
1600–1700	−6.1	67.1	335	2.8
1699–1701	−10.0	68.0		
1779–1781	−15.5	66.25		
1849–1851	−12.3	63.5		
1874–1876	−9.3	62.7		
1899–1901	−7.1	62.3		
1949–1951	−.44	63.55		
1989–1991	1.6	64.0		

* AD unless otherwise stated.
Also contained in this table are seven data selected from the results of direct observations.

up to 30–40% (the percentage IRM anisotropy is defined as [100 × (max. − min. axis of IRM ellipsoid) / average] (see Stephenson *et al.* 1986). The maximum deviation of the IRM vector from the magnetizing field linearly increases from zero with a gradient of 0.29° per 1% increment of anisotropy. The minimum axis of the IRM ellipsoid was found to be always perpendicular to the plane of the floor and would result in shallowing of the inclination if the anisotropy

properties of the IRM copied those of the NRM (TRM) in these structures. However, this certainly is not the case because the low field IRM ellipsoid is dominated by the properties of the multidomain particles present while the TRM of our samples is carried predominantly by single domain grains (as shown by the large Q values mentioned earlier). Thus, in the present case, the results of the low field IRM experiments cannot be used directly for correction of the direction of the TRM.

For such a correction the anisotropy field of the single domain particles present is needed. This can be detected, in principle, by IRM measurements provided that the field is sufficiently high (50–60 mT). The method is that of sifting, i.e. giving the IRM in a field B and then demagnetizing it in an alternating field with a peak value less than B. The higher the amplitude of the alternating field the greater will be the high coercivity (single domain) contribution to the IRM (Stephenson *et al.* 1986) left after demagnetization. In the few experiments that we have carried out to determine the high field anisotropy of the IRM we used 50 mT or 60 mT fields to obtain the IRM and 20 mT or 40 mT peak alternating fields to demagnetize it. The results were unequivocal, showing a drastical decrease in the anisotropy of IRM in high fields relative to that given in low fields. The reduction varied between 50 and 80% of the low field anisotropy, i.e. in these samples the high coercivity fraction was strikingly less anisotropic than the low coercivity fraction. Thus it can be concluded preliminarily that the correction of the direction of the TRM for magnetic anisotropy may generally not be necessary, but clearly many more experiments are needed before a valid conclusion can be drawn for each particular feature. For the time being, however, the archaeomagnetic site mean directions (Table 1) can be analysed further without any correction for any of these three factors.

Data analysis

As a first step, directions for any dated observations with identical age ranges were combined and the means were referred to the centre of the respective age range. Twenty one mean directions and the 13 remaining data singlets (Table 1) gave 34 'independent' observations for the period between 200 BC and AD 1650 (Table 2). Considering the gaps in the observations for part of the 'Dark Ages' and late Medieval times this amounts to roughly two data per century. For the last 300 years, which is not represented in the archaeomagnetic record, seven data were

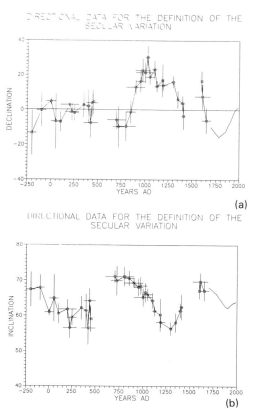

(a)

(b)

Fig. 2. Geomagnetic declination (**a**) and inclination (**b**) data from Table 2. Error bars show age ranges and directional equivalents of α_{95}.

chosen from the direct observations with approximately the same sampling frequency, i.e. two data per century and added to the database (Table 2 and Fig. 2). No smoothing was applied, except that inherent in the averaging of site mean directions within the same age range. These provide all the information presently available about the geomagnetic secular variation for these two millenia.

Both directional elements show definite trends as follows. (a) The declination is moderately scattered about zero before AD 500, shows westerly values during the Avar age and very rapid increase till the eleventh century when it attains its maximum easterly declination of about 30°E. This is followed by an uneven decrease until about AD 1400 when the value is about zero. The direct observations show that the declination is already easterly (Fig. 1) and increasing during the fifteenth and sixteenth centuries, attaining another maximum at about 15°E in the second half of the sixteenth century, after which it rapidly decreases to its minimum value of 16°W in 1792. During the last 200 years

the declination increased again and passed zero at 1950; (b) The variation of the inclination is rather sinusoidal. The main minima occur during the third and thirteenth centuries and the maxima during the eighth century, at about AD 1600 and perhaps during the second century BC. The maximum change of the inclination during these times is about 15°.

A more objective definition of the secular variation can be achieved by the application of some smoothing of the original data (Table 1). For this study, the magnetic directions were smoothed using a moving-window technique. Owing to the favourable properties of the Gaussian smoothing (Véges 1970), each datum of Table 1 was weighted according to its distance from the centre of a particular time window using Gaussian weights, i.e.

$$ w = \exp\left[-\left(\frac{r\pi}{4} \right)^2 \right] $$

where w is the weighting function and r is the distance in time between the centre of the window and that of the age range of the datum normalized by half of the window size. The precision parameter, k, was used as a separate weight to reduce the influence of less precisely determined archaeomagnetic directions. Outliers, if any, were dealt with as follows. For each non-overlapping window the feature overlapping it and farthest from the mean direction for that window was determined. If it had less than 0.5% probability of being a member of the same population as the other features in that interval it was deleted as an outlier (McFadden 1980, eqn. 20; Sternberg 1989). After a number of trials to find a sufficiently smooth variation with good resolution, the 'final' result was obtained somewhat subjectively for a window length of 100 years and with weighting by precision parameter and Gaussian weights. The increment between successive windows was also 100 years, i.e. successive windows do not overlap. The data points for the secular variation curve for Budapest corresponding to this smoothing were then connected with a best fit line calculated by computer using a cubic spline approximation with stretching factor 2.0 (Fig. 3).

The spectral content of the secular variation was studied by computing peridograms for both the declination and inclination records (Fig. 3). To avoid aliasing, the records were first smoothed by a $\cos^2$ window, then the tapered data were used to compute the periodograms at L discrete points for the interval $0 < \tau/T < 0.5$ where τ is the sampling distance and T is the period searched. Here $\tau = 100$ years and $L = 500$. On this analysis, the declination record

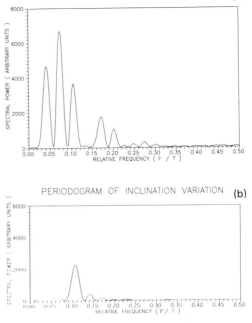

Fig. 4. Periodiograms of: (**a**) declination; and (**b**) inclination obtained after tapering the directional data of Fig. 4a by $\cos^2\pi\,(\tau/T)$, where $\tau = 100$ years and $T \geqslant 200$ years.

exhibits periods of 2381 years, 1351 years, 961 years (and perhaps 585 years and 498 years), while the inclination record is apparently monochromatic with a (dominant) period of 943 years. The same analysis on another data set which was obtained identically with the former 'final' one, except with successive windows overlapping by 50 years, yielded periods 2272 years, 1315 years, 925 years (and 574 years and 481 years) for declination and 909 years for inclination. It appears therefore that an oscillation with a period of somewhat above 900 years is present in both records and an approximately 580 years oscillation might also be common.

Comparison with other archaeomagnetic results

In general, the current knowledge about secular variation for Hungary for the past two thousand years (Fig. 3) compares well with those for other European regions such as the Balkans (Kovacheva 1983), the Ukraine (Zagnii & Rusakov 1982), Sicily (Rolph & Shaw 1986), France (Thellier 1981) and Britain (Aitken & Weaver 1965; Clarke *et al.* 1988). The agreement with the French data is especially good except for a few centuries, during the so-called Avar age,

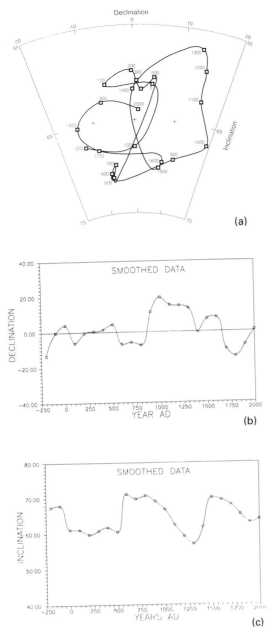

Fig. 3. Present secular variation model for Hungary (Budapest) obtained after smoothing the archaeomagnetic data of Table 1 combined with the direct observations selected in Table 2 by the moving-window technique. Both window length and step size (increment between successive windows) are 100 years (non-overlapping windows). Gaussian and precision weighting was applied (see text). Best fit lines were calculated by computer using cubic spline approximations with stretching factor 2. (**a**) stereographic projection of directional secular variation. Labels are dates in years; (**b**) declination; and (**c**) inclination.

when the declinations for Hungary are always westerly whereas those for France, though decreasing, are always easterly. However, the validity of these westerly declinations in the Hungarian record is evidenced by both of the data sets for East Europe. Thus the occurrence of westerly declinations during the first half of the Dark Ages appears to be specific to East Europe. Later in the Hungarian record, another local minimum of declination occurs at about AD 1400 that is also present in the French and Sicilian data and, some 50 years earlier, in the British data. This is not present in either of the East European data sets. The lack of this feature of the secular variation east of Hungary might be real or might just be unresolved by the present data.

Archaeomagnetic dating

The variations (Fig. 3) provide master curves that can be used for the determination of archaeomagnetic ages. However, the usefulness of the master curves of the archaeosecular variation for dating is different for different intervals of time. As there was apparently little change in the secular variation during the first half of the first millenium and also between AD 600 and 800 precision dating for these intervals is presently not possible, though the two intervals are clearly distinguishable, e.g. on the basis of the inclinations. From AD 800 on, good magnetic dating can be made using the sharp decrease of the inclination until about AD 1300, which is followed by a sharp increase until about AD 1500. There is also enough change in the declination to improve the precision of the dating. Except for the sixteenth century, which is unresolvable on the basis of the present data, precise dating appears to be feasible again for the more recent periods.

Conclusion

A secular variation record has been derived from archaeomagnetic and direct observations for Hungary for the past two thousand years. It has been argued that the effects of shape anisotropy on the archaeomagnetic directions are negligible. Magnetic anisotropy effects due to particle alignment have also been considered and it has been concluded tentatively that they may not be significant for the archaeomagnetic samples in which the TRM is carried by single domain grains as indicated by high Q values. The secular variation record has been obtained using an objective smoothing of the data which allows for imprecision both in age and magnetic

direction. The general pattern of direction change is dominated by an approximately 900 year oscillation and is in agreement with that for France, Sicily, Britain, the Ukraine and the Balkans. The secular variation has sufficient resolution for dating purposes for most of the period under consideration.

I wish to thank the editors for making improvements to the manuscript.

References

AITKEN, M. J. & WEAVER, G. H. 1965. Recent archaeomagnetic results in England. *Journal of Geomagnetism and Geoelectricity*, **17**, 391–394.

BARTA, GY. 1957. *Földmágnesség*. Akadémiai Kiadó. Budapest.

BURLATSKAYA, S. P. & MÁRTON, P. 1986. The variation of ancient geomagnetic field intensity for the territory of Hungary. *Journal of Geomagnetism and Geoelectricity*, **38**, 1369–1372.

CLARK, A. J., TARLING, D. H. & NOËL, M. 1988. Developments in archaeomagnetic dating in Britain. *Journal of Archaeological Science*, **15**, 645–667.

COE, R. S. 1979. The effect of shape anisotropy on TRM direction. *Geophysical Journal of the Royal Astronomical Society*, **56**, 369–383.

FISHER, R. A. 1953. Dispersion on a sphere. *Proceedings of the Royal Society, London*, **A217**, 295–305.

HANSTEEN, C. 1819. *Untersuchungen über den Magnetismus der Erde*. Christiana.

KOVACHEVA, M. 1983. Archaeomagnetic data from Bulgaria and South Eastern Yugoslavia. *In*: CREER, K. M., TUCHOLKA, P. & BARTON, C. E. (eds) *Geomagnetism of Baked Clays and Recent Sediments*, Elsevier. Amsterdam, 106–110.

MÁRTON, P. 1990. Archaeomagnetic directional data from Hungary: Some new results. *In*: PERNICKA, E. & WAGNER, G. (eds) *Archaeometry '90*. Birkhäuser Verlag, Basel, 569–576.

McFADDEN, P. L. 1980. Determination of the angle in a Fisher distribution which will be exceeded with a given probability. *Geophysical Journal of the Royal Astronomical Society*, **60**, 391–396.

ROLPH, T. C. & SHAW, J. 1986. Variation of the Geomagnetic Field in Sicily. *Journal of Geomagnetism and Geoelectricity*, **38**, 1269–1277.

SCHENZL, G. 1881. *Adalékok a Magyar Koronához Tartozó Országok Földmágnességi Viszonyainak Ismeretéhez*. Magyar Királyi Természettudonányi Társulat. Budapest.

STERNBERG, R. S. 1989. Secular variation of archaeomagnetic direction in the American Southwest, AD. 750–1425. *Journal of Geophysical Research*, **94**, 527–546.

STEPHENSON, A., SADIKUN, S. & POTTER, D. K. 1986. A theoretical and experimental comparison of the anisotropies of magnetic susceptibility and remanence in rocks and minerals. *Geophysical Journal of the Royal Astronomical Society*, **84**, 185–200.

TARLING, D. H. 1988. Secular variations of the geomagnetic Field – The archaeomagnetic record. *In*: STEVENSON, F. R. & WOLFENDALE, A. W. (eds) *Secular Solar and Geomagnetic Variations in the Last 10,000 Years*. Kluwer Academic Publishers, 349–365.

THELLIER, E. 1981. The direction of the Earth's magnetic field, in France, during the last 2000 years. *Physics of the Earth and Planetary Interiors*, **24**, 89–132.

VÉGES, I. 1970. Map plotting with weighted average on the surface of a circular disc. *Pure and Applied Geophysics*, **78**, 5–17.

ZAGNII, G. P. & RUSAKOV, O. M. 1982. *Archaeosecular variation of the geomagnetic field in the Southwest of the Soviet Union* [in Russian]. Naukova Dumka, Kiev.

Glossary of basic palaeomagnetic and rock magnetic terms

A. MORRIS

Department of Geological Sciences, University of Plymouth, Drake Circus, Plymouth PL48AA, UK

The basic principles of palaeomagnetism can be stated straightforwardly as follows. A rock unit can acquire magnetization components at various stages in its history, from the time of its formation until the present day. Each component can record the direction of the ambient geomagnetic field at the location of the rock unit at the time of magnetization. Under ideal conditions, the direction and intensity of each of these magnetic components can be determined by laboratory study, and their relative ages and relationships to geological events established. The inclination (dip) of each magnetization component is related directly to the geographic latitude of the rock at the time of magnetization, whereas the declination (azimuth) of the magnetic vector indicates the subsequent rotation of the rock unit with respect to geographic north. Hence, information on latitudinal and rotational movements of the rock unit can be derived. This underlying simplicity is, of course, complicated by a host of factors affecting the magnetization recorded by a rock, and by a wide range of experimental procedures and difficulties. These give rise to a complex terminology, which can make extraction of geologically useful information from a palaeomagnetic study difficult for the non-specialist. The papers in this volume are necessarily full of technical palaeomagnetic and rock magnetic terms. This glossary aims to provide a reference point for readers who are not familiar with detailed palaeomagnetic practice, but who would like to know more about the concepts behind the subject. It will hopefully increase the usefulness of both this volume and other palaeomagnetic literature to non-specialists.

The glossary is not exhaustive, but attempts to cover most of the terms used in the papers in this volume. For full and detailed treatments of palaeomagnetic methodology and practice, the reader is referred to comprehensive texts such as Collinson (1983), Tarling (1983), O'Reilly (1984), Piper (1987) and Butler (1992). Several entries in this glossary draw heavily upon descriptions given in these texts. On a more basic level, Cox & Hart (1986) provide an introduction to the fundamentals of palaeomagnetism, while a clear introduction to palaeomagnetic applications in geology is given by Park (1983).

Glossary

Words in upright bold type refer to other glossary entries.

α_{95}: The semi-angle of the cone of 95% confidence surrounding a **mean direction of magnetization** or **pole positions** (see **Fisherian statistics**).

AF demagnetization: see **Alternating field demagnetization**.

Alternating field (AF) demagnetization: This is carried out by subjecting a specimen to an alternating magnetic field of gradually decreasing magnitude in the presence of a zero direct magnetic field. The alternating field is produced by passing an alternating current through a coil. Maximum obtainable fields are usually around 100 mT. The current ramps up to produce the selected peak alternating field. All magnetic grains with **coercivities** less than the peak applied field will have their magnetizations pulled into alignment with the alternating magnetic field. As the magnitude of the applied field decreases during each alternating cycle, a fraction of the magnetic grains present in the specimen will cease to be affected by the field. Approximately half of these grains will be left with their magnetizations aligned along their preferred axes (see **Domains**) with a component along the axis of the demagnetizer coil, with the other half having a component in the opposite direction. The total magnetic moments of these grains will approximately cancel out. Subsequent cycles result in cancelling out of the magnetizations of successively lower and lower coercivity fractions. After a few minutes, when the alternating field has reduced to zero, the net result is the effective demagnetization of all grains with coercivities less than the peak applied field.

From Morris, A. & Tarling, D. H. (eds), 1996, *Palaeomagnetism and Tectonics of the Mediterranean Region*, Geological Society Special Publication No. 105, pp. 401–415.

This technique can be applied in two different ways. Static AF demagnetization of the three orthogonal axes of a specimen can be carried out in turn. A more efficient method, however, involves tumbling a specimen about two or three axes using nested gears. Such **tumblers** are designed to present in sequence all axes of the specimen to the axis of the demagnetizer coil, thereby demagnetizing all specimen axes during a single treatment and randomizing the magnetization carried by all grains with coercivities less than the peak applied field. In either case, each specimen is usually subjected to **stepwise demagnetization**.

AF demagnetization is most effective for rocks in which **magnetite** or **titanomagnetite** is the dominant ferromagnetic mineral present. An advantage of the technique is that it does not produce chemical alteration in a specimen, which is a common problem with **thermal demagnetization**. AF treatment is, however, ineffective in demagnetizing rocks where the remanence is carried by **hematite** or **goethite**, which have coercivities which exceed those of most AF demagnetization systems.

AMS: see **Anisotropy of magnetic susceptibility**.

Anhysteretic remanent magnetization (ARM): A magnetic remanence acquired when a ferromagnetic grain (*s.l.*) is subjected simultaneously to alternating and direct magnetic fields. ARMs can be produced during **AF demagnetization** if either: (i) the Earth's magnetic field is not adequately cancelled out by the magnetic shielding of the AF demagnetizer; or (ii) the alternating current passed through the demagnetizer coil does not have a very pure waveform. In either case the resulting small direct magnetic field introduces spurious ARM components into a specimen, which produce a noisy demagnetization path and can obscure the natural remanence (particularly at high demagnetizing fields of $>50 \, \text{mT}$). Both these effects can be reduced by using a **tumbler** during AF demagnetization. ARMs can be deliberately produced to investigate the variation in magnetic grain size within a sample collection (see Fig. 2e of Feinberg, Saddiqi & Michard, this volume.)

Anisotropy of magnetic susceptibility (AMS): A property of a material whereby identical magnetic fields applied in different directions produce different intensities of induced magnetization. AMS reflects the statistical alignment of platy or elongate magnetic (usually ferromagnetic) grains. AMS is defined in terms of the magnetic susceptibility ellipsoid, which has principal axes along the directions of maximum (K_1), intermediate (K_2) and minimum

(K_3) susceptibility. If $K_1 = K_2 = K_3$, the ellipsoid is spherical and the specimen has an isotropic magnetic susceptibility. If $K_1 \approx K_2 > K_3$, the ellipsoid is oblate (disc-shaped). If $K_1 > K_2 \approx K_3$, the ellipsoid is prolate (cigar-shaped). Oblate susceptibility ellipsoids are commonly observed in sedimentary rocks and in rocks with a significant foliation, with K_3 oriented perpendicular to the bedding and foliation, respectively. Prolate ellipsoids can be observed in volcanic lava flows and current-deposited sediments, where K_1 is aligned parallel to the palaeoflow direction. Significant AMS can also be produced during straining of rocks, and has been used to infer the orientation of the strain ellipsoid (e.g. Kligfield *et al.* 1983). Anisotropy of remanent magnetization is also of interest in palaeomagnetic studies. A full treatment of anisotropies of susceptibility and remanent magnetization is given in Tarling & Hrouda (1993).

Antiferromagnetism: Describes the behaviour of solids with antiparallel coupling between adjacent layers of atomic magnetic moments, where opposing layers have equal magnetic moment and thus produce no net magnetization.

Apparent polar wander path (APWP): A plot of sequential positions of **palaeomagnetic poles** from a particular lithospheric plate or tectonostratigraphic terrane, usually presented on the present-day geographic grid. APWPs allow presentation of palaeomagnetic data covering significant periods of geological time. Comparison of APWPs from different continents and terranes allows documentation of the timing of continental and terrane separation and collision (docking). APWPs also form the basis for one method of magnetic dating, whereby poles obtained from a unit of uncertain age are compared with well-dated poles from an appropriate reference APWP (e.g. Najman *et al.* 1994).

The APWP for the African plate is not well-defined, particularly for the Mesozoic. This has been partially overcome by rotating palaeomagnetic data from other continents (particularly North American data) into African coordinates to produce synthetic African APWPs (see Channell, this volume, for a discussion related to analysis of data from Adria). The African and Eurasian APWPs which are most widely used in palaeomagnetic studies in the Mediterranean region are those given by Westphal *et al.* (1986) and Besse & Courtillot (1991). North American and Indian APWPs are given by Besse & Courtillot (1991). An earlier synthesis of APWPs for North

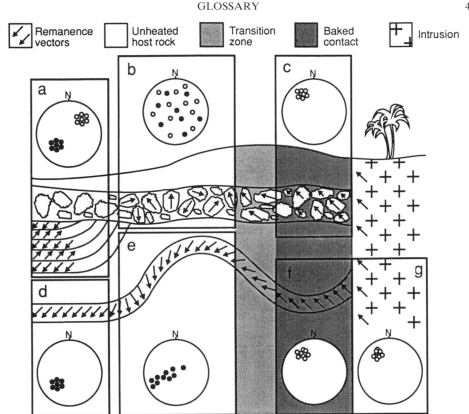

Fig. 1. Schematic geological section illustrating four field tests of palaeomagnetic stability: (a) the **reversal test**; (b, c) the **conglomerate test**; (d, e) the **fold test**; and (d, f, g) the **baked contact test**. See glossary entries for details. Solid symbols on stereonets are downward remanence directions, open symbols are upward directions (modified after Park 1983).

America, Northern Eurasia, Africa, South America, Australia and India was given by Irving & Irving (1982).

ARM: see **Anhysteretic remanent magnetization**.

Baked contact test: A **field test** of palaeomagnetic stability (Fig. 1d, f, g). Samples are collected from an igneous intrusion, and from its baked contact and host rock. If the host rock is much older than the igneous rock, then **ChRM** directions from the host (Fig. 1d) and igneous (Fig. 1g) rocks may be expected to be significantly different. The baked contact, however, should have the same direction as the igneous rock since its magnetization should be thermally reset during intrusion (Fig. 1f). A positive test is where the igneous rock and baked contact have similar magnetization directions which are different from that of the host rock far from the intrusive contact, suggesting that the primary **TRM** has been isolated. A negative test (uniform directions throughout) suggests that all units have experienced a **remagnetization** event.

Bedding (or tilt) correction: If a sampled rock unit is no longer in the attitude in which it acquired its magnetization because of tilting, it is necessary to apply a correction to restore the sampled unit back to its presumed initially horizontal attitude (**palaeohorizontal**). This usually involves rotating bedding planes back to horizontal about a line of strike. This can, however, lead to erroneous palaeomagnetic directions if the deformation involved tilting around inclined axes or more than one phase of tilting, or if fold axes are plunging (see **plunge correction**). The bedding correction is potentially the largest source of error in palaeomagnetic studies.

Blocking temperature: Magnetic **relaxation time** is exponentially inversely proportional to temperature. For any given volume of a **SD** grain of a certain composition, there is a temperature above which relaxation times are so short that the grain can not hold a fixed magnetization direction on a laboratory timescale and displays **superparamagnetism**. As the grain cools below this temperature, it exhibits stable SD behaviour. This transition temperature is called the *blocking temperature*. At temperatures between the **Curie temperature** and the blocking temperature, the grain is **ferromagnetic** but the **remanent magnetization** of an assemblage of such grains will quickly decay to zero. Below the blocking temperature, the remanent magnetization rapidly becomes increasingly stable as relaxation times increase dramatically with decreasing temperature. Relaxation time and blocking temperature are fundamental to theories of **thermoremanent magnetization**.

Blocking temperature spectra: The range of **blocking temperatures** present within a specimen, arising from a distribution of **ferromagnetic** grain sizes, shapes and compositions.

Blocking volume: Magnetic **relaxation time** is directly proportional to grain volume. Above the blocking volume, a grain exhibits SD behaviour and the **remanent magnetization** of an assemblage of such grains can be stable. Relaxation time and blocking volume are fundamental to theories of **chemical remanent magnetization**.

Bulk susceptibility: The magnetic **susceptibility** due to all minerals in a specimen, but usually dominated by the ferromagnetic minerals (Butler 1992).

Characteristic Remanent Magnetization (ChRM): The highest-stability component of **NRM** isolated during demagnetization. Unlike the term **primary remanence**, ChRM does not imply a time of acquisition of the magnetic component.

Chemical Remanent Magnetization (CRM): A magnetic remanence acquired as a ferromagnetic grain nucleates and grows below the **Curie temperature** in the presence of a magnetic field. For a palaeomagnetically useful stable remanence to be acquired, grains must only grow to stable **SD** or **PSD** size and no greater. Reactions which produce a CRM include: (i) alteration of an existing mineral to a ferromagnetic mineral; and (ii) direct precipitation of a ferromagnetic mineral from solution (Butler 1992).

ChRM: see **Characteristic Remanent Magnetization**.

Cleaning: The process of removing secondary components of magnetization from a specimen in order to recover the primary component of remanence. The two most frequently used cleaning methods are **stepwise thermal** and **AF demagnetization**. It is good practice to apply both thermal and AF demagnetization methods to a sample collection. Complete overlap of **coercivity** or **unblocking temperature spectra** of

two **magnetic components** in a rock can yield an AF or thermal demagnetization path which appears to show only a single component. Total reliance on a single demagnetization method could, therefore, produce misleading results. It is unlikely, however, that both the coercivity and unblocking temperature spectra of individual magnetic components overlap completely.

Coercivity: Magnetic grains are magnetized along 'easy' axes (see **Domains**). The coercivity of a SD grain is the magnetic field which must be applied to force its magnetization to flip direction by 180° (i.e. to result in a 'permanent' change in direction of magnetization upon removal of the field).

Coercivity spectra: The range of **coercivities** present within a specimen, arising from a distribution of **ferromagnetic** grain sizes, shapes and compositions.

Confidence limits on magnetic poles (dp, dm): The circular confidence limit, α_{95}, on a given **mean direction of magnetization** maps onto a confidence oval around the corresponding magnetic pole position. This confidence oval has a semi-major axis, dm, and semi-minor axis, dp, given by:

$$dm = \alpha_{95}\left(\frac{\sin p}{\cos I}\right) \text{ and } dp = 2\alpha_{95}\left(\frac{1}{1 + 3\cos^2 I}\right)$$

where I = the **inclination** of the site mean direction; α_{95} = the 95% confidence limit on the site mean direction; and p = the angular distance from the reference pole to the sampling site (see **Reference direction**).

Conglomerate test: A **field test** of palaeomagnetic stability (Fig. 1b, c). Samples are collected from clasts within a conglomerate, and **ChRM** directions are determined for each clast by demagnetization experiments. Uniform directions of magnetization within individual clasts but a random distribution between clasts indicates that the magnetization of the source rock has been stable since at least the time of formation of the conglomerate (positive test; Fig. 1b). Uniform directions between clasts indicate that the ChRM of the conglomerate was formed after deposition (negative test; Fig. 1c). A positive test carried out on an intraformational conglomerate strongly suggests that the ChRM of the source formation is a **primary remanence**.

CRM: see **Chemical Remanent Magnetization**.

Curie temperature: The temperature above which **ferro-** and **ferrimagnetic** substances behave paramagnetically. Since interatomic distances increase upon heating, the strength of exchange coupling between atomic magnetic moments decreases with increasing temperature, thereby reducing the resultant magnetization. At the Curie temperature, interatomic distances become so great that exchange coupling breaks down, the atomic magnetic moments become independent, and the material exhibits **paramagnetism**. Upon cooling below the Curie temperature exchange coupling and ferro- or ferrimagnetism reappear. Curie temperatures of the two most common palaeomagnetic carriers are 580°C for **magnetite** and 680°C for **hematite**.

Declination: The horizontal angle between either a magnetization vector or the Earth's magnetic field and geographic north.

Demagnetization: see **Cleaning**; **AF demagnetization**; **Thermal demagnetization**.

Detrital remanent magnetization (DRM): A **remanent magnetization** acquired during deposition and lithification of sedimentary rocks. The most common **ferromagnetic** carrier of DRM is detrital **magnetite**. DRM acquisition is a complex process affected by depositional environment and post-depositional disturbances such as bioturbation. Detrital remanences can be subdivided into those produced by physical alignment of ferromagnetic particles with the ambient geomagnetic field during deposition (referred to as depositional detrital remanent magnetizations), and those arising from post-depositional alignment (referred to as post-depositional detrital remanent magnetizations or **PDRMs**). The total DRM observed in a sedimentary rock usually results from a combination of depositional and post-depositional alignments. Fine-grained sediments are more accurate recorders of the geomagnetic field direction during (or soon after) deposition since they contain a high proportion of stable **single-domain** and **pseudo-single-domain** grains, and these finer grains are likely to have strong magnetizations and are hence more effectively aligned by the geomagnetic field. Larger **multi-domain** grains are less stable. They are also less likely to move freely within pore spaces in a water-saturated sediment, and are therefore less effectively aligned by either depositional or post-depositional processes. For these reasons, fine-grained sediments such as siltstones are widely sampled for palaeomagnetic study.

Diamagnetism: In diamagnetic substances, the application of an external magnetic field produces a small induced magnetization in the opposite direction to the applied field. The magnetization is proportional to the applied field, and decays to zero when the field is removed. Magnetic susceptibility for a diamagnetic substance is negative. Examples of diamagnetic minerals are quartz, calcite and dolomite.

Domains: The magnetization within a small region within a ferromagnetic grain is uniform in direction and has a preferred orientation, aligned either along specific crystallographic axes, known as magnetocrystalline 'easy' axes or along the length of the grain (for small, elongate grains). These axes are hereafter referred to as 'preferred axes'. In larger grains (e.g. >10 μm), however, a number of volume elements are present, each of which has its magnetization aligned along a preferred axis. These volume elements are called *magnetic domains*. The domains form an arrangement which minimizes the total magnetic energy of the grain.

Three types of different magnetic behaviour are observed, depending on grain size.
(a) Single-domain behaviour: SD grains have a high coercivity and their magnetization can be stable over geological time periods, and are thus efficient palaeomagnetic carriers. SD grains of cubic magnetite are smaller than 0.1 μm, whereas elongate SD magnetite grains can be up to 1 μm in length.
(b) Multi-domain behaviour: MD grains have a low coercive force, and their magnetization decays with time. MD grains are, therefore, less effective palaeomagnetic recorders than SD grains.
(c) Pseudo-single-domain behaviour: The presence of crystal lattice imperfections in some MD grains prevents simple interaction between adjacent domains. The resulting *pseudo-single-domain* (PSD) magnetic behaviour is then closer to that of SD grains. PSD behaviour also results from the intermediate size of some grains. Here the magnetic moments form a vortex pattern which produces high coercivity grains but of lower total magnetic moments. The PSD grain-size interval for magnetite is approximately 0.1–1.0 μm. PSD grains exhibit significant time-stability of remanent magnetization, and can be important palaeomagnetic carriers. (Butler 1992; Tarling & Hrouda 1993).

dm(δm): The semi-major axis of the confidence oval on a magnetic pole (see **Confidence limits on magnetic poles**).

dp(δp): The semi-minor axis of the confidence oval on a magnetic pole (see **Confidence limits on magnetic poles**).

DRM: see **Detrital Remanent Magnetization**.

Ferrimagnetism: Describes the behaviour of solids with antiparallel coupling between adjacent layers of atomic magnetic moments, where opposing layers have unequal magnetic moment and thus produce a net magnetization in the direction of the dominant layer.

Ferromagnetism: Ferromagnetism (*sensu lato*) refers to all solids with exchange coupling of atomic magnetic moments, including materials displaying **antiferromagnetism** and **ferrimagnetism**. Ferromagnetism (*sensu stricto*) describes the behaviour of solids with parallel coupling between adjacent layers of atomic magnetic moments, producing a strong magnetization (even in the absence of an external magnetic field).

Field tests: Various methods of determining the timing of ChRM acquisition, which generally require carefully designed sampling strategies. See **fold test**; **reversal test**; **baked contact test**; **conglomerate test**; **syn-folding remanence**; and Fig. 1.

Fisherian statistics: Fisher (1953) derived a probability density function for vectors considered as points on a sphere, known as the Fisher distribution, which is used to statistically define the dispersion of a set of magnetization vectors around the mean direction and to perform statistical tests. The theoretical precision parameter, κ, for the Fisher distribution varies from zero if all vectors in the total population are randomly distributed to infinity if they are all identical to the mean. The best estimate, k, of this precision parameter (based on the finite number of samples drawn from the total population) is simply given by:

$$k = \frac{N-1}{N-R} \qquad \text{for } \mathbf{N} > 7 \text{ and } \kappa > 3.$$

where R is the length of the resultant vector of the N individual magnetization vectors (see **Mean direction of magnetization**).
Values of $k > 10$ indicate that the observed mean direction is close to the true mean of the total population.

The Fisherian confidence limit associated with a calculated mean direction of magnetization is usually quoted for the 0.95 probability level and is given by:

$$\alpha_{95} = \cos^{-1}\left\{1 - \frac{N-R}{R}\left[\left(20^{\frac{1}{N-1}}\right) - 1\right]\right\}$$

which, if $k \geqslant 7$, can be approximated by:

$$\alpha_{95} \approx \frac{140}{\sqrt{kN}}$$

There is a 95% probability that the true mean direction of the total population of magnetic

vectors (from which the sampled population of magnetic vectors was drawn) will lie within this cone.

A well-defined mean direction of magnetization will have a high value of k (>10) and a small α_{95} angle ($<15°$).

Fold test: A **field test** of palaeomagnetic stability (Fig. 1d, e). Samples are collected from beds with different present structural orientations (e.g. from around a fold). The distributions of **ChRM** directions before and after correcting for bedding tilt are compared. If ChRM directions become more closely grouped after untilting (compare Fig. 1d and 1e), this indicates that the ChRM was acquired before folding (positive fold test). Increased dispersion of ChRM directions after untilting indicates that the magnetization is post-folding in age (negative fold test). Care must be taken to ensure that appropriate corrections for bedding tilt are applied, and that any plunge of the fold axis is also corrected (see **bedding** (or **tilt**) **correction** and **syn-folding remanence**). A plethora of statistical tests have been devised to determine the significance of fold tests. The most widely used tests are those of McElhinny (1964) and McFadden & Jones (1981).

Geocentric axial dipole (GAD): A fundamental assumption of the palaeomagnetic method is that the time-averaged geomagnetic field can be modelled by a single magnetic dipole at the centre of the Earth which is aligned along the rotation axis, the geocentric axial dipole (GAD) model. The magnetic field **inclination** in this model is related to geographic latitude by the dipole equation: $tan\,\mathrm{I} = 2\,tan\,\lambda$. The **declination** of the field is everywhere zero. This model does *not* describe the present geomagnetic field, which is more closely modelled by a geocentric dipole inclined at 11.5° to the rotation axis. This best fitting inclined dipole accounts for approximately 90% of the present geomagnetic field at the surface. The remaining *c.* 5% is called the non-dipole field. However, palaeomagnetic records spanning the last 5 million years show that the average position of the geomagnetic pole is indistinguishable from the rotation axis. Thus, over periods sufficient to average out **secular variation** (*c.* 10^5 years) the geomagnetic field appears to be adequately described by the GAD model. The GAD assumption is equally valid for periods of normal and reversed polarity of the geomagnetic field, but does not apply to periods when the field is transitional between the two polarity states. The GAD model can not be used to interpret components of remanence acquired during such transitions.

Geographic coordinates: Magnetization directions given before **bedding correction**.

Goethite: An oxyhydroxide of iron with the composition $\alpha FeOOH$ which displays imperfect **antiferromagnetism** or weak **ferromagnetism**. It has a very high **coercivity** of $>5\,T$ with a low maximum **unblocking temperature** of 80–120°C. Natural dehydration of goethite produces **hematite** which acquires a **chemical remanent magnetization** (CRM) by grain growth, and is an important process in the magnetization of redbeds. Goethite is common in limestones, and forms by either direct precipitation from seawater or by diagenetic alteration or sub-aerial weathering of pyrite. The dehydration of goethite to hematite at 300–400°C produces complications during laboratory **thermal demagnetization**. The presence of goethite is easy to establish using IRM acquisition experiments (see **Isothermal remanent magnetization**) in high applied fields, since it has a higher coercivity than hematite. It can also be distinguished by a rapid intensity decrease by 120°C during thermal demagnetization of NRM or IRM.

Great circle analysis: If two components of magnetization are present in a **specimen** and their **coercivity** and/or **unblocking temperature spectra** do not completely overlap, then **demagnetization** results in a movement of the resultant specimen magnetization along a great circle path on a stereonet. Demagnetization can fail to isolate the direction of the higher coercivity/unblocking temperature component (e.g. because of overlap at the higher end of the demagnetization spectra, intensity decreasing below the noise level of the magnetometer, or chemical alteration during **thermal demagnetization**). However, if other great circle paths are available from other specimens at the **site** and these great circles converge, then an estimate of the final magnetization direction may be obtained. Early methods of calculating the final direction used the intersection points of the converging great circles. A more statistically rigorous method was devised by McFadden & McElhinny (1988), which is based upon maximum likelihood analysis, and is now in general use. Their method allows information from specimens where the final direct was successfully isolated (direct observations) to be combined with information from great circle paths. The mean direction of the direction observations is calculated and used as an initial estimate for the maximum likelihood estimate μ of the true final mean direction (in the absence of direct observations a guess at the final direction is

used). An iterative procedure is then followed, whereby the point on each great circle closest to μ is found, the mean direction of these points and the direct observations is calculated and used as a new value of μ, and the process repeated until successive iterations produce no significant change in the direction of μ. The confidence limits associated with the final maximum likelihood estimate direction can also be calculated. The McFadden & McElhinny (1988) method has provided a powerful tool for palaeomagnetic analysis, and has resulted in a significant increase in the number of useful data recovered by laboratory demagnetization experiments.

Hematite: A mineral with the composition αFe_2O_3 with hexagonal structure. Atomic magnetic moments of Fe^{3+} cations are parallel coupled within basal planes, but adjacent layers of cations are approximately anti-parallel coupled. A net magnetization in the basal plane arises from this imperfect anti-parallel coupling. The resulting imperfect **antiferromagnetism** is referred to as *canted antiferromagnetism*. In addition, some naturally occurring hematite has a defect **ferromagnetism** caused by lattice defects or impurities. The overall effect is one of weak but very stable ferromagnetism. Hematite has a maximum **unblocking temperature** of 675°C and a maximum **coercivity** of 1.5–5.0 T (O'Reilly 1984; Lowrie 1990). Hematite can be the dominant ferromagnetic material in highly silicic and/or highly oxidized igneous rocks, and is nearly always the dominant ferromagnetic material in red beds (Butler 1992).

In situ directions: Magnetization directions given before **bedding correction**.

Inclination: The angle between either a magnetization vector or the Earth's magnetic field and the horizontal plane.

Inclination shallowing: Inclinations which are shallower than those of the expected **reference direction** for a sampled unit are frequently encountered. In sedimentary rocks this is often attributed to the effects of post-depositional compaction or gravitational forces acting during deposition (see Garcés *et al.* this volume). Other potential causes of shallow inclinations are the biasing effect of a strong magnetic **anisotropy** on the magnetic remanence, and the terrain effect (Baag *et al.* 1995) in strongly magnetized lavas.

IRM: see **Isothermal remanent magnetization**.

Isothermal remanent magnetization (IRM): An artificial magnetization imparted by subjecting a specimen to a direct magnetic field in the laboratory. IRMs are also produced naturally by lightning strikes. Laboratory IRMs are used extensively to determine the nature of the magnetic minerals which are capable of carrying a natural remanence in a specimen. The standard procedure is to apply progressively increasing magnetic fields to a specimen, measuring the IRM produced after each field application. The shape of the resulting graph of IRM against applied field is characteristic for different ferromagnetic minerals. For example, rapid increases in IRM and subsequent flattening off of the curve (saturation) by applied fields of 100–300 mT indicate the presence of **magnetite, titanomagnetite** or **maghemite**. In contrast, **hematite** does not reach saturation until 1.5–5.0 T whereas **goethite** only saturates in applied fields greater than 5.0 T. The interpretation of these curves can be improved by **stepwise thermal demagnetization** of the acquired IRM, particularly if three different fields are applied along the three orthogonal axes of a specimen (Lowrie 1990), thereby allowing **unblocking temperature spectra** to be deduced for different coercivity fractions of IRM.

k: The Fisherian precision parameter which measures the dispersion of a set of magnetization vectors around the **mean direction of magnetization**. High values of *k* indicate that the observed mean direction is close to the true mean of the total population (from which the sampled population of magnetic vectors was drawn). See **Fisherian statistics**.

Locality: Generally an area of less than 1 km^2 from which more than one site has been collected.

MAD: see **Principal component analysis**.

Maghemite: A **ferromagnetic** mineral (γFe_2O_3) with the composition of **hematite** but the cubic (spinel) structure of **magnetite**. It forms by low-temperature (<200°C) oxidation of magnetite during subaerial or subaqueous weathering. It has a maximum **coercivity** of 300 mT (O'Reilly 1984), equivalent to that of magnetite. It is destroyed by heating to 350°C when it inverts to hematite.

Magnetic components: A rock may record one or more components of magnetization acquired at different times during its history. The total **natural remanent magnetization** of a rock is the vector sum of all magnetic components. The direction and intensity of each component can be determined using **demagnetization** experiments, **principal component analysis**, and **great circle analysis**. Magnetic components with completely overlapping **coercivity** and **unblocking**

temperature spectra can not be separated (see **cleaning**).

Magnetic intensity or magnetization (SI unit: A m^{-1}): The net magnetic moment per unit volume. Magnetometers measure the vector sum of magnetic moments within a **specimen**. The magnetic moment (SI unit: A m^2) is converted simply to magnetization by dividing by the specimen volume.

Magnetic susceptibility: see **Susceptibility**.

Magnetite: see **Titanomagnetites**.

Magnetization: see **Magnetic intensity**.

Magnetometers:
(a) Cryogenic: These are the most sensitive magnetometers available. They employ a magnetic field sensor called a SQUID (acronym for Superconducting QUantum Interference Device) which operates at liquid helium temperatures. They are capable of measuring **magnetic moments** of the order of 10^{-10} A m^2. The development of cryogenic systems in the early 1970's extended the range of rock types which could be subjected to palaeomagnetic analysis, and led to the routine analysis of weakly magnetic carbonates. Full details of the principles of cryogenic magnetometers are given by Collinson (1983). High temperature (i.e. liquid nitrogen) SQUIDS are now commercially available.
(b) Fluxgate: The most widely used type of magnetometer. A magnetic field sensor (fluxgate) detects the magnetic field of the magnetized **specimen**. Sensitivity is increased by slowly spinning the specimen. Measurements are typically made using a sequence of four or six specimen orientations, in order to average out inhomogeneities. Data from the complete sequence of six positions are used to calculate the mean direction of magnetization and its reliability. Full details of the principles of fluxgate magnetometers are given by Collinson (1983).

MD: Abbreviation for multi-domain; see **Domains**.

Mean direction of magnetization: The mean direction of a number, *N*, of palaeomagnetic directions (e.g. samples at a site, or sites at a locality) is found by simple vector summation. The direction cosines of each individual vector are calculated from:

$$l_i = \cos I_i \cos D_i \qquad m_i = \cos I_i \sin D_i \qquad n_i = \sin I_i$$

where D_i and I_i are the declination and inclination of the *i*th vector, and l_i, m_i, n_i are the directions cosines of the *i*th vector with respect to north, east and down directions respectively. The resultant vector, *R*, of the *N* individual vectors is then calculated from:

$$R^2 = \left(\sum_{i=1}^{N} l_i \right)^2 + \left(\sum_{i=1}^{N} m_i \right)^2 + \left(\sum_{i=1}^{N} n_i \right)^2$$

and the direction cosines of the mean direction found from:

$$l = \left(\sum_{i=1}^{N} l_i \right) \bigg/ R \qquad m = \left(\sum_{i=1}^{N} m_i \right) \bigg/ R \qquad n = \left(\sum_{i=1}^{N} n_i \right) \bigg/ R$$

The declination, D_m, and inclination, I_m, of the mean direction are then given by:

$$D_m = \tan^{-1}(m/l) \quad I_m = \sin^{-1} n$$

The length of the resultant vector, R, forms the basis for analyses of the degree of clustering of the N magnetization vectors using **Fisherian statistics**. For a perfectly aligned set of vectors, $R = N$. Values of $R \ll N$ indicate widely scattered unit magnetization vectors.

Multi-domain grains: see **Domains**.

Natural remanent magnetization (NRM): The summation of all components of magnetic remanence acquired by natural processes. The NRM of a specimen can consist of several components (a multi-component remanence) acquired at different times during its history. For example, the NRM of a lava may comprise a primary **thermoremanent magnetization**, a **secondary magnetization** acquired during low-grade metamorphism, and a **viscous remanent magnetization** acquired in the present day field. Components of magnetization are separated in laboratory studies using **stepwise** (progressive) **demagnetization**.

Net tectonic rotation: The single rotation (usually about an inclined axis) which restores the *in situ* magnetization direction to the appropriate **reference direction** and simultaneously restores the present bedding pole to vertical (in the case of presumed initially horizontal beds). A full treatment is given by MacDonald (1980). See also Kirker & McClelland (this volume).

NRM: see **Natural remanent magnetization**.

Orthogonal demagnetization diagram: The most common method of displaying the variations in intensity and direction of magnetization of a sample resulting from progressive **demagnetization**. The data from each demagnetization step are plotted as points on two sets of superimposed axes. The N and E Cartesian components of magnetization are plotted on N–S/E–W axes. The projection of the vertical Cartesian component of magnetization Z is plotted on to either N–S/up–down or E–W/up–down axes. Successive points are usually joined by straight lines. The angle subtended by each point with the N axis is the **declination**. The angle between each

point and the horizontal in the selected vertical plane gives the apparent inclination I_{app} which is related to the true **inclination** by:

$$\tan I = \tan I_{app} |\cos D| \quad \text{for the N–S vertical plane}$$

$$\tan I = \tan I_{app} |\sin D| \quad \text{for the E–W vertical plane}$$

The distance of each point from the origin is proportional to the intensity of the component of magnetization plotted on to that plane. A linear segment in the demagnetization path defined by a number of successive points on these plots indicates demagnetization of a single component of magnetization with a constant direction (or conceivably two components of magnetization with identical **unblocking temperature** or **coercivity spectra**). The declination and inclination of successively removed components can be easily calculated. Partial overlap of unblocking temperature or coercivity spectra of two components produces curved demagnetization paths, and may require the use of **great circle analysis** to determine the magnetization direction(s).

Overprint: See **Secondary magnetization**.

Palaeohorizontal: A surface in a rock unit which can be assumed to initially have been horizontal before deformation (e.g. a bedding plane). Recovery of the pre-deformation orientation of a magnetic remanence vector by application of suitable structural corrections relies upon good palaeohorizontal control.

Palaeointensity: The past intensity of the geomagnetic field, usually determined by analysis of the thermoremanent magnetization carried by an igneous rock. The standard procedure for palaeointensity determination involves comparison of **TRM** acquired by a specimen when cooled in a known laboratory magnetic field with the natural TRM carried by the specimen. The technique is described more fully by Bardot *et al.* (this volume).

Palaeolatitude: The latitude of a sampled unit at the time it acquired its magnetization. The palaeolatitude (λ) and magnetic inclination (I) are related by the equation: $\tan I = 2 \tan \lambda$. Potential errors in determining inclinations and hence palaeolatitudes include: (i) misinterpretation of the age of a remanence (i.e. using tilt corrected vectors when the remanence is post-folding in age); (ii) the application of inappropriate **bedding (or tilt) corrections** (for example in volcanic terrains where lava flows can be deposited on significant palaeoslopes); and (iii) biasing due to **inclination shallowing**.

Palaeomagnetic pole: A pole position found by

averaging a number of site mean **virtual geomagnetic poles**, in order to average out **secular variation** of the dipole and non-dipole components of the geomagnetic field. A set of VGPs derived from palaeomagnetic sites magnetized over approximately 10^4 to 10^5 years provides adequate sampling of secular variation. The resultant palaeomagnetic pole gives the position of the Earth's rotation axis with respect to the sampling area at the time of magnetization.

Palaeomagnetic rotation: The angular difference between the declination of the remanence vector at a site and that of the appropriate **reference direction**.

Paramagnetism: Describes the behaviour of solids containing atoms with atomic magnetic moments but where no interaction occurs between adjacent atomic moments. The atomic magnetic moments oscillate rapidly and randomly in orientation at any temperature above absolute zero, producing no net magnetization in the absence of an applied magnetic field. Application of a magnetic field exerts an aligning torque on the atomic magnetic moments, but this is effectively overcome by thermal energy (even at room temperature) and only a small net magnetization is produced. The magnetization disappears when the field is removed. Paramagnetic minerals have a positive susceptibility. Examples are pyroxenes and olivine.

pca: see **Principal component analysis**.

PDRM: see **Post-depositional remanent magnetization**.

Plunge correction: If a fold axis has a significant plunge, then the application of standard **bedding (or tilt) corrections** to restore the fold limbs back to horizontal can produce anomalous magnetic declinations. In such situations, it is necessary to first restore the fold axis to the horizontal before unfolding the limbs. Errors produced by plunge of fold axes are insignificant for bedding dips of less than 30° (see **Tarling 1983**).

Pole position: see **Virtual geomagnetic pole** and **Palaeomagnetic pole**.

Post-depositional remanent magnetization (PDRM): A magnetic remanence acquired by sediments after deposition but prior to final induration. PDRMs are usually produced by a combination of physical rotation of interstitial magnetic grains into alignment with the ambient magnetic field and chemical changes as sediments consolidate. PDRM is common in deep-sea and lake sediments. See also **Detrital remanent magnetization**.

Precision parameter: see **Fisherian statistics**.

Primary magnetization: That component of magnetisation acquired at the time of formation of a rock unit. For igneous rocks the primary magnetization is the **TRM** acquired during initial cooling, whereas for sedimentary rocks it is the **DRM** acquired during deposition. In general it is difficult to prove whether a characteristic magnetization (**ChRM**) isolated by demagnetization is the true primary remanence in a sampled rock unit. **Fold tests** can only be used to identify the timing of remanence acquisition relative to deformation (pre- or post-folding), and cases of pre-folding remagnetization have been reported (see Villalaín *et al.* this volume). Most interpretations of ChRM directions as primary magnetizations are therefore based on assumption. A unique case where this assumption appears to be justified is that of a dyke (or other igneous intrusion) where it can be demonstrated that: (i) the magnetization direction of the dyke and its baked margin is different from that of the host rock far from the dyke (see **baked contact test**); (ii) the unblocking temperature of the TRM component imparted in the host rock by heating at the time of dyke emplacement decreases with increasing distance from the dyke margin; and (iii) the dyke rock yields a realistic **palaeointensity**.

Principal component analysis (pca): A statistically rigorous, quantitative technique used to determine the direction of a best-fit line through a set of magnetization directions obtained by **stepwise demagnetization** (as displayed on an **orthogonal demagnetization diagram**). A quantitative measure of the precision of the calculated best-fit line is given by the maximum angular deviation (MAD). There are three ways in which pca can be applied to demagnetization data: (i) the line may be anchored to the origin; (ii) the origin may be used as a separate data point; or (iii) the origin may be ignored to produce a free line fit. Care must be taken if anchored line fits are used since information from low intensity, high stability components may potentially be lost.

PSD: Abbreviation for pseudo-single-domain; see **Domains**.

Pseudo-single-domain grains: see **Domains**.

PTRM: See **Thermoremanent magnetization**.

Pyrrhotite: A ferrimagnetic iron sulphide with monoclinic crystal structure and composition in the range Fe_7S_8 to Fe_9S_{10}. It has a maximum unblocking temperature of 325°C and a maximum coercivity of 0.5–1.0 T. Pyrrhotite forms during diagenesis of marine sediments and in contact metamorphic aureoles.

Reference direction: The expected magnetization direction at a **site** or **locality**, usually derived from a coeval reference pole obtained outside the area of interest (normally from a stable region outside the deformed zone under study). Reference directions in the Mediterranean region are usually derived from the apparent polar wander paths of the African (Gondwanan) or Eurasian plates, or from other smaller lithospheric units (e.g. Stable Iberia). If the reference pole has latitude λ_p and longitude ϕ_p and the site under consideration is at latitude λ_s and longitude ϕ_s, then the inclination, I_x, and declination, D_x, of the reference direction are given by:

$$D_x = \cos^{-1}\left(\frac{\sin \lambda_p - \sin \lambda_s \cos p}{\cos \lambda_s \sin p}\right)$$

$$I_x = \tan^{-1}(2 \cot p)$$

where p = the angular distance from the reference pole to the sampling site

$$= \cos^{-1}[\sin \lambda_p \sin \lambda_s + \cos \lambda_p \cos \lambda_s \cos(\phi_p - \phi_s)]$$

The 95% confidence limit, c, about D_x, I_x is given by:

$$c = \sin^{-1}\left[\cos I_x\left(\frac{\sin A_{95}}{\sin p}\right)\right]$$

where A_{95} = the 95% cone of confidence on the reference pole.
(Equations from Butler 1992)

Relaxation time: Thermal vibrations of atomic magnetic moments can overcome internal (magnetostatic and magnetocrystalline) energy barriers in a **single-domain** grain and cause the magnetization to flip backwards and forwards along the 'preferred axes' (see **Domains**). In an assemblage of SD grains whose preferred axes are randomly distributed, thermal processes will result in a randomization and cancelling out of the individual magnetizations, thereby producing no net externally observed magnetization. The **remanent magnetization** of an assemblage of SD grains therefore decays exponentially with time. Multi-domain systems behave in a similar way, as thermal vibrations allow domain walls to move past energy barriers. The relaxation time, τ, is defined as the time taken for the initial remanent magnetization, J_o, to decay to J_o/e. Relaxation times vary over many orders of magnitude. Effective palaeomagnetic recorders must have relaxation times on the scale of geological time. SD grains can have relaxation

times greater than the age of the Earth and so are excellent geomagnetic recorders. See also **Blocking temperature** and **Blocking volume**.

Remagnetization: An event which produces a **secondary magnetization** which can partially or completely obscure the **primary magnetization** or an earlier secondary magnetization.

Remagnetization circles: see **Great circle analysis**.

Remanent magnetization: A permanent magnetization of a material which persists after removal of the magnetizing field. The remanence of a rock records information on the direction of the geomagnetic field at the time of magnetization acquisition. The measurement and analysis of components of remanent magnetization within suitable rock types forms the core of palaeomagnetism. For details of the main types of remanent magnetization see: **Thermoremanent magnetization**; **Detrital remanent magnetization**; **Chemical remanent magnetization**; **Viscous remanent magnetization**.

Reversal test: A **field test** of palaeomagnetic stability. In a rock sequence formed over a period during which the polarity of the geomagnetic field underwent one or more reversals, different horizons may carry either normal or reverse characteristic components of magnetization (ChRMs). A palaeomagnetic dataset obtained from such a sequence will then contain antiparallel magnetization directions, providing that the characteristic remanence has been successfully isolated and that **secular variation** has been adequately sampled (Fig. 1a). Such a dataset passes the reversal test if the angular difference between the mean directions of the normal and reversed groups is not significantly less than 180°. Any residual components of **secondary magnetization** will introduce a bias towards one polarity or the other and produce imperfectly antiparallel groupings, resulting in a negative reversal test. Unlike other field tests, the reversal test does not give information about the age of remanence relative to a geological event (such as folding or intrusion). A positive fold test does not indicate that the ChRM is a primary magnetization, but only that it is stable and free of later secondary components and the effects of secular variation. **Remagnetization** events spanning a field reversal can also give rise to normal and reverse ChRMs which may pass a reversal test. The statistical basis of the test is described by McFadden & McElhinny (1990).

Sample: An independently oriented volume of material collected for palaeomagnetic analysis. Samples are usually collected either by *in situ* coring in the field using petrol or electrically driven rock drills or by hand sampling (in which case cores are drilled from each sample in the laboratory). Sample **ChRM** directions of magnetization are combined to give site mean directions.

SD: Abbreviation for single-domain; see **Domains**.

Secondary magnetization (or overprint): Any component of magnetization which was acquired subsequent to initial formation of a rock unit. Secondary magnetizations can be produced by a wide variety of mechanisms, including: reheating events (e.g. burial and subsequent exhumation) giving rise to thermoviscous remanent magnetizations; chemical alteration (e.g. by orogenic fluids or during weathering) producing **CRMs**; lightning strikes, which impart an **IRM**; and acquisition of viscous magnetization (**VRM**) by exposure to the geomagnetic field. Secondary magnetizations may be distinguished using a variety of **field tests** (e.g. Morris & Robertson 1993). Demagnetization (magnetic cleaning) of the natural remanence (**NRM**) can be used to remove such secondary components, as they will normally have a different stability to **thermal** or **AF demagnetization** than the **primary magnetization** component.

Secular variation (SV): The intensity, inclination and declination of the Earth's magnetic field all exhibit temporal variations. Temporal variations occur both on a daily basis (diurnal variation) and over periods of 1 year to 10^5 years. These longer-term changes, or secular variations, are due to convective movements in the outer core. For example, direct observations of the geomagnetic field in London over the past 400 years show that the declination has varied almost from 20°E to 20°W and nearly back to zero. SV records for pre-observatory times are derived from analysis of archaeological materials (e.g. oven walls, hearth floors etc.; see P. Márton, this volume, and Evans, this volume). SV reference curves allow dating of archaeological and Recent geological materials (e.g. historical lava flows; see Incoronato, this volume). Longer-term (c. 10 ka to 100 ka) SV records can be obtained by analysis of lake sediments, loess deposits and stalactites. In tectonic studies it is important that adequate statistical sampling of SV is achieved at a site so that SV is averaged out when calculating the site **mean direction of magnetization**.

Single-domain grains: see **Domains**.

Site: A collection of separately oriented samples spread over c. 5–10 m^2 of outcrop. The precision

of a mean direction of magnetization determined from the samples collected at a site initially increases as the number of samples, n, increases (i.e. as n increases, the precision parameter, k, increases and the α_{95} value decreases; see **Fisherian statistics**). Precision tends to become constant, however, for values of $n > 8$. Larger sample numbers produce no significant improvement in precision for the extra effort involved in their collection and analysis. Sites usually, therefore, consist of 6–10 samples.

Specimen: Standard palaeomagnetic specimens are cylinders of 22 mm height and 25 mm diameter, cut from individual samples collected in the field. This ratio of height to diameter provides the closest approximation to a theoretically ideal, but impractical, spherical specimen. Specimen **ChRM** directions of magnetization are combined to give sample mean directions.

Stepwise demagnetization: The most widely used demagnetization procedure, in which a specimen is demagnetized at progressively increasing temperatures or alternating fields, with the magnetic remanence (and the susceptibility in the case of thermal treatment) being measured after each demagnetization step. The procedure is repeated until either the remanence is reduced to the noise level of the magnetometer being used, the temperature exceeds that of the **Curie temperature** of hematite (680°C), or the maximum alternating field of the AF demagnetizer is reached (usually 100 mT).

Stratigraphic coordinates: Magnetization directions given after **bedding correction**.

Sun compass: A device used to orient palaeomagnetic cores or block samples in the field, consisting of a circular plate marked in degrees with a central pin. The sun compass can be mounted on a pivoting orientation table attached to a hollow tube, which can be placed over the *in situ* core. The axis of the core corresponds to 0°. After levelling the table, the angle made by the shadow of the pin is noted, along with the time of the reading (to within 2 minutes) and the location of the sampling **site** (to within 0.25° of latitude and longitude). The azimuth of the core axis with respect to geographic north can then be calculated. The pivoting table has a built-in dip meter, which is used to measure the hade of the core axis with respect to the vertical.

Superparamagnetism: Describes the magnetic behaviour of very small grains (such as <0.01 μm **magnetite** grains and <0.03 μm **hematite** grains) in which atomic magnetic moments align in the presence of a magnetic field to produce a strong magnetization, but where the magnetization is rapidly destroyed by thermal vibrations very soon after the field is removed. Superparamagnetic grains can have high magnetic **susceptibilities**.

Susceptibility: A measure of the ease with which a material can be magnetized. In magnetic fields as weak as the Earth's, the magnetization J_i induced in a material is directly proportional to the field strength H. The constant of proportionality is called the magnetic susceptibility χ (i.e. $\chi = J_i/H$), and is dimensionless in SI units. The determination of susceptibility is useful for an estimate of the total magnetic content of a specimen (see Thompson & Oldfield 1986), and as an important monitor of thermochemical changes during **thermal demagnetization**. See also **Anisotropy of magnetic susceptibility**.

SV: see **Secular variation**.

Syn-folding remanence: Standard palaeomagnetic structural tilt corrections (bedding corrections) rotate the limbs of a fold back to the horizontal (after application of a **plunge correction** where necessary). Occasionally, the best clustering of magnetization directions is obtained before the fold limbs have been completely corrected for tilt (i.e. by partial unfolding). This implies that the magnetization may have been acquired during formation of the fold, and the magnetization may be interpreted as a syn-folding remanence. The degree of clustering at different stages of unfolding is assessed using **Fisherian statistics**. Due to the potential errors introduced by standard tilt corrections (see **bedding (or tilt) correction**), apparent syn-folding remanences formed at less than approximately 20% or more than approximately 80% of unfolding should be treated with caution, since they may represent post- or pre-folding remanences respectively. Care must be taken to ensure that a combination of pre- and post-folding remanence is not being analysed.

Tesla (T): The unit of magnetic field strength commonly used by palaeomagnetists. In SI units, B in tesla (fundamental units: $kg\,s^{-1}\,C^{-1}$) is strictly the magnetic induction, whereas H is measured in $A\,m^{-1}$ (fundamental units: $C\,s^{-1}\,m^{-1}$) and is the magnetic field.

Thermal demagnetization: Involves heating a **specimen** to an elevated temperature and then cooling to room temperature in zero magnetic field. The magnetizations of all **ferromagnetic** grains within the specimen with **unblocking temperatures** less than or equal to the demagnetization temperature are randomised upon heating. In the absence of a magnetic field, the magnetizations retain this random distribution upon subsequent cooling. A **stepwise demagnetization** procedure is usually followed, with successively increasing temperatures demagnetizing successively higher unblocking temperature fractions. The zero ($<15\,nT$) magnetic field is usually provided by a set of magnetic shields (μ-metal), although arrangements of Helmholtz coils are used to cancel the geomagnetic field in some apparatus.

In addition to being able to demagnetize **magnetite** and **titanomagnetite**, thermal demagnetization is effective in removing magnetization components carried by **hematite** and **goethite**, which usually have **coercivities** exceeding the maximum field produced by **AF demagnetization** apparatus. Problems can arise from thermochemical alteration of specimens at high temperatures, with resulting production of new magnetic phases (e.g. pyrite altering to magnetite at 350–500°C). To monitor such changes, the magnetic **susceptibility** of specimens is routinely measured after each heating step, as any change in susceptibility indicates the destruction or creation of magnetic minerals.

Thermoremanent magnetization (TRM): A **remanent magnetization** acquired upon cooling from temperatures above the **Curie temperature** in the presence of a magnetic field. Most igneous rocks acquire a TRM during initial cooling. A range of **blocking temperatures**, distributed downward from the Curie temperature, will be present in the rock due to a distribution of **ferromagnetic** grain sizes and compositions. Each individual grain experiences a rapid increase in **relaxation time** as the temperature decreases through its blocking temperature, and the grain then holds a stable magnetization. Upon cooling through the lowest blocking temperature of the assemblage of ferromagnetic grains within the rock, the total TRM of the rock is blocked in. The proportion of the total TRM acquired in a distinct temperature interval during cooling is known as a *partial TRM* or *PTRM*.

Tilt corrected directions: Magnetization directions given after **bedding correction**.

Titanomagnetites: **Ferrimagnetic** minerals of the composition $Fe_{3-x}Ti_xO_4$ (where $0 \geqslant x \geqslant 1$), ranging from magnetite (Fe_3O_4) to ulvöspinel (Fe_2TiO_4). They have a cubic (spinel) structure at room temperature and form a solid-solution series at temperatures above 800°C. Under slow cooling conditions, the high-temperature solid solution unmixes or exsolves into fairly pure magnetite and either ilmenite or ulvöspinel, of which only magnetite is magnetic at room temperature (Tarling 1983). Rapid cooling can, however, preserve intermediate titano-magnetite compositions. The endmember magnetite has a maximum **unblocking temperature** of 575°C and a maximum **coercivity** of 300 mT. Maximum unblocking temperatures and coercivities of titanomagnetites both decrease with increasing Ti content, being 350°C and 200 mT respectively for $x = 0.3$, and 150°C and 100 mT respectively for $x = 0.6$ (O'Reilly 1984; Lowrie 1990).

TRM: see **Thermoremanent magnetization**.

Tumbler: An apparatus which rotates a specimen about two or three axes within the demagnetizing coil during **AF demagnetization**. The tumbler presents in sequence all axes of the specimen to the axis of the coil, thereby allowing demagnetization of the specimen during a single treatment. Tumbling also reduces the risk of introducing spurious laboratory remanences (e.g. **ARM**) into a specimen during AF demagnetization.

TVRM: see **Viscous remanent magnetization**.

Unblocking temperature: The temperature at which a component of magnetization in a specimen becomes thermally demagnetized in a laboratory experiment. Unblocking occurs during laboratory heating when the **relaxation times** of the grains carrying the magnetization become equivalent to the length of time at which the specimen is held at elevated temperature.

A **TRM** component acquired at a particular **blocking temperature** does not necessarily unblock at the same temperature during laboratory heating. The difference between blocking and unblocking temperatures is a function of the rate of cooling from the initial blocking temperature, and the difference tends towards zero for components which became blocked in at temperatures close to the **Curie temperature** of the magnetic carrier (Dodson & McClelland Brown 1980).

Unblocking temperatures of other remanent magnetizations (e.g. **DRM**, **CRM**) do not relate to any temperature conditions during acquisition of magnetization. Laboratory unblocking temperatures of thermoviscous remanences (**TVRMs**) can, however, be used to determine a range of temperature-time conditions over which the remanence may have been acquired (e.g. Pullaiah *et al.* 1975). Maximum unblocking temperatures of all remanence types can be used to characterise the **ferromagnetic** minerals carrying the remanence (see **Titanomagnetites**;

Hematite; **Goethite**; and **Pyrrhotite** for the corresponding maximum unblocking temperatures).

Unblocking temperature spectra: The range of **unblocking temperatures** present within a specimen.

Vector component diagram: see **Orthogonal demagnetization diagram**.

VGP: see **Virtual Geomagnetic Pole**.

Virtual Geomagnetic Pole (VGP): The position of a pole of a geocentric dipole that can account for a mean ChRM direction of magnetization observed at a single sampling site (Butler 1992). The latitude, λ_p, and longitude, ϕ_p, of the VGP corresponding to a **mean direction of magnetization** with declination, D, and inclination, I, observed at a site at latitude, λ_s, and longitude, ϕ_s, is given by:

$$\lambda_p = \sin^{-1}(\sin \lambda_s \sin \lambda + \cos \lambda_s \cos \lambda \cos D)$$

$$\phi_p = \phi_s + [\sin^{-1}(\cos \lambda \sin D/\cos \lambda_p)]$$

where λ = magnetic latitude of the sampling location = $\tan^{-1}[(\tan I)/2]$

The α_{95} circle of confidence about the site mean direction of magnetization maps into an ellipse of confidence about the VGP (see **Confidence limits on magnetic poles**).

Individual VGPs rarely correspond to the actual geomagnetic pole position at the time of magnetization of the sampled unit even if **secular variation** of the dipolar geomagnetic field has been cancelled out, because of the influence of non-dipole components (see also **Palaeomagnetic pole**).

Viscous remanent magnetization: A **remanent magnetization** gradually acquired during exposure to weak magnetic fields. VRM acquisition can be considered as the inverse of magnetic relaxation (Butler 1992; see **Relaxation time**), and involves alignment of magnetic moments of grains with short relaxation time, τ. Exposure of an assemblage of grains to the geomagnetic field for a time, T, will result in effective unblocking of all grains with $\tau \leqslant T$. The magnetic moments of these grains can align with the geomagnetic field. The intensity of the resulting VRM increases with increasing exposure time as the relaxation times of more grains are exceeded. The rate of VRM acquisition increases with temperature, as higher thermal energy aids rotation of magnetic moments past energy barriers. A VRM acquired at an elevated temperature is referred to as a *thermoviscous remanent magnetization* (TVRM). Natural VRM in rocks commonly results in present field direction overprints. **Single-domain** and **pseudo-single-domain grains** are less likely to acquire VRM than **multi-domain grains** because of their much longer relaxation times.

VRM: see **Viscous remanent magnetization**.

Zijderveld plot: see **Orthogonal demagnetization diagram**.

My thanks to Roy Thompson and Graeme Taylor for critical reviews of this glossary, and to Darren Randall for helpful suggestions.

References

BAAG, C., HELSLEY, C. E., XU, S. Z. & LEINERT, B. R. 1995. Deflection of palaeomagnetic directions due to magnetization of the underlying terrain. *Journal of Geophysical Research*, **100**, 10013–10027.

BESSE, J. & COURTILLOT, V. 1991. Revised and synthetic apparent polar wander paths of the African, Eurasian, North American and Indian Plates, and true polar wander since 200 Ma. *Journal of Geophysical Research*, **96**, 4029–4050.

BUTLER, R. F. 1992. *Paleomagnetism: magnetic domains to geologic terranes*. Blackwell Scientific Publications, Boston.

COLLINSON, D. W. 1983. *Methods in Palaeomagnetism and Rock Magnetism*. Chapman & Hall, London.

COX, A. & HART, R. B. 1986. *Plate tectonics: how it works*. Blackwell Scientific Publications.

DODSON, M. H. & McCLELLAND BROWN, E. 1980. Magnetic blocking temperatures of single-domain grains during slow cooling. *Journal of Geophysical Research*, **85**, 2625–2637.

IRVING, E. & IRVING, G. A. 1982. Apparent polar wander paths from Carboniferous through Cenozoic and the assembly of Gondwana. *Geophysical Surveying*, **5**, 141–188.

KLIGFIELD, R., LOWRIE, W., HIRT, A. M. & SIDDANS, A. W. B. 1983. Effect of progressive deformation on remanent magnetization of Permian redbeds from the Alpes Maritimes (France). *Tectonophysics*, **97**, 59–85.

LOWRIE, W. 1990. Identification of ferromagnetic minerals in a rock by coercivity and unblocking temperature properties. *Geophysical Research Letters*, **17**, 159–162.

MACDONALD, W. D. 1980. Net tectonic rotation, apparent tectonic rotation, and the structural tilt correction in palaeomagnetic studies. *Journal of Geophysical Research*, **85**, 3659–3669.

McELHINNY, M. W. 1964. Statistical significance of the fold test in palaeomagnetism. *Geophysical Journal of the Royal Astronomical Society*, **8**, 338–340.

McFADDEN, P. L. & JONES, D. L. 1981. The fold test in palaeomagnetism. *Geophysical Journal of the Royal Astronomical Society*, **67**, 53–58.

—— & McELHINNY, M. W. 1988. The combined analysis of remagnetization circles and direct observations in palaeomagnetism. *Earth and Planetary Science Letters*, **87**, 161–172.

—— & —— 1990. Classification of the reversal test in palaeomagnetism. *Geophysical Journal International*, **103**, 725–729.

MORRIS, A. & ROBERTSON, A. H. F. 1993. Miocene remagnetisation of carbonate platform and Antalya Complex units within the Isparta Angle, SW Turkey. *Tectonophysics*, **220**, 243–266.

NAJMAN, Y. M. R., ENKIN, R. J., JOHNSON, M. R. W., ROBERTSON, A. H. F. & BAKER, J. 1994. Palaeomagnetic dating of the earliest continental Himalayan foredeep sediments: implications for Himalayan evolution. *Earth and Planetary Science Letters*, **128**, 713–718.

O'REILLY, W. 1984. *Rock and mineral magnetism*. Blackie, Glasgow.

PARK, J. K. 1983. Paleomagnetism for geologists. *Geoscience Canada*, **10**, 180–188.

PIPER, J. D. A. 1987. *Palaeomagnetism and the continental crust*. Open University Press, Milton Keynes.

PULLAIAH, G., IRVING, E., BUCHAN, K. L. & DUNLOP, D. J. 1975. Magnetization changes caused by burial and uplift. *Earth and Planetary Science Letters*, **28**, 133–143.

TARLING, D. H. 1983. *Palaeomagnetism: principles and applications in geology, geophysics and archaeology*. Chapman & Hall, London.

—— & HROUDA, F. 1993. The magnetic anisotropy of rocks. Chapman & Hall, London.

THOMPSON, R. & OLDFIELD, F. 1986. *Environmental magnetism*. Allen & Unwin, London.

WESTPHAL, M., BAZHENOV, M. L., LAUER, J. P., PECHERSKY, D. M. & SIBUET, J.-C. 1986. Palaeomagnetic implications on the evolution of the Tethys Belt from the Atlantic Ocean to the Pamirs since the Triassic. *Tectonophysics*, **123**, 37–82.

Index

Page numbers given in **bold** refer to glossary definitions. Page numbers given in *italics* refer to Figures or Tables.